Probability   ·   Bayes' theorem

$$P(B_i | A) = \frac{P(B_i) \cdot P(A|B_i)}{P(B_1) \cdot P(A|B_1) + P(B_2) \cdot P(A|B_2) + \cdots + P(B_k) \cdot P(A|B_k)}$$

Conditional probability

$$P(A|B) = \frac{P(A \cap B)}{P(B)}$$

General addition rule

$$P(A \cup B) = P(A) + P(B) - P(A \cap B)$$

General multiplication rule

$$P(A \cap B) = P(B) \cdot P(A|B) \qquad \text{or} \qquad P(A \cap B) = P(A) \cdot P(B|A)$$

Mathematical expectation

$$E = a_1 p_1 + a_2 p_2 + \cdots + a_k p_k$$

Probability   Binomial distribution
Distributions

$$f(x) = \binom{n}{x} p^x (1 - p)^{n - x}$$

Mean of probability distribution

$$\mu = \sum x \cdot f(x)$$

Standard deviation of probability distribution

$$\sigma = \sqrt{\sum (x - \mu)^2 \cdot f(x)}$$

Continued inside back cover

MODERN
ELEMENTARY
STATISTICS

MODERN ELEMENTARY STATISTICS

Tenth Edition

JOHN E. FREUND

Arizona State University

PRENTICE HALL, Upper Saddle River, New Jersey 07458

Library of Congress Cataloging-in-Publication Data

Freund, John E.
  Modern elementary statistics.—10th ed./John E. Freund.
    p. cm.
  Statistics.
  Includes bibliographical references and index.
  ISBN 0-13-017701-6
    I. Title.
QA276.12.F738 2001
519.5—dc21   00-029792

Acquisition Editor: **Kathleen Boothby Sestak**
Special Projects Manager: **Ann Heath**
Editorial/Production Supervision: **Bayani Mendoza de Leon**
Assistant Vice President of Production and Manufacturing: **David W. Riccardi**
Senior Managing Editor: **Linda Mihatov Behrens**
Executive Managing Editor: **Kathleen Schiaparelli**
Manufacturing Buyer: **Alan Fischer**
Manufacturing Manager: **Trudy Pisciotti**
Director of Marketing: **John Tweeddale**
Marketing Manager: **Angela Battle**
Marketing Assistant: **Vince Jansen**
Associate Editor, Mathematics/Statistics Media: **Audra J. Walsh**
Director of Creative Services: **Paul Belfanti**
Art Director: **Maureen Eide**
Assistant to the Art Director: **John Christiana**
Art Manager: **Gus Vibal**
Art Editor: **Grace Hazeldine**
Cover Designer: **Stacey Abraham**
Interior Designer: **Donna Wickes**
Editorial Assistant: **Joanne Wendelken**
Cover Image: **Bruce Roberts/Photo Researcher, Inc.**

©2001, 1997, 1992, 1988, 1984, 1979, 1973, 1967, 1960, 1952 by Prentice-Hall, Inc.
Upper Saddle River, New Jersey 07458

MINITAB is a registered trademark of MINITAB statistical software.

Printed in the United States of America

10 9 8 7 6 5 4 3 2 1

ISBN 0-13-017701-6

Prentice-Hall International (UK) Limited, *London*
Prentice-Hall of Australia Pty. Limited, *Sydney*
Prentice-Hall Canada Inc., *Toronto*
Prentice-Hall Hispanoamericana, S.A., *Mexico*
Prentice-Hall of India Private Limited, *New Delhi*
Prentice-Hall of Japan, Inc., *Tokyo*
Pearson Education Asia Pte. Ltd.
Editora Prentice-Hall do Brasil, Ltda., *Rio de Janeiro*

CONTENTS†

PREFACE xi

1 INTRODUCTION   1

1.1   The Growth of Modern Statistics   2
1.2   The Study of Statistics   3
1.3   Descriptive Statistics and Statistical Inference   4
1.4   The Nature of Statistical Data   7
1.5   Checklist of Key Terms   10
1.6   References   10

2 SUMMARIZING DATA: LISTING AND GROUPING   12

2.1   Listing Numerical Data   13
2.2   Stem-and-Leaf Displays   16
2.3   Frequency Distributions   20
2.4   Graphical Presentations   30
*2.5   Summarizing Two-Variable Data   37
2.6   Checklist for Key Terms   45
2.7   References   45

3 SUMMARIZING DATA: MEASURES OF LOCATION   47

3.1   Populations and Samples   48
3.2   The Mean   49
3.3   The Weighted Mean   53
3.4   The Median   58
3.5   Other Fractiles   62
3.6   The Mode   66

†All sections marked * are optional; they can be omitted without loss of continuity.

v

*3.7   The Description of Grouped Data   70
3.8   Technical Note (Summations)   76
3.9   Checklist of Key Terms   79
3.10   References   79

4 SUMMARIZING DATA: MEASURES OF VARIATION   80

4.1   The Range   81
4.2   The Variance and the Standard Deviation   82
4.3   Applications of the Standard Deviation   85
*4.4   The Description of Grouped Data   92
4.5   Some Further Descriptions   94
4.6   Checklist of Key Terms   100
4.7   References   101

REVIEW EXERCISES FOR CHAPTERS 1, 2, 3, AND 4   102

5 POSSIBILITIES AND PROBABILITIES   108

5.1   Counting   109
5.2   Permutations   113
5.3   Combinations   116
5.4   Probability   123
5.5   Checklist of Key Terms   132
5.6   References   133

6 SOME RULES OF PROBABILITY   134

6.1   Sample Spaces and Events   135
6.2   The Postulates of Probability   145
6.3   Probabilities and Odds   148
6.4   Addition Rules   153
6.5   Conditional Probability   160
6.6   Multiplication Rules   163
*6.7   Bayes' Theorem   170
6.8   Checklist of Key Terms   175
6.9   References   175

7 EXPECTATIONS AND DECISIONS   177

7.1   Mathematical Expectation   178
*7.2   Decision Making   184
*7.3   Statistical Decision Problems   188
7.4   Checklist of Key Terms   190
7.5   References   191

REVIEW EXERCISES FOR CHAPTERS 5, 6, AND 7   192

8 PROBABILITY DISTRIBUTIONS   198

8.1   Random Variables   199
8.2   Probability Distributions   200
8.3   The Binomial Distribution   202
8.4   The Hypergeometric Distribution   211
8.5   The Poisson Distribution   215
*8.6   The Multinomial Distribution   219
8.7   The Mean of a Probability Distribution   223
8.8   The Standard Deviation of a Probability Distribution   225
8.9   Checklist of Key Terms   232
8.10   References   232

9 THE NORMAL DISTRIBUTION   233

9.1   Continuous Distributions   234
9.2   The Normal Distribution   237
*9.3   A Check for Normality   247
9.4   Applications of the Normal Distribution   250
9.5   The Normal Approximation to the Binomial
       Distribution   253
9.6   Checklist of Key Terms   259
9.7   References   260

10 SAMPLING AND SAMPLING DISTRIBUTIONS   261

10.1   Random Sampling   262
*10.2 Sample Designs   269
*10.3 Systematic Sampling   269

*10.4   Stratified Sampling   270
*10.5   Cluster Sampling   272
10.6   Sampling Distributions   276
10.7   The Standard Error of the Mean   279
10.8   The Central Limit Theorem   282
10.9   Some Further Considerations   285
*10.10   Technical Note (Simulation)   290
10.11   Checklist of Key Terms   293
10.12   References   293

REVIEW EXERCISES FOR CHAPTERS 8, 9, AND 10   295

11 PROBLEMS OF ESTIMATION   302

11.1   The Estimation of Means   303
11.2   The Estimation of Means (σ Unknown)   308
11.3   The Estimation of Standard Deviations   316
11.4   The Estimation of Proportions   323
11.5   Checklist of Key Terms   329
11.6   References   330

12 TESTS OF HYPOTHESES: MEANS   331

12.1   Tests of Hypotheses   332
12.2   Significance Tests   337
12.3   Tests Concerning Means   345
12.4   Tests Concerning Means (σ Unknown)   349
12.5   Differences between Means   354
12.6   Differences between Means (σ's Unknown)   357
12.7   Differences between Means (Paired Data)   359
12.8   Checklist of Key Terms   365
12.9   References   365

13 TESTS OF HYPOTHESES: STANDARD DEVIATIONS   366

13.1   Tests Concerning Standard Deviations   367
13.2   Tests Concerning Two Standard Deviations   370
13.3   Checklist of Key Terms   375
13.4   References   376

14 TESTS OF HYPOTHESES BASED ON COUNT DATA   377

14.1   Tests Concerning Proportions   378
14.2   Tests Concerning Proportions (Large Samples)   379
14.3   Differences between Proportions   382
14.4   The Analysis of an $r \times c$ Table   386
14.5   Goodness of Fit   401
14.6   Checklist of Key Terms   408
14.7   References   408

REVIEW EXERCISES FOR CHAPTERS 11, 12, 13, AND 14   410

15 ANALYSIS OF VARIANCE   417

15.1   Differences among k Means: An Example   418
15.2   The Design of Experiments: Randomization   422
15.3   One-Way Analysis of Variance   424
15.4   Multiple Comparisons   430
15.5   The Design of Experiments: Blocking   436
15.6   Two-Way Analysis of Variance   438
15.7   Two-Way Analysis of Variance without Interaction   438
15.8   The Design of Experiments: Replication   443
15.9   Two-Way Analysis of Variance with Interaction   443
15.10   The Design of Experiments: Further Considerations   450
15.11   Checklist of Key Terms   457
15.12   References   458

16 REGRESSION   459

16.1   Curve Fitting   460
16.2   The Method of Least Squares   462
16.3   Regression Analysis   473
*16.4   Multiple Regression   482
*16.5   Nonlinear Regression   486
16.6   Checklist of Key Terms   496
16.7   References   496

17 CORRELATION 498

17.1   The Coefficient of Correlation   499
17.2   The Interpretation of r 506
17.3   Correlation Analysis   511
*17.4   Multiple and Partial Correlation   516
17.5   Checklist of Key Terms   519
17.6   References   519

18 NONPARAMETRIC TESTS 520

18.1   The Sign Test   521
18.2   The Sign Test (Large Samples)   523
*18.3   The Signed-Rank Test   527
*18.4   The Signed-Rank Test (Large Samples)   531
18.5   The U-Test   535
18.6   The U-Test (Large Samples)   540
18.7   The H-Test   541
18.8   Tests of Randomness: Runs   545
18.9   Tests of Randomness: Runs (Large Samples)   547
18.10   Tests of Randomness: Runs Above and
        Below the Median   548
18.11   Rank Correlation   551
18.12   Some Further Considerations   555
18.13   Summary   555
18.14   Checklist of Key Terms   556
18.15   References   556

REVIEW EXERCISES FOR CHAPTERS 15, 16, 17, AND 18   558

STATISTICAL TABLES 567

ANSWERS TO ODD-NUMBERED EXERCISES 593

INDEX 611

PREFACE

The first edition of this book was keynoted by the statement that "statistical methods in general are nothing but a refinement of everyday thinking." These words paraphrase the statement made by Albert Einstein, the renowned scientist, that "the whole of science is nothing more than a refinement of everyday thinking." When asked whether he could be quoted on that, Dr. Einstein declined, saying that he could not claim any originality for this remark, which he made only out of the conviction that it is so.

*I*n spite of some of the changes that have been advocated about the teaching of statistics and also about the scope and the limitations of the basic subject matter of statistics itself, our views have remained unchanged. We present this revision out of the conviction that this is how statistics should be taught as a vital part of one's general education.

NEW TO THIS EDITION

Most of the innovations in this edition are based on suggestions of colleagues and students, who generously took the time to share their experiences and their thoughts. Among other things, this led to more emphasis on the use of *p*-values; the rearrangement of the basic material on statistical inference; with some regrets the deletion of the material on Bayesian estimation; and much more emphasis on the conditions (assumptions) underlying the various techniques. There are also some changes in terminology brought about by the widespread access to computers and other technology. For instance, without the limitations imposed by the size of statistical tables, some methods are no longer referred to as small-sample or as large-sample techniques.

There are also changes in format. For easy reference, the examples in the text are numbered consecutively throughout each chapter. Furthermore, the review exercises found at the end of each chapter in preceding editions are replaced by review exercises for sets of related chapters.

Computer Printouts

In particular, a greater effort is made to make the reader aware of the technologies that are available for work in statistics. Included are 36 new printouts generated by means of Release 10 Xtra for Macintosh of MINITAB software, with some of them modified slightly to delete material that is of no immediate relevance and to avoid the repetitious listing of the data. The most recent version of MINITAB is MINITAB for Windows Release 13. Completely new are 16 reproductions from the display screen of a TI-83 graphing calculator, a technology that has become increasingly popular in the teaching of elementary statistics.

Let us make it clear, however, that neither computers nor graphing calculators are required for the use of our text. Indeed, the book can be used effectively by readers who do not possess or have easy access to computers and statistical software, or to graphing calculators. Some of the exercises are labeled with the special icon 🖳 for the use of a computer and/or the icon 🖩 for the use of a graphing calculator, but these are optional. To repeat what we indicated in the preceding paragraph and shall repeat again later, the purpose of the computer printouts and the graphing calculator reproductions is to make the reader aware of the existence of these technologies for work in statistics.

Exercises

Finally, many of the more than 1,300 exercises are new or updated from previous editions. Of course, the reader is not expected to work each one, but there is an adequate variety to provide exercise material for just about everyone, regardless of his or her primary area of interest.

SUPPLEMENTS FOR TEACHING AND LEARNING

Instructor Supplements

Instructor's Solutions Manual (ISBN 0-13-018468-3)   Worked-out solutions to all of the exercises are provided in this manual. Careful attention has been paid to ensure that all methods of solution and notation are consistent with those used in the text.

Test Item File, prepared by Laurel Technical Services (ISBN 0-13-018469-1) Revised to reflect this edition's changes, the Test Item File includes three types of questions—true/false, multiple-choice, and free response—that correlate to problems in the text.

Windows PH Custom Test (ISBN 0-13-018460-8)   Incorporates three levels of test creation: 1.) selection of questions from a test item file, 2.) addition of new questions with the ability to import test and graphics files from WordPerfect, Microsoft Word, and Wordstar, and 3.) algorithmic generation of multiple questions from a single-question template. The PH Custom Test has a full-featured graphics editor supporting the complex formulas and graphics required by the statistics discipline.

Student Supplements

Data Disk   Available with the text, this tool includes the larger data sets used in the text problems and exercises, eliminating extensive data entry. The selected problems included on the data disk are indicated by the icon 🖳 .

Student Solutions Manual (ISBN 0-13-018472-1)   The Student Solutions Manual contains worked-out solutions to all of the text's odd-numbered exercises—a great resource for students as they study and work through the problem material.

For Instructors and Students

Prentice Hall Companion Web site—www.prenhall.com/freund   Designed to complement and expand upon the text, the text web site offers a variety of interactive teaching and learning tools, including links to related web sites, quizzes, Syllabus Builder, and more! For more information, contact your local Prentice Hall representative.

For additional information on the supplements and special statistical software packages available from Prentice Hall, visit us online at *http://www.prenhall.com/*

ACKNOWLEDGMENTS

Appreciation goes to my colleagues and students for their helpful suggestions and reviews of the present as well as preceding editions of this text. For this revision, I am grateful to:

Michael B. Dollinger, Pacific Lutheran University;

Kathleen Ebert, Alfred State College;

Lynn R. Eisenberg, Durham Technical Community College;

J.S. Huang, Columbia College;

Bob Jensen, Pacific Lutheran University;

Lionel Mordecai, Southwestern Community College; and especially Dr. Benjamin M. Perles of Suffolk University for his careful review of the entire manuscript.

The author is indebted to Prentice Hall for permission to reproduce the material in Table II; to the *Biometrika* trustees to reproduce the material in Tables III and IV; to the American Cyanamid Company to reproduce the material in Table VI; to the Addison-Wesley Publishing Company to reproduce the material in Table VII; to the editor of the *Annals of Mathematical Statistics* to reproduce the material in Table VIII; and to the Aerospace Research Laboratories, U.S. Air Force, to reproduce the material in Table IX.

Thanks also to Laurel Technical Services for its accuracy check of the book and preparation of the Test Item File. Finally, gratitude goes to our Prentice-Hall team: Kathy Boothby Sestak, Joanne Wendelken, Linda Behrens, Maureen Eide, and Bayani DeLeon.

John E. Freund
Paradise Valley, Arizona

MODERN
ELEMENTARY
STATISTICS

1

INTRODUCTION

1.1 The Growth of Modern Statistics 2

1.2 The Study of Statistics 3

1.3 Descriptive Statistics and Statistical Inference 4

1.4 The Nature of Statistical Data 7

1.5 Checklist of Key Terms 10

1.6 References 10

*E*verything that deals even remotely with the collection, processing, interpretation, and presentation of data belongs to the domain of statistics, and so does the detailed planning that precedes all these activities. Indeed, statistics includes such diversified tasks as calculating the batting averages of baseball players; collecting and recording data on births, marriages, and deaths; evaluating the effectiveness of commercial products; and forecasting the weather. Even one of the most advanced branches of atomic physics goes by the name of quantum statistics.

The word "statistics" itself is used in various ways. It can be used, for example, to denote the mere tabulation of numerical data, as in reports of stock market transactions and in publications such as the *Statistical Abstract of the United States* or the *World Almanac*. It can also be used to denote the totality of methods that are employed in the collection, processing, and analysis of data, numerical and otherwise, and it is in this sense that "statistics" is used in the title of this book.

The word "statistician" is also used in several ways. It can be applied to those who simply collect information, as well as to those who prepare analyses or interpretations, and it is also applied to scholars who develop the mathematical theory on which the whole subject is based. Finally, the word "statistic" in the singular is used to denote a particular measure or formula, such as an average, a range of values, a growth rate such as an economic indicator, or a measure of the correlation (or relationship) between variables.

In Sections 1.1 and 1.2 we discuss the recent growth of statistics with its ever-widening range of applications and the need for its study as part of one's specialized (professional) training as well as one's general education. In Section 1.3 we explain the distinction between the two major branches of statistics—descriptive statistics and statistical inference—and in Section 1.4 we discuss the nature of different kinds of data and thus alert the reader to the indiscriminate application of some methods used in the analysis of data.

1.1   THE GROWTH OF MODERN STATISTICS

There are several reasons why the scope of statistics and the need to study statistics have grown enormously in the last fifty or so years. One reason is the increasingly quantitative approach employed in all the sciences, as well as in business and many other activities that directly affect our lives. This includes the use of mathematical techniques in the evaluation of antipollution controls, in inventory planning, in the analysis of traffic patterns, in the study of the effects of various kinds of medications, in the evaluation of teaching techniques, in the analysis of competitive behavior of business managers and governments, in the study of diet and longevity, and so forth. Also, the availability of computers has greatly increased our ability to deal with numerical information, so that sophisticated statistical work can be done even by small businesses and by college and high school students.

The other reason is that the amount of data that is collected, processed, and disseminated to the public for one reason or another has increased almost beyond comprehension, and what part is good statistics and what part is bad statistics is anybody's guess. To act as watchdogs, more and more persons with some knowledge of statistics are needed to take an active part in the collection of the data, in the analysis of the data, and, what is equally important, in all of the preliminary planning. Without the latter, it is frightening to think of all the things that can go wrong in the compilation of statistical data. The results of costly surveys can be useless if questions are ambiguous or asked in the wrong way, if they are asked of the wrong persons, in the wrong place, or at the wrong time. Much of this is just common sense, as is illustrated by the following examples.

EXAMPLE   1.1   To determine public sentiment about the continuation of a government program, an interviewer asks, "Do you feel that this wasteful program should be stopped?" Explain why this question will probably not evoke a fair (objective) response.

Solution   The interviewer is *begging the question* by suggesting, in fact, that the program is wasteful.   ∎

EXAMPLE   1.2   To study consumer reaction to a new convenience food, a house-to-house survey is conducted during weekday mornings, with no provisions for return visits in case no one is home. Explain why this approach may well yield misleading information.

Solution   This survey will fail to reach those most likely to use the product: single persons and married couples with both spouses employed.   ∎

Although much of the aforementioned growth of statistics began prior to the "computer revolution," the widespread availability and use of computers has greatly accelerated the process. In particular, computers enable one to handle, analyze, and dissect large masses of data and to perform calculations that previously had been too cumbersome to contemplate. Let us point out, though, that access to a computer is not imperative for the study of statistics so long as one's main goal is to gain an understanding of the subject. Some computer uses are illustrated in this textbook, but they are intended merely to make the reader aware of the technology that is available for work in statistics. Thus, computers are not required for the use of this textbook, but for those familiar with statistical software, there are some special exercises marked with an appropriate icon. Otherwise, none of the exercises require more than a simple hand-held calculator.

1.2   THE STUDY OF STATISTICS

The subject of statistics can be presented at various levels of mathematical difficulty, and it may be directed toward applications in various fields of inquiry. Accordingly, many textbooks have been written on business statistics, educational statistics, medical statistics, psychological statistics, ... , and even on statistics for historians. Although problems arising in these various disciplines will sometimes require special statistical techniques, none of the basic methods discussed in this text is restricted to any particular field of application. In the same way in which $2 + 2 = 4$ regardless of whether we are adding dollar amounts, horses, or trees, the methods we shall present provide **statistical models** that apply regardless of whether the data are IQs, tax payments, reaction times, humidity readings, test scores, and so on. To illustrate this further, consider the following illustration (reproduced from Example 14.1 on page 378):

> It has been claimed that more than 70% of the students attending a large state university are opposed to a plan to increase student fees in order to build new parking facilities. If 15 of 18 students selected at random at that university are opposed to the plan, test the claim at the 0.05 level of significance.

Except for the reference to the "level of significance," which is a technical term, the question asked here should be clear, and it should also be apparent that the answer would be of interest primarily to students attending this university and its administrators. However, if we had wanted to give an example that is of special interest to growers of fruit and vegetables, engineers, doctors, and ecologists, we could have rephrased Example 14.1 as follows:

It has been claimed that more than 70% of the citrus trees in a Florida county were severely damaged by a recent frost. If 15 of 18 citrus trees selected at random in that county were severely damaged by that frost, test the claim at the 0.05 level of significance.

It has been claimed that more than 70% of certain airplanes have cracks in their tail rudders due to metal fatigue. If 15 of 18 such airplanes selected at random had cracks in their tail rudders due to metal fatigue, test the claim at the 0.05 level of significance.

It has been claimed that more than 70% of all doctors associated with HMOs are dissatisfied with their fees. If 15 of 18 doctors selected at random from among those associated with HMOs are dissatisfied with their fees, test the claim at the 0.05 level of significance.

It has been claimed that more than 70% of all cars of a certain model year emit excessive amounts of pollutants. If 15 of 18 such cars selected at random were found to emit excessive amounts of pollutants, test the claim at the 0.05 level of significance.

Insofar as the work in this book is concerned, the statistical treatment of all these versions of Example 14.1 is the same, and with some imagination the reader should be able to rephrase it for virtually any field of specialization. As some authors do, we could present, and so designate, special problems for readers with special interests, but this would defeat our goal of impressing upon the reader the importance of statistics in all of science, business, and everyday life. To attain this goal, we have included in this text exercises covering a wide spectrum of interests.

To avoid the possibility of misleading anyone with the different versions of Example 14.1 given above, let us make it clear that all statistical problems cannot be squeezed into the same mold. Although the methods we shall study in this book are all widely applicable, it is always important to make sure that the statistical model we are using is the right one.

1.3 DESCRIPTIVE STATISTICS AND STATISTICAL INFERENCE

The origin of modern statistics can be traced to two areas of interest that, on the surface, have very little in common: government (political science) and games of chance.

Governments have long used censuses to count persons and property, and the problem of describing, summarizing, and analyzing census data has led to the development of methods that, until recently, constituted about all there was to the subject of statistics. These methods, which at first consisted primarily of presenting data in the form of tables and charts, make up what we now call **descriptive statistics.** This includes anything done to data that is designed to summarize, or describe, without going any further; that is, without attempting to infer anything that goes beyond the data themselves. For instance, if tests performed on six compact cars imported in 1999 showed that they were able to accelerate from 0 to 60 mph (miles per hour) in 12.9, 16.5, 11.3, 15.2, 18.2, and 17.7

seconds, and we report that half of them accelerated from 0 to 60 mph in less than 16.0 seconds, our work belongs to the domain of descriptive statistics. This would also be the case if we claim that these cars averaged

$$\frac{12.9 + 16.5 + 11.3 + 15.2 + 18.2 + 17.7}{6} = 15.3$$

seconds but not if we concluded that half of *all* compact cars imported that year could accelerate from 0 to 60 mph in less than 16.0 seconds.

Although descriptive statistics is an important branch of statistics and it continues to be widely used, statistical information usually arises from samples (from observations made on only part of a large set of items), and this means that its analysis requires generalizations that go beyond the data. As a result, the most important feature of the recent growth of statistics has been a shift in emphasis from methods that merely describe to methods that serve to make generalizations; that is, a shift in emphasis from descriptive statistics to the methods of **statistical inference.**

Such methods are required, for instance, to predict the operating life span of a hand-held calculator (on the basis of the performance of several such calculators); to estimate the 2005 assessed value of all privately owned property in Orange County, California (on the basis of business trends, population projections, and so forth); to compare the effectiveness of two reducing diets (on the basis of the weight losses of persons who have been on the diets); to determine the most effective dose of a new medication (on the basis of tests performed with volunteer patients from selected hospitals); or to predict the flow of traffic on a freeway that has not yet been built (on the basis of past traffic counts on alternative routes).

In each of the situations described in the preceding paragraph, there are uncertainties because there is only partial, incomplete, or indirect information; therefore, the methods of statistical inference are needed to judge the merits of our results, to choose a "most promising" prediction, or to select a "most reasonable" (perhaps a "potentially most profitable") course of action.

In view of the uncertainties, we handle problems like these with statistical methods that find their origin in games of chance. Although the mathematical study of games of chance dates back to the seventeenth century, it was not until the early part of the nineteenth century that the theory developed for "heads or tails," for example, or "red or black" or "even or odd," was applied also to real-life situations where the outcomes were "boy or girl," "life or death," "pass or fail," and so forth. Thus, **probability theory** was applied to many problems in the behavioral, natural, and social sciences and at present provides an important tool for the analysis of any situation (in science, in business, or in everyday life) that in some way involves an element of uncertainty or chance. In particular, it provides the basis for the methods that we use when we generalize from observed data, namely, when we use the methods of statistical inference.

In recent years it has been suggested that the emphasis has swung too far from descriptive statistics to statistical inference, and that more attention should be paid to the treatment of problems requiring only descriptive techniques. To accommodate these needs, some new descriptive methods have been developed under the general heading of **exploratory data analysis.** Two of these will be presented in Sections 2.2 and 4.5.

■ E x e r c i s e s

1.1 Rephrase Example 14.1 referred to on page 3 so that it would be of special interest to
(a) an insurance salesperson;
(b) a travel agent.

1.2 Rephrase Example 14.1 referred to on page 3 so that it would be of special interest to
(a) a biologist;
(b) an architect.

1.3 Bad statistics may well result from asking questions in the wrong way or of the wrong persons. Explain why the following may lead to useless data:
(a) To study business executives' reaction to photocopying machines, the Xerox corporation hires a research organization to ask business executives the question, "How do you like using Xerox copiers?"
(b) To determine what the average person spends on a vacation, a researcher interviews persons disembarking from a Concorde jet.

1.4 Bad statistics may well result from asking questions in the wrong place or at the wrong time. Explain why the following may lead to useless data:
(a) To predict an election, a poll taker interviews persons coming out of the building that houses the headquarters of a political party.
(b) To study the spending patterns of individuals, a survey is conducted during the first three weeks of December.

1.5 Explain why each of the following studies may fail to yield the desired information:
(a) To ascertain facts about personal habits, a sample of adults are asked how often they take a bath or a shower.
(b) To determine the average annual income of its graduates 10 years after graduation, the alumni office of a university sent questionnaires in 2002 to all members of the class of 1992.

1.6 In four French vocabulary tests a student received successive grades of 56, 62, 70, and 78. Which of the following conclusions can be obtained from these data by purely descriptive methods and which require generalizations? Explain your answers.
(a) Only three of the grades exceeded 60.
(b) The student's grades increased from each test to the next.
(c) The student must have studied harder for each successive test.
(d) In the fourth test the student must have been lucky that the questions covered material he had studied the day before the test.

1.7 Ben and Jim are avid readers. In a recent month Ben read four fiction books and two nonfiction books, while Jim read three fiction books and three nonfiction books. Which of the following conclusions can be obtained from these figures by purely descriptive methods and which require generalizations? Explain your answers.
(a) In the given month Ben and Jim read equally many books.
(b) Ben always reads more fiction books than Jim.
(c) Over a year, Jim averages three nonfiction books per month.
(d) Ben and Jim's reading speed is just about the same.

1.8   In 1990, 1991, 1992, 1993, and 1994 the amount of money spent on radio advertising totaled 8,726, 8,476, 8,654, 9,457, and 10,295 million dollars. Which of the following conclusions can be obtained from these data by purely descriptive methods and which require generalizations? Explain your answers.
(a) In each of these years the total exceeded 8,000 million dollars.
(b) From 1990 through 1994 the total increased from year to year.
(c) The amount of money spent on radio advertising in 1995 must have exceeded 10,000 million dollars.
(d) In all these years, the total spent on radio advertising was less than that spent on direct mail advertising.

1.9   Driving the same model truck, five persons averaged 15.5, 14.7, 16.0, 15.5, and 14.8 miles per gallon. Which of the following conclusions can be obtained from these data by purely descriptive methods and which require generalizations? Explain your answers.
(a) The third driver must have driven mostly on rural roads.
(b) The second driver must have driven faster than the other four.
(c) More often than any other figure, the drivers averaged 15.5 miles per gallon.
(d) None of the drivers averaged better than 16.0 miles per gallon.

1.10   With reference to Exercise 1.9, can we conclude that the five drivers averaged 15.3 miles per gallon?

1.11   A statistically minded broker has her office on the third floor of a very tall office building, and whenever she leaves her office she records whether the first elevator that stops is going up or going down. Having done this for some time, she discovers that the vast majority of the time the first elevator that stops is going down. Comment on the following "conclusions":
(a) Fewer elevators are going up than are going down.
(b) The next time she leaves her office the first elevator that stops will be going down.

1.4   THE NATURE OF STATISTICAL DATA

Essentially, there are two kinds of statistical data: **numerical data** and **categorical data.** The former are obtained by measuring or counting, and they are also referred to as **quantitative data.** Such data may consist, for example, of the weights of the guinea pigs used in an experiment (obtained by measuring) or the daily absences from a class throughout a school year (obtained by counting). In contrast, categorical data result from descriptions, and they may exist, for example, of the blood types of hospital patients, their marital status, or their religious affiliation. Categorical data are also referred to as **qualitative data.** For ease in manipulating (recording or sorting) categorical data, they are often **coded** by assigning numbers to the different categories, and thus converting the categorical data to numerical data in a trivial sense. For example, marital status might be coded by letting 1, 2, 3, and 4 denote a person's being single (never married), married, widowed, or divorced.

Numerical data are classified further as being **nominal data, ordinal data, interval data,** or **ratio data.** Nominal data are numerical in name only, as typified by the preceding example, where the numbers 1, 2, 3, and 4 were used to denote a person's being single (never married), married, widowed, or divorced. By "nominal data are numerical in name only" we mean that they do not share any of the properties of the numbers we deal with in ordinary arithmetic. With regard to the codes for marital status, we cannot write $3 > 1$ or $2 < 4$, and we cannot write $2 - 1 = 4 - 3$, $1 + 3 = 4$, or $4 \div 2 = 2$. This illustrates how important it is always to check whether the mathematical treatment of statistical data is really legitimate.

Let us now consider some examples where data share some, but not necessarily all, of the properties of the numbers we deal with in ordinary arithmetic. For instance, in mineralogy the hardness of solids is sometimes determined by observing "what scratches what." If one mineral can scratch another it receives a higher hardness number, and on the Mohs scale the numbers from 1 to 10 are assigned, respectively, to talc, gypsum, calcite, fluorite, apatite, feldspar, quartz, topaz, sapphire, and diamond. With these numbers we can write $6 > 3$, for example, or $7 < 9$, since feldspar is harder than calcite and quartz is softer than sapphire. On the other hand, we cannot write $10 - 9 = 2 - 1$, for example, because the difference in hardness between diamond and sapphire is actually much greater than that between gypsum and talc. Also, it would be meaningless to say that topaz is twice as hard as fluorite simply because their respective hardness numbers on the Mohs scale are 8 and 4.

If we cannot do anything except set up inequalities, as was the case in the preceding example, we refer to the data as ordinal data. In connection with ordinal data, $>$ does not necessarily mean "greater than." It may be used to denote "happier than," "preferred to," "more difficult than," "tastier than," and so forth.

If we can also form differences, but not multiply or divide, we refer to the data as interval data. To give an example, suppose we are given the following temperature readings in degrees Fahrenheit: $63°$, $68°$, $91°$, $107°$, $126°$, and $131°$. Here we can write $107° > 68°$ or $91° < 131°$, which simply means that $107°$ is warmer than $68°$ and that $91°$ is colder than $131°$. Also, we can write $68° - 63° = 131° - 126°$, since equal temperature differences are equal in the sense that the same amount of heat is required to raise the temperature of an object from $63°$ to $68°$ as from $126°$ to $131°$. On the other hand, it would not mean much if we say that $126°$ is twice as hot as $63°$, even though $126 \div 63 = 2$. To show why, we have only to change to the Celsius scale, where the first temperature becomes $\frac{5}{9}(126 - 32) = 52.2°$, the second temperature becomes $\frac{5}{9}(63 - 32) = 17.2°$, and the first figure is now more than three times the second. This difficulty arises because the Fahrenheit and Celsius scales both have artificial origins (zeros); in other words, the number 0 of neither scale is indicative of the absence of whatever quantity we are trying to measure.

If we can also form quotients, we refer to the data as ratio data, and such data are not difficult to find. They include all the usual measurements (or determinations) of length, height, money amounts, weight, volume, area, pressure, elapsed time (though not calendar time), sound intensity, density, brightness, velocity, and so on.

The distinction we have made here between nominal, ordinal, interval, and ratio data is important, for as we shall see, the nature of a set of data may suggest the use of particular statistical techniques. To emphasize the point that what we can and cannot do arithmetically with a given set of data depends on the nature of the data, consider the following scores that four students obtained in the three parts of a comprehensive history test:

	American history	European history	Ancient history
Linda	89	51	40
Tom	61	56	54
Henry	40	70	55
Rose	13	77	72

The totals for the four students are 180, 171, 165, and 162, so that Linda scored highest, followed by Tom, Henry, and Rose.

Suppose now that somebody proposes that instead of adding the scores obtained in the three parts of the test, we compare the overall performance of the four students by ranking their scores from high to low for each part of the test and then average their ranks (that is, add them and divide by 3). What we get is shown in the following table:

	American history	European history	Ancient history	Average rank
Linda	1	4	4	3
Tom	2	3	3	$2\frac{2}{3}$
Henry	3	2	2	$2\frac{1}{3}$
Rose	4	1	1	2

Now, if we look at the average ranks, we find that Rose came out best, followed by Henry, Tom, and Linda, so that the order has been reversed from what it was before. How can this be? Well, strange things can happen when we average ranks. For instance, when it comes to their ranks, Linda's outscoring Tom by 28 points in American history counts just as much as Tom's outscoring her by 5 points in European history, and Tom's outscoring Henry by 21 points in American history counts just as much as Henry's outscoring him by a single point in ancient history. We conclude that, perhaps, we should not have averaged the ranks, but it might also be pointed out that, perhaps, we should not even have totaled the original scores. The variation of the American history scores, which go from 13 to 89, is much greater than that of the other two kinds of scores, and this strongly affects the total scores and suggests a possible shortcoming of the procedure. These comments sound interesting, but we shall not follow up on them as it has been our goal merely to alert the reader against the indiscriminate use of statistical techniques; that is, to show how the choice of a statistical technique may be dictated by the nature of the data.

Exercises

1.12 Do we get nominal data or ordinal data if mechanics are asked whether replacing the spark plugs on a new model car is very difficult, difficult, easy, or very easy, and these alternatives are coded 1, 2, 3, and 4?

1.13 Do we get nominal data or ordinal data if the religion of patients in a hospital are recorded as 1, 2, 3, 4, or 5, representing Protestant, Catholic, Jewish, other, and none?

1.14 Are the following nominal data, ordinal data, interval data, or ratio data? Explain your answers.
(a) The number of passengers on a flight from San Francisco to Chicago.
(b) Social security numbers.
(c) Academic rank of faculty members coded as 1, 2, 3, or 4 depending on whether they are professors, associate professors, assistant professors, or instructors.

1.15 Are the following nominal data, ordinal data, interval data, or ratio data? Explain your answers.
(a) Leap years.
(b) The numbers on the uniforms of football players.

1.16 In two major golf tournaments one professional golfer finished second and ninth, while another finished sixth and fifth. Comment on the argument that since $2 + 9 = 6 + 5$, the overall performance of the two golfers in these two tournaments was equally good.

1.5 Checklist of Key Terms *(with page references to their definitions)*

Categorical data, 7
Descriptive statistics, 4
Exploratory data analysis, 6
Interval data, 7
Nominal data, 7
Numerical data, 7
Ordinal data, 7

Probability theory, 5
Qualitative data, 7
Quantitative data, 7
Ratio data, 7
Scaling, 11
Statistical inference, 5
Statistical model, 3

1.6 References

Brief and informal discussions of what statistics is and what statisticians do may be found in pamphlets titled Careers in Statistics *and* Statistics as a Career: Women at Work, *which are published by the American Statistical Association. They may be obtained by writing to this organization at 1429 Duke Street, Alexandria, Virginia 22314-3402.*

Among the few books on the history of statistics, on the elementary level is

WALKER, H. M., *Studies in the History of Statistical Method.* Baltimore: The Williams & Wilkins Company, 1929.

and on the more advanced level

KENDALL, M. G., and PLACKETT, R. L., eds. *Studies in the History of Statistics and Probability,* Vol. II. New York: Macmillan Publishing Co., Inc., 1977.

PEARSON, E. S., and KENDALL, M. G., eds. *Studies in the History of Statistics and Probability.* New York: Hafner Press, 1970.

STIGLER, S. M., *The History of Statistics.* Cambridge, Mass.: Harvard University Press, 1986.

*A more detailed discussion of the nature of statistical data and the general problem of **scaling** (namely, the problem of constructing scales of measurement or assigning scale scores) may be found in*

HILDEBRAND, D. K., LAING, J. D., and ROSENTHAL, H., *Analysis of Ordinal Data.* Beverly Hills, Calif.: Sage Publications, Inc., 1977.

REYNOLDS, H. T., *Analysis of Nominal Data.* Beverly Hills, Calif.: Sage Publications, Inc., 1977.

SIEGEL, S., *Nonparametric Statistics for the Behavioral Sciences.* New York: McGraw-Hill Book Company, 1956.

The following are some titles from the ever-growing list of books on statistics that are written for the layperson:

BROOK, R. J., ARNOLD, G. C., HASSARD, T. H., and PRINGLE, R. M., eds., *The Fascination of Statistics.* New York: Marcel Dekker, Inc., 1986.

FEDERER, W. T., *Statistics and Society.* New York: Marcel Dekker, Inc., 1991.

GONICK, L., and SMITH, W., *The Cartoon Guide to Statistics.* New York: HarperCollins Publishers, Inc., 1993.

HOLLANDER, M., and PROSCHAN, F., *The Statistical Exorcist: Dispelling Statistics Anxiety.* New York: Marcel Dekker, Inc., 1984.

HOOKE, R., *How to Tell the Liars from the Statisticians.* New York: Marcel Dekker, Inc., 1983.

KIMBLE, G. A., *How to Use (and Misuse) Statistics.* Upper Saddle River, N.J.: Prentice Hall, Inc., 1978.

LARSEN, R. J., and STROUP, D. F., *Statistics in the Real World.* New York: Macmillan Publishing Co., Inc., 1976.

RUNYON, R. P., *Winning with Statistics.* Reading, Mass.: Addison-Wesley Publishing Company, Inc., 1977.

TANUR, J. M., ed., *Statistics, A Guide to the Unknown.* San Francisco: Holden-Day, Inc., 1972.

WANG, C., *Sense and Nonsense of Statistical Inference.* New York: Marcel Dekker, Inc., 1993.

2 SUMMARIZING DATA: LISTING AND GROUPING†

2.1   Listing Numerical Data   13

2.2   Stem-and-Leaf Displays   16

2.3   Frequency Distributions   20

2.4   Graphical Presentations   30

*2.5   Summarizing Two-Variable Data   37

2.6   Checklist of Key Terms   45

2.7   References   45

*I*n recent years the collection of statistical data has grown at such a rate that it would be impossible to keep up with even a small part of the things that directly affect our lives unless this information is disseminated in "predigested" or summarized form. The whole matter of putting large masses of data into a usable form has always been important, but it has multiplied greatly in the last few decades. This has been due partly to the development of computers, which now make it possible to accomplish in minutes what was previously left undone because it would have taken months or years, and partly to the deluge of data generated by

†Since computer printouts and a reproduction from the display screen of a graphing calculator appear first in this chapter, let us repeat from the Preface that the purpose of the printouts and the graphing calculator reproductions is to make the reader aware of the existence of these technologies for work in statistics. Let us make it clear, however, that neither computers nor graphing calculators are required for the use of our text. Indeed, this book can be used effectively by readers who do not possess or have easy access to computers and statistical software or to graphing calculators. Some of the exercises are labeled with special icons for the use of a computer or a graphing calculator, but these exercises are optional.

the increasingly quantitative approach of the sciences, especially the behavioral and social sciences, where nearly every aspect of human life is nowadays measured in one way or another.

The most common method of summarizing data is to present them in condensed form in tables or charts, and at one time this took up the better part of an elementary course in statistics. Nowadays, there is so much else to learn in statistics that very little time is devoted to this kind of work. In a way this is unfortunate, because one does not have to look far in newspapers, magazines, and even professional journals to find unintentionally or intentionally misleading statistical charts.

In Sections 2.1 and 2.2 we shall present ways of listing data so that they present a good overall picture and, hence, are easy to use. By **listing** we are referring to any kind of treatment that preserves the identity of each value (or item). In other words, we rearrange but do not change. A speed of 63 mph remains a speed of 63 mph, a salary of $75,000 remains a salary of $75,000, and when sampling public opinion, a Republican remains a Republican and a Democrat remains a Democrat. In Sections 2.3 and 2.4 we shall discuss ways of **grouping** data into a number of classes, intervals, or categories and presenting the result in the form of a table or a chart. This will leave us with data in a relatively compact and easy-to-use form, but it does entail a substantial loss of information. Instead of a person's weight, we may know only that he or she weighs anywhere from 160 to 169 pounds, and instead of an actual pollen count we may know only that it is *medium* (11–25 parts per cubic meter).

2.1 LISTING NUMERICAL DATA

Listing and, thus, organizing the data is usually the first task in any kind of statistical analysis. As a typical situation, consider the following data, representing the lengths (in centimeters) of 60 sea trout caught by a commercial trawler in Delaware Bay:

19.2	19.6	17.3	19.3	19.5	20.4	23.5	19.0	19.4	18.4
19.4	21.8	20.4	21.0	21.4	19.8	19.6	21.5	20.2	20.1
20.3	19.7	19.5	22.9	20.7	20.3	20.8	19.8	19.4	19.3
19.5	19.8	18.9	20.4	20.2	21.5	19.9	21.7	19.5	20.9
18.1	20.5	18.3	19.5	18.3	19.0	18.2	21.9	17.0	19.7
20.7	21.1	20.6	16.6	19.4	18.6	22.7	18.5	20.1	18.6

The mere gathering of this information is no small task, but it should be clear that more must be done to make the numbers comprehensible.

What can be done to make this mass of information more usable? Some persons find it interesting to locate the extreme values, which are 16.6 and 23.5 for this list. Occasionally, it is useful to sort the data in an ascending or descending order. The following list gives the lengths of the trout arranged in an ascending order:

16.6	17.0	17.3	18.1	18.2	18.3	18.3	18.4	18.5	18.6
18.6	18.9	19.0	19.0	19.2	19.3	19.3	19.4	19.4	19.4
19.4	19.5	19.5	19.5	19.5	19.5	19.6	19.6	19.7	19.7
19.8	19.8	19.8	19.9	20.1	20.1	20.2	20.2	20.3	20.3
20.4	20.4	20.4	20.5	20.6	20.7	20.7	20.8	20.9	21.0
21.1	21.4	21.5	21.5	21.7	21.8	21.9	22.7	22.9	23.5

Sorting a large set of numbers in an ascending or descending order can be a surprisingly difficult task. It is simple, though, if we can use a computer or a graphing calculator. In that case, entering the data is the most tedious part. Then, with a graphing calculator we press **STAT** and **2,** fill in the list where we put the data, press **ENTER,** and the display screen spells out **DONE.**

If a set of data consists of relatively few values, many of which are repeated, we simply count how many times each value occurs and then present the result in the form of a table or a **dot diagram.** In such a diagram we indicate by means of dots how many times each value occurs.

EXAMPLE 2.1   An audit of twenty tax returns revealed 0, 2, 0, 0, 1, 3, 0, 0, 0, 1, 0, 1, 0, 0, 2, 1, 0, 0, 1, and 0 mistakes in arithmetic.

(a) Construct a table showing the number of tax returns with 0, 1, 2, and 3 mistakes in arithmetic.
(b) Draw a dot diagram displaying the same information.

Solution   Counting the number of 0's, 1's, 2's, and 3's, we find that there are, respectively, 12, 5, 2, and 1. This information is displayed as follows, in tabular form on the left and in graphical form on the right.

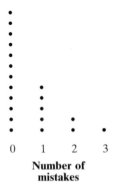

Number of mistakes	Number of tax returns
0	12
1	5
2	2
3	1

There are various ways in which dot diagrams can be modified; for instance, instead of dots we can use other symbols such as ×'s, *'s, ★'s, or ◇'s. Also, we could align the dots horizontally rather than vertically.

The methods we used to display relatively few numerical values, many of which are repeated, can also be used to display categorical data.   ∎

EXAMPLE   2.2

The faculty of a university's mathematics department consists of four professors, six associate professors, eleven assistant professors, and nine instructors. Display this information in the form of a horizontally aligned dot diagram.

 olution

Faculty rank	
Professor	☆ ☆ ☆ ☆
Associate professor	☆ ☆ ☆ ☆ ☆ ☆
Assistant professor	☆ ☆ ☆ ☆ ☆ ☆ ☆ ☆ ☆ ☆ ☆
Instructor	☆ ☆ ☆ ☆ ☆ ☆ ☆ ☆ ☆

Another way of modifying dot diagrams is to replace the numbers of dots with rectangles, whose lengths are proportional to the respective numbers of dots. Such diagrams are referred to as **bar charts,** and the rectangles are often supplemented with the corresponding frequencies (numbers of symbols) as in Figure 2.1. ∎

EXAMPLE   2.3

Draw a bar chart for the data of Example 2.1; that is, for the numbers of mistakes in arithmetic in the twenty tax returns.

Solution

FIGURE 2.1
Bar chart of mistakes in arithmetic in tax returns.

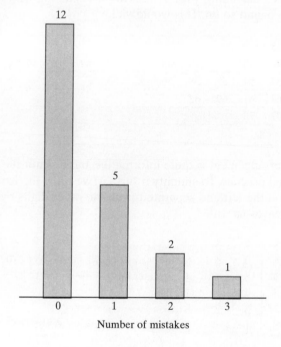

∎

2.2 STEM-AND-LEAF DISPLAYS

Dot diagrams are impractical and ineffective when a set of data contains many different values or categories, or when some of the values or categories require too many dots to yield a coherent picture. To give an example, consider the first-round scores in a PGA tournament, where the lowest score was a 62, the highest score was an 88, and 27 of the 126 golfers shot a par 72. This illustrates both of the reasons cited previously for not using dot diagrams. There are too many different values from 62 to 88, and at least one of them, 72, requires too many dots.

In recent years, an alternative method of listing data has been proposed for the exploration of relatively small sets of numerical data. It is called a **stem-and-leaf display** and it also yields a good overall picture of the data without any appreciable loss of information. Again, each value retains its identity, and the only information we lose is the order in which the data were obtained.

To illustrate this technique, consider the following data on the number of rooms occupied each day in a resort hotel during a recent month of June:

55	49	37	57	46	40	64	35	73	62
61	43	72	48	54	69	45	78	46	59
40	58	56	52	49	42	62	53	46	81

The smallest and largest values are 35 and 81, so that a dot diagram would require that we allow for 47 possible values. Actually, only 25 of the values occur, but in order to avoid having to allow for that many possibilities, let us combine all the values beginning with a 3, all those beginning with a 4, all those beginning with a 5, and so on. This would yield

37	35									
49	46	40	43	48	45	46	40	49	42	46
55	57	54	59	58	56	52	53			
64	62	61	69	62						
73	72	78								
81										

This arrangement is quite informative, but it is not the kind of diagram we use in actual practice. To simplify it further, we show the first digit only once for each row, on the left and separated from the other digits by means of a vertical line. This leaves us with

3	7	5							
4	9	6	0	3	8	5	6	0	9	2	6
5	5	7	4	9	8	6	2	3		
6	4	2	1	9	2					
7	3	2	8							
8	1									

and this is what we refer to as a stem-and-leaf display. In this arrangement, each row is called a **stem,** each number on a stem to the left of the vertical line is called a **stem label,** and each number on a stem to the right of the vertical line is called a **leaf.** As we shall see later, there is a certain advantage to arranging the leaves on each stem according to size, and for our data this would yield

3	5	7								
4	0	0	2	3	5	6	6	6	8	9	9
5	2	3	4	5	6	7	8	9		
6	1	2	2	4	9					
7	2	3	8							
8	1									

A stem-and-leaf display is actually a hybrid kind of arrangement, obtained in part by grouping and in part by listing. The values are grouped into the six stems, and yet each value retains its identity. Thus, from the preceding stem-and-leaf display, we can reconstruct the original data as 35, 37, 40, 40, 42, 43, 45, 46, 46, 46, 48, 49, 49, 52, 53, . . . , and 81, though not in their original order.

There are various ways in which stem-and-leaf displays can be modified. For instance, the stem labels or the leaves could be two-digit numbers, so that

$$24 \mid 0 \quad 2 \quad 5 \quad 8 \quad 9$$

would represent the numbers 240, 242, 245, 248, and 249, and

$$2 \mid 31 \quad 45 \quad 70 \quad 88$$

would represent the numbers 231, 245, 270, and 288.

Now suppose that in the room occupancy example we had wanted to use more than six stems. Using each stem label twice, if necessary, once to hold the leaves from 0 to 4 and once to hold the leaves from 5 to 9, we would get

3	5	7					
4	0	0	2	3			
4	5	6	6	6	8	9	9
5	2	3	4				
5	5	6	7	8	9		
6	1	2	2	4			
6	9						
7	2	3					
7	8						
8	1						

and this is called a **double-stem display.** Another modification of stem-and-leaf displays is mentioned in Exercise 2.22.

We shall not discuss stem-and-leaf displays in any great detail, as it has been our objective mainly to present one of the relatively new techniques, which come under the general heading of **exploratory data analysis.** These techniques are

```
MTB > Stem-and-Leaf cl.


Stem-and-leaf of C1            N  = 30
Leaf Unit = 1.0


     2       3  57
     6       4  0023
    13       4  5666899
    (3)      5  234
    14       5  56789
     9       6  1224
     5       6  9
     4       7  23
     2       7  8
     1       8  1
```

FIGURE 2.2

Computer printout of a double-stem display of the room occupancy data.

used primarily to explore data without or before using the more traditional methods of statistical analysis. Although the construction of stem-and-leaf displays is straightforward and quite easy, the work can be simplified further by using a computer and appropriate software. This is illustrated by Figure 2.2.

■ Exercises

2.1   In ten recent years, there were 1, 6, 2, 1, 1, 3, 0, 1, 1, and 1 hurricanes that reached U.S. shores. Construct a dot diagram.

2.2   The number of passengers in 25 four-door sedans observed on I-17 north of Phoenix, Arizona, were 1, 3, 0, 0, 2, 4, 1, 1, 2, 0, 1, 3, 1, 0, 2, 2, 3, 3, 0, 1, 0, 2, 0, 1, and 1. Construct a dot diagram.

2.3   On 40 business days, a pharmacy filled 7, 4, 6, 9, 5, 8, 8, 7, 6, 10, 7, 7, 6, 9, 6, 8, 4, 9, 8, 7, 5, 8, 7, 5, 8, 10, 6, 9, 7, 7, 8, 10, 6, 6, 7, 8, 7, 9, 7, and 8 prescriptions for AMBIEN sleeping pills.
      (a) Construct a table showing on how many days the pharmacy filled 4, 5, 6, 7, 8, 9, and 10 prescriptions for this sleeping pill.
      (b) Construct a dot diagram for these data, using asterisks instead of dots.

2.4   In a special sale, a Buick dealer advertised the following secondhand cars: 95 LeSabre, 98 Regal, 97 LeSabre, 97 Park Avenue, 94 Regal, 95 Skylark, 97 LeSabre, 94 Skylark, 94 Century, 98 Skylark, 97 Skylark, 98 LeSabre, 96 Regal, 97 Skylark, 96 Century, 98 Regal, 97 LeSabre, 96 LeSabre, 98 Park Avenue, 97 Skylark, 97 Riviera, 94 LeSabre, 96 Regal, 96 Century, and 98 LeSabre.
      (a) Construct a dot diagram showing how these cars are distributed according to model year.
      (b) Construct a dot diagram showing how these cars are distributed according to model name.

2.5   At a dog show, an interviewer asked 30 persons to name their favorite breed of dog in the Hound Group. Their replies were dachshund, greyhound, basset, beagle, afghan, afghan, beagle, dachshund, beagle, afghan, dachshund, greyhound, beagle, greyhound, dachshund, dachshund, afghan, dachshund, greyhound, beagle, afghan, greyhound, beagle, dachshund, dachshund, beagle, bloodhound, greyhound, basset, and beagle. Construct a dot diagram for these categorical data.

2.6   Draw a bar chart with horizontal bars for the categorical data of Exercise 2.5.

2.7   On Wednesdays, mutual funds are denoted by the letter A in the financial pages of the Arizona Republic if they are in the top 20 percent among funds with the same investment objectives, by the letter B if they are in the next 20 percent, . . . , and by the letter E if they are in the bottom 20 percent. On the third Wednesday in June, 1998, seventeen Vanguard Index Funds were denoted by

A A D B B A A A A B C B A A E B C

Construct a dot diagram of this information.

2.8   If the categories in a dot diagram are arranged in descending order according to their frequencies (numbers of dots), such a dot diagram is also referred to as a **Pareto diagram.** Present the data of Exercise 2.5 in the form of a Pareto diagram.

2.9   Pareto diagrams are often used in industrial quality control to illustrate the relative importance of different kinds of defects. Denoting broken parts, paint defects, missing parts, faulty connections, and all other defects by the codes 0, 1, 2, 3, and 4, a quality control inspector observed the following kinds of defects in a large production run of cellular phones:

3   3   2   3   2   2   0   3   3   4   1   3   2   0   2   0   3   3
2   0   1   2   3   4   3   3   0   2   3   3   1   3   2   3   3

Present these defects in the form of a Pareto diagram.

2.10   List the data that correspond to the following stems of stem-and-leaf displays:
(a) 1 | 4 7 0 1 5;
(b) 4 | 2 0 3 9 8;
(c) 7 | 3 5 1 1 6.

2.11   List the data that correspond to the following stems of stem-and-leaf displays:
(a) 3 | 6 1 7 5 2;
(b) 4 | 15 38 50 77;
(c) 25 | 4 4 0 3 9.

2.12   List the data that correspond to the following stems of double-stem displays:
(a) 5 | 3 0 4 4 1 2
    5 | 9 9 7 5 8 6
(b) 6 | 7 8 5 9 6
    7 | 1 1 0 4 3
    7 | 5 5 8 9 6

2.13   Following are the lengths of young-of-the-year fresh-water drum (in millimeters), caught near Rattlesnake Island in Lake Erie: 79, 77, 65, 78, 71, 66, 95, 86, 84, 83, 88, 72, 81, 64, 71, 58, 60, 81, 73, 67, 85, 89, 75, 80, and 56. Construct a stem-and-leaf display with the stem labels 5, 6, 7, 8, and 9.

2.14   Convert the stem-and-leaf display obtained in Exercise 2.13 into a double-stem display.

2.15   Following are the scores that 50 college students obtained in a religious literacy test, in which the maximum score was 60:

35   31   54   34   41   30   38   36   43   40
50   31   36   34   44   35   49   43   39   56

26   36   30   43   36   25   40   41   39   51
25   39   48   37   29   31   33   30   43   45
46   44   38   38   53   34   51   41   36   42

Construct a stem-and-leaf display with the stem labels 2, 3, 4, and 5.

2.16   Convert the stem-and-leaf display obtained in Exercise 2.15 into a double-stem display.

2.17   Following are the lifetimes of 25 electronic components sampled from a production lot: 834, 919, 784, 865, 839, 912, 888, 783, 655, 831, 886, 842, 760, 854, 939, 961, 826, 954, 866, 675, 760, 865, 901, 632, and 718. Construct a stem-and-leaf display with one-digit stem labels and two-digit leaves. (Data are in hours of continuous use.)

2.18   Following are the low temperatures recorded at the Phoenix Sky Harbor Airport during a recent month of February: 46, 43, 54, 53, 43, 42, 47, 46, 46, 45, 43, 39, 52, 51, 48, 42, 43, 47, 49, 54, 53, 45, 50, 52, 53, 49, 35, and 34. Construct a double-stem display.

2.19   Following are measurements (to the nearest hundredth of a second) of the time required for sound to travel between two points: 1.53, 1.66, 1.42, 1.54, 1.37, 1.44, 1.60, 1.68, 1.72, 1.59, 1.54, 1.63, 1.58, 1.46, 1.52, 1.58, 1.53, 1.50, 1.49, and 1.62. Construct a stem-and-leaf display with the stem labels 1.3, 1.4, 1.5, 1.6, and 1.7, and one-digit leaves.

2.20   Convert the stem-and-leaf display obtained in Exercise 2.19 into a double-stem display.

2.21   Following are the IQs of 24 persons empaneled for jury duty by a municipal court: 108, 97, 103, 122, 84, 105, 101, 113, 127, 103, 124, 97, 88, 109, 103, 115, 96, 110, 104, 92, 105, 106, 93, and 99. Construct a stem-and-leaf display with the stem labels 8, 9, 10, 11, and 12, and one-digit leaves.

2.22   To get more stems, we sometimes repeat each stem label five times, with the first holding the leaves 0 and 1, the second holding the leaves 2 and 3, . . . , and the fifth holding the leaves 8 and 9. Construct this kind of stem-and-leaf display for the following numbers of operations performed at a hospital during 80 weeks:

42	50	49	44	41	54	47	38	45	44
46	33	40	36	39	53	42	48	41	52
57	44	42	48	45	46	40	59	41	44
41	48	39	43	45	34	47	48	36	49
36	55	48	45	42	57	50	49	47	43
52	60	46	35	49	37	33	38	51	47
40	52	57	56	46	45	48	37	50	55
43	56	55	46	48	37	62	61	57	53

2.3   FREQUENCY DISTRIBUTIONS

When we deal with large sets of data, and sometimes even when we deal with not so large sets of data, it can be quite a problem to get a clear picture of the information that they convey. As we saw in Sections 2.1 and 2.2, this usually requires that we rearrange and/or display the **raw** (untreated) **data** in some special form. Traditionally, this involves a **frequency distribution** or one of its **graphical presentations,** where we group or classify the data into a number of categories or classes.

Following are two examples. A recent study of their total billings (rounded to the nearest dollar) yielded data for a sample of 4,757 law firms. Rather than providing printouts of the 4,757 values, the information is disseminated by means of the following table:

Total billings	Number of law firms
Less than $300,000	2,405
$300,000 to $499,999	1,088
$500,000 to $749,999	271
$750,000 to $999,999	315
$1,000,000 or more	678
Total	4,757

This distribution does not show much detail, but it may well be adequate for most practical purposes. This should also be the case in connection with the following table, which summarizes the 2,439 complaints received by an airline about comfort-related characteristics of its airplanes:

Nature of complaint	Number of complaints
Inadequate leg room	719
Uncomfortable seats	914
Narrow aisles	146
Insufficient carry-on facilities	218
Insufficient restrooms	58
Miscellaneous other complaints	384
Total	2,439

When data are grouped according to numerical size, as in the first example, the resulting table is called a **numerical** or **quantitative distribution.** When they are grouped into nonnumerical categories, as in the second example, the resulting table is called a **categorical** or **qualitative distribution.**

Frequency distributions present data in a relatively compact form, give a good overall picture, and contain information that is adequate for many purposes, but, as we said previously, there is some loss of information. Some things that can be determined from the original data cannot be determined from a distribution. For instance, in the first example the distribution does not tell us the exact size of the lowest and the highest billings, nor does it provide the total of the billings of the 4,757 law firms. Similarly, in the second example we cannot tell how many of the complaints about uncomfortable seats pertained to their width or how many complains about insufficient carry-on facilities applied to particular size luggage. Nevertheless, frequency distributions present information in a generally more usable form, and the price we pay for this—the loss of certain information—is usually a fair exchange.

The construction of a frequency distribution consists essentially of three steps:

1. Choosing the **classes** (intervals or categories)
2. Sorting or tallying the data into these classes
3. Counting the number of items in each class

Since the second and third steps are purely mechanical, we concentrate here on the first, namely, that of choosing a suitable classification.

For numerical distributions, this consists of deciding how many classes we are going to use and from where to where each class should go. Both of these choices are essentially arbitrary, but the following rules are usually observed:

We seldom use fewer than 5 or more than 15 classes; the exact number we use in a given situation depends largely on how many measurements or observations there are.

Clearly, we would lose more than we gain if we group five observations into 12 classes with most of them empty, and we would probably discard too much information if we group a thousand measurements into three classes.

We always make sure that each item (measurement or observation) goes into one and only one class.

To this end we must make sure that the smallest and largest values fall within the classification, that none of the values can fall into a gap between successive classes, and that the classes do not overlap, namely, that successive classes have no values in common.

Whenever possible, we make the classes cover equal ranges of values.

Also, if we can, we make these ranges multiples of numbers that are easy to work with, such as 5, 10, or 100, since this will tend to facilitate the construction and the use of a distribution.

If we assume that the law firm billings were all rounded to the nearest dollar, only the third of these rules was violated in the construction of the distribution on page 21. However, had the billings been given to the nearest cent, then a billing of, say, $499,999.54 would have fallen between the second class and the third class, and we would also have violated the second rule. The third rule was violated because the classes do not all cover equal ranges of values; in fact, the first class and the last class have, respectively, no specified lower and upper limits.

Classes of the "less than," "or less," "more than," or "or more" variety are referred to as **open classes,** and they are used to reduce the number of classes that are needed when some of the values are much smaller than or much greater than the rest. Generally, open classes should be avoided, however, because they make it impossible to calculate certain values of interest, such as averages or totals (see Exercise 3.62).

Insofar as the second rule is concerned, we have to watch whether the data are given to the nearest dollar or to the nearest cent, whether they are given to the nearest inch or to the nearest tenth of an inch, whether they are given to the nearest ounce or to the nearest hundredth of an ounce, and so on. For instance, if we want to group the weights of certain animals, we might use the first of the following classifications when the weights are given to the nearest kilogram, the second when the weights are given to the nearest tenth of a kilogram, and the third when the weights are given to the nearest hundredth of a kilogram:

Weight (kilograms)	Weight (kilograms)	Weight (kilograms)
10–14	10.0–14.9	10.00–14.99
15–19	15.0–19.9	15.00–19.99
20–24	20.0–24.9	20.00–24.99
25–29	25.0–29.9	25.00–29.99
30–34	30.0–34.9	30.00–34.99
etc.	etc.	etc.

To illustrate what we have been discussing in this section, let us now go through the actual steps of grouping a set of data into a frequency distribution.

EXAMPLE   2.4   Based on 1997 figures, the following are 110 "waiting times" (in minutes) between eruptions of the Old Faithful Geyser in Yellowstone National Park:

81	83	94	73	78	94	73	89	112	80
94	89	35	80	74	91	89	83	80	82
91	80	83	91	89	82	118	105	64	56
76	69	78	42	76	82	82	60	73	69
91	83	67	85	60	65	69	85	65	82
53	83	62	107	60	85	69	92	40	71
82	89	76	55	98	74	89	98	69	87
74	98	94	82	82	80	71	73	74	80
60	69	78	74	64	80	83	82	65	67
94	73	33	87	73	85	78	73	74	83
83	51	67	73	87	85	98	91	73	108

Construct a frequency distribution.

Solution   Since the smallest value is 33 and the largest value is 118, we have to cover an interval of 86 values and a convenient choice would be to use the nine classes 30–39, 40–49, 50–59, 60–69, 70–79, 80–89, 90–99, 100–109, and 110–119. These classes will accommodate all of the data, they do not overlap, and they are all of the same size. There are other possibilities (for instance, 25–34, 35–44, 45–54, 55–64, 65–74, 75–84, 85–94, 95–104, 105–114, and 115–124), but it should be apparent that our first choice will facilitate the tally.

We now tally the 110 values and get the result shown in the following table:

Waiting times between eruptions (minutes)	Tally	Frequency
30–39	||	2
40–49	||	2
50–59	||||	4
60–69	⊞⊞ ⊞⊞ ⊞⊞ ||||	19
70–79	⊞⊞ ⊞⊞ ⊞⊞ ⊞⊞ ||||	24
80–89	⊞⊞ ⊞⊞ ⊞⊞ ⊞⊞ ⊞⊞ ⊞⊞ ⊞⊞ ||||	39
90–99	⊞⊞ ⊞⊞ ⊞⊞	15
100–109	|||	3
110–119	||	2
	Total	110

The numbers given in the right-hand column of this table, which show how many values fall into each class, are called the **class frequencies.** The smallest and largest values that can go into any given class are called its **class limits,** and for the distribution of the waiting times between eruptions they are 30 and 39, 40 and 49, 50 and 59, . . . , and 110 and 119. More specifically, 30, 40, 50, . . . , and 110 are called the **lower class limits,** and 39, 49, 59, . . . , and 119 are called the **upper class limits.**

The amounts of time that we grouped in our example were all given to the nearest minute, so that 30 actually includes everything from 29.5 to 30.5, 39 includes everything from 38.5 to 39.5, and the class 30–39 includes everything from 29.5 to 39.5. Similarly, the second class includes everything from 39.5 to 49.5, . . . , and the class at the bottom of the distribution includes everything from 109.5 to 119.5. It is customary to refer to 29.5, 39.5, 49.5, . . . , and 119.5 as the **class boundaries** or the **real class limits** of the distribution. Although 39.5 is the **upper boundary** of the first class and also the **lower boundary** of the second class, 49.5 is the upper boundary of the second class and also the lower boundary of the third class, and so forth, there is no cause for alarm. The class boundaries are by choice *impossible values* that cannot occur among the data being grouped. If we assume again that the law firm billings grouped in the distribution on page 21 were all rounded to the nearest dollar, the class boundaries $299,999.50, $499,999.50, $749,999.50, and $999,999.50 are also impossible values.

We emphasize this point because, to avoid gaps in the continuous number scale, some statistics texts, some widely used computer programs, and some graphing calculators (MINITAB, for example, and the TI-83) include in each class its lower boundary, and the highest class also includes its upper boundary. They would include 29.5 but not 39.5 in the first class of the preceding distribution of waiting times between eruptions of Old Faithful. Similarly, they would include 39.5 but not 49.5 in the second class, . . . , but 109.5 as well as 119.5 in the highest class of the distribution. All this is immaterial, of course, so long as the class boundaries are impossible values that cannot occur among the data being grouped. Especially for this reason, the use of impossible class boundaries cannot be overemphasized (see also discussion of Figure 10.4 on page 278).

Numerical distributions also have what we call **class marks** and **class intervals.** Class marks are simply the midpoints of the classes, and they are found by adding the lower and upper limits of a class (or its lower and upper boundaries) and dividing by 2. A class interval is merely the length of a class, or the range of values it can contain, and it is given by the difference between its boundaries. If the classes of a distribution are all equal in length, their common class interval, which we call the **class interval of the distribution,** is also given by the difference between any two successive class marks. Thus, the class marks of the waiting-time distribution are 34.5, 44.5, 54.5, . . . , and 114.5, and the class intervals and the class interval of the distribution are all equal to 10.

There are essentially two ways in which frequency distributions can be modified to suit particular needs. One way is to convert a distribution into a **percentage distribution** by dividing each class frequency by the total number of items grouped, and then multiplying by 100.

EXAMPLE   2.5   Convert the waiting-time distribution of Example 2.4 into a percentage distribution.

Solution   The first class contains $\frac{2}{110} \cdot 100 = 1.82\%$ of the data (rounded to two decimals), and so does the second class. The third class contains $\frac{4}{110} \cdot 100 = 3.64\%$ of the data, the fourth class contains $\frac{19}{110} \cdot 100 = 17.27\%$ of the data, . . . , and the bottom class again contains 1.82% of the data. These results are shown in the following table:

Waiting times between eruptions (minutes)	Percentage
30–39	1.82
40–49	1.82
50–59	3.64
60–69	17.27
70–79	21.82
80–89	35.45
90–99	13.64
110–109	2.73
110–119	1.82

The percentages total 100.01, with the difference, of course, due to rounding.   ∎

The other way of modifying a frequency distribution is to convert it into a "less than," "or less," "more than," or "or more" **cumulative distribution.** To construct a cumulative distribution, we simply add the class frequencies, starting either at the top or at the bottom of the distribution.

EXAMPLE 2.6   Convert the waiting-time distribution of Example 2.4 into a cumulative "less than" distribution.

Solution   Since none of the values is less than 30, 2 of the values are less than 40, $2 + 2 = 4$ of the values are less than 50, $2 + 2 + 4 = 8$ of the values are less than 60, . . . , and all 110 of the values are less than 120, we get

Waiting times between eruptions (minutes)	Cumulative frequency
Less than 30	0
Less than 40	2
Less than 50	4
Less than 60	8
Less than 70	27
Less than 80	51
Less than 90	90
Less than 100	105
Less than 110	108
Less than 120	110

Note that instead of "less than 30" we could have written "29 or less," instead of "less than 40" we could have written "39 or less," instead of "less than 50" we could have written "49 or less," and so forth.   ■

In the same way we can also convert a percentage distribution into a **cumulative percentage distribution.** We simply add the percentages instead of the frequencies, starting either at the top or at the bottom of the distribution.

So far we have discussed only the construction of numerical distributions, but the general problem of constructing categorical (or qualitative) distributions is about the same. Here again we must decide how many categories (classes) to use and what kind of items each category is to contain, making sure that all the items are accommodated and that there are no ambiguities. Since the categories must often be chosen before any data are actually collected, it is usually prudent to include a category labeled "others" or "miscellaneous."

For categorical distributions, we do not have to worry about such mathematical details as class limits, class boundaries, and class marks. On the other hand, there is often a serious problem with ambiguities and we must be very careful and explicit in defining what each category is to contain. For instance, if we had to classify items sold at a supermarket into "meats," "frozen foods," "baked goods," and so forth, it would be difficult to decide, for example, where to put frozen beef pies. Similarly, if we had to classify occupations, it would be difficult to decide where to put a farm manager, if our table contained (without qualification) the two categories "farmers" and "managers." For this reason, it is advisable, where possible, to use standard categories developed by the Bureau of the Census and other government agencies.

■ E x e r c i s e s

2.23 The weights of 150 rats used in medical research vary from 268 to 395 grams. Show the class limits of a table into which these weights could conveniently be grouped.

2.24 The burning times of certain solid-fuel rockets given to the nearest tenth of a second vary from 3.2 to 5.9 seconds. Show the class limits of a table into which these burning times could conveniently be grouped.

2.25 The monthly electricity bills of the residents of a retirement community vary from $27.45 to $174.69, with the large variation being due to the high cost of airconditioning during the summer months. Show the class limits of a table with not more than ten classes into which these monthly bills could conveniently be grouped.

2.26 Decide for each of the following whether it can be determined from the distribution of law firm billings on page 21; if possible, give a numerical answer:
(a) The number of law firms with billings exceeding $300,000.
(b) The number of law firms with billings exceeding $749,999.
(c) The number of law firms with billings less than $250,000.
(d) The number of law firms with billings less than $500,000.

2.27 Following is the distribution of the weights of 133 mineral specimens collected on a field trip:

Weight (grams)	Number of specimens
5.0–19.9	8
20.0–34.9	27
35.0–49.9	42
50.0–64.9	31
65.0–79.9	17
80.0–94.9	8

Find
(a) the lower class limits;
(b) the upper class limits;
(c) the class boundaries;
(d) the class intervals.

2.28 To group data on the number of rainy days reported by a newspaper for the month of May, we plan to use the classes 1–9, 10–19, 20–25, and 25–30. Explain where difficulties might arise.

2.29 To group sales invoices ranging from $12.00 to $79.00, a department store's accountant uses the following classes: 10.00–29.99, 30.00–49.99, 60.00–79.99, and 70.00–99.99. Explain where difficulties might arise.

2.30 Difficulties can also arise when we choose inappropriate classes for grouping categorical data. If men's shirts are classified according to the fibers of which they are made, explain where difficulties might arise if we include only the three categories: wool, silk, and synthetic fibers.

2.31 Explain where difficulties might arise if, in a study of the nutritional value of desserts, we use the categories pie, cake, fruit, pudding, and ice cream.

2.32 Temperature readings, rounded to the nearest degree Fahrenheit, are grouped into a distribution with the classes 55–60, 61–66, 67–72, 73–78, and 79–84. Find
(a) the class boundaries of this distribution:
(b) the class marks.

2.33 Measurements given to the nearest centimeter are grouped into a table having the class boundaries 19.5, 24.5, 29.5, 34.5, 39.5, and 44.5. Find
(a) the class limits of these five classes;
(b) their class marks;
(c) their class intervals.

2.34 The class marks of a distribution of retail food prices (in cents) are 27, 42, 57, 72, 87, and 102. Find the corresponding
(a) class boundaries;
(b) class limits.

2.35 The wingspans of certain birds are grouped into a distribution with the class boundaries 59.95, 74.95, 89.95, 104.95, 119.95, and 134.95 centimeters. Find the corresponding
(a) class limits;
(b) class marks.

2.36 Following are the percent shrinkages on drying of 40 plastic clay specimens:

20.3	16.8	21.7	19.4	15.9	18.3	22.3	17.1
19.6	21.5	21.5	17.9	19.5	19.7	13.3	20.5
24.4	19.8	19.3	17.9	18.9	18.1	23.5	19.8
20.4	18.4	19.5	18.5	17.4	18.7	18.3	20.4
20.1	18.5	17.8	17.3	20.0	19.2	18.4	19.0

Group these percentages into a frequency distribution with the classes 13.0–14.9, 15.0–16.9, 17.0–18.9, 19.0–20.9, 21.0–22.9, and 23.0–24.9.

2.37 Convert the distribution obtained in Exercise 2.36 into a percentage distribution.

2.38 Convert the distribution obtained in Exercise 2.36 into a cumulative "less than" distribution.

2.39 Following are 60 measurements (in 0.00001 inch) of the thickness of an aluminum alloy plating obtained in the analysis of an anodizing process:

24	24	41	36	32	33	22	34	39	25	21	32
36	26	43	28	30	27	38	25	33	42	30	32
31	34	21	27	35	48	35	26	21	30	37	39
25	33	36	27	29	28	26	22	23	30	43	20
31	22	37	23	30	29	31	28	36	38	20	24

Group these thicknesses into a distribution with the classes 20–24, 25–29, 30–34, 35–39, 40–44, and 45–49.

2.40   Convert the distribution obtained in Exercise 2.39 into a percentage distribution.

2.41   Convert the distribution obtained in Exercise 2.40 into a cumulative "or less" percentage distribution.

2.42   Following are the lengths of root penetrations (in feet) of 120 crested wheatgrass seedlings one month after planting:

0.95	0.88	0.90	1.23	0.83	0.67	1.41	1.04	1.01	0.81
0.78	1.21	0.80	1.43	1.27	1.16	1.06	0.86	0.70	0.80
0.71	0.93	1.00	0.62	0.80	0.81	0.75	1.25	0.86	1.15
0.91	0.62	0.84	1.08	0.99	1.38	0.98	0.93	0.80	1.25
0.82	0.97	0.85	0.79	0.90	0.84	0.53	0.83	0.83	0.60
0.95	0.68	1.27	0.97	0.80	1.13	0.89	0.83	1.47	0.96
1.34	0.87	0.75	0.95	1.13	0.95	0.85	1.00	0.73	1.36
0.94	0.80	1.33	0.91	1.03	0.93	1.34	0.82	0.82	0.95
1.11	1.02	1.21	0.90	0.80	0.92	1.06	1.17	0.85	1.00
0.88	0.86	0.64	0.96	0.88	0.95	0.74	0.57	0.96	0.78
0.89	0.81	0.89	0.88	0.73	1.08	0.87	0.83	1.19	0.84
0.94	0.70	0.76	0.85	0.97	0.86	0.94	1.06	1.27	1.09

Group these lengths into a table having the classes 0.50–0.59, 0.60–0.69, 0.70–0.79, ..., and 1.40–1.49.

2.43   Convert the distribution obtained in Exercise 2.42 into a cumulative "more than" distribution.

2.44   Regroup the distribution of the waiting times of Example 2.4 into a distribution with the classes 25–34, 35–44, 45–54, 55–64, ..., and 115–124.

2.45   Following are the scores that 180 students obtained in a twelfth-grade achievement test in the social sciences:

60	74	65	61	70	51	81	63	73	74	56	69	76	55	39
55	60	71	65	75	68	62	75	68	91	65	78	82	53	67
68	76	63	66	74	86	69	57	74	70	89	37	68	74	52
82	63	77	79	67	62	63	78	74	50	79	95	72	66	45
58	62	46	73	55	73	70	83	63	66	54	72	76	60	58
57	69	75	70	72	58	43	66	52	84	72	67	91	82	67
67	62	73	58	38	74	78	94	77	62	68	58	76	64	82
63	63	64	77	67	71	75	88	90	86	73	68	61	74	59
70	65	71	79	74	72	59	73	65	55	60	84	42	75	72
44	59	65	71	69	94	72	57	84	71	65	73	79	50	68
61	63	67	55	75	64	66	81	64	85	54	57	65	90	73
93	50	76	69	51	71	56	72	48	50	81	70	46	56	78

Group these scores into convenient classes with an interval of 10.

2.46 Convert the distribution obtained in Exercise 2.45 into a cumulative "or more" distribution.

2.47 Convert the distribution obtained in Exercise 2.46 into a cumulative "or more" percentage distribution.

2.4 GRAPHICAL PRESENTATIONS

When frequency distributions are constructed mainly to condense large sets of data and present them in an "easy to digest" form, it is usually most effective to display them graphically. As the saying goes, a picture speaks louder than a thousand words, and this was true even before the current proliferation of computer graphics. Nowadays, each statistical software package strives to outdo its competitors by means of more and more elaborate pictorial presentations of statistical data.

For frequency distributions, the most common form of graphical presentation is the **histogram,** like the one shown in Figure 2.3. Histograms are constructed by representing the measurements or observations that are grouped (in Figure 2.3 the waiting times between eruptions of Old Faithful) on a horizontal scale, the class frequencies on a vertical scale, and drawing rectangles whose bases equal the class intervals and whose heights are the corresponding class frequencies.

The markings on the horizontal scale of a histogram can be the class limits as in Figure 2.3, the class marks, the class boundaries, or arbitrary key values. For practical reasons, it is usually preferable to show the class limits, even though the rectangles actually go from one class boundary to the next. After all, they tell us *what values go into each class.* Note that histograms cannot be drawn for distributions with open classes and that they require special care when the class intervals are not all equal (see Exercise 2.57 on page 35).

The data that led to Figure 2.3 were easy to group because there were only 110 values in the sample. For really large sets of data, it may be convenient to construct histograms directly from raw data by using a suitable computer package or a graphing calculator. An example of the latter is shown in Figure 4.5. We said that it *may be* convenient to use a computer package or a graphing calcu-

FIGURE 2.3
Histogram of the waiting times between eruptions of Old Faithful geyser.

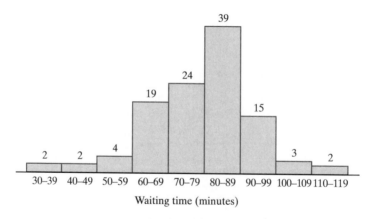

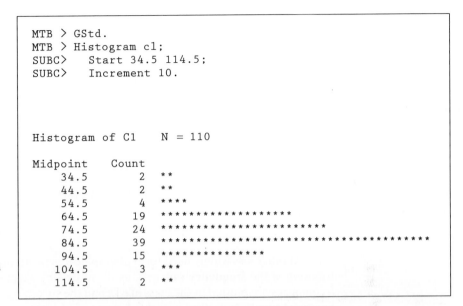

```
MTB > GStd.
MTB > Histogram c1;
SUBC>    Start 34.5 114.5;
SUBC>    Increment 10.



Histogram of C1    N = 110

Midpoint    Count
      34.5      2    **
      44.5      2    **
      54.5      4    ****
      64.5     19    *******************
      74.5     24    ************************
      84.5     39    ***************************************
      94.5     15    ***************
     104.5      3    ***
     114.5      2    **
```

FIGURE 2.4
Computer printout of a histogram for the waiting times between eruptions of Old Faithful geyser.

lator—in actual practice, just entering the data in a computer or a calculator can be more work than tallying the data manually and drawing the rectangles.

Our definition of "histogram" may be referred to as traditional, for nowadays the term is applied much more loosely to all sorts of graphical presentations of frequency distributions, where the class frequencies are not necessarily represented by rectangles. For instance, Figure 2.4 shows an older MINITAB generated histogram, which really looks more like a dot diagram (see Section 2.1), except that the dots aligned at the class marks represent the various values in the corresponding classes rather than repeated identical values.

Also referred to at times as histograms are bar charts (see Section 2.1), such as the one shown in Figure 2.5. The heights of the rectangles, or bars, again represent the class frequencies, but there is no pretense of having a continuous horizontal scale.

FIGURE 2.5
Bar chart of the distribution of waiting times between eruptions of Old Faithful geyser.

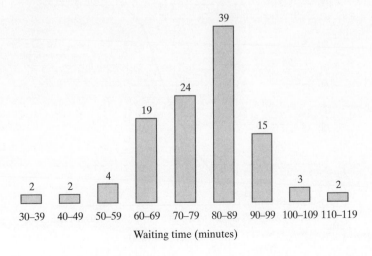

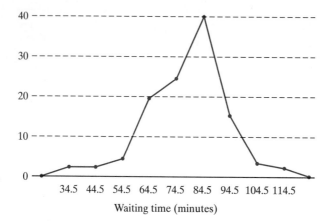

FIGURE 2.6
Frequency polygon of the distribution of waiting times between eruptions of Old Faithful geyser.

Another less widely used form of graphical presentation of a frequency distribution is the **frequency polygon,** as illustrated by Figure 2.6. Here, the class frequencies are plotted at the class marks and the successive points are connected by straight lines. Note that we added classes with zero frequencies at both ends of the distribution to "tie down" the graph to the horizontal scale. If we apply a similar technique, to a cumulative distribution, usually a "less than" distribution, we obtain what is called an **ogive** (which rhymes with five or jive). However, in an ogive the cumulative frequencies are plotted at the class boundaries instead of the class marks—it stands to reason that the cumulative frequency corresponding to, say, "less than 60" should be plotted at the class boundary 59.5, since "less than 60" actually includes everything up to 59.5. Figure 2.7 shows an ogive of the "less than" distribution of the waiting times obtained on page 26.

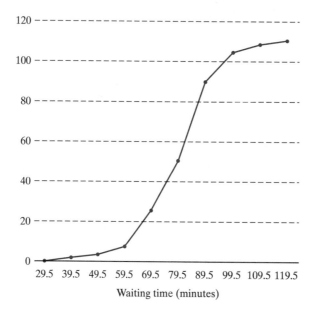

FIGURE 2.7
Ogive of the distribution of waiting times between eruptions of Old Faithful geyser.

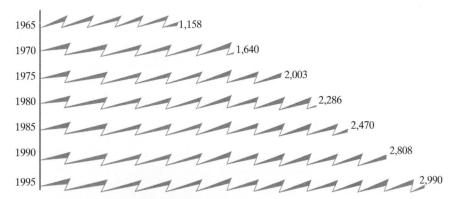

FIGURE 2.8
Net generation of electric energy in the United States (billions of kilowatt-hours).

Although the visual appeal of histograms, bar charts, frequency polygons, and ogives is a marked improvement over that of mere tables, there are various ways in which distributions can be presented even more dramatically and often more effectively. An example of such a pictorial presentation (often seen in newspapers, magazines, and reports of various sorts) is the **pictogram** shown in Figure 2.8.

Categorical distributions are often presented graphically as **pie charts,** like the one shown in Figure 2.9, where a circle is divided into sectors—pie-shaped pieces—that are proportional in size to the corresponding frequencies or percentages. To construct a pie chart, we first convert the distribution into a percentage distribution. Then, since a complete circle corresponds to 360 degrees, we obtain the central angles of the various sectors by multiplying the percentages by 3.6.

EXAMPLE 2.7 The following table shows the educational attainment of women who had a child in 1992 (in thousands).

Less than 4 years of high school	12,159
High school diploma	19,063
College: No degree	12,422
Associate's degree	3,982
Bachelor's degree	8,173
Graduate or Professional degree	2,812
	58,611

Construct a pie chart.

Solution The percentages corresponding to the six categories are $\frac{12,159}{58,611} \cdot 100\% = 20.75\%$,

$\frac{19,063}{58,611} \cdot 100\% = 32.52\%$, $\quad \frac{12,422}{58,611} \cdot 100\% = 21.19\%$, $\quad \frac{3,982}{58,611} \cdot 100\% = 6.79\%$,

$$\frac{8{,}173}{58{,}611} \cdot 100\% = 13.94\%, \text{ and } \frac{2{,}812}{58{,}611} \cdot 100\% = 4.80\% \text{ rounded to two decimals.}$$

Multiplying these percentages by 3.6, we find that the central angles of the six sectors are 74.7, 117.1, 76.3, 24.4, 50.2, and 17.3 degrees. Rounding the angles to the nearest degree and using a protractor, we get the pie chart shown in Figure 2.9. ■

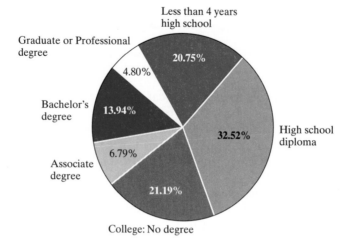

FIGURE 2.9
Educational attainment of women who had a child in 1992.

Many computers are preprogrammed so that, once the data have been entered, a simple command will produce a pie chart, or a variation thereof. Some computer-generated pie charts use color, some are three dimensional, some cut out sectors (like pieces of pie) for emphasis, and some shade or tint the various sectors.

■ E x e r c i s e s

2.48   Draw a histogram of the following distribution of the frequencies with which rifle shots hit the respective distances from the center of a target.

Distance (centimeters)	Frequency
0.0–1.9	23
2.0–3.9	18
4.0–5.9	12
6.0–7.9	9
8.0–9.9	5
10.0–11.9	2
12.0–13.9	1

2.49 Convert the distribution of Exercise 2.48 into a cumulative "less than" distribution and draw an ogive.

2.50 Construct a bar chart of the distribution of the weights of mineral specimens given in Exercise 2.27.

2.51 Construct a histogram of whichever data you grouped among those of Exercises 2.36, 2.39, or 2.42.

2.52 Construct a frequency polygon of whichever data you grouped among those of Exercises 2.36, 2.39, or 2.42.

2.53 Following is the distribution of the number of fish tacos served for lunch by a Mexican restaurant on 60 weekdays:

Number of fish tacos	Number of weekdays
30–39	4
40–49	23
50–59	28
60–69	5

Construct a
(a) histogram;
(b) bar chart;
(c) frequency polygon;
(d) ogive.

2.54 Following are 80 measurements of the iron-solution index of tin-plate specimens, designed to measure the corrosion resistance of tin-plated steel:

0.78	0.65	0.48	0.83	1.43	0.92	0.92	0.72	0.48	0.96
0.72	0.48	0.83	0.49	0.78	0.96	1.06	0.83	0.78	0.82
1.12	0.78	1.03	0.88	1.23	0.28	0.95	1.16	0.47	0.55
0.97	1.20	0.77	0.72	0.45	1.36	0.65	0.73	0.39	0.94
0.79	1.26	1.06	0.90	0.77	0.45	0.78	0.77	1.09	0.73
0.64	0.91	0.95	0.71	1.20	0.88	0.83	0.78	1.04	1.33
0.52	0.32	0.54	0.63	0.44	0.92	1.00	0.79	0.63	1.23
0.65	0.64	0.48	0.79	0.99	0.57	0.91	1.12	0.70	1.05

Group these measurements into a table with the class interval 0.20 and draw its histogram.

2.55 Convert the distribution obtained in Exercise 2.54 into a cumulative "less than" distribution and draw its ogive.

2.56 Draw a frequency polygon of the distribution obtained in Exercise 2.54.

2.57 Figure 2.10 shows the distribution of the scores of 80 incoming college freshmen on a French language placement examination. Explain why it might easily give a misleading impression and indicate how it might be improved.

2.58 Combine the second and third classes of the distribution of Exercise 2.48 and draw a histogram in which the areas of the rectangles are proportional to the class frequencies.

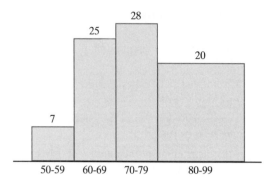

FIGURE 2.10
Distribution of scores on a French language placement examination.

Scores on French language placement examination

2.59   Construct a pie chart of the distribution of complaints received by an airline about comfort-related characteristics of its airplanes given on page 21.

2.60   Construct a pie chart for the categorical data of Exercise 2.5.

2.61   Construct a pie chart of the following distribution that shows how the dogs entered in a dog show are distributed according to A.K.C. classifications:

Group	Number
Sporting dogs	13
Hounds	24
Working dogs	50
Terriers	31
Toys	17
Nonsporting dogs	15

2.62   Asked to rate the maneuverability of a new model car as excellent, very good, good, fair, or poor, 50 drivers responded as follows: very good, good, good, fair, excellent, good, good, good, very good, poor, good, good, good, good, very good, good, fair, good, good, poor, very good, fair, good, good, excellent, very good, good, good, good, fair, fair, very good, good, very good, excellent, very good, fair, good, very good, good, fair, good, good, excellent, very good, fair, fair, good, very good, and good. Construct a pie chart showing the percentages corresponding to these ratings.

2.63   The pictogram of Figure 2.11 is intended to illustrate that per capita personal income in the United States has doubled from $10,000 in 1980 to $20,000 in 1992. Explain why this pictogram conveys a misleading impression and indicate how it might be modified.

 2.64   Use a computer package or a graphing calculator to construct a histogram for the lengths of the 60 sea trout on page 13, using the classes 16.0–16.9, 17.0–17.9, 18.0–18.9, 19.0–19.9, 20.0–20.9, 21.0–21.9, 22.0–22.9, and 23.0–23.9.

 2.65   Following are the scores which 150 applicants for secretarial positions in a government agency obtained in an achievement test:

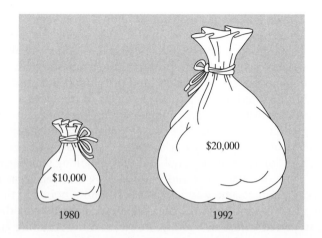

FIGURE 2.11
Per capita personal
income.

62	37	49	56	89	52	41	70	80	28	54	45	95	52	66
43	59	56	70	64	55	62	79	48	26	61	56	62	49	71
58	77	74	63	37	68	41	52	60	69	58	73	14	60	84
55	44	63	47	28	83	46	55	53	72	54	83	70	61	36
46	50	35	56	43	61	76	63	66	42	50	65	41	62	74
45	60	47	72	87	54	67	45	76	52	57	32	55	70	44
81	72	54	57	92	61	42	30	57	58	62	86	45	63	28
57	40	44	55	36	55	44	40	57	28	63	45	86	61	51
68	56	47	86	52	70	59	40	71	56	34	62	81	58	43
46	60	45	69	74	42	55	46	50	53	77	70	49	58	63

Use a computer package or a graphing calculator to construct a histogram of the
distribution of these data with the classes 10–19, 20–29, 30–39, 40–49, 50–59, 60–
69, 70–79, 80–89, and 90–99.

*2.5   SUMMARIZING TWO-VARIABLE DATA[†]

So far we have dealt only with situations involving one variable—the room oc-
cupancies in Section 2.2, the waiting times between eruptions of Old Faithful in
Example 2.4, the plating thicknesses of Exercise 2.39, the root penetrations of
Exercise 2.42, and so on. In actual practice, many statistical methods apply to situations involving two variables, and some of them apply even when the number
of variables cannot be counted on one's fingers and toes. Not quite so extreme
would be a problem in which we want to study the values of one-family homes,
taking into consideration their age, their location, the number of bedrooms, the
number of baths, the size of the garage, the type of roof, the number of fireplaces,
the lot size, the value of nearby properties, and the accessibility of schools.

[†]As indicated in the Table of Contents, all sections marked with an asterisk are optional. They
may be omitted without loss of continuity.

Leaving some of this work to later chapters and, in fact, most of it to advanced courses in statistics, we shall treat here only the display, listing, and grouping of data involving two variables; that is, problems dealing with the display of paired data. In most of these problems the main objective is to see whether there is a relationship, and if so what kind of relationship, so that we can predict one variable, denoted by the letter y, in terms of the other variable, denoted by the letter x. For instance, the x's might be family incomes and the y's might be family expenditures on medical care, they might be annealing temperatures and the hardness of steel, or they might be the time that has elapsed since the chemical treatment of a swimming pool and the remaining concentration of chlorine.

Pairs of values of x and y are usually referred to as **data points,** denoted by (x, y), in the same way in which we denote points in the plane, with x and y being their x- and y-coordinates. When we actually plot the points corresponding to paired values of x and y, we refer to the resulting graph as a **scatter diagram,** a **scatter plot,** or a **scattergram.** As their name implies, such graphs are useful tools in the analysis of whatever relationship there may exist between the x's and the y's, namely, in judging whether there are any discernible patterns.

EXAMPLE   2.8   Raw materials used in the production of synthetic fiber are stored in a place that has no humidity control. Following are measurements of the relative humidity in the storage place, x, and the moisture content of a sample of the raw material, y, on 15 days:

x (Percent)	y (Percent)
36	12
27	11
24	10
50	17
31	10
23	12
45	18
44	16
43	14
32	13
19	11
34	12
38	17
21	8
16	7

Construct a scattergram.

Solution   Scattergrams are easy enough to draw, yet the work can be simplified by using appropriate computer software or a graphing calculator. The one shown in Figure 2.12 was reproduced from the display screen of a TI-83 graphing calculator.

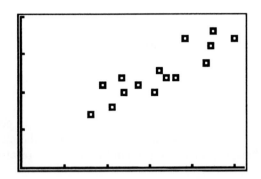

FIGURE 2.12
Scattergram of the humidity and water content data.

As can be seen from the diagram, the points are fairly widely scattered, yet there is evidence of an upward trend; that is, increases in the water content of the raw material seem to go with increases in humidity. In Figure 2.12 the dots are squares with their centers removed, but they can also be circles, ×'s, dots, or other kinds of symbols. (The units are not marked on either scale, but on the horizontal axis the tick marks are at 10, 20, 30, 40, and 50, and on the vertical axis they are at 5, 10, 15, and 20.) ∎

Some difficulties arise when two or more of the data points are identical. In that case, the TI-83 graphing calculator shows only one point and so do some of the printouts obtained with statistical software. However, MINITAB has a special scattergram to take care of situations like this. Its so-called *character plot* prints the number 2 instead of the symbol × or ★ to indicate that there are two identical data points, and it would print a 3 if there were three. This is illustrated by the following example.

EXAMPLE 2.9 The following data were obtained in a study of the relationship between the resistance (in ohms) and the failure time (in minutes) of certain overloaded resistors.

Resistance x	Failure time y
33	39
36	36
30	34
44	51
34	36
25	21
40	45
28	25
40	45
46	36
42	39

48	41
47	45
25	21

Construct a scattergram.

Solution   As can be seen, there are two duplicates among the data points; (40, 45) appears twice, and so does (25, 21). A scattergram that shows the number 2 instead of the symbol ★ at these two points is given in the computer printout of Figure 2.13.   ■

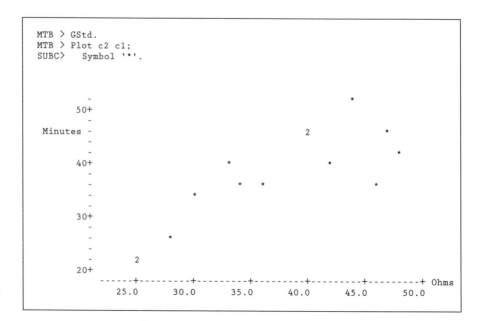

FIGURE 2.13
Scattergram of the resistance and failure time data.

To illustrate the steps needed to group paired data into a **two-way frequency distribution** and then draw a **three-dimensional histogram,** let us use data obtained in checking the reliability of an achievement test in elementary statistics. In general, a test is considered to be reliable if good students to whom it is administered will consistently score high while poor students will consistently score low. Rather than have students repeatedly take the test, a more convenient way of checking on reliability is to divide the test into two parts, usually the even-numbered problems and the odd-numbered problems, and then compare the scores that students received on both halves of the test.

EXAMPLE 2.10   Following are the scores which 40 students obtained on both parts of the test, with the scores on the even-numbered problems denoted by x and the scores on the odd-numbered problems denoted by y:

x	y	x	y	x	y	x	y
40	39	32	23	37	34	32	28
45	45	45	35	41	38	40	34
27	24	42	36	35	33	37	37
42	39	44	42	34	30	47	45
42	29	41	35	38	40	44	40
49	40	48	45	42	34	35	35
36	28	44	39	32	35	44	35
39	39	40	28	38	27	43	38
43	38	50	48	36	37	37	35
39	34	37	39	43	42	43	33

Choosing the five classes 26–30, 31–35, 36–40, 41–45, and 46–50 for x and the six classes 21–25, 26–30, 31–35, 36–40, 41–45, and 46–50 for y, group these data into a two-way frequency distribution and draw a three-dimensional histogram.

Solution   Performing the tally, we find that the first pair of values, 40 and 39, goes into the cell belonging to the third column and the fourth row, the second pair of values, 45 and 45, goes into the cell belonging to the fourth column and the fifth row, and so on. We thus get

	x				
y	26–30	31–35	36–40	41–45	46–50
21–25	|	|			
26–30		||	|||	|	
31–35		|||	||||	┼┼┼	
36–40			┼┼┼ |	┼┼┼ ||	|
41–45				|||	||
46–50					|

and, hence, the following two-way frequency distribution:

	x				
y	26–30	31–35	36–40	41–45	46–50
21–25	1	1			
26–30		2	3	1	
31–35		3	4	5	
36–40			6	7	1
41–45				3	2
46–50					1

In Figure 2.14, the three-dimensional histogram of this distribution, the heights of the blocks are proportional to the corresponding frequencies.   ■

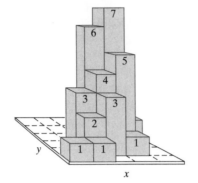

FIGURE 2.14
Histogram of the
two-way frequency
distribution.

Exercises[†]

*2.66 The following data pertain to the chlorine residual in a swimming pool at various times after it has been treated with chemicals:

Number of hours	Chlorine residual (parts per million)
2	1.8
4	1.5
6	1.4
8	1.1
10	1.1
12	0.9

Draw a scatter diagram and describe what relationship, if any, it seems to indicate.

*2.67 Following are the high school averages, x, and the first-year college grade-point indexes, y, of ten students:

x	y
3.0	2.6
2.7	2.4
3.8	3.9
2.6	2.1
3.2	2.6

[†]All exercises marked with an asterisk pertain to optional material.

3.4	3.3
2.8	2.2
3.1	3.2
3.5	2.8
3.3	2.5

Draw a scattergram and describe what kind of relationship, if any, it seems to display.

*2.68 Following are the drying times of a certain varnish and the amount of a certain chemical that has been added:

Amount of Additive (Grams) x	Drying Time (Hours) y
1	7.2
2	6.7
3	4.7
4	3.7
5	4.7
6	4.2
7	5.2
8	5.7

Draw a scatter diagram and describe what kind of relationship, if any, it seems to display.

 *2.69 In a study of the growth of saguaro cacti, an experiment was performed to determine how well the height of such cacti can be estimated from aerial photographs. Following are the heights of 36 saguaros (in inches) estimated from aerial photographs, x, and measured on the ground, y:

x	y	x	y
118	103	163	163
166	160	124	137
141	143	171	173
164	187	165	112
150	111	123	132
151	134	142	151
133	121	144	148
122	143	130	117
165	141	135	165
168	149	139	112
109	125	161	121
153	128	170	189

135	101	148	156
158	136	136	158
104	117	174	182
183	121	186	161
173	156	194	153
125	130	181	183

Draw a scatter plot and describe what kind of relationship, if any, it seems to display.

 *2.70   Following are the numbers of minutes it took 30 students to memorize two lists of Spanish verbs, one in the morning and one in the late afternoon.

Morning x	Afternoon y	Morning x	Afternoon y
15	16	18	21
21	28	25	23
17	22	13	19
23	23	20	24
23	17	16	26
12	17	24	25
28	25	18	27
23	26	22	28
16	24	27	25
25	29	19	18
22	21	23	22
21	20	14	22
17	15	26	26
21	29	24	21
22	18	19	23

Draw a scattergram and use it to describe what kind of relationship, if any, it seems to display.

*2.71   Group the data of Exercise 2.69 into a two-way frequency distribution with the classes 100–119, 120–139, 140–159, 160–179, and 180–199 for both variables.

*2.72   Use the distribution obtained in Exercise 2.71 to draw a three-dimensional histogram.

*2.73   Group the data of Exercise 2.70 into a two-way frequency distribution with the classes 10–14, 15–19, 20–24, and 25–29 for x, and the classes 15–19, 20–24, and 25–29 for y.

*2.74   Use the distribution obtained in Exercise 2.73 to draw a three-dimensional histogram.

2.6 Checklist of Key Terms *(with page references to their definitions)*†

Bar chart, 15
Categorical distribution, 21
Class, 22
  boundary, 24
  frequency, 24
  interval, 25
  limit, 24
  mark, 25
Cumulative distribution, 26
Cumulative percentage distribution, 26
Data point, 38
Dot diagram, 14
Double-stem display, 17
Exploratory data analysis, 17
Frequency distribution, 20
  two-way, 40
Frequency polygon, 32
Graphical presentation, 20
Grouping, 13
Histogram, 30
  three-dimensional, 40
Interval of distribution, 25
Leaf, 17

Listing, 13
Lower boundary, 24
Lower class limit, 24
Numerical distribution, 21
Ogive, 32
Open class, 22
Pareto diagram, 19
Percentage distribution, 25
Pictogram, 33
Pie chart, 33
Qualitative distribution, 21
Quantitative distribution, 21
Raw data, 20
Real class limits, 24
Scattergram, 38
Stem, 17
Stem-and-leaf display, 16
Stem label, 17
Three-dimensional histogram, 40
Two-way frequency distribution, 40
Upper boundary, 24
Upper class limit, 24

2.7 References

Detailed information about statistical charts may be found in

CLEVELAND, W. S., *The Elements of Graphing Data.* Monterey, Calif.: Wadsworth Advanced Books and Software, 1985.

SCHMID, C. F., *Statistical Graphics: Design Principles and Practices.* New York: John Wiley & Sons, Inc., 1983.

TUFTE, E. R., *The Visual Display of Quantitative Information.* Cheshire, Conn.: Graphics Press, 1985.

and some interesting information about the history of the graphical presentation of statistical data is given in an article by E. Royston in

PEARSON, E. S. and KENDALL, M. G., eds., *Studies in the History of Statistics and Probability.* New York: Hafner Press, 1970.

Discussions of what not to do in the presentation of statistical data may be found in

CAMPBELL, S. K., *Flaws and Fallacies in Statistical Thinking.* Upper Saddle River, N.J.: Prentice Hall, Inc., 1974.

HUFF, D., *How to Lie with Statistics.* New York: W.W. Norton & Company, Inc., 1954.

†Terms in italics were introduced in exercises.

REICHMAN, W. J., *Use and Abuse of Statistics.* New York: Penguin Books, 1971.

SPIRER, H. E., SPIRER, L., and JAFFE, A. J., *Misused Statistics,* 2nd ed. New York: Marcel Dekker, Inc., 1998.

Useful references to lists of standard categories are given in

HAUSER, P. M., and LEONARD, W. R., *Government Statistics for Business Use,* 2nd ed. New York: John Wiley & Sons, Inc., 1956.

For further information about exploratory data analysis and stem-and-leaf displays in particular, see

HARTWIG, F., and DEARING, B. E., *Exploratory Data Analysis.* Beverly Hills, Calif.: Sage Publications, Inc., 1979.

HOAGLIN, D. C., MOSTELLER, F., and TUKEY, J. W., *Understanding Robust and Exploratory Data Analysis.* New York: John Wiley & Sons, Inc., 1983.

KOOPMANS, L. H., *An Introduction to Contemporary Statistics.* North Scituate, Mass.: Duxbury Press, 1981.

TUKEY, J. W., *Exploratory Data Analysis.* Reading, Mass.: Addison-Wesley Publishing Company, Inc., 1977.

VELLEMAN, P. F., and HOAGLIN, D. C., *Applications, Basics, and Computing for Exploratory Data Analysis.* North Scituate, Mass.: Duxbury Press, 1980.

3

SUMMARIZING DATA: MEASURES OF LOCATION

3.1 Populations and Samples 48

3.2 The Mean 49

3.3 The Weighted Mean 53

3.4 The Median 58

3.5 Other Fractiles 62

3.6 The Mode 66

*3.7 The Description of Grouped Data 70

3.8 Technical Note (Summations) 76

3.9 Checklist of Key Terms 79

3.10 References 79

*W*hen we are about to describe a set of data, it is sound advice to say neither too little nor too much. Thus, depending on the nature of the data and the purpose we have in mind, statistical descriptions can be very brief or very elaborate. Sometimes we present data just as they are and let them speak for themselves; on other occasions we may just group the data and present their distribution in tabular or graphical form. Most of the time, though, we have to describe data in various other ways.

It is often appropriate to summarize data by means of a few well-chosen numbers that, in their way, are descriptive of the entire set. Exactly what sort of numbers we choose depends on the particular characteristics we want to describe. In one study we may be interested in a value that somehow describes the middle or the most typical of a set of data; in another we may be interested in the value that is exceeded only by 25% of the data; and in still another we may be interested in the length of the interval between the smallest and the largest values among the data. The statistical measures cited in the first two situations come under the heading of **measures of location** and the one cited in the third situation fits the definition of a **measure of variation.**

In this chapter we shall concentrate on measures of location, and in particular on **measures of central location,** which in some way describe the center or the middle of a set of data. Measures of variation and some other kinds of statistical descriptions will be discussed in Chapter 4.

3.1   POPULATIONS AND SAMPLES

When we stated that the choice of a statistical description may depend on the nature of the data, we were referring among other things to the following distinction:

> **If a set of data consists of all conceivably possible (or hypothetically possible) observations of a given phenomenon, we call it a population; if a set of data consists of only a part of these observations, we call it a sample.**

Here we added the phrase "hypothetically possible" to take care of such clearly hypothetical situations as where we look at the outcomes (heads or tails) of 12 flips of a coin as a sample from the potentially unlimited number of flips of the coin, where we look at the weights of ten 30-day-old lambs as a sample of the weights of all (past, present, and future) 30-day-old lambs raised at a certain farm, or where we look at four determinations of the uranium content of an ore as a sample of the many determinations that could conceivably be made. In fact, we often look at the results of an experiment as a sample of what we might get if the experiment were repeated over and over again.

Originally, statistics dealt with the description of human populations, census counts, and the like, but as it grew in scope, the term "population" took on the much wider connotation given to it in the preceding distinction between populations and samples. Whether or not it sounds strange to refer to the heights of all the trees in a forest or the speeds of all the cars passing a checkpoint as populations is beside the point—in statistics, "population" is a technical term with a meaning of its own.

Although we are free to call any group of items a population, what we do in practice depends on the context in which the items are to be viewed. Suppose, for instance, that we are offered a lot of 400 ceramic tiles, which we may or may not buy depending on their strength. If we measure the breaking strength of 20 of these tiles in order to estimate the average breaking strength of all the tiles, these 20 measurements are a sample from the population that consists of the breaking strengths of the 400 tiles. In another context, however, if we consider entering into a long-term contract calling for the delivery of tens of thousands of such tiles, we would look upon the breaking strengths of the original 400 tiles only as a sample. Similarly, the complete figures for a recent year, giving the elapsed times between the filing and disposition of divorce suits in San Diego County, can be looked upon as either a population or a sample. If we are interested only in San Diego County and that particular year, we would look upon the data as a population; on the other hand, if we want to generalize about the time that is required for the disposition of divorce suits in the entire United States, in some other county, or in some other year, we would look upon the data as a sample.

As we have used it here, the word "sample" has very much the same meaning as it has in everyday language. A newspaper considers the attitudes of 150 readers toward a proposed school bond to be a sample of the attitudes of all its readers toward the bond; and a consumer considers a box of Mrs. See's candy a sample of the firm's product. Later, we shall use the word "sample" only when referring to data that can reasonably serve as the basis for valid generalizations about the populations from which they came; in this more technical sense, many sets of data that are popularly called samples are not samples at all.

In this chapter and in Chapter 4 we shall describe things statistically without making any generalizations. For future reference, though, it is important to distinguish even here between populations and samples. Thus, we shall use different symbols depending on whether we are describing populations or samples.

3.2   THE MEAN

The most popular measure of central location is what the layperson calls an "average" and what the statistician calls an **arithmetic mean,** or simply a **mean.**[†] It is defined as follows:

The mean of n numbers is their sum divided by n.

It is all right to use the word "average," and on occasion we shall use it ourselves, but there are other kinds of averages in statistics and we cannot afford to speak loosely when there is any risk of ambiguity.

EXAMPLE 3.1   From 1990 through 1994, the combined seizure of heroin by the Drug Enforcement Administration, the F.B.I., and the U.S. Custom's Service added up to 1,794, 3,030, 2,551, 3,514, and 2,824 pounds. Find the mean seizure of heroin by these government agencies for the given five-year period.

Solution   The total for the five years is $1{,}794 + 3{,}030 + 2{,}551 + 3{,}514 + 2{,}824 = 13{,}713$ pounds, so that the mean is $\dfrac{13{,}713}{5} = 2{,}742.6$ pounds.   ∎

EXAMPLE 3.2   In the 97th through 104th Congress of the United States, there were, respectively, 67, 71, 78, 82, 96, 110, 104, and 92 Representatives at least 60 years old at the beginning of the first session. Find the mean.

Solution   The total of these figures is $67 + 71 + 78 + 82 + 96 + 110 + 104 + 92 = 700$. Hence, the mean is $\dfrac{700}{8} = 87.5$.   ∎

[†]The term "arithmetic mean" is used mainly to distinguish the mean from the **geometric mean** and the **harmonic mean,** two other kinds of averages used only in very special situations (see Exercises 3.15 and 3.16).

Since we shall have occasion to calculate the means of many different sets of sample data, it will be convenient to have a simple formula that is always applicable. This requires that we represent the figures to be averaged by some general symbol such as *x, y,* or *z;* the number of values in a sample, the **sample size,** is usually denoted by the letter *n.* Choosing the letter *x,* we can refer to the *n* values in a sample as $x_1, x_2, \ldots,$ and x_n (which read "*x* sub-one," "*x* sub-two," … , and "*x* sub-*n*"), and write

$$\text{sample mean} = \frac{x_1 + x_2 + x_3 + \cdots + x_n}{n}$$

This formula will take care of any set of sample data, but it can be made more compact by assigning the sample mean the symbol $\bar{x}$ (which reads "*x* bar") and using the Σ notation. The symbol Σ is capital *sigma,* the Greek letter for S. In this notation we let Σx stand for "the sum of the *x*'s" (that is, $\Sigma x = x_1 + x_2 + \cdots + x_n$), and we can write

*Sample
mean*

$$\bar{x} = \frac{\Sigma x}{n}$$

If we refer to the measurements as *y*'s or *z*'s, we write their mean as $\bar{y}$ or $\bar{z}$. In the formula for $\bar{x}$, the term Σx does not state explicitly which values of *x* are added; let it be understood, however, that Σx always refers to the sum of all the *x*'s under consideration in a given situation. In Section 3.8, the use of the sigma notation will be discussed in some detail.

The number of values in a population, the **population size,** is usually denoted by *N.* The mean of a population of *N* items is defined in the same way as the mean of a sample. It is the sum of the *N* items, $x_1 + x_2 + \cdots + x_N$, or Σx, divided by *N.*

Assigning the population mean the symbol μ (*mu,* the Greek letter for lowercase *m*), we write

*Population
mean*

$$\mu = \frac{\Sigma x}{N}$$

with the reminder that Σx is now the sum of all *N* values of *x* that constitute the population.[†]

Also, to distinguish between descriptions of populations and descriptions of samples, we not only use different symbols such as μ and $\bar{x}$, but we refer to a description of a population as a **parameter** and a description of a sample as a **statistic.** Parameters are usually denoted by Greek letters.

[†]When the population size is unlimited, as discussed in the beginning of Section 3.1, population means cannot be defined in this way. Definitions of the mean of infinite populations can be found in most textbooks on mathematical statistics.

To illustrate the terminology and notation just introduced, suppose that we are interested in the mean lifetime of a production lot of $N = 40{,}000$ light bulbs. Obviously, we cannot test all of the light bulbs for there would be none left to use or sell, so we take a sample, calculate $\bar{x}$, and use this quantity as an estimate of μ.

EXAMPLE   3.3   If $n = 5$ and the light bulbs in the sample last 967, 949, 952, 940, and 922 hours, what can we conclude about the mean lifetime of the 40,000 light bulbs in the production lot?

Solution   The mean of this sample is

$$\bar{x} = \frac{967 + 949 + 952 + 940 + 922}{5} = 946 \text{ hours}$$

and if we can assume that the data constitute a sample in the technical sense (namely, a set of data from which valid generalizations can be made), we estimate the mean of all 40,000 light bulbs as $\mu = 946$ hours.   ■

For nonnegative data, the mean not only describes their middle, but it also puts some limitation on their size. If we multiply by n on both sides of the equation $\bar{x} = \dfrac{\Sigma x}{n}$, we find that $\Sigma x = n \cdot \bar{x}$ and, hence, that no part, or subset of the data can exceed $n \cdot \bar{x}$.

EXAMPLE   3.4   If the mean salary paid to three NBA players for the 1998–1999 season is $2,450,000, can

    (a) any one of them receive an annual salary of $4,000,000;

    (b) any two of them receive an annual salary of $4,000,000?

Solution   The combined salaries of the three players total $3(2{,}450{,}000) = \$7{,}350{,}000$.

    (a) If one of them receives an annual salary of $4,000,000, this would leave $7{,}350{,}000 - 4{,}000{,}000 = \$3{,}350{,}000$ for the other two players, so this could be the case.

    (b) For two of them to receive an annual salary of $4,000,000 would require $2(4{,}000{,}000) = \$8{,}000{,}000$, which exceeds the total paid to the three players. Hence, this cannot be the case.   ■

EXAMPLE   3.5   If six high school juniors averaged 57 on the verbal part of the PSAT/MSQT test, at most how many of them could have scored 72 or better on the test?

Solution   Since $n = 6$ and $\bar{x} = 57$, it follows that their combined scores total $6(57) = 342$. Since $342 = 4 \cdot 72 + 54$, we find that at most four of the six students could have scored 72 or more.   ■

The popularity of the mean as a measure of the "middle" or "center" of a set of data is not accidental. Anytime we use a single number to describe some aspect of a set of data, there are certain requirements, or desirable features, that should be kept in mind. Aside from the fact that the mean is a simple and familiar measure, the following are some of its noteworthy properties:

(1) The mean can be calculated for any set of numerical data, so it always exists.

(2) Any set of numerical data has one and only one mean, so it is always unique.

(3) The mean lends itself to further statistical treatment; for instance, as we shall see, the means of several sets of data can always be combined into the overall mean of all the data.

(4) The mean is relatively reliable in the sense that means of repeated samples drawn from the same population usually do not fluctuate, or vary, as widely as other statistical measures used to estimate the mean of a population.

The fourth of these properties is of fundamental importance in statistical inference, and we shall study it in some detail in Chapter 10.

Finally, let us consider another property of the mean that, on the surface, seems desirable.

(5) The mean takes into account each item in a set of data.

Note, however, that samples may contain very small or very large values that are so far removed from the main body of the data that the appropriateness of including them in the sample is questionable. Such values may be due to chance, they may be due to gross errors in recording the data, gross errors in calculations, malfunctioning of equipment, or other identifiable sources of contamination. In any case, when such values are averaged in with the other values, they can affect the mean to such an extent that it is debatable whether it really provides a useful, or meaningful, description of the "middle" of the data.

EXAMPLE   3.6   The editor of a book on nutritional values needs a figure for the calorie count of a slice of a 12-inch pepperoni pizza. Letting a laboratory with a calorimeter do the job, she gets the following figures for the pizza from six different fast-food chains: 265, 332, 340, 225, 238, and 346.

   (a) Calculate the mean, which the editor will report in her book.
   (b) Suppose that when calculating the mean, the editor makes the mistake of entering 832 instead of 238 in her calculator. How much of an error would this make in the figure that she reports in her book?

Solution   **(a)** The correct mean is $\bar{x} = \dfrac{265 + 332 + 340 + 225 + 238 + 346}{6} = 291.$

(b) The incorrect mean is $\bar{x} = \dfrac{265 + 332 + 340 + 225 + 832 + 346}{6} =$ 390, so that her error would be a disastrous $390 - 291 = 99$.   ∎

EXAMPLE   3.7   The ages of six students who went on a geology field trip are 16, 17, 15, 19, 16, and 17, and the age of the instructor who went with them is 54. Find the mean age of these seven persons.

Solution   The mean is $\bar{x} = \dfrac{16 + 17 + 15 + 19 + 16 + 17 + 54}{7} = 22$, but any statement to the effect that the average age of the group is 22 could easily be misinterpreted. We might well infer incorrectly that most of the persons who went on the field trip are in their low twenties.   ∎

To avoid the possibility of being misled by a mean affected by a very small value or a very large value, we sometimes find it preferable to describe the middle or center of a set of data with a statistical measure other than the mean; perhaps, with the **median,** which we shall discuss in Section 3.4.

3.3   THE WEIGHTED MEAN

When we calculate a mean, we may be making a serious mistake if we overlook the fact that the quantities we are averaging are not all of equal importance with reference to the situation being described. Consider, for example, a cruise line that advertizes the following fares for single-occupancy cabins on an 11-day Caribbean cruise:

Cabin category	Fare
Ultra deluxe (outside)	$7,870
Deluxe (outside)	$7,080
Outside	$5,470
Outside (shower only)	$4,250
Inside (shower only)	$3,460

The mean of these five fares is $\bar{x} = \dfrac{7,870 + 7,080 + 5,470 + 4,250 + 3,460}{5} =$ $5,626, but we cannot very well say that the average fare for one of these single-occupancy cabins is $5,626. To get that figure, we would also have to know how many cabins there are in each of the categories. Referring to the ship's deck plan, where the cabins are color-coded by category, we find that there are, respectively, 6, 4, 8, 13, and 22 cabins available in these five categories. If it can be assumed that these 53 cabins will all be occupied, the cruise line can expect to receive a total of

$$6(7{,}870) + 4(7{,}080) + 8(5{,}470) + 13(4{,}250) + 22(3{,}460) = \$250{,}670$$

for the 53 cabins and, hence, on the average $\dfrac{250{,}670}{53} \approx \$4{,}729.62$ per cabin.

To give quantities being averaged their proper degree of importance, it is necessary to assign them (relative importance) **weights** and then calculate a **weighted mean.** In general, the weighted mean $\bar{x}_w$ of a set of numbers x_1, x_2, $x_3, \ldots$, and x_n, whose relative importance is expressed numerically by a corresponding set of numbers $w_1, w_2, w_3, \ldots$, and w_n, is given by

Weighted mean

$$\bar{x}_w = \frac{w_1 x_1 + w_2 x_2 + \cdots + w_n x_n}{w_1 + w_2 + \cdots + w_n} = \frac{\Sigma w \cdot x}{\Sigma w}$$

Here $\Sigma w \cdot x$ is the sum of the products obtained by multiplying each x by the corresponding weight, and Σw is simply the sum of the weights. Note that when the weights are all equal, the formula for the weighted mean reduces to that for the ordinary (arithmetic) mean.

EXAMPLE   3.8   The following table shows the number of households in the five Pacific states in 1990, and the corresponding percentage changes in the number of households 1990–1994:

	Number of households (1,000)	Percentage change
Washington	1,872	9.1
Oregon	1,103	8.3
California	10,381	4.5
Alaska	189	10.3
Hawaii	356	7.1

Calculate the weighted mean of the percentage changes using the 1990 numbers of households as weights.

Solution   Substituting $x_1 = 9.1$, $x_2 = 8.3$, $x_3 = 4.5$, $x_4 = 10.3$, $x_5 = 7.1$, $w_1 = 1{,}872$, $w_2 = 1{,}103$, $w_3 = 10{,}381$, $w_4 = 189$, and $w_5 = 356$ into the formula for the weighted mean, we get

$$\frac{9.1(1,872) + 8.3(1,103) + 4.5(10,381) + 10.3(189) + 7.1(356)}{1,872 + 1,103 + 10,381 + 189 + 356}$$

$$= \frac{77,378.9}{13,901} \approx 5.6\%$$

Note that we used the symbol $\approx$ to mean "approximately equal to." We use this symbol only for steps where numerical rounding occurs. ∎

A special application of the formula for the weighted mean arises when we must find the overall mean, or **grand mean,** of k sets of data having the means $\bar{x}_1, \bar{x}_2, \bar{x}_3, \ldots,$ and $\bar{x}_k$, and consisting of $n_1, n_2, n_3, \ldots,$ and n_k measurements or observations. The result is given by

Grand mean of combined data

$$\bar{\bar{x}} = \frac{n_1\bar{x}_1 + n_2\bar{x}_2 + \cdots + n_k\bar{x}_k}{n_1 + n_2 + \cdots + n_k} = \frac{\sum n \cdot \bar{x}}{\sum n}$$

where the weights are the sizes of the samples, the numerator is the total of all the measurements or observations, and the denominator is the number of items in the combined samples.

EXAMPLE 3.9 There are three sections of a course in European history, with 19 students in the section meeting MWF at 9 A.M., 27 in the section meeting MWF at 11 A.M., and 24 in the section meeting MWF at 1 P.M. If the students in the 9 A.M. section averaged 66 in the midterm examination, those in the 11 A.M. section averaged 71, and those in the 1 P.M. section averaged 63, what is the mean score for all three sections combined?

Solution Substituting $n_1 = 19$, $n_2 = 27$, $n_3 = 24$, $\bar{x}_1 = 66$, $\bar{x}_2 = 71$, and $\bar{x}_3 = 63$ into the formula for the grand mean of combined data, we get

$$\bar{\bar{x}} = \frac{19 \cdot 66 + 27 \cdot 71 + 24 \cdot 63}{19 + 27 + 24}$$

$$= \frac{4,683}{70}$$

$$= 66.9$$

or 67 rounded to the nearest integer. ∎

■ Exercises

3.1 Suppose we are given the numbers of home runs hit in each game of the College World Series in Omaha in 1999. Give one illustration each of a situation where these figures would be looked upon as
(a) a population;
(b) a sample.

3.2   Suppose that the final election returns from a given city show that the three candidates for a certain office received, respectively, 15,873, 13,499, and 2,580 votes.
What office might these candidates be running for so that these figures would constitute
(a) a population;
(b) a sample?

3.3   If we determine the length of the delay of the departure of eight TWA flights from
LAX, give two different examples of a population from which we might be sampling.

3.4   Suppose we are given the high and low temperatures recorded each day of August
1999, in Seattle, Washington. Decide in each case whether this information constitutes a sample or a population, if we want to
(a) calculate the difference between the high and low temperatures for each of
these days;
(b) establish norms for these days;
(c) predict the high and low temperatures for August 2009;
(d) determine for each of these days whether the high temperature exceeded an
established norm or whether the low temperature fell below an established
norm.

3.5   Following are the numbers of seconds that 12 insects survived after having been
sprayed with a certain insecticide: 112, 83, 102, 84, 105, 121, 76, 110, 98, 91, 103, and
85. Find the mean of these survival times.

3.6   Following are the numbers of twists that were required to break ten forged alloy
bars: 23, 14, 37, 25, 29, 45, 19, 30, 36, and 42. Calculate the mean of this set of data.

3.7   A 10-ml pipette was calibrated and found to deliver the following volumes: 9.96,
9.98, 9.92, 9.98, and 9.96. Find the mean of these data and use it to estimate by how
much the calibration is off.

3.8   Find the means of the eight values in each of the five lines in the listing of percent
shrinkages in Exercise 2.36 on page 28.

3.9   With reference to Exercise 2.42 on page 29, find the mean of the subsample obtained by beginning with the fourth value, 1.23, and then taking every fifth value
among the 120 root penetrations.

3.10   Find the mean of the 60 lengths of sea trouts given on page 13.

3.11   Following are the speeds (miles per hour) at which 20 cars were timed on I-17 in
Arizona in late-evening traffic: 77, 69, 82, 76, 69, 71, 80, 66, 70, 77, 72, 73, 80, 86, 74,
77, 69, 89, 74, and 75.
(a) Find the mean of these speeds.
(b) Find the mean of these speeds after subtracting 75 from each value and then
adding 75 to the result. What general simplification does this suggest for the
calculation of means?

3.12   An elevator in a department store is designed to carry a maximum load of 3,200
pounds. If it is loaded with 18 persons having a mean weight of 166 pounds, is there
any danger of it being overloaded?

3.13 Generalizing the argument of Examples 3.4 and 3.5, it can be shown that for any set of nonnegative data with the mean $\bar{x}$, the fraction of the data that are greater than or equal to any positive constant k cannot exceed $\bar{x}/k$. Use this result, called **Markov's theorem**, in the following problems:
   (a) If the mean breaking strength of certain linen threads is 33.5 ounces, at most what fraction of the threads can have a breaking strength of 50.0 ounces or more?
   (b) If the diameters of the orange trees in an orchard have a mean of 17.2 cm, at most what fraction of the trees can have a diameter of 20.0 cm or more?

3.14 Records show that in Phoenix, Arizona, the normal daily maximum temperature for each month is 65, 69, 74, 84, 93, 102, 105, 102, 98, 88, 74, and 66 degrees Fahrenheit. Verify that the mean of these figures is 85 and comment on the claim that, in Phoenix, the average daily maximum temperature is a very comfortable 85 degrees.

3.15 The **geometric mean** of n positive numbers is the nth root of their product. For example, the geometric mean of 3 and 12 is $\sqrt{3 \cdot 12} = \sqrt{36} = 6$ and the geometric mean of 1, 3, and 243 is $\sqrt[3]{1 \cdot 3 \cdot 243} = \sqrt[3]{729} = 9$.
   (a) Find the geometric mean of 9 and 36.
   (b) Find the geometric mean of 1, 2, 8, and 81.
   (c) During a flu epidemic, 12 cases were reported on the first day, 18 on the second day, and 48 on the third day. Thus, from the first day to the second day the number of cases reported was multiplied by $\dfrac{18}{12} = \dfrac{3}{2}$, and from the second day to the third day the number of cases was multiplied by $\dfrac{48}{18} = \dfrac{8}{3}$. Find the geometric mean of these two growth rates and, assuming that the growth pattern continues, predict the number of cases that will be reported on the fourth and fifth days.

3.16 The **harmonic mean** of n positive numbers $x_1, x_2, x_3, \ldots,$ and x_n is defined as the reciprocal of the mean of their reciprocals. Its usefulness is limited, but it is appropriate in some special situations. For instance, if someone drives 10 miles on a highway at 60 mph and on the way back at 30 mph, he will not have averaged $\dfrac{60 + 30}{2} =$ 45 mph. He will have driven 20 miles in 30 minutes, which makes his average speed 40 mph.
   (a) Verify that the harmonic mean of 60 and 30 is 40, so that it gives the appropriate average for the preceding example.
   (b) If an investor buys $18,000 worth of a company's stock at $45 a share and another $18,000 at $36 a share, she is buying

$$\frac{18{,}000}{45} + \frac{18{,}000}{36} = 900$$

   shares for $36,000 and she is paying $\dfrac{36{,}000}{900} = \$40$ per share. Verify that 40 is, in fact, the harmonic mean of 45 and 36.

3.17 A Dodge dealer advertised the following prices as part of a "stock reduction" sale:

1998 Neon	$7,988
1999 Stratus/Breeze	$10,988
1999 Stratus ES	$13,988

He has, respectively, 24, 18, and 15 of these cars in stock. Using these in-stock figures as weights, find the weighted mean of the prices of these three kinds of cars.

3.18 An instructor counts the final examination in a course four times as much as each of three one-hour examinations. Which of two students has a higher weighted average score, the one who received scores of 72, 80, and 65 in the one-hour examinations and an 82 in the final examination, or the one who received scores of 81, 87, and 75 in the one-hour examinations and a 78 in the final examination?

3.19 A person invests $5,000 in a municipal bond paying 4.75%, $10,000 in a money market fund paying 4.95%, and $2,000 in a certificate of deposit paying 3.25%. Using the respective amounts invested as weights, find the weighted average return on these investments. Does this figure equal the investor's actual return on these three investments?

3.20 Among the students receiving bachelor's degrees from a certain university in 1993, 382 majoring in the humanities had salary offers averaging $24,373, 450 majoring in the social sciences had salary offers averaging $22,684, and 113 majoring in computer science had salary offers averaging $31,329. What was the average salary offered to these 945 graduates?

3.21 The 401 thousand male students taking the ACT mathematics test in 1999 averaged 20.8 (maximum score 36), and the 489 thousand female students averaged 19.6. What was their combined average score?

3.22 A home appliance center advertised the following refrigerators, of which it had, respectively, 19, 12, 8, 15, and 26 in stock:

Brand	Size	Price
Magic Chef	15 cu. ft.	$379
Magic Chef	21 cu. ft.	$499
Jenn-Air	19 cu. ft.	$549
Jenn-Air	21 cu. ft.	$649
Maytag	24 cu. ft.	$799

(a) What is the average size of these refrigerators?
(b) What is the average price of these refrigerators?

3.4 THE MEDIAN

To avoid the possibility of being misled by one or a few very small or very large values, we sometimes describe the "middle" or "center" of a set of data with statistical measures other than the mean. One of these, the **median** of n values requires that we first arrange the data according to size. Then it is defined as follows:

The median is the value of the middle item when n is odd, and the mean of the two middle items when n is even.

In either case, when no two values are alike, the median is exceeded by as many values as it exceeds. When some of the values are alike, this may not be the case.

EXAMPLE 3.10   In five recent weeks, a town reported 36, 29, 42, 25, and 29 burglaries. Find the median number of burglaries for these weeks.

Solution   The median is not 42, the third (or middle) item, because the data must first be arranged according to size. Thus, we get

$$25 \quad 29 \quad 29 \quad 36 \quad 42$$

and it can be seen that the middle one, the median, is 29.   ■

Note that in this example there are two 29's among the data and that we did not refer to either of them as *the median*—the median is a number and not necessarily a particular measurement or observation.

EXAMPLE 3.11   In some cities, persons cited for minor traffic violations can attend a class in defensive driving in lieu of paying a fine. Given that 12 such classes in Phoenix, Arizona, were attended by 37, 32, 28, 40, 35, 38, 40, 24, 30, 37, 32, and 40 persons, find the median of these data.

Solution   Ranking these attendance figures according to size, from low to high, we get

$$24 \quad 28 \quad 30 \quad 32 \quad 32 \quad 35 \quad 37 \quad 37 \quad 38 \quad 40 \quad 40 \quad 40$$

and we find that the median is the mean of the two values nearest the middle, namely, $\dfrac{35 + 37}{2} = 36$.   ■

Some of the values were alike in this example, but this did not affect the median, which exceeds six of the values and is exceeded by equally many. The situation is quite different, however, in the example that follows.

EXAMPLE 3.12   On the seventh hole of a certain golf course, a par four, nine golfers scored par, birdie (one below par), par, par, bogey (one above par), eagle (two below par), par, birdie, birdie. Find the median.

Solution   Ranking these figures according to size, from low to high, we get

$$2 \quad 3 \quad 3 \quad 3 \quad 4 \quad 4 \quad 4 \quad 4 \quad 5$$

and it can be seen that the fifth value, the median, is equal to par 4.   ■

This time the median exceeds four of the values but is exceeded by only one, and it may well be misleading to think of the median, 4, as the middle of the nine scores. It is not exceeded by as many values as it exceeds, but *by definition* the median *is* 4.

The symbol that we use for the median of n sample values $x_1, x_2, x_3, \ldots,$ and x_n is $\tilde{x}$ (and, hence, $\tilde{y}$ or $\tilde{z}$ if we refer to the values of y's or z's). If a set of data constitutes a population, we denote its median by $\tilde{\mu}$.

Thus, we have a symbol for the median, but no formula; there is only a formula for the **median position.** Referring again to data arranged according to size, usually ranked from low to high, we can write

Median position

> *The median is the value of the* $\dfrac{n+1}{2}$ *th item.*

EXAMPLE 3.13   Find the median position for

(a) $n = 17$;   (b) $n = 41$.

Solution   With the data arranged according to size (and counting from either end)

(a) $\dfrac{n+1}{2} = \dfrac{17+1}{2} = 9$ and the median is the value of the 9th item;

(b) $\dfrac{n+1}{2} = \dfrac{41+1}{2} = 21$ and the median is the value of the 21st item. ∎

EXAMPLE 3.14   Find the median position for

(a) $n = 16$;   (b) $n = 50$.

Solution   With the data arranged according to size (and counting from either end)

(a) $\dfrac{n+1}{2} = \dfrac{16+1}{2} = 8.5$ and the median is the mean of the values of the 8th and 9th items;

(b) $\dfrac{n+1}{2} = \dfrac{50+1}{2} = 25.5$ and the median is the mean of the values of the 25th and 26th items. ∎

It is important to remember that $\dfrac{n+1}{2}$ is the formula for the median position and not a formula for the median, itself. It is also worth mentioning that determining the median can usually be simplified, especially for large sets of data, by first presenting the data in the form of a stem-and-leaf display.

EXAMPLE 3.15   On page 16 we gave data on the number of rooms occupied each day in a resort hotel during the month of June, and in Figure 2.2 we displayed these data as follows:

2	3	57
6	4	0023
13	4	5666899
(3)	5	234
14	5	56789
9	6	1224
5	6	9
4	7	23
2	7	8
1	8	1

Use this double-stem display to find the median of these room-occupancy data.

Solution   When we gave this display in Figure 2.2, we did not explain the significance of the figures in the column to the left of the stem labels. As can easily be verified, they are simply the accumulated numbers of leaves counted from either end. Furthermore, the parentheses around the 3 are meant to tell us that the median of the data is on that stem (or else is the mean of two values on that stem).

Since $n = 30$ for the given table, the median position is $\dfrac{30 + 1}{2} = 15.5$, so that the median is the mean of the fifteenth and sixteenth largest values among the data. Since $2 + 4 + 7 = 13$ of the values are represented by leaves on the first three stems, the median is the mean of the values represented by the second and third leaves on the fourth stem. These are 53 and 54, and hence the median of the room-occupancy data is $\dfrac{53 + 54}{2} = 53.5$. Note that this illustrates why we said that it is generally advisable to arrange the leaves on each stem, so that they are ranked from low to high.   ∎

As a matter of interest, let us also mention that the mean of the room-occupancy data is 55.7. It really should not come as a surprise that the median does not equal the mean—it defines the middle of a set of data in a different way. The median is average in the sense that it splits the data into two parts so that, unless there are duplicates, there are equally many values above and below the median. The mean, on the other hand, is average in the sense that if each value is replaced by some constant k while the total remains unchanged, this number k will have to be the mean. (This follows directly from the relationship $n \cdot \overline{x} = \Sigma x$.) In this sense, the mean has also been likened to a center of gravity.

The median shares some, but not all, of the properties of the mean listed on page 52. Like the mean, the median always exists and it is unique for any set of data. Also like the mean, the median is simple enough to find once the data have been arranged according to size, but as we indicated earlier, sorting a set of data manually can be a surprisingly difficult task.

Unlike the mean, the medians of several sets of data cannot generally be combined into an overall median of all the data, and in problems of statistical inference the median is usually less reliable than the mean. This is meant to say that the medians of repeated samples from the same population will usually vary more widely than the corresponding means (see Exercises 3.32 and 4.20). On the other hand, sometimes the median may be preferable to the mean because it is not so easily, or not at all, affected by extreme (very small or very large) values. For instance, in Example 3.6 we showed that incorrectly entering 832 instead of 238 into a calculator caused an error of 99 in the mean. As the reader will be asked to verify in Exercise 3.29, the corresponding error in the median would have been only 37.5.

Finally, also unlike the mean, the median can be used to define the middle of a number of objects, properties, or qualities that can be ranked, namely, when we deal with ordinal data. For instance, we might rank a number of tasks according to their difficulty and then describe the middle (or median) one as being of "average difficulty." Also, we might rank samples of chocolate fudge according to their consistency and then describe the middle (or median) one as having "average consistency."

Besides the median and the mean there are several other measures of central location; for example, the **midrange** described in Exercise 3.35 and the **midquartile** defined on page 64. Each describes the "middle" or "center" of a set of data in its own way, and it should not come as a surprise that their values may well all be different. Then there is also the **mode** described in Section 3.6.

3.5   OTHER FRACTILES

The median is but one of many **fractiles** that divide data into two or more parts, as nearly equal as they can be made. Among them we also find **quartiles, deciles,** and **percentiles,** which are intended to divide data into four, ten, and a hundred parts. Until recently, fractiles were determined mainly for distributions of large sets of data, and in this connection we shall study them in Section 3.7.

In this section we shall concern ourselves mainly with a problem that has arisen in **exploratory data analysis**—in the preliminary analysis of relatively small sets of data. It is the problem of dividing such data into four nearly equal parts, where we say "nearly equal" because there is no way in which we can divide a set of data into four equal parts for, say, $n = 27$ or $n = 33$. Statistical measures designed for this purpose have traditionally been referred to as the three **quartiles,** Q_1, Q_2, and Q_3, and there is no argument about Q_2, which is simply the median. On the other hand, there is some disagreement about the definition of Q_1 and Q_3.

As we shall define them, the quartiles divide a set of data into four parts such that there are as many values less than Q_1 as there are between Q_1 and Q_2, between Q_2 and Q_3, and greater than Q_3. Assuming that no two values are alike, this is accomplished by letting

Q_1 be the median of all the values less than the median of the whole set of data,

and

Q_3 be the median of all the values greater than the median of the whole set of data.

EXAMPLE 3.16   Following are the high-temperature readings in twelve European capitals on a recent day in the month of June: 90, 75, 86, 77, 85, 72, 78, 79, 94, 82, 74, and 93. Find Q_1, Q_2 (the median), and Q_3.

Solution   For $n = 12$ the median position is $\dfrac{12 + 1}{2} = 6.5$ and, after arranging the data according to size, we find that the sixth and seventh values among

$$72 \quad 74 \quad 75 \quad 77 \quad 78 \quad 79 \quad 82 \quad 85 \quad 86 \quad 90 \quad 93 \quad 94$$

are 79 and 82. Hence the median is $\dfrac{79 + 82}{2} = 80.5$. For the six values below 80.5 the median position is $\dfrac{6 + 1}{2} = 3.5$, and since the third and fourth values are 75 and 77, $Q_1 = \dfrac{75 + 77}{2} = 76$. Counting from the other end, the third and fourth values are 90 and 86, and $Q_3 = \dfrac{90 + 86}{2} = 88$. As can be seen from the data and also from Figure 3.1, there are three values below 76, three values between 76 and 80.5, three values between 80.5 and 88, and three values above 88.   ■

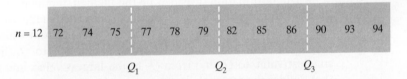

FIGURE 3.1
The three quartiles of
Example 3.16.

Everything worked nicely in this example, but $n = 12$ happened to be a multiple of 4, which raises the question whether our definition of Q_1 and Q_3 will work also when this is not the case.

EXAMPLE 3.17   Suppose that the city where the high temperature was 77 failed to report, so that we are left with the following 11 numbers arranged according to size:

$$72 \quad 74 \quad 75 \quad 78 \quad 79 \quad 82 \quad 85 \quad 86 \quad 90 \quad 93 \quad 94$$

Find Q_1, Q_2, and Q_3.

Solution   For $n = 11$ the median position is $\dfrac{11 + 1}{2} = 6$ and, referring to the preceding data, which are already arranged according to size, we find that the median is 82. For the five values below 82 the median position is $\dfrac{5 + 1}{2} = 3$, and Q_1, the third value, equals 75. Counting from the other end, Q_3, the third value, equals 90. As can be seen from the data and also from Figure 3.2, there are two values below 75, two values between 75 and 82, two values between 82 and 90, and two values above 90. Again, this satisfies the requirement for the three quartiles, $Q_1, Q_2,$ and Q_3. ∎

FIGURE 3.2
The three quartiles of Example 3.17.

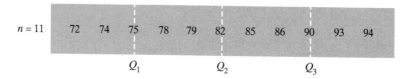

If some of the values are alike, we modify the definitions of Q_1 and Q_3 by replacing "less than the median" by "to the left of the median position" and "greater than the median" by "to the right of the median position." For instance, for Example 3.12 we already showed that the median, the fifth value, equals 4. Now, the median of the four values to the left of the median position, Q_1, equals 3, and the median of the four values to the right of the median position, Q_3, equals 4.

Quartiles are not meant to be descriptive of the "middle" or "center" of a set of data, and we have given them here mainly because, like the median, they are fractiles and they are determined in more or less the same way. The **midquartile** $\frac{Q_1 + Q_3}{2}$, has been used on occasion as another measure of central location. An alternative definition of Q_1 and Q_3 is provided in Exercise 3.52.

The information provided by the median, the quartiles Q_1 and Q_3, and the smallest and largest values is sometimes presented in the form of a **boxplot.** Originally referred to somewhat whimsically as a **box-and-whisker plot,** such a display consists of a rectangle that extends from Q_1 to Q_3, lines drawn from the smallest value to Q_1 and from Q_3 to the largest value, and a line at the median that divides the rectangle into two parts. In practice, boxplots are sometimes embellished with other features, but the simple form shown here is adequate for most purposes.

EXAMPLE 3.18   In Example 3.15 we used the following double-stem display to show that the median of the room occupancy data, originally given on page 16, is 53.5:

2	3	57
6	4	0023
13	4	5666899
(3)	5	234
14	5	56789
9	6	1224
5	6	9

4	7	23
2	7	8
1	8	1

(a) Find the smallest and largest values.
(b) Find Q_1 and Q_3.
(c) Draw a boxplot.

Solution

(a) As can be seen by inspection the smallest value is 35 and the largest value is 81.

(b) For $n = 30$ the median position is $\dfrac{30 + 1}{2} = 15.5$ and, hence, for the 15 values below 53.5 the median position is $\dfrac{15 + 1}{2} = 8$. It follows that Q_1, the eighth value, is 46. Similarly, Q_3, the eighth value from the other end, is 62.

(c) Combining all this information, we obtain the boxplot shown in Figure 3.3. ∎

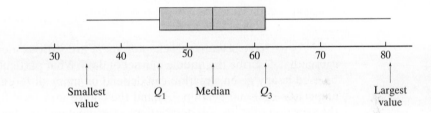

FIGURE 3.3
Boxplot of the room occupancy data.

Boxplots can also be constructed with appropriate computer software or a graphing calculator. Using the same data as in Example 3.18, we reproduced the one shown in Figure 3.4 from the display screen of a TI-83 graphing calculator.

FIGURE 3.4
Boxplot of the room occupancy data.

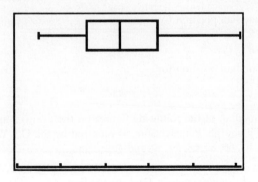

3.6 THE MODE

Another measure that is sometimes used to describe the middle or center of a set of data is the **mode,** which is defined simply as the value that occurs with the highest frequency and more than once. Its two main advantages are that it requires no calculations, only counting, and it can be determined for qualitative, or nominal, data.

EXAMPLE 3.19   The 20 meetings of a square dance club were attended by 22, 24, 23, 24, 27, 25, 20, 24, 26, 28, 26, 23, 21, 24, 24, 25, 23, 28, 26, and 25 of its members. Find the mode.

Solution   Among these numbers, 20, 21, 22, and 27 each occurs once, 28 occurs twice, 23, 25, and 26 each occurs three times, and 24 occurs 5 times. Thus, the modal attendance is 24.   ∎

EXAMPLE 3.20   In Example 3.12 we gave the scores of nine golfers on a par-four hole as 2, 3, 3, 3, 4, 4, 4, 4, and 5. Find the mode.

Solution   Since these data are already arranged according to size, it can easily be seen that 4, which occurs four times, is the modal score.   ∎

As we have seen in this chapter, there are various measures of central location that describe the middle of a set of data. What particular "average" should be used in any given situation can depend on many different things (see, for example, the optional Section 7.3) and the choice may be difficult to make. Since the selection of statistical descriptions often contains an element of arbitrariness, some persons believe that the magic of statistics can be used to prove nearly anything. Indeed, a famous nineteenth-century British statesman is often quoted as saying that there are three kinds of lies: *lies, damned lies, and statistics.* Exercises 3.34 and 3.35 describe a situation where this kind of criticism is well justified.

■ Exercises

3.23 Find the median position for
(a) $n = 25$;
(b) $n = 18$.

3.24 Find the median position for
(a) $n = 13$;
(b) $n = 32$.

3.25 Find the median of the following figures on the average monthly percent of sunshine in Pittsburgh, Pennsylvania, as reported by the U.S. Weather Bureau: 38, 40, 53, 53, 57, 65, 66, 63, 68, 59, 50, and 40.

3.26 On fifteen days, a restaurant served breakfast to 38, 50, 53, 36, 38, 56, 46, 54, 54, 58, 35, 61, 44, 48, and 59 persons. Find the median.

3.27 Thirty-two NBA games lasted 138, 142, 113, 164, 159, 157, 135, 122, 126, 139, 140, 142, 157, 121, 143, 140, 169, 130, 142, 146, 155, 117, 158, 148, 145, 151, 137, 128, 133, 150, 134, and 147 minutes. Determine the median length of these games.

3.28 In a study of the stopping ability of standard passenger cars on dry, clean, level pavement, twenty-one drivers going 30 mph were able to stop in 78, 69, 79, 91, 66, 72, 74, 85, 84, 66, 76, 67, 79, 83, 70, 77, 67, 79, 79, 77, and 67 feet. Find the median of these stopping distances.

3.29 With reference to Example 3.6, suppose that the editor of the book on nutrition had used the median instead of the mean to average the respective calorie counts. Show that with the median the error of using 832 instead of 238 would have been only 37.5.

3.30 Following are the miles per gallon obtained with 30 tankfuls of gas:

24.1	24.9	25.2	23.8	24.7	22.9	25.0	24.1	23.6	24.5
23.7	24.4	24.7	23.9	25.1	24.6	23.3	24.3	24.8	22.8
23.9	24.2	24.7	24.9	25.0	24.8	24.5	23.4	24.6	25.3

Construct a double-stem display and use it to determine the median of these data.

3.31 Following are the body weights of 60 small lizards used in a study of vitamin deficiencies (in grams):

125	128	106	111	116	123	119	114	117	143
136	92	115	118	121	137	132	120	104	125
119	115	101	129	87	108	110	133	135	126
127	103	110	126	118	82	104	137	120	95
146	126	119	113	105	132	126	118	100	113
106	125	117	102	146	129	124	113	95	148

Construct a stem-and-leaf display with the stem labels 8, 9, 10, 11, 12, 13, and 14, and use it to determine the median of these data.

3.32 To verify the claim that the mean is generally more reliable than the median (namely, that the mean is subject to smaller chance fluctuations), a student conducts an experiment consisting of 12 tosses of three dice. The following are his results:

2, 4, and 6,   5, 3, and 5,   4, 5, and 3,   5, 2, and 3,   6, 1, and 5,   3, 2, and 1,
3, 1, and 4,   5, 5, and 2,   3, 3, and 4,   1, 6, and 2,   3, 3, and 3,   4, 5, and 3

(a) Calculate the 12 medians and the 12 means.
(b) Group the medians and the means obtained in part (a) into separate distributions having the classes

$$1.5 - 2.5, \quad 2.5 - 3.5, \quad 3.5 - 4.5, \quad 4.5 - 5.5$$

(Note that there will be no ambiguities since the medians of three whole numbers and the means of three whole numbers cannot equal 2.5, 3.5, or 4.5.)
(c) Draw histograms of the two distributions obtained in part (b) and explain how they illustrate the claim that the mean is generally more reliable than the median.

3.33 Repeat Exercise 3.32 with your own data by repeatedly rolling three dice (or one die three times) and constructing distributions of the medians and the means.

3.34 A consumer testing service obtains the following mileages per gallon during five test runs performed with each of three compact cars:

Car A:	27.9	30.4	30.6	31.4	31.7
Car B:	31.2	28.7	31.3	28.7	31.3
Car C:	28.6	29.1	28.5	32.1	29.7

(a) If the manufacturers of car A want to advertise that their car performed best in this test, which of the "averages" discussed so far in this chapter could they use to substantiate this claim?

(b) If the manufacturers of car B want to advertise that their car performed best in this test, which of the "averages" discussed so far in this chapter could they use to substantiate this claim?

3.35 With reference to Exercise 3.34, suppose that the manufacturers of car C hire an unscrupulous statistician and instruct him to find some kind of "average" that will substantiate that their car performed best in the test. Show that the **midrange** (the mean of the smallest and largest values) will serve their purpose.

3.36 Find the positions of the median, Q_1, and Q_3 for
(a) $n = 16$;
(b) $n = 27$.

3.37 Find the positions of the median, Q_1, and Q_3 for
(a) $n = 31$;
(b) $n = 28$.

3.38 Find the positions of the median, Q_1, and Q_3 for $n = 29$ and verify that when no two values are alike, there are as many values below Q_1 as there are between Q_1 and the median, between the median and Q_3, and above Q_3.

3.39 Find the positions of the median, Q_1, and Q_3 for $n = 34$ and verify that even though some of the values may be alike, there are as many values to the left of the Q_1 position as there are to the right of the Q_1 position and to the left of the median position, to the right of the median position and to the left of the Q_3 position, and to the right of the Q_3 position.

3.40 Following are the thicknesses of grease coatings (in microns) produced on twenty steel rods: 41, 51, 63, 57, 57, 66, 63, 60, 44, 41, 46, 43, 53, 55, 48, 49, 65, 61, 58, and 66. Find the median, Q_1, and Q_3. Also verify that there are as many values below Q_1 as there are between Q_1 and the median, between the median and Q_3, and above Q_3.

3.41 With reference to Exercise 3.40, determine the smallest and largest values among the data, and use the information obtained here and in Exercise 3.40 to draw a boxplot.

3.42 Following are fourteen temperature readings taken at different locations in a large kiln: 409, 412, 439, 411, 432, 432, 405, 411, 422, 417, 440, 427, 411, and 417. Find the median, Q_1, and Q_3. Also verify that there are as many values to the left of the Q_1 position as there to the right of the Q_1 position and to the left of the median position, to the right of the median position and to the left of the Q_3 position, and to the right of the Q_3 position.

3.43 With reference to Exercise 3.42, determine the smallest and largest values among the data, and use the information obtained here and in Exercise 3.42 to draw a boxplot.

3.44 With reference to Example 2.4, use a computer package or a graphing calculator to sort the $n = 110$ waiting times between eruptions of Old Faithful in ascending order, and use this arrangement to determine the median, Q_1, and Q_3.

3.45 Use the arrangement obtained in Exercise 3.44 to read off the smallest and largest values among the waiting times, and use the information obtained here and in Exercise 3.44 to draw a boxplot.

3.46 With reference to Exercise 3.31, use a computer package or a graphing calculator to sort the data, ranked from low to high, and use this arrangement to determine the median, Q_1, and Q_3 for the 60 body weights.

3.47 Use the arrangement obtained in Exercise 3.46 to read off the smallest and largest among the body weights, and use the information obtained here and in Exercise 3.46 to draw a boxplot.

3.48 With reference to Exercise 2.42, use a computer package or a graphing calculator to rearrange the $n = 120$ lengths of root penetrations, ranked from low to high, and use this arrangement to determine the median, Q_1, and Q_3.

3.49 Use the arrangement obtained in Exercise 3.48 to read off the smallest and largest of the lengths of the root penetrations, and use the information obtained here and in Exercise 3.48 to draw a boxplot.

3.50 Use the double-stem display obtained in Exercise 3.30 to
(a) read off the smallest and largest values;
(b) determine Q_1 and Q_3.

3.51 Use the results obtained in Exercises 3.30 and 3.50 to draw a boxplot for the mileage data.

3.52 An alternative definition of the first and third quartiles defines the **lower hinge** as the median of all the values less than *or equal to* the median of the whole set of data, and the **upper hinge** as the median of all the values greater than *or equal to* the median of the whole set of data.[†] It is assumed here again that no two values are alike. Rework Example 3.16, finding the two hinges instead of Q_1 and Q_3, and draw a diagram like the one shown in Figure 3.1.

3.53 Rework Example 3.17, finding the two hinges (as defined in Exercise 3.52) instead of Q_1 and Q_3, and draw a diagram like the one shown in Figure 3.2.

3.54 Using the diagrams obtained in Exercises 3.52 and 3.53, decide in each case whether there are as many values below the lower hinge as there are between the lower hinge and the median, between the median and the upper hinge, and above the upper hinge.

3.55 Find the mode of each of the following data, provided, of course, that it exists:
(a) 6, 8, 6, 5, 5, 7, 7, 9, 7, 6, 8, 4, and 7;
(b) 57, 39, 54, 30, 46, 22, 48, 35, 27, 31, and 23;
(c) 11, 15, 13, 14, 13, 12, 10, 11, 12, 13, 11, and 13.

3.56 Following are the numbers of chicken dinners served by a restaurant on 40 Sundays: 41, 52, 46, 42, 46, 36, 46, 61, 58, 44, 49, 48, 48, 52, 50, 45, 68, 45, 48, 47, 49, 57, 44, 48, 49, 45, 47, 48, 43, 45, 45, 56, 48, 54, 51, 47, 42, 53, 48, and 41. Find the mode.

3.57 Following are the numbers of blossoms on 50 cacti in a desert botanical garden: 1, 0, 3, 0, 4, 1, 0, 1, 0, 0, 1, 6, 1, 0, 0, 0, 3, 3, 0, 1, 1, 5, 0, 2, 0, 3, 1, 1, 0, 4, 0, 0, 1, 2, 1, 1, 2, 0, 1, 0, 3, 0, 0, 1, 5, 3, 0, 0, 1, and 0. Find the mode.

3.58 On a seven-day cruise, twenty passengers complained of seasickness on 0, 4, 5, 1, 0, 0, 5, 4, 5, 5, 0, 2, 0, 0, 6, 5, 4, 1, 3, and 2 days. Find the mode and explain why it may very well give a misleading picture of the actual situation.

[†]The use of hinges as an alternative to quartiles was once popular in exploratory data analysis, but that popularity has diminished.

3.59   When there is more than one mode, this is often construed as an indication that the data actually consist of a combination of several distinct sets of data. Reanalyze the data of Exercise 3.58 after changing the fourth value from 1 to 5.

3.60   Asked whether they ever go to the opera, 40 persons in the age group from 20 to 29 replied as follows: rarely, occasionally, never, occasionally, occasionally, occasionally, rarely, rarely, never, occasionally, never, rarely, occasionally, frequently, occasionally, rarely, never, occasionally, occasionally, rarely, rarely, never, occasionally, occasionally, rarely, frequently, rarely, occasionally, occasionally, never, rarely, frequently, never, rarely, occasionally, occasionally, rarely, rarely, occasionally and never. What is their modal reply?

3.61   With reference to Exercise 2.5, what is the modal reply of the 30 persons interviewed at the dog show?

*3.7   THE DESCRIPTION OF GROUPED DATA

In the past, considerable attention was paid to the description of grouped data, because it usually simplified matters to group large sets of data before calculating various statistical measures. This is no longer the case, since the necessary calculations can now be made in a matter of seconds with the use of computers or even hand-held calculators. Nevertheless, we shall devote this section and Section 4.4 to the description of grouped data, since many kinds of data (for example, those reported in government publications) are available only in the form of frequency distributions.

As we have already seen in Chapter 2, the grouping of data entails some loss of information. Each item loses its identity, so to speak; we know only how many values there are in each class or in each category. This means that we shall have to be satisfied with approximations. Sometimes we treat our data as if all the values falling into a class were equal to the corresponding class mark, and we shall do so to define the mean of a frequency distribution. Sometimes we treat our data as if all the values falling into a class are spread evenly throughout the corresponding class interval, and we shall do so to define the median of a frequency distribution. In either case, we get good approximations since the resulting errors will tend to average out.

To give a general formula for the mean of a distribution with k classes, let us denote the successive class marks by $x_1, x_2, \ldots,$ and x_k, and the corresponding class frequencies by $f_1, f_2, \ldots,$ and f_k. Then, the sum of all the measurements is approximated by

$$x_1 \cdot f_1 + x_2 \cdot f_2 + \cdots + x_k \cdot f_k = \sum x \cdot f$$

and the mean of the distribution is given by

*Mean of
grouped data*

$$\bar{x} = \frac{\sum x \cdot f}{n}$$

Here n is the size of the sample, $f_1 + f_2 + \cdots + f_k$, and to write a corresponding formula for the mean of a population we substitute μ for $\bar{x}$ and N for n.

EXAMPLE 3.21   Find the mean for the distribution of the waiting times between eruptions of Old Faithful geyser that was obtained in Example 2.4.

Solution   To get $\Sigma x \cdot f$, we perform the calculations shown in the following table, where the first column contains the class marks, the second column consists of the class frequencies shown on page 24, and the third column contains the products $x \cdot f$:

Class mark x	Frequency f	$x \cdot f$
34.5	2	69.0
44.5	2	89.0
54.5	4	218.0
64.5	19	1,225.5
74.5	24	1,788.0
84.5	39	3,295.5
94.5	15	1,417.5
104.5	3	313.5
114.5	2	229.0
	110	8,645.0

Then, substitution into the formula yields $\bar{x} = \dfrac{8,645.0}{110} = 78.59$ rounded to two decimals.   ∎

To check on the **grouping error,** namely, the error introduced by replacing each value within a class by the corresponding class mark, we can calculate $\bar{x}$ for the original data given on page 23, or use the same computer software that led to Figure 2.4. Having already entered the data, we simply change the command to MEAN C1 and we get 78.273, or 78.27 rounded to two decimals. Thus, the grouping error is only $78.59 - 78.27 = 0.32$, which is fairly small.

When dealing with grouped data, we can determine most other statistical measures besides the mean, but we may have to make different assumptions and/or modify the definitions. For instance, for the median of a distribution we use the second of the assumptions mentioned on page 70 (namely, the assumption that the values within a class are spread evenly throughout the corresponding class interval). Thus, with reference to a histogram

The median of a distribution is such that the total area of the rectangles to its left equals the total area of the rectangles to its right.

To find the dividing line between the two halves of a histogram (each of which represents $\frac{n}{2}$ of the items grouped), we must count $\frac{n}{2}$ of the items starting at either end of the distribution. How this is done is illustrated by the following example and Figure 3.5.

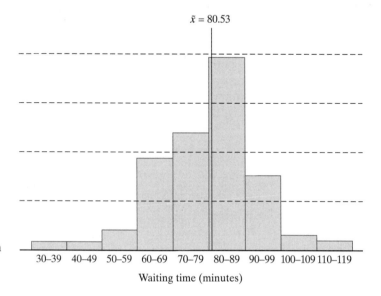

$\tilde{x} = 80.53$

FIGURE 3.5
The median of the distribution of the waiting times between eruptions of Old Faithful.

30–39 40–49 50–59 60–69 70–79 80–89 90–99 100–109 110–119

Waiting time (minutes)

EXAMPLE 3.22   Find the median of the distribution of the waiting times between eruptions of Old Faithful.

Solution   Since $\frac{n}{2} = \frac{110}{2} = 55$, we must count 55 of the items starting at either end. Starting at the bottom of the distribution (that is, beginning with the smallest values), we find that $2 + 2 + 4 + 19 + 24 = 51$ of the values fall into the first five classes. Therefore, we must count $55 - 51 = 4$ more values from among the values in the sixth class. Based on the assumption that the 39 values in the sixth class are spread evenly throughout that class, we accomplish this by adding $\frac{4}{39}$ of the class interval of 10 to 79.5, which is its lower class boundary. This yields

$$\tilde{x} = 79.5 + \frac{4}{39} \cdot 10 = 80.53$$

rounded to two decimals.   ■

In general, if L is the lower boundary of the class into which the median must fall, f is its frequency, c is its class interval, and j is the number of items we still lack when we reach L, then the median of the distribution is given by

*Median of
grouped data*

$$\tilde{x} = L + \frac{j}{f} \cdot c$$

If we prefer, we can find the median of a distribution by starting to count at the other end (beginning with the largest values) and subtracting an appropriate fraction of the class interval from the upper boundary of the class into which the median must fall.

EXAMPLE 3.23   Use this alternative approach to find the median of the waiting times between eruptions of Old Faithful.

Solution   Since $2 + 3 + 15 = 20$ of the values fall above 89.5, we need $55 - 20 = 35$ of the 39 values in the next class to reach the median. Thus, we write

$$\tilde{x} = 89.5 - \frac{35}{39} \cdot 10 = 80.53$$

and the result is, of course, the same.   ∎

Note that the median of a distribution can be found regardless of whether the class intervals are all equal. In fact, it can be found even when either or both classes at the top and at the bottom of a distribution are open, so long as the median does not fall into either class (see Exercise 3.62).

The method by which we found the median of a distribution can also be used to determine other fractiles. For instance, Q_1 and Q_3 are defined for grouped data so that 25% of the total area of the rectangles of the histogram lies to the left of Q_1 and 25% lies to the right of Q_3. Similarly, the nine deciles (which are intended to divide a set of data into ten equal parts) are defined for grouped data so that 10 percent of the total area of the rectangles of the histogram lies to the left of D_1, 10 percent lies between D_1 and $D_2, \ldots$, and 10 percent lies to the right of D_9. And finally, the ninety-nine percentiles (which are intended to divide a set of data into a hundred equal parts) are defined for grouped data so that 1 percent of the total area of the rectangles of the histogram lies to the left of P_1, 1 percent lies between P_1 and $P_2, \ldots$, and 1 percent lies to the right of P_{99}. Note that D_5 and P_{50} are equal to the median and that P_{25} equals Q_1 and P_{75} equals Q_3.

EXAMPLE 3.24   Find Q_1 and Q_3 for the distribution of the waiting times between eruptions of Old Faithful.

Solution   To find Q_1 we must count $\frac{110}{4} = 27.5$ of the items starting at the bottom of the distribution. Since there are $2 + 2 + 4 + 19 = 27$ values in the first four classes, we must count $27.5 - 27 = 0.5$ of the 24 values in the fifth class to reach Q_1. This yields

$$Q_1 = 69.5 + \frac{0.5}{24} \cdot 10 \approx 69.71$$

Since $2 + 3 + 15 = 20$ of the values fall into the last three classes, we must count $27.5 - 20 = 7.5$ of the 39 values in the next class to reach Q_3. Thus, we write

$$Q_3 = 89.5 - \frac{7.5}{39} \cdot 10 \approx 87.58 \qquad \blacksquare$$

EXAMPLE 3.25   Find D_2 and P_8 for the distribution of the waiting times between eruptions of Old Faithful.

Solution   To find D_2 we must count $110 \cdot \frac{2}{10} = 22$ of the items starting at the bottom of the distribution. Since there are $2 + 2 + 4 = 8$ values in the first three classes, we must count $22 - 8 = 14$ of the 19 values in the fourth class to reach D_2. This yields

$$D_2 = 59.5 + \frac{14}{19} \cdot 10 \approx 66.87$$

Since $2 + 3 + 15 = 20$ of the values fall into the last three classes, we must count $22 - 20 = 2$ of the 39 values in the next class to reach P_8. Thus, we write

$$P_8 = 89.5 - \frac{2}{39} \cdot 10 \approx 88.99 \qquad \blacksquare$$

Note that when we determine a fractile of a distribution, the number of items we have to count and the quantity j in the formula on page 73 need not be a whole number.

■ E x e r c i s e s

*3.62   For each of the following distributions determine whether it is possible to find the mean and/or the median:

(a) Grade	Frequency		(b)	IQ	Frequency
40–49	5			Less than 90	3
50–59	18			90–99	14
60–69	27			100–109	22
70–79	15			110–119	19
80–89	6			More than 119	7

(c)	Weight	Frequency
	100 or less	41
	101–110	13
	111–120	8
	121–130	3
	131–140	1

*3.63 Following is the distribution of the percentages of the students who are bilingual in 50 elementary schools:

Percentage	Number of schools
0–4	18
5–9	15
10–14	9
15–19	7
20–24	1

Find the mean and the median.

*3.64 With reference to the distribution of Exercise 3.63, find
(a) Q_1 and Q_3;
(b) the deciles D_3 and D_7;
(c) the percentiles P_5 and P_{95}.

*3.65 Following is a distribution of the compressive strengths (in 1,000 psi) of 120 concrete samples:

Compressive strength	Frequency
4.20–4.39	6
4.40–4.59	12
4.60–4.79	23
4.80–4.99	40
5.00–5.19	24
5.20–5.39	11
5.40–5.59	4
	120

Find the mean and the median.

*3.66 With reference to the distribution of Exercise 3.65, find
(a) Q_1 and Q_3;
(b) D_1 and D_9;
(c) P_{15} and P_{85}.

*3.67   With reference to the distribution of Exercise 2.27, find the mean and the median of the weights of the mineral specimens.

*3.68   With reference to the distribution of Exercise 2.27, find the first and third quartiles of the weights of the mineral specimens.

*3.69   With reference to the distribution of Exercise 2.53, find the mean and the median for the numbers of fish tacos served by the restaurant.

*3.70   With reference to the distribution of Exercise 2.53, find Q_1 and Q_3 for the numbers of fish tacos served by the restaurant.

*3.71   Use the distribution you obtained in Exercise 2.36, 2.39, or 2.42 to calculate
(a) the mean;
(b) the median;
(c) Q_1 and Q_3.

*3.72   The calculation of the mean of a distribution can usually be simplified by replacing the class marks with consecutive integers. This process is referred to as **coding**; when the class intervals are all equal, and only then, we assign the value 0 to a class mark near the middle of the distribution and code the class marks ..., −3, −2, −1, 0, 1, 2, 3, ... Denoting the coded class marks by the letter u, we then use the formula

Mean of grouped data (with coding)

$$\bar{x} = x_0 + \frac{\sum u \cdot f}{n} \cdot c$$

where x_0 is the class mark in the original scale to which we assign 0 in the new scale, c is the class interval, n is the total number of items grouped, and $\sum u \cdot f$ is the sum of the products obtained by multiplying each of the new class marks by the corresponding class frequency. Use this kind of coding to recalculate
(a) the mean for the distribution of Exercise 3.65;
(b) the mean of the distribution of the waiting times between eruptions of Old Faithful obtained in Example 3.21.

3.8   TECHNICAL NOTE (SUMMATIONS)

In the notation introduced on page 50, $\sum x$ does not tell us which, or how many, values of x we must add. This is taken care of by the more explicit notation

$$\sum_{i=1}^{n} x_i = x_1 + x_2 + \cdots + x_n$$

where it is made clear that we are adding the x's whose subscripts i are $1, 2, \ldots$, and n. We are not using the more explicit notation in this text to simplify the overall appearance of the formulas, assuming that it is clear in each case what x's we are referring to and how many there are.

Using the Σ notation, we shall also have occasion to write such expressions as Σx^2, Σxy, $\Sigma x^2 f$, ..., which (more explicitly) represent the sums

$$\sum_{i=1}^{n} x_i^2 = x_1^2 + x_2^2 + x_3^2 + \cdots + x_n^2$$

$$\sum_{j=1}^{m} x_j y_j = x_1 y_1 + x_2 y_2 + \cdots + x_m y_m$$

$$\sum_{i=1}^{n} x_i^2 f_i = x_1^2 f_1 + x_2^2 f_2 + \cdots + x_n^2 f_n$$

Working with two subscripts, we shall also have the occasion to evaluate **double summations** such as

$$\sum_{j=1}^{3} \sum_{i=1}^{4} x_{ij} = \sum_{j=1}^{3} (x_{1j} + x_{2j} + x_{3j} + x_{4j})$$

$$= x_{11} + x_{21} + x_{31} + x_{41} + x_{12} + x_{22} + x_{32} + x_{42}$$

$$+ x_{13} + x_{23} + x_{33} + x_{43}$$

To verify some of the formulas involving summations that are stated but not proved in the text, the reader will need the following rules:

Rules for summations

Rule A: $\displaystyle\sum_{i=1}^{n} (x_i \pm y_i) = \sum_{i=1}^{n} x_i \pm \sum_{i=1}^{n} y_i$

Rule B: $\displaystyle\sum_{i=1}^{n} k \cdot x_i = k \cdot \sum_{i=1}^{n} x_i$

Rule C: $\displaystyle\sum_{i=1}^{n} k = k \cdot n$

The first of these rules states that the summation of the sum (or difference) of two terms equals the sum (or difference) of the individual summations, and it can be extended to the sum or difference of more than two terms. The second rule states that we can, so to speak, factor a constant out of a summation, and the third rule states that the summation of a constant is simply n times that constant. All of these rules can be proved by actually writing out in full what each of the summation represents.

Exercises

3.73 Write each of the following in full; that is, without summation signs:

(a) $\displaystyle\sum_{i=1}^{6} x_i;$

(d) $\displaystyle\sum_{j=1}^{8} x_j f_j;$

(b) $\sum_{i=1}^{5} y_i$;

(e) $\sum_{i=3}^{7} x_i^2$;

(c) $\sum_{i=1}^{3} x_i y_i$;

(f) $\sum_{j=1}^{4} (x_j + y_j)$.

3.74 Write each of the following as a summation; that is, in the Σ notation:
(a) $z_1 + z_2 + z_3 + z_4 + z_5$;
(b) $x_5 + x_6 + x_7 + x_8 + x_9 + x_{10} + x_{11} + x_{12}$;
(c) $x_1 f_1 + x_2 f_2 + x_3 f_3 + x_4 f_4 + x_5 f_5 + x_6 f_6$;
(d) $y_1^2 + y_2^2 + y_3^2$;
(e) $2x_1 + 2x_2 + 2x_3 + 2x_4 + 2x_5 + 2x_6 + 2x_7$;
(f) $(x_2 - y_2) + (x_3 - y_3) + (x_4 - y_4)$;
(g) $(z_2 + 3) + (z_3 + 3) + (z_4 + 3) + (z_5 + 3)$;
(h) $x_1 y_1 f_1 + x_2 y_2 f_2 + x_3 y_3 f_3 + x_4 y_4 f_4$.

3.75 Given $x_1 = 3$, $x_2 = 2$, $x_3 = -2$, $x_4 = 5$, $x_5 = -1$, $x_6 = 3$, $x_7 = 2$, and $x_8 = 4$, find
(a) $\sum_{i=1}^{8} x_i$; (b) $\sum_{i=1}^{8} x_i^2$.

3.76 Given $x_1 = 2$, $x_2 = 3$, $x_3 = 4$, $x_4 = 5$, $x_5 = 6$, $f_1 = 2$, $f_2 = 8$, $f_3 = 9$, $f_4 = 3$, and $f_5 = 2$, find
(a) $\sum_{i=1}^{5} x_i$;

(c) $\sum_{i=1}^{5} x_i f_i$;

(b) $\sum_{i=1}^{5} f_i$;

(d) $\sum_{i=1}^{5} x_i^2 f_i$.

3.77 Given $x_1 = 4$, $x_2 = -2$, $x_3 = 3$, $x_4 = -1$, $y_1 = 5$, $y_2 = -2$, $y_3 = 4$, and $y_4 = -1$, find
(a) $\sum_{i=1}^{4} x_i$;

(d) $\sum_{i=1}^{4} y_i^2$;

(b) $\sum_{i=1}^{4} y_i$;

(e) $\sum_{i=1}^{4} x_i y_i$.

(c) $\sum_{i=1}^{4} x_i^2$;

3.78 Given $x_{11} = 4$, $x_{12} = 2$, $x_{13} = -1$, $x_{14} = 3$, $x_{21} = 2$, $x_{22} = 5$, $x_{23} = -1$, $x_{24} = 6$, $x_{31} = 4$, $x_{32} = -1$, $x_{33} = 3$, and $x_{34} = 4$, find
(a) $\sum_{i=1}^{3} x_{ij}$ separately for $j = 1, 2, 3$, and 4;

(b) $\sum_{j=1}^{4} x_{ij}$ separately for $i = 1, 2$, and 3.

3.79 With reference to Exercise 3.78, evaluate the double summation $\sum_{i=1}^{3} \sum_{j=1}^{4} x_{ij}$ using

(a) the results of part (a) of that exercise;
(b) the results of part (b) of that exercise.

3.80 Show that $\sum_{i=1}^{n} (x - \bar{x}) = 0$ for any set of x's whose mean is $\bar{x}$.

3.81 Is it true in general that $\left(\sum_{i=1}^{n} x_i \right)^2 = \sum_{i=1}^{n} x_i^2$? (*Hint:* Check whether the equation holds for $n = 2$.)

3.9 Checklist of Key Terms *(with page references to their definitions)*[†]

Arithmetic mean, 49	Median, 58, 71
Boxplot, 64	Median position, 60
Box-and-whisker plot, 64	Midquartile, 64
*Coding, 76	*Midrange, 68
Decile, 62, 73	Mode, 66
Double summation, 77	Parameter, 50
Fractile, 62	Percentile, 62, 73
*Geometric mean, 57	Population, 48
Grand mean, 55	Population size, 50
*Grouping error, 71	Quartile, 62, 73
*Harmonic mean, 57	Sample, 48
*Lower hinge, 69	Sample size, 50
*Markov's theorem, 57	Statistic, 50
Mean, 49, 70	Summation notation, 50, 76
Measures of central location, 48	*Upper hinge, 69
Measures of location, 47	Weight, 54
Measures of variation, 47	Weighted mean, 54

3.10 References

Informal discussions of the ethics involved in choosing among averages and other questions of ethics in statistics in general are given in

HOOKE, R., *How to Tell the Liars from the Statisticians.* New York: Marcel Dekker, Inc., 1983.

HUFF, D., *How to Lie with Statistics.* New York: W. W. Norton & Company, Inc., 1954.

For further information about the use and interpretation of hinges, see the books on exploratory data analysis referred to on page 46.

[†]Terms in italics were introduced in exercises.

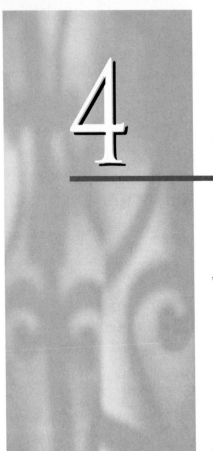

4 SUMMARIZING DATA: MEASURES OF VARIATION

4.1   The Range 81

4.2   The Variance and the Standard Deviation 82

4.3   Applications of the Standard Deviation 85

*4.4   The Description of Grouped Data 92

4.5   Some Further Descriptions 94

4.6   Checklist of Key Terms 100

4.7   References 101

*O*ne aspect of most sets of data is that the values are not all alike; indeed, the extent to which they are unalike, or vary among themselves, is of basic importance in statistics. Consider the following examples:

In a hospital where each patient's pulse rate is taken three times a day, that of patient *A* is 72, 76, and 74, while that of patient *B* is 72, 91, and 59. The mean pulse rate of the two patients is the same, 74, but observe the difference in variability. Whereas patient *A*'s pulse rate is stable, that of patient *B* fluctuates widely.

A supermarket stocks certain 1-pound bags of mixed nuts, which on the average contain 12 almonds per bag. If all the bags contain anywhere from 10 to 14 almonds, the product is consistent and satisfactory, but the situation is quite different if some of the bags have no almonds while others have 20 or more.

Measuring variability is of special importance in statistical inference. Suppose, for instance, that we have a coin that is slightly bent and we wonder whether there is still a fifty–fifty chance for heads. What if we toss the coin 100 times and get 28 heads and 72 tails? Does the shortage of heads—only 28 where we might have expected 50—imply that the count is not "fair?" To answer such questions we must have some idea about the magnitude of the fluctuations, or variations, that are brought about by chance when coins are tossed 100 times.

We have given these three examples to show the need for measuring the extent to which data are dispersed, or spread out; the corresponding measures that provide this information are called **measures of variation.** In Sections 4.1 through 4.3 we present the most widely used measures of variation and some of their special applications. Some statistical descriptions other than measures of location and measures of variation are discussed in Section 4.5.

4.1   THE RANGE

To introduce a simple way of measuring variability, let us refer to the first of the three examples cited previously, where the pulse rate of patient A varied from 72 to 76 while that of patient B varied from 59 to 91. These extreme (smallest and largest) values are indicative of the variability of the two sets of data, and just about the same information is conveyed if we take the differences between the respective extremes. So, let us make the following definition:

> **The range of a set of data is the difference between the largest value and the smallest.**

For patient A the pulse rates had a range of $76 - 72 = 4$ and for patient B they had a range of $91 - 59 = 32$. Also, for the lengths of the trout on page 13 the range was $23.5 - 16.6 = 6.9$ centimeters and for the waiting times between eruptions of Old Faithful in Example 2.4, the range was $118 - 33 = 85$ minutes.

Conceptually, the range is easy to understand, its calculation is very easy, and there is a natural curiosity about the smallest and largest values. Nevertheless, it is not a very useful measure of variation—its main shortcoming being that it does not tell us anything about the dispersion of the values that fall between the two extremes. For example, each of the following three sets of data

Set A:	5	18	18	18	18	18	18	18	18	18
Set B:	5	5	5	5	5	18	18	18	18	18
Set C:	5	6	8	9	10	12	14	15	17	18

has a range of $18 - 5 = 13$, but their dispersions between the first and last values are totally different.

In actual practice, the range is used mainly as a "quick and easy" measure of variability; for instance, in industrial quality control it is used to keep a close check on raw materials and products on the basis of small samples taken at regular intervals of time.

Whereas the range covers all the values in a sample, a similar measure of variation covers (more or less) the middle 50 percent. It is the **interquartile range** $Q_3 - Q_1$, where Q_1 and Q_3 may be defined as in Section 3.5, in Section 3.7, or in Exercise 3.52. For instance, for the twelve temperature readings in Example 3.16 we might use $88 - 76 = 12$ and for the grouped data in Example 3.24 we might use $87.58 - 69.71 = 17.87$. Some statisticians also use the **semi-interquartile range** $\frac{1}{2}(Q_3 - Q_1)$, which is sometimes referred to as the **quartile deviation.**

4.2   THE VARIANCE AND THE STANDARD DEVIATION

To define the **standard deviation,** by far the most generally useful measure of variation, let us observe that the dispersion of a set of data is small if the values are closely bunched about their mean, and that it is large if the values are scattered widely about their mean. It would seem reasonable, therefore, to measure the variation of a set of data in terms of the amounts by which the values deviate from their mean. If a set of numbers

$$x_1, \quad x_2, \quad x_3, \quad \dots, \quad \text{and} \quad x_n$$

constitutes a sample with the mean $\bar{x}$, then the differences

$$x_1 - \bar{x}, \quad x_2 - \bar{x}, \quad x_3 - \bar{x}, \quad \dots, \quad \text{and} \quad x_n - \bar{x}$$

are called the **deviations from the mean,** and we might use their average (that is, their mean) as a measure of the variability of the sample. Unfortunately, this will not do. Unless the x's are all equal, some of the deviations from the mean will be positive, some will be negative, and as the reader was asked to show in Exercise 3.80, the sum of the deviations from the mean, $\Sigma(x - \bar{x})$, and hence also their mean, is always equal to zero.

Since we are really interested in the magnitude of the deviations, and not in whether they are positive or negative, we might simply ignore the signs and define a measure of variation in terms of the absolute values of the deviations from the mean. Indeed, if we add the deviations from the mean as if they were all positive or zero and divide by n, we obtain the statistical measure that is called the **mean deviation.** This measure has intuitive appeal, but because of the absolute values it leads to serious theoretical difficulties in problems of inference, and it is rarely used.

An alternative approach is to work with the squares of the deviations from the mean, as this will also eliminate the effect of the signs. Squares of real numbers cannot be negative; in fact, squares of the deviations from a mean are all positive unless a value happens to coincide with the mean. Then, if we average the squared deviations from the mean and take the square root of the result (to compensate for the fact that the deviations were squared), we get

$$\sqrt{\frac{\Sigma(x - \bar{x})^2}{n}}$$

and this is how, traditionally, the standard deviation used to be defined. Expressing literally what we have done here mathematically, it is also called the **root-mean-square deviation.**

Nowadays, it is customary to modify this formula by dividing the sum of the squared deviations from the mean by $n - 1$ instead of n. Following this practice, which will be explained later, let us define the **sample standard deviation,** denoted by s, as

Sample standard deviation

$$s = \sqrt{\frac{\Sigma(x - \bar{x})^2}{n - 1}}$$

and its square, the **sample variance,** as

Sample
variance

$$s^2 = \frac{\sum(x - \bar{x})^2}{n - 1}$$

These formulas for the standard deviation and the variance apply to samples, but if we substitute μ for $\bar{x}$ and N for n, we obtain analogous formulas for the standard deviation and the variance of a population. It is customary to denote the **population standard deviation** by σ (*sigma*, the Greek letter for lowercase s) when dividing by N, and by S when dividing by $N - 1$. Thus, for σ we write

Population
standard
deviation

$$\sigma = \sqrt{\frac{\sum(x - \mu)^2}{N}}$$

and the **population variance** is σ^2.

Ordinarily, the purpose of calculating a sample statistic (such as the mean, the standard deviation, or the variance) is to estimate the corresponding population parameter. If we actually took many samples from a population that has the mean μ, calculated the sample means $\bar{x}$, and then averaged all these estimates of μ, we should find that their average is very close to μ. However, if we calculated the variance of each sample by means of the formula $\dfrac{\sum(x - \bar{x})^2}{n}$ and then averaged all these supposed estimates of σ^2, we would probably find that their average is less than σ^2. Theoretically, it can be shown that we can compensate for this by dividing by $n - 1$ instead of n in the formula for s^2. Estimators having the desirable property that their values will on the average equal the quantity they are supposed to estimate are said to be **unbiased;** otherwise, they are said to be **biased.** So, we say that $\bar{x}$ is an unbiased estimator of the population mean μ, and that s^2 is an unbiased estimator of the population variance σ^2. It does not follow from this that s is also an unbiased estimator of σ, but when n is large the bias is small and can usually be ignored.

In calculating the sample standard deviation using the formula by which it is defined, we must (1) find $\bar{x}$, (2) determine the n deviations from the mean $x - \bar{x}$, (3) square these deviations, (4) add all the squared deviations, (5) divide by $n - 1$, and (6) take the square root of the result arrived at in step 5. In actual practice, this formula is rarely used—there are various shortcuts—but we shall illustrate it here to emphasize what is really measured by a standard deviation.

EXAMPLE   4.1   A bacteriologist found 8, 11, 7, 13, 10, 11, 7, and 9 microorganisms of a certain kind in eight cultures. Calculate *s*.

Solution   First calculating the mean, we get

$$\bar{x} = \frac{8 + 11 + 7 + 13 + 10 + 11 + 7 + 9}{8} = 9.5$$

and then the work required to find $\Sigma(x - \bar{x})^2$ may be arranged as in the following table:

x	$x - \bar{x}$	$(x - \bar{x})^2$
8	−1.5	2.25
11	1.5	2.25
7	−2.5	6.25
13	3.5	12.25
10	0.5	0.25
11	1.5	2.25
7	−2.5	6.25
9	−0.5	0.25
	0.0	32.00

Finally, dividing 32.00 by $8 - 1 = 7$ and taking the square root (using a simple handheld calculator), we get

$$s = \sqrt{\frac{32.00}{7}} \approx \sqrt{4.57} = 2.14$$

rounded to two decimals.   ∎

Note in the preceding table that the total for the middle column is zero; since this must always be the case, it provides a convenient check on the calculations.
It was easy to calculate s in this example because the data were whole numbers and the mean was exact to one decimal. Otherwise, the calculations required by the formula defining s can be quite tedious, and, unless we can get s directly with a statistical calculator or a computer, it helps to use the formula

Computing formula for the sample standard deviation

$$s = \sqrt{\frac{S_{xx}}{n - 1}} \qquad \text{where} \qquad S_{xx} = \Sigma x^2 - \frac{(\Sigma x)^2}{n}$$

EXAMPLE 4.2   Use this computing formula to rework Example 4.1.

Solution   First we calculate Σx and Σx^2, getting

$$\Sigma x = 8 + 11 + 7 + 13 + 10 + 11 + 7 + 9$$

$$= 76$$

and

$$\Sigma x^2 = 64 + 121 + 49 + 169 + 100 + 121 + 49 + 81$$

$$= 754$$

Then, substituting these totals and $n = 8$ into the formula for S_{xx}, and $n - 1 = 7$ and the value obtained for S_{xx} into the formula for s, we get

$$S_{xx} = 754 - \frac{(76)^2}{8} = 32$$

and, hence, $s = \sqrt{\frac{32}{7}} = 2.14$ rounded to two decimals. This agrees, as it should, with the result obtained before.   ∎

As should have been apparent from these two examples, the advantage of the computing formula is that we got the result without having to determine $\bar{x}$ and work with the deviations from the mean. Incidentally, the computing formula can also be used to find σ, with the n in the formula for S_{xx} and the $n - 1$ in the formula for s replaced by N.

4.3   APPLICATIONS OF THE STANDARD DEVIATION

In subsequent chapters, sample standard deviations will be used primarily to estimate population standard deviations in problems of inference. Meanwhile, to provide the reader with more of a feeling of what a standard deviation really measures, we shall devote this section to some applications.

In the argument that led to the definition of the standard deviation, we observed that the dispersion of a set of data is small if the values are bunched closely about their mean, and that it is large if the values are scattered widely about their mean. Correspondingly, we can now say that if the standard deviation of a set of data is small, the values are concentrated near the mean, and if the standard deviation is large, the values are scattered widely about the mean. This idea is expressed more formally by the following theorem, called **Chebyshev's theorem** after the Russian mathematician P. L. Chebyshev (1821–1894):

Chebyshev's theorem

> *For any set of data (population or sample) and any constant k greater than 1, the proportion of the data that must lie within k standard deviations on either side of the mean is at least*
>
> $$1 - \frac{1}{k^2}$$

It may be surprising that we can make such definite statements, but it is a certainty that at least $1 - \frac{1}{2^2} = \frac{3}{4}$, or 75%, of the values in *any* set of data must lie within two standard deviations on either side of the mean, at least $1 - \frac{1}{5^2} = \frac{24}{25}$, or 96%, must lie within five standard deviations on either side of the mean, and at least $1 - \dfrac{1}{10^2} = \dfrac{99}{100}$, or 99%, must lie within ten standard deviations on either side of the mean. Here we arbitrarily let $k = 2, 5,$ and 10.

EXAMPLE   4.3   A study of the nutritional value of a certain kind of reduced-fat cheese showed that on the average a one-ounce slice contains 3.50 grams of fat with a standard deviation of 0.04 gram of fat.

(a) According to Chebyshev's theorem, at least what percent of the one-ounce slices of this kind of cheese must have a fat content between 3.38 and 3.62 grams of fat?

(b) According to Chebyshev's theorem, between what values must be the fat content of at least 93.75% of the one-ounce slices of this kind of cheese?

Solution   (a) Since $3.62 - 3.50 = 3.50 - 3.38 = 0.12$, we find that $k(0.04) = 0.12$ and, hence, $k = \dfrac{0.12}{0.04} = 3$. It follows that at least $1 - \dfrac{1}{3^2} = \dfrac{8}{9}$, or approximately 88.9%, of the one-ounce slices of the cheese have a fat content between 3.38 and 3.62 grams of fat.

(b) Since $1 - \dfrac{1}{k^2} = 0.9375$, we find that $\dfrac{1}{k^2} = 1 - 0.9375 = 0.0625$,

$$k^2 = \dfrac{1}{0.0625} = 16, \text{ and } k = 4.$$ It follows that 93.75% of the one-ounce slices of this cheese contain between $3.50 - 4(0.04) = 3.34$ and $3.50 + 4(0.04) = 3.66$ grams of fat.   ∎

Chebyshev's theorem applies to any kind of data, but it has its shortcomings. Since it tells us merely "at least what proportion" of a set of data must lie between certain limits (that is, it provides only a lower limit to the actual proportion), it has few practical applications. Indeed, we have presented it here only to provide the reader with some idea of how to relate the standard deviation to the spread, or dispersion, of a set of data, and vice versa.

For distributions having the general shape of the cross section of a bell (see Figure 4.1), we can make the following much stronger statements:

About 68 percent of the values will lie within one standard deviation of the mean, that is, between $\bar{x} - s$ and $\bar{x} + s$.

About 95 percent of the values will lie within two standard deviations of the mean, that is between $\bar{x} - 2s$ and $\bar{x} + 2s$.

About 99.7 percent of the values will lie within three standard deviations of the mean, that is, between $\bar{x} - 3s$ and $\bar{x} + 3s$.

This result is sometimes referred to as the **empirical rule,** presumably because such percentages are observed in practice. Actually, it is a theoretical result based on the normal distribution, which we shall study in Chapter 9 (in particular, see Exercise 9.10).

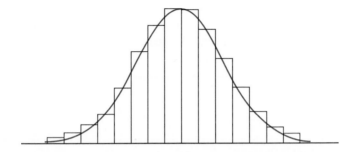

FIGURE 4.1
Bell-shaped
distribution.

EXAMPLE   4.4   In Example 3.21 we calculated the mean of the grouped waiting times between eruptions of Old Faithful, getting 78.59, and in Example 4.7 we will show that the corresponding standard deviation is 14.35. Use these figures to determine from the original data given in Example 2.4 what percentage of the values falls within three standard deviations of the mean.

Solution   Since $\bar{x} = 78.59$ and $s = 14.35$, we shall have to determine what percentage of the values falls between $78.59 - 3(14.35) = 35.54$ and $78.59 + 3(14.35) = 121.64$. Counting two of the values, 33 and 35, below 35.54 and none above 121.64, we find that $110 - 2 = 108$ of the values and, hence

$$\frac{108}{110} \cdot 100 = 98.2\%$$

of the original waiting times fall within three standard deviations of the mean. This is fairly close to the expected 99.7%, but then the distribution of the waiting times is not really perfectly bell shaped.   ■

In the introduction to this chapter we gave three examples in which knowledge about the variability of the data was of special importance. This is also the case when we want to compare numbers belonging to different sets of data. To illustrate, suppose that the final examination in a French course consists of two parts, vocabulary and grammar, and that a certain student scored 66 points in the vocabulary part and 80 points in the grammar part. At first glance it would seem that the student did much better in grammar than in vocabulary, but suppose that all the students in the class averaged 51 points in the vocabulary part with a standard deviation of 12, and 72 points in the grammar part with a standard deviation of 16. Thus, we can argue that the student's score in the vocabulary part is $\dfrac{66 - 51}{12} = 1.25$ standard deviations above the average for the class, while her score in the grammar part is only $\dfrac{80 - 72}{16} = 0.50$ standard deviation above the average for the class. Whereas the original scores cannot be meaningfully compared, these new scores, expressed in terms of standard deviations, can. Clearly, the given student rates much higher on her command of French vocabulary than on her knowledge of French grammar, compared to the rest of the class.

What we have done here consists of converting the grades into **standard units** or **z-scores.** In general, if x is a measurement belonging to a set of data having the mean $\bar{x}$ (or μ) and the standard deviation s (or σ), then its value in standard units, denoted by z, is

Formula for converting to standard units

$$z = \frac{x - \bar{x}}{s} \quad or \quad z = \frac{x - \mu}{\sigma}$$

depending on whether the data constitute a sample or a population. In these units, z tells us how many standard deviations a value lies above or below the mean of the set of data to which it belongs. Standard units will be used frequently in later chapters.

EXAMPLE   4.5   Mrs. Clark belongs to an age group for which the mean weight is 112 pounds with a standard deviation of 11 pounds, and Mr. Clark, her husband, belongs to an age group for which the mean weight is 163 pounds with a standard deviation of 18 pounds. If Mrs. Clark weighs 132 pounds and Mr. Clark weighs 193 pounds, which of the two is relatively more overweight compared to his/her age group?

Solution   Mr. Clark's weight is $193 - 163 = 30$ pounds above average while Mrs. Clark's weight is "only" $132 - 112 = 20$ pounds above average, yet in standard units we get $\dfrac{193 - 163}{18} \approx 1.67$ for Mr. Clark and $\dfrac{132 - 112}{11} \approx 1.82$ for Mrs. Clark. Thus, relative to their age groups Mrs. Clark is somewhat more overweight than Mr. Clark.   ∎

A serious disadvantage of the standard deviation as a measure of variation is that it depends on the units of measurement. For instance, the weights of certain objects may have a standard deviation of 0.10 ounce, but this really does not tell us whether it reflects a great deal of variation or very little variation. If we are weighing the eggs of quails, a standard deviation of 0.10 ounce would reflect a considerable amount of variation, but this would not be the case if we are weighing, say, 100-pound bags of potatoes. What we need in a situation like this is a **measure of relative variation** such as the **coefficient of variation,** defined by the following formula:

Coefficient of variation

$$V = \frac{s}{\bar{x}} \cdot 100\% \quad or \quad V = \frac{\sigma}{\mu} \cdot 100\%$$

The coefficient of variation expresses the standard deviation as a percentage of what is being measured, at least on the average.

EXAMPLE   4.6   Several measurements of the diameter of a ball bearing made with one micrometer had a mean of 2.49 mm and a standard deviation of 0.012 mm, and several measurements of the unstretched length of a spring made with another micrometer had a mean of 0.75 in. with a standard deviation of 0.002 in. Which of the two micrometers is relatively more precise?

Solution   Calculating the two coefficients of variation, we get

$$\frac{0.012}{2.49} \cdot 100\% \approx 0.48\% \qquad \text{and} \qquad \frac{0.002}{0.75} \cdot 100\% \approx 0.27\%$$

Thus, the measurements of the length of the spring are relatively less variable, which means that the second micrometer is more precise.   ∎

■ E x e r c i s e s

4.1   Experimenting with a new cake mix, a home economist found that cakes prepared with five packages of the mix had heights of 4.5, 5.8, 6.0, 4.3, and 4.4 cm. Find
(a) the range;
(b) the standard deviation (using the formula that defines s).

4.2   Following are four measurements of the specific gravity of aluminum: 2.65, 2.69, 2.66, and 2.64. Find
(a) the range;
(b) the standard deviation (using the formula that defines s).

4.3   It has been claimed that for samples of size $n = 4$, the range should be roughly twice as large as the standard deviation. Use the results of Exercise 4.2 to check this claim.

4.4   The ten employees of a machine shop, given a course in CPR, scored 17, 20, 12, 16, 18, 16, 17, 16, 18, and 19 in a test administered after the completion of the course. Find
(a) the range;
(b) the standard deviation (using the formula that defines s).

4.5   It has been claimed that for samples of size $n = 10$, the range should be roughly three times as large as the standard deviation. Use the results of Exercise 4.4 to check this claim.

4.6   With reference to the scores of Exercise 4.4, find
(a) the range [or use the result obtained in part (a) of Exercise 4.4];
(b) the interquartile range, $Q_3 - Q_1$.
Also, explain why it should not come as a surprise that the range is more than double the interquartile range.

4.7   Find the range of the measurements of the aluminum alloy platings given in Exercise 2.39.

4.8   Find the range of the lengths of the root penetrations of the crested wheatgrass seedlings given in Exercise 2.42.

4.9   Use the double-stem display of Figure 2.2 to determine the range of the room-occupancy data.

4.10   In Exercise 3.5 we gave the numbers of seconds that 12 insects survived after having been sprayed with a certain insecticide as 112, 83, 102, 84, 105, 121, 76, 110, 98, 91, 103, and 85. Find the standard deviation using
(a) the formula that defines s;
(b) the shortcut formula for s.

4.11   Ten meetings of the board of directors of a cheritable organization were attended by 8, 12, 9, 12, 10, 7, 11, 10, 11, and 8 members. Time yourself on calculating the standard deviation using
(a) the formula that defines s;
(b) the shortcut formula for s.

4.12   In 1992 there were 7, 8, 4, 11, 13, 15, 6, and 4 Sunday newspapers in the eight Mountain States. Time yourself calculating σ for these data using
(a) the formula that defines σ;
(b) the shortcut formula for s adapted for the calculation of σ.

4.13   In Exercises 3.7, a 10-ml pipette was found to deliver 9.96, 9.98, 9.92, 9.98, and 9.96 mls.
(a) Calculate s for these five measurements.
(b) Substract 9.90 from each of the five measurements and then calculate s for the resulting data.
What simplification does this suggest for the calculation of a standard deviation?

4.14   In Exercise 3.42 we gave the following temperature readings taken at different locations in a large kiln: 409, 412, 439, 411, 432, 432, 405, 411, 422, 417, 440, 427, 411, and 417 degrees Fahrenheit.
(a) Calculate s for these fourteen temperature readings.
(b) Subtract 400 from each of these temperature readings and then calculate s for the resulting data.
What simplification does this suggest for the calculation of a standard deviation?

4.15   On a rainy day during the monsoon season, 0.13, 0.05, 0.26, 0.41, 0.57, 0.02, 0.25, 0.10, 0.60, and 0.18 inches of rain were recorded in ten cities in Arizona.
(a) Calculate s for these data.
(b) Multiply each of these figures by 100, calculate s for the resulting data, and compare the result with that obtained in part (a).
What possible simplification does this suggest for the calculation of a standard deviation?

4.16   Calculate the variance of the thicknesses of the twenty grease coatings given in Exercise 3.40.

 4.17   Use a computer package or a graphing calculator to verify that the standard deviation of the waiting times between eruptions of Old Faithful given in Example 2.4 is $s = 14.666$ rounded to three decimals.

 4.18   Use a computer package or a graphing calculator to determine the standard deviation of the lengths of the 60 sea trout given in the beginning of Section 2.1.

 4.19   With reference to Exercise 2.45, use a computer package or a graphing calculator to calculate the standard deviation of the scores that the 180 students obtained in the achievement test in the social sciences.

4.20 Find the standard deviation of the twelve means and also that of the twelve medians obtained in part (a) of Exercise 3.32. What is illustrated by the difference in the size of these two standard deviations?

4.21 According to Chebyshev's theorem, what can we assert about the proportion of any set of data that must lie within k standard deviations on either side of the mean when
(a) $k = 6$;
(b) $k = 12$?

4.22 According to Chebyshev's theorem, what can we assert about the proportion of any set of data that must lie within k standard deviations on either side of the mean when
(a) $k = 2.5$;
(b) $k = 9$?

4.23 Hospital records show that a certain surgical procedure takes on the average 111.6 minutes with a standard deviation of 2.8 minutes. At least what percentage of these surgical procedures take anywhere between
(a) 106.0 and 117.2 minutes;
(b) 97.6 and 125.6 minutes?

4.24 With reference to Exercise 4.23, between how many minutes must be the length of
(a) at least $\frac{35}{36}$ of these surgical procedures;
(b) at least 99% of these surgical procedures?

4.25 Having kept records for several months, Ms. Lewis knows that it takes her on the average 47.7 minutes with a standard deviation of 2.46 minutes to drive to work from her suburban home. If she always starts out exactly one hour before she has to arrive at work, at most what percentage of the time will she arrive late?

 4.26 Following are the amounts of sulfur oxides (in tons) emitted by an industrial plant on eighty days:

15.8	26.4	17.3	11.2	23.9	24.8	18.7	13.9	9.0	13.2
22.7	9.8	6.2	14.7	17.5	26.1	12.8	28.6	17.6	23.7
26.8	22.7	18.0	20.5	11.0	20.9	15.5	19.4	16.7	10.7
19.1	15.2	22.9	26.6	20.4	21.4	19.2	21.6	16.9	19.0
18.5	23.0	24.6	20.1	16.2	18.0	7.7	13.5	23.5	14.5
14.4	29.6	19.4	17.0	20.8	24.3	22.5	24.6	18.4	18.1
8.3	21.9	12.3	22.3	13.3	11.8	19.3	20.0	25.7	31.8
25.9	10.5	15.9	27.5	18.1	17.9	9.4	24.1	20.1	28.5

(a) Group these data into the classes 5.0–8.9, 9.0–12.9, 13.0–16.9, 17.0–20.9, 21.0–24.9, 25.0–28.9, and 29.0–32.9. Also, draw a histogram of this distribution and judge whether it may well be described as bell shaped.
(b) Use a computer package or a graphing calculator to determine the values of $\bar{x}$ and s for the ungrouped data.
(c) Use the results of part (b) to determine the values of $\bar{x} \pm s, \bar{x} \pm 2s,$ and $\bar{x} \pm 3s$.
(d) Use the results of part (c) to determine what percent of the original data falls within one standard deviation on either side of the mean, what percent falls within two standard deviations on either side of the mean, and what percent falls within three standard deviations on either side of the mean.
(e) Compare the percentages obtained in part (d) with the 68, 95, and 99.7 percent claimed by the empirical rule.

4.27 The applicants to one branch of a state university have a mean ACT English score of 19.4 with a standard deviation of 3.1, while the applicants to another branch of the state university have a mean ACT English score of 20.1 with a standard deviation of 2.8. If an applicant applied to both branches, with respect to which of the two branches is he or she in a relatively better position with
(a) an ACT English score of 24;
(b) an ACT English score of 29?

4.28 An investment service reports for each stock on its list the price at which it is currently selling, its average price over the last six months, and a measure of its variability. Stock A, it reports, is currently selling at $76.75 and averaged $58.25 over the last six months with a standard deviation of $11.00. Stock B is currently selling at $49.50 and averaged $37.50 over the last six months with a standard deviation of $4.00. Leaving all other considerations aside, which of the two stocks is relatively more overpriced at this time?

4.29 In ten rounds of golf, one golfer averaged 76.2 with a standard deviation of 2.4, while another golfer averaged 84.9 with a standard deviation of 3.5. Which of the two golfers is relatively more consistent?

4.30 If five specimens of hard yellow brass had shearing strengths of 49, 52, 51, 53, and 55 thousand psi, and on four Sundays the rainfall at a marina amounted to 0.22, 0.18, 0.16, and 0.24 inches, which of these two sets of data is relatively more variable?

4.31 According to their medical records, one person's blood glucose level, measured before breakfast over several months, averaged 118.2 with a standard deviation of 4.8, while that of another person, also measured before breakfast over several months, averaged 109.7 with a standard deviation of 4.7. Which of the two persons' blood glucose level was relatively more variable?

4.32 An alternative measure of relative variation is the **coefficient of quartile variation,** which is defined as the ratio of the semi-interquartile range to the midquartile multiplied by 100, namely, as

$$\frac{Q_3 - Q_1}{Q_3 + Q_1} \cdot 100$$

Find the coefficient of quartile variation for the test scores given in Exercise 4.4.

*4.4   THE DESCRIPTION OF GROUPED DATA

As we saw in Chapter 2 and then again in Section 3.7, the grouping of data entails some loss of information. Each item has lost its identity and we know only how many values there are in each class or in each category. To define the standard deviation of a distribution we shall have to be satisfied with an approximation and, as we did in connection with the mean, we shall treat our data as if all the values falling into a class were equal to the corresponding class mark. Thus, letting $x_1, x_2, \ldots,$ and x_k denote the class marks, and $f_1, f_2, \ldots,$ and f_k the corresponding class frequencies, we approximate the actual sum of all the measurements or observations with

$$\Sigma x \cdot f = x_1 f_1 + x_2 f_2 + \cdots + x_k f_k$$

and the sum of their squares with

$$\Sigma x^2 \cdot f = x_1^2 f_1 + x_2^2 f_2 + \cdots + x_k^2 f_k$$

Then, we write the computing formula for the standard deviation of grouped sample data as

$$s = \sqrt{\frac{S_{xx}}{n-1}} \qquad \text{where} \qquad S_{xx} = \Sigma x^2 \cdot f - \frac{(\Sigma x \cdot f)^2}{n}$$

which is very similar to the corresponding computing formula for s for ungrouped data. To obtain a corresponding computing formula for σ, we replace n by N in the formula for S_{xx} and $n-1$ by N in the formula for s.

When the class marks are large numbers or given to several decimals, we can simplify things further by using the coding suggested in Exercise 3.72. When the class intervals are all equal, and only then, we replace the class marks with consecutive integers, preferably with 0 at or near the middle of the distribution. Denoting the coded class marks by the letter u, we then calculate S_{uu} and substitute into the formula

$$s_u = \sqrt{\frac{S_{uu}}{n-1}}$$

This kind of coding is illustrated by Figure 4.2, where we find that if u varies (is increased or decreased) by 1, the corresponding value of x varies (is increased or decreased) by the class interval c. Thus, to change s_u from the u-scale to the original scale of measurement, the x-scale, we multiply it by c.

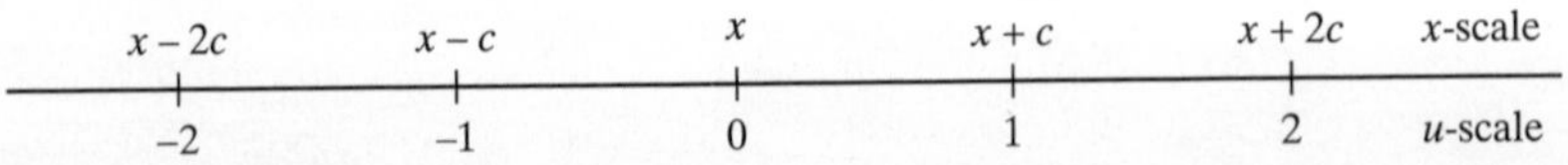

FIGURE 4.2
Coding the class marks
of a distribution.

$x - 2c$	$x - c$	x	$x + c$	$x + 2c$	x-scale
-2	-1	0	1	2	u-scale

EXAMPLE   4.7   With reference to the distribution of the waiting times between eruptions of Old Faithful shown in Example 2.4 and also in Example 3.21, calculate its standard deviation

    (a) without coding;
    (b) with coding.

Solution   **(a)**

x	f	$x \cdot f$	$x^2 \cdot f$
34.5	2	69	2,380.5
44.5	2	89	3,960.5
54.5	4	218	11,881
64.5	19	1,225.5	79,044.75
74.5	24	1,788	133,206
84.5	39	3,295.5	278,469.75
94.5	15	1,417.5	133,953.75
104.5	3	313.5	32,760.75
114.5	2	229	26,220.5
	110	8,645	701,877.5

so that

$$S_{xx} = 701,877.5 - \frac{(8,645)^2}{110} = 22,459.1$$

and

$$s = \sqrt{\frac{22,459.1}{109}} \approx 14.35$$

(b)	u	f	$u \cdot f$	$u^2 \cdot f$
	-4	2	-8	32
	-3	2	-6	18
	-2	4	-8	16
	-1	19	-19	19
	0	24	0	0
	1	39	39	39
	2	15	30	60
	3	3	9	27
	4	2	8	32
		110	45	243

so that

$$S_{uu} = 243 - \frac{(45)^2}{110} \approx 224.59$$

and

$$s_u = \sqrt{\frac{224.59}{109}} \approx 1.435$$

Finally, $s = 10(1.435) = 14.35$, which agrees, as it should, with the result obtained in part (a). This clearly demonstrates how the coding simplified the calculations.

■

4.5   SOME FURTHER DESCRIPTIONS

So far we have discussed only statistical descriptions that come under the general heading of measures of location or measures of variation. Actually, there is no limit to the number of ways in which statistical data can be described, and statisticians continually develop new methods of describing characteristics of numerical data that are of interest in particular problems. In this section we shall consider briefly the problem of describing the overall shape of a distribution.

Although frequency distributions can take on almost any shape or form, most of the distributions we meet in practice can be described fairly well by one or another of a few standard types. Among these, foremost in importance is the aptly described symmetrical **bell-shaped distribution** shown in Figure 4.1. The two distributions shown in Figure 4.3 can, by a stretch of the imagination, be described as bell shaped, but they are not symmetrical. Distributions like these, having a "tail" on one side or the other, are said to be **skewed;** if the tail is on the left we say that they are **negatively skewed** and if the tail is on the right we say that they are **positively skewed.** Distributions of incomes or wages are often positively skewed because of the presence of some relatively high values that are not offset by correspondingly low values.

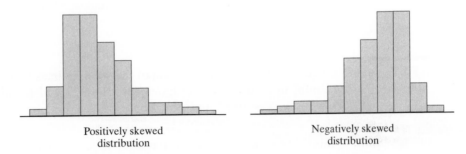

FIGURE 4.3
Skewed distributions.

Positively skewed
distribution

Negatively skewed
distribution

The concepts of symmetry and skewness apply to any kind of data, not only distributions. Of course, for a large set of data we may just group the data and draw and study a histogram, but if that is not enough, we can use any one of several statistical **measures of skewness.** A relatively easy one is based on the fact that when there is perfect symmetry, as in the distribution shown in Figure 4.1, the mean and the median will coincide. When there is positive skewness and some of the high values are not offset by correspondingly low values, as in Figure 4.4, the mean will be greater than the median; when there is a negative skewness and some of the low values are not offset by correspondingly high values, the mean will be smaller than the median.

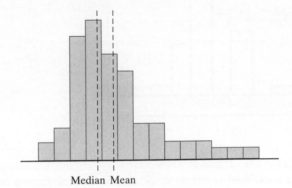

FIGURE 4.4
Mean and median of
positively skewed
distribution.

Median Mean

This relationship between the median and the mean can be used to define a relatively simple measure of skewness, called the **Pearsonian coefficient of skewness.** It is given by

Pearsonian
coefficient
of skewness

$$SK = \frac{3(mean - median)}{standard\ deviation}$$

For a perfectly symmetrical distribution, such as the one pictured in Figure 4.1, the mean and the median coincide and $SK = 0$. In general, values of the Pearsonian coefficient of skewness must fall between -3 and 3, and it should be noted that division by the standard deviation makes SK independent of the scale of measurement.

EXAMPLE   4.8   Calculate SK for the distribution of the waiting times between eruptions of Old Faithful, using the results of Examples 3.21, 3.22, and 4.7, where we showed that $\bar{x} = 78.59$, $\tilde{x} = 80.53$, and $s = 14.35$.

Solution   Substituting these values into the formula for SK, we get

$$SK = \frac{3(78.59 - 80.53)}{14.35} \approx -0.41$$

which shows that there is a definite, though modest, negative skewness. This is also apparent from the histogram of the distribution, shown originally in Figure 2.3 and here again in Figure 4.5, reproduced from the display screen of a TI-83 graphing calculator.   ■

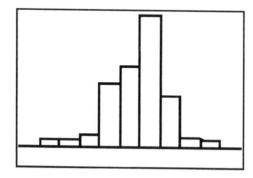

FIGURE 4.5
Histogram of the
distribution of the
waiting times between
eruptions of Old
Faithful.

When a set of data is so small that we cannot meaningfully construct a histogram, a good deal about its shape can be learned from a boxplot (defined originally on page 64). Whereas the Pearsonian coefficient is based on the difference between the mean and the median, with a boxplot we judge the symmetry or skewness of a set of data on the basis of the position of the median relative to the two quartiles, Q_1 and Q_3. In particular, if the line at the median is at or near the center of the box, this is an indication of the symmetry of the data; if it is appreciably to the left of center, this is an indication that the data are positively skewed; and if it is appreciably to the right of center, this is an indication that the data are negatively skewed. The relative length of the two "whiskers," extending from the smallest value to Q_1 and from Q_3 to the largest value, can also be used as an indication of symmetry or skewness.

EXAMPLE   4.9   Following are the annual incomes of fifteen CPAs in thousands of dollars: 88, 77, 70, 80, 74, 82, 85, 96, 76, 67, 80, 75, 73, 93, and 72. Draw a boxplot and use it to judge the symmetry or skewness of the data.

Solution   Arranging the data according to size, we get

67	70	72	73	74	75	76	77
80	80	82	85	88	93	96	

and it can be seen that the smallest value is 67; the largest value is 96; the median is the eighth value from either side, which is 77; Q_1 is the fourth value from the left, which is 73; and Q_3 is the fourth value from the right, which is 85. All this information is summarized by the MINITAB printout of the boxplot shown in Figure 4.6. As can be seen, there is a strong indication that the data are positively

FIGURE 4.6
Box plot of the incomes of the CPAs.

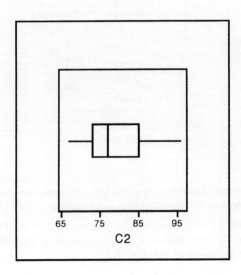

skewed. The line at the median is well to the left of the center of the box and the "whisker" on the right is quite a bit longer than the one on the left.   ■

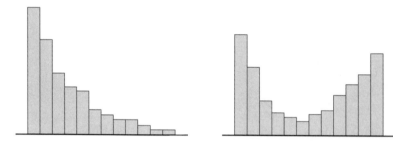

FIGURE 4.7
Reverse J-shaped and U-shaped distributions.

Besides the distributions we have discussed in this section, two others sometimes met in practice are the **reverse J-shaped** and **U-shaped** distributions shown in Figure 4.7. As can be seen from this figure, the names of these distributions literally describe their shape. Examples of such distributions may be found in Exercises 4.45 and 4.47.

Exercises

*4.33   In a factory or office, the time during working hours in which a machine is not operating as a result of breakage or failure is called a *downtime*. The following distribution shows a sample of the length of the downtimes of a certain machine (rounded to the nearest minute).

Downtime (minutes)	Frequency
0–9	2
10–19	15
20–29	17
30–39	13
40–49	3

Find the standard deviation of this distribution
(a) without coding;
(b) with coding.

*4.34   With reference to Exercise 3.63, find the standard deviation of the distribution of the percentages of the students who are bilingual in fifty elementary schools.

*4.35   Use the results of Exercises 3.63 and 4.34 to calculate the Pearsonian coefficient of skewness for the distribution of the percentages of bilingual students. Discuss the symmetry or skewness of this distribution.

*4.36 With reference to Exercise 3.65, find the standard deviation of the distribution of the compressive strength of the 120 concrete samples.

*4.37 Use the results of Exercises 3.65 and 4.36 to calculate the Pearsonian coefficient of skewness for the distribution of compressive strengths.

*4.38 Following is the distribution of the grades that 500 students received in a geography test:

Grade	Number of students
10–24	44
25–39	70
40–54	92
55–69	147
70–84	115
85–99	32

Calculate
(a) the mean and the median;
(b) the standard deviation.

*4.39 Use the results of Exercise 4.38 to calculate the Pearsonian coefficient of skewness for the given data, and discuss the symmetry or skewness of their distribution.

4.40 Following are the numbers of accidents that occurred in July of 1999 in a certain town at 20 intersections without left-turn arrows:

25	30	32	22	26	10	2	32	6	13
27	22	18	12	28	35	8	29	31	8

Find the median, Q_1, and Q_3.

4.41 Use the results of Exercise 4.40 to construct a boxplot, and use it to discuss the symmetry or skewness of these accident data.

4.42 Following are the response times of 30 integrated circuits (in picoseconds):

3.7	4.1	4.5	4.6	4.4	4.8	4.3	4.4	5.1	3.9
3.3	3.4	3.7	4.1	4.7	4.6	4.2	3.7	4.6	3.4
4.6	3.7	4.1	4.5	6.0	4.0	4.1	5.6	6.0	3.4

Construct a stem-and-leaf display using the units digits as the stem labels and the tenths digits as the leaves. Use this stem-and-leaf display to judge the symmetry, or the lack of it, of these data.

4.43 With reference to Exercise 4.42, find all the information that is required to draw a boxplot. Then use it to draw a boxplot and judge the symmetry or skewness of the response times of these integrated circuits.

4.44   With reference to Exercise 3.27, find all the information that is necessary to draw a boxplot. Then use it to draw a boxplot and judge the symmetry or skewness of the lengths of the NBA games.

4.45   Following are the numbers of 3's obtained in fifty rolls of four dice: 0, 0, 1, 0, 0, 0, 2, 0, 0, 1, 0, 0, 0, 0, 1, 1, 0, 1, 2, 0, 0, 1, 0, 0, 0, 1, 1, 0, 1, 0, 0, 1, 2, 1, 0, 0, 3, 1, 1, 0, 4, 0, 0, 1, 2, 1, 0, 0, 1, and 1. Construct a frequency distribution and use it to judge the overall shape of the data.

4.46   With reference to Exercise 4.45, find all the information that is needed to draw a boxplot. Then draw the boxplot and use it to describe the symmetry, or the lack of it, of the data. Also discuss the overall shape of the data.

4.47   If a coin is flipped five times in a row, the result can be represented by means of a sequence of H's and T's (for example, HHTTH), where H stands for heads and T for tails. Having obtained such a sequence of H's and T's, we can then check after each successive flip whether the number of heads exceeds the number of tails. For instance, for HHTTH heads is ahead after the first flip, after the second flip, after the third flip, not after the fourth flip, but again after the fifth flip. Altogether, heads is ahead four times. Actually we repeated this "experiment" sixty times and found that heads was ahead

1	1	5	0	0	5	0	1	2	0	1	0	5	1	0
0	5	0	0	0	0	1	0	0	5	0	2	0	1	0
5	5	0	5	4	3	5	0	5	0	1	5	0	1	5
3	1	5	5	2	1	2	4	2	3	0	5	5	0	0

times. Construct a frequency distribution and discuss the overall shape of the data.

4.48   With reference to Exercise 4.47, find all the information that is needed to construct a boxplot. What features of the boxplot suggest that the data have a very unusual shape?

4.6 *Checklist of Key Terms* (with page references to their definitions)

Bell-shaped distribution, 87, 95
Biased estimator, 83
Chebyshev's theorem, 85
Coefficient of quartile variation, 92
Coefficient of variation, 88
Deviation from mean, 82
Empirical rule, 87
Interquartile range, 81
Mean deviation, 82
Measures of relative variation, 88
Measures of skewness, 95
Measures of variation, 81
Negatively skewed distribution, 95
Pearsonian coefficient of skewness, 95
Population standard deviation, 83
Population variance, 83

Positively skewed distribution, 95
Quartile deviation, 81
Range, 81
Reverse J-shaped distribution, 98
Root-mean-square deviation, 82
Sample standard deviation, 82
Sample variance, 83
Semi-interquartile range, 81
Skewed distribution, 95
Standard deviation, 82
Standard units, 86
Unbiased estimator, 83
U-shaped distribution, 98
Variance, 83
z-scores, 88

4.7 References

A proof that division by $n - 1$ makes the sample variance an unbiased estimator of the population variance may be found in most textbooks on mathematical statistics; for instance, in

MILLER, I., and MILLER, M., *John E. Freund's Mathematical Statistics*, 6th ed. Upper Saddle River, N.J.: Prentice Hall, 1998.

Some information about the effect of grouping on the calculation of various statistical descriptions may be found in some of the older textbooks on statistics; for instance, in

MILLS, F. C., *Introduction to Statistics*. New York: Holt, Rinehart and Winston, 1956.

R E V I E W   E X E R C I S E S   for Chapters 1, 2, 3, and 4

R.1   The number of artifacts uncovered each day at an archaeological dig are to be grouped into a table with the classes 0–4, 5–14, 15–24, 23–35, and 40 or more. Explain where difficulties might arise.

R.2   On four days a building inspector reported 12, 8, 22, and 6 violations of a city's building code. Which of the following conclusions can be obtained from these data by purely descriptive methods and which ones require generalizations? Explain your answers.

(a) Altogether, the building inspector reported 48 violations of the city's building code on these four days.

(b) This building inspector rarely, if ever, reports more than 25 violations of the city's building code on any one day.

(c) The building inspector reported only 6 violations on the fourth day because it started snowing and he went home from work right after lunch.

(d) Probably, the third figure was recorded incorrectly and should have been 12 instead of 22.

(e) On the average, the building inspector reported 12 violations of the city's building code.

R.3   List the measurements that are grouped in the following stem-and-leaf display with unit leaves.

```
12   5   3
13   7   0   4   8
14   1   4   6   6   9   3
15   2   2   8   0   5
16   7   1
```

R.4   Twenty pilots were tested in a flight simulator. Following are the times (in seconds) it took them to take corrective action to an emergency situation: 4.9, 10.1, 6.3, 8.5, 7.7, 6.3, 3.9, 6.5, 6.8, 9.0, 11.3, 7.5, 5.8, 10.4, 8.2, 7.4, 4.6, 5.3, 9.7, and 7.3. Find

(a) the median;

(b) Q_1 and Q_3.

R.5   Use the results obtained in Exercise R.4 to construct a boxplot for the given data.

R.6   Explain why it is impossible to have $n = 6$, $\Sigma x = 18$, and $\Sigma x^2 = 47$ for a given set of data.

*R.7   Following is a distribution of the numbers of mistakes that 80 graduate students made in translating a passage from French to English as part of the language requirement for an advanced degree:

Number of mistakes	Number of students
0–4	34
5–9	20
10–14	15
15–19	9
20–24	2

Calculate

(a) the mean;

(b) the median;

(c) the standard deviation;

(d) the Pearsonian coefficient of skewness.

R.8 Convert the distribution of Exercise R.7 into a cumulative "or less" distribution.

R.9 A fishery expert found the following concentrations of mercury, in parts per million, in 32 fish caught in a certain stream:

0.045	0.063	0.049	0.062	0.065	0.054	0.050	0.048
0.072	0.060	0.062	0.054	0.049	0.055	0.058	0.067
0.055	0.058	0.061	0.047	0.063	0.068	0.056	0.057
0.072	0.052	0.058	0.046	0.052	0.057	0.066	0.054

(a) Construct a stem-and-leaf display with the stem labels 0.04, 0.05, 0.06, and 0.07.

(b) Use the stem-and-leaf display obtained in part (a) to determine the median, Q_1, and Q_3.

(c) Draw a boxplot and use it to describe the overall shape of the given data.

R.10 According to Chebyshev's theorem, what can we assert about the percentage of any set of data that must lie within k standard deviations of the mean for (a) $k = 3.5$; (b) $k = 4.5$?

R.11 A meteorologist has complete data for the last ten years on how many days in June the maximum temperature in Palm Springs, California, exceeded 110 degrees. Give one example each of a situation in which the meteorologist would look upon these data as

(a) a population;

(b) a sample.

R.12 IQ scores are sometimes looked upon as interval data. What assumption would this entail about the differences in intelligence of three persons with IQs of 95, 100, and 115? Is this assumption reasonable?

R.13 Rephrase Example 14.1, referred to on page 3, so that it would be of special interest to

(a) a lawyer;

(b) a writer of mystery stories.

R.14 Following are the numbers of articles that forty college professors have published in professional journals: 12, 8, 22, 45, 3, 27, 18, 12, 6, 32, 15, 17, 4, 19, 10, 2, 9, 16, 21, 17, 18, 11, 15, 2, 13, 15, 27, 16, 1, 5, 6, 15, 11, 32, 16, 10, 18, 4, 18, and 19. Determine

(a) the mean;

(b) the median.

R.15 Certain mass-produced metal shafts have a mean diameter of 24.00 mm (as required by specifications) with a standard deviation of 0.03 mm. At least what percentage of the shafts have diameters between 23.91 and 24.09 mm?

R.16 Among the students graduating from a university, 45 majoring in computer science had job offers averaging $31,100 (rounded to the nearest $100), 63 majoring in mathematics had job offers averaging $30,700, 112 majoring in engineering had job offers averaging $35,000, and 35 majoring in chemistry had job offers averaging $30,400. Find the mean job offer received by these 255 students.

R.17 From the distribution of Exercise R.7, can we determine how many of the 80 graduate students made

(a) more than 14 mistakes;

(b) anywhere from 5 to 19 mistakes;

(c) exactly 17 mistakes;

(d) anywhere from 10 to 20 mistakes?

If possible, give a numerical answer.

R.18 Following are the numbers of whales seen breaching on sixty whale-watching trips
off the cost of Baja California:

10	18	14	9	7	3	14	16	15	8	12	18
13	6	11	22	18	8	22	13	10	14	8	5
8	12	16	21	13	10	7	3	15	24	16	18
12	18	10	8	6	13	12	9	18	23	15	11
19	10	11	15	12	6	4	10	13	27	14	6

Determine the mean, the median, and the standard deviation of these data.

R.19 Use the results of Exercise R.18 to calculate the coefficient of variation.

R.20 Group the data of Exercise R.18 into a distribution with the classes 0–4, 5–9,
10–14, 15–19, 20–24, and 25–29. Also, draw a histogram of this distribution.

R.21 With reference to the data of Example 2.4, the waiting times between eruptions of
Old Faithful, regroup them into a distribution with the classes 25–39, 40–54, 55–69,
70–84, 85–99, 100–114, and 115–129. Also, draw a histogram of this distribution and
compare it with the one shown in Figure 2.3.

R.22 The average hourly wages paid by the deli department of a supermarket are, respectively, $8.20 for its male employees and $8.00 for its female employees. For the
produce department, the corresponding figures are $7.80 and $7.60, so that in both
departments the men are paid better than the women. Is it at all possible that for
both departments *combined* the women are paid better than the men?

R.23 The class limits of a distribution of weights (in ounces) are 10–29, 30–49, 50–69,
70–89, and 90–109. Find

(a) the class boundaries;

(b) the class marks;

(c) the class interval of the distribution.

R.24 For a given set of data, the smallest value is 5.0, the largest value is 65.0, the median is 15.0, the first quartile is 11.5, and the third quartile is 43.5. Draw a boxplot
and discuss the symmetry or skewness of the set of data.

R.25 Given $x_1 = 3.5$, $x_2 = 7.2$, $x_3 = 4.4$, and $x_4 = 2.0$, find

(a) Σx;

(b) Σx^2;

(c) $(\Sigma x)^2$.

R.26 If a population consists of the integers $1, 2, 3, \dots$, and k, its variance is $\sigma^2 = \dfrac{k^2 - 1}{12}$.
Verify this formula for

(a) $k = 3$;

(b) $k = 5$.

R.27 For a given set of data, the mean is 19.5 and the coefficient of variation is 32%. Find the standard deviation.

R.28 Will we get nominal data or ordinal data if
  (a) consumers must say whether they prefer Brand X to Brand Y, like them equally, or prefer Brand Y to Brand X;
  (b) consumers must say whether they prefer Brand X to Brand Y, like them equally, prefer Brand Y to Brand X, or have no opinion?

R.29 In the construction of a categorical distribution, men's shirts are classified according to whether they are made of wool, silk, linen, or synthetic fibers. Explain where difficulties might arise.

R.30 On thirty days, the numbers of registered nurses present at a nursing home were 2, 3, 1, 1, 3, 0, 0, 2, 1, 2, 2, 3, 0, 1, 2, 3, 2, 2, 2, 1, 1, 0, 2, 3, 2, 2, 2, 1, 0, and 2. Construct a dot diagram.

R.31 The daily numbers of persons attending an art museum are grouped into a distribution with the classes 0–29, 30–59, 60–89, and 90 or more. Can the resulting distribution be used to determine on how many days
  (a) at least 89 persons attended the museum;
  (b) more than 89 persons attended the museum;
  (c) anywhere from 30 to 89 persons attended the museum;
  (d) more than 100 persons attended the museum?

R.32 If a set of measurements has the mean $\bar{x} = 45$ and the standard deviation $s = 8$, convert each of the following values of x into standard units:
  (a) $x = 65$;
  (b) $x = 39$;
  (c) $x = 55$.

R.33 Explain why each of the following data may well fail to yield the desired information.
  (a) To determine public sentiment about certain import restrictions, an interviewer asks voters, "Do you feel that this unfair practice should be stopped?"
  (b) To predict an election for the governor of a state, a public opinion poll interviews persons selected haphazardly from a city's telephone directory.

R.34 Following are the numbers of false alarms that a security service received on twenty evenings: 9, 8, 4, 12, 15, 5, 5, 9, 3, 2, 6, 12, 5, 17, 6, 3, 7, 10, 8, and 4. Construct a boxplot and discuss the symmetry or skewness of the data.

R.35 The following distribution was obtained in a two-week study of the productivity of 100 workers:

Number of acceptable pieces produced	Number of workers
15–29	3
30–44	14
45–59	18
60–74	26
75–89	20
90–104	12
105–119	7

Find

(a) the class boundaries;

(b) the class marks;

(c) the class interval.

R.36 Draw a histogram of the distribution of Exercise R.35.

R.37 Convert the distribution of Exercise R.35 into a cumulative "less than" distribution and draw an ogive.

*R.38 Calculate the mean, the median, and the standard deviation of the distribution of Exercise R.35. Also, determine the Pearsonian coefficient of skewness.

R.39 Following are the systolic blood pressures of twenty-two hospital patients:

151	173	142	154	165	124	153	155	146	172	162
182	162	135	159	204	130	162	156	158	149	130

Construct a stem-and-leaf display with the stem labels 12, 13, 14, ... , and 20.

R.40 Use the stem-and-leaf display obtained in Exercise R.39 to get the information needed for the construction of a boxplot. Draw a boxplot and discuss the symmetry or skewness of the data.

R.41 Based on past experience, it is known that the bus that leaves downtown Phoenix at 8:05 A.M. takes 42 minutes with a standard deviation of 2.5 minutes to reach the Arizona State University campus. At least what percentage of the time will the bus reach the A.S.U. campus between 8:37 A.M. and 8:57 A.M.?

R.42 An official of a symphony orchestra reported that its five concerts were attended by 462, 480, 1,455, 417, and 432 patrons.

(a) Calculate the mean and the median of these attendance figures.

(b) Discovering that the third value was printed incorrectly and should have been just 455, recalculate the mean and the median of the corrected attendance figures.

(c) Compare the effect of this printing error on the mean and on the median.

R.43 The 30 pages of a preliminary printout of a manuscript were proofread for typographical errors, yielding the following numbers of mistakes:

2	0	3	1	0	0	0	5	0	1	2	1	4	0	1
0	1	3	1	2	0	1	0	3	1	2	0	1	0	2

Construct a dot diagram.

R.44 An experiment was performed by scientists to estimate the average (mean) increase in the pulse rate of astronauts performing a certain task in outer space. Simulating weightlessness, they obtained the following data (increase in pulse rate in beats per minute) for 33 persons who performed the given task:

34	26	22	24	23	18	21	27	33	26	31
28	29	25	13	22	21	15	30	24	23	37
26	22	27	31	25	28	20	25	27	24	18

Calculate

(a) the mean and the median;

(b) the standard deviation;

(c) The Pearsonian coefficient of skewness.

R.45 In an air pollution study, eight different samples of air yielded 2.2, 1.8, 3.1, 2.0, 2.4, 2.0, 2.1, and 1.2 micrograms of suspended benzene-soluble organic matter per cubic meter. Calculate the coefficient of variation for these data.

 *R.46 Following are the scores obtained by 44 cadets firing at a target from a kneeling position, *x*, and from a standing position, *y*:

x	*y*	*x*	*y*	*x*	*y*	*x*	*y*
81	83	81	76	94	86	77	83
93	88	96	81	86	76	97	86
76	78	86	91	91	90	83	78
86	83	91	76	85	87	86	89
99	94	90	81	93	84	98	91
98	87	87	85	83	87	93	82
82	77	90	89	83	81	88	78
92	94	98	91	99	97	90	93
95	94	94	94	90	96	97	92
98	84	75	76	96	86	89	87
91	83	88	88	85	84	88	92

Use a computer or a graphing calculator to produce a scattergram and describe the relationship, if any, between the cadets' scores in the two positions.

R.47 If students calculate their grade-point indexes (that is, average their grades) by counting A, B, C, D, and F as 1, 2, 3, 4, and 5, what does this assume about the nature of the grades?

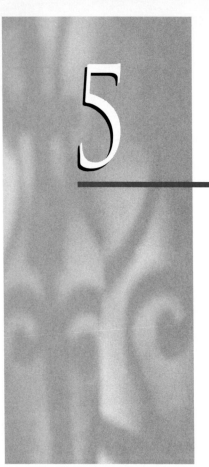

5

POSSIBILITIES AND PROBABILITIES

5.1   Counting 109
5.2   Permutations 113
5.3   Combinations 116
5.4   Probability 123
5.5   Checklist of Key Terms 132
5.6   References 133

Since statistics may be called the art or science of "how to live with uncertainty and chance," it deals with problems that are as old as humankind. Until the last few centuries, everything relating to chance was looked upon as divine intent, and it was considered impious, or even sacreligious, to try to analyze the "mechanics" of the supernatural through mathematics. This explains the slow development of the mathematics of probability. It was not until the advent of scientific thought, with its emphasis on observation and experimentation, that it even occurred to anyone that **probability theory** might be used in the study of the laws of nature, or that it might be applied to the solution of simple problems of everyday life.

In this chapter we shall see how uncertainties can actually be measured, how they can be assigned numbers, and how these numbers are to be interpreted. In subsequent chapters we shall see how these numbers, called **probabilities,** can be used to live with uncertainties—how they can be used to make choices or decisions that promise to be most profitable, or otherwise most desirable.

In Sections 5.1 through 5.3 we present mathematical preliminaries dealing with the question of "what is possible" in given situations. After all, we can hardly predict the outcome of a football game unless we know what teams are playing, and we cannot very well predict what will happen in an election unless we know what candidates are running for office. Then, in Section 5.4 we shall learn how to judge also "what is probable"; that is, we shall learn about several ways in which probabilities are defined, or interpreted, and their values are determined.

5.1   COUNTING

In contrast to the high-powered methods used nowadays in science, in business, and even in everyday life, the simple process of counting still plays an important role. We still have to count $1, 2, 3, 4, 5, \ldots$, for example, to determine how many persons take part in a demonstration, the size of the response to a questionnaire, the number of damaged cases in a shipment of wines from Portugal, or when preparing a report showing how many times the temperature in Phoenix, Arizona, went over 100 degrees in a given month. Sometimes, the process of counting can be simplified by using mechanical devices (for instance, when counting spectators passing through turnstiles), or by performing counts indirectly (for instance, by subtracting the serial numbers of invoices to determine the total number of sales). At other times, the process of counting can be simplified greatly by means of special mathematical techniques, such as the ones given in this section.

In the study of "what is possible," there are essentially two kinds of problems. There is the problem of listing everything that can happen in a given situation, and then there is the problem of determining how many different things can happen (without actually constructing a complete list). The second kind of problem is especially important, because there are many situations in which we do not need a complete list, and hence, can save ourselves a great deal of work. Although the first kind of problem may seem straightforward and easy, the following example illustrates that this is not always the case:

EXAMPLE   5.1   A restaurant offers three kinds of house wine by the glass—a red wine, a white wine, and a rosé. List the number of ways in which three dinner guests can order three glasses of wine, without taking into account who gets which wine.

Solution   Evidently, there are many different possibilities. Guests might order three glasses of red wine; they might order two glasses of red wine and one glass of rosé; they might order one glass of white wine and two glasses of rosé; they might order one glass of each kind; and so forth. Continuing this way carefully, we may be able to list all ten possibilities, but there is a good chance that we will miss one or two.   ∎

Problems like this can be handled systematically by drawing a **tree diagram** such as the one pictured in Figure 5.1. This diagram shows that there are four possibilities (four branches) corresponding to 0, 1, 2, or 3 glasses of red wine. Then, for white wine there are four branches coming from the top branch (0 glasses of red wine), three branches coming from the next branch (1 glass of red wine), two branches coming from the following branch (2 glasses of red wine), and only one branch coming from the bottom branch (3 glasses of red wine). After that, there is only one possibility for the number of glasses of rosé, since the numbers of glasses must always add up to three. Thus, we find that altogether there are ten possibilities.   ■

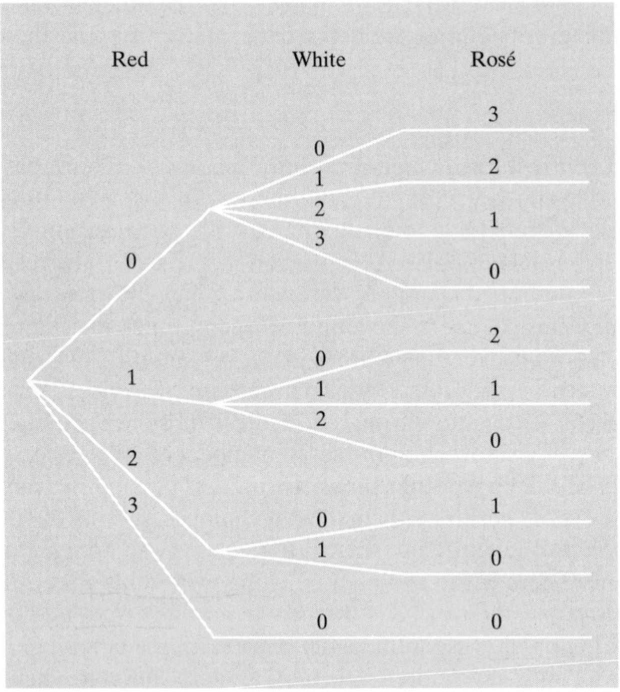

FIGURE 5.1
Tree diagram for
Example 5.1.

EXAMPLE   5.2   In a medical study, patients are classified according to whether they have blood type A, B, AB, or O, and also according to whether their blood pressure is low, normal, or high. In how many different ways can a patient thus be classified?

Solution   As is apparent from the tree diagram of Figure 5.2, the answer is 12. Starting at the top, the first path along the "branches" corresponds to a patient having blood type A and low blood pressure, the second path corresponds to a patient having blood type A and normal blood pressure, . . . , and the twelfth path corresponds to a patient having blood type O and high blood pressure.   ■

The answer we got in Example 5.2 is $4 \cdot 3 = 12$, namely, the product of the number of blood types and the number of blood pressure levels. Generalizing from this example, let us state the following rule:

Multiplication of choices

> *If a choice consists of two steps, of which the first can be made in m ways and for each of these the second can be made in n ways, then the whole choice can be made in $m \cdot n$ ways.*

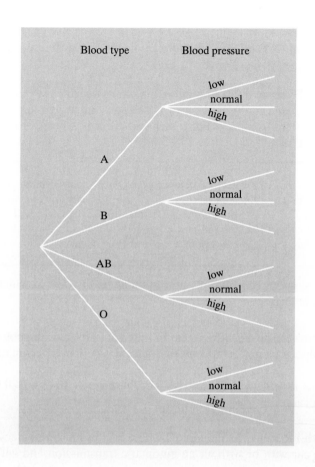

FIGURE 5.2
Tree diagram for
Example 5.2.

To prove this we have only to draw a tree diagram like that of Figure 5.2. First there are m branches corresponding to the m possibilities in the first step, and then there are n branches emanating from each of these branches corresponding to the n possibilities in the second step. This leads to $m \cdot n$ paths along the branches of the tree diagram, and hence to $m \cdot n$ possibilities.

EXAMPLE 5.3 If a research worker wants to experiment with one of 12 new medications for sinusitis, trying it on mice, guinea pigs, or rats, in how many different ways can she choose one of the medications and one of the three kinds of laboratory animals?

Solution Since $m = 12$ and $n = 3$, there are $12 \cdot 3 = 36$ different ways in which the experiment can be arranged. ∎

EXAMPLE 5.4 If a physics department schedules four lecture sections and sixteen laboratory sections for its basic course, in how many different ways can a student choose one of each?

Solution   Since $m = 4$ and $n = 16$, there are $4 \cdot 16 = 64$ different ways in which a student can choose one of each. ∎

By using appropriate tree diagrams, we can easily generalize the preceding rule for the multiplication of choices so that it will apply to choices involving more than two steps. For k steps, where k is a positive integer, we get the following rule:

Multiplication of choices (generalized)

> *If a choice consists of k steps, of which the first can be made in n_1 ways, for each of these the second can be made in n_2 ways, for each combination of choices made in the first two steps the third can be made in n_3 ways, ..., and for each combination of choices made in the first $k - 1$ steps the kth can be made in n_k ways, then the whole choice can be made in $n_1 \cdot n_2 \cdot n_3 \cdot \cdots \cdot n_k$ ways.*

We simply keep multiplying the numbers of ways in which the different steps can be made.

EXAMPLE   5.5   A new-car dealer offers a car in four body styles, in ten colors, and with a choice of three engines. In how many different ways can a person order one of the cars?

Solution   Since $n_1 = 4$, $n_2 = 10$, and $n_3 = 3$, there are $4 \cdot 10 \cdot 3 = 120$ ways in which a person can order one of the cars. ∎

EXAMPLE   5.6   With reference to Example 5.5, if a person must also decide whether to order the car with or without an automatic transmission and with or without aircon-ditioning, in how many different ways can he or she order one of the cars?

Solution   Since $n_1 = 4$, $n_2 = 10$, $n_3 = 3$, $n_4 = 2$, and $n_5 = 2$, there are $4 \cdot 10 \cdot 3 \cdot 2 \cdot 2 = 480$ ways in which the person can order one of the cars. ∎

EXAMPLE   5.7   A test consists of fifteen multiple choice questions, with each question having four possible answers. In how many different ways can a student check off one answer to each question?

Solution   Since $n_1 = n_2 = n_3 = \cdots = n_{15} = 4$, there are altogether

$$4 \cdot 4 \cdot 4 \cdot 4 \cdot 4 \cdot 4 \cdot 4 \cdot 4 \cdot 4 \cdot 4 \cdot 4 \cdot 4 \cdot 4 \cdot 4 \cdot 4 = 1,073,741,824$$

different ways in which a student can check off one answer to each question. (Note that in only one of the 1,073,741,824 possibilities all the answers are correct, and in

$$3 \cdot 3 \cdot 3 \cdot 3 \cdot 3 \cdot 3 \cdot 3 \cdot 3 \cdot 3 \cdot 3 \cdot 3 \cdot 3 \cdot 3 \cdot 3 \cdot 3 = 14,348,907$$

of them all the answers are wrong.) ∎

5.2   PERMUTATIONS

The rule for the multiplication of choices and its generalization are often used when several choices are made from one and the same set and we are concerned with the order in which they are made.

EXAMPLE   5.8   If 20 women entered the Miss Arizona contest, in how many different ways can the judges choose the winner and the first runner up?

Solution   Since the winner can be chosen in $m = 20$ ways and the first runner up must be one of the other $n = 19$ contestants, there are altogether $20 \cdot 19 = 380$ ways in which the judges can make their selection.   ∎

EXAMPLE   5.9   In how many different ways can the 52 members of a labor union choose a president, a vice-president, a secretary, and a treasurer?

Solution   Since $n_1 = 52$, $n_2 = 51$, $n_3 = 50$, and $n_4 = 49$ (regardless of which officer is chosen first, second, third, and fourth), there are altogether $52 \cdot 51 \cdot 50 \cdot 49 = 6{,}497{,}400$ different possibilities.   ∎

In general, if r objects are selected from a set of n distinct objects, any particular arrangement (order) of these objects is called a **permutation.** For instance, 4 1 2 3 is a permutation of the first four positive integers; Maine, Vermont, and Connecticut is a permutation (a particular ordered arrangement) of three of the six New England states; and

UCLA, Stanford, USC, and Washington

ASU, Oregon, WSU, and California

are two different permutations (ordered arrangements) of four of the ten universities with football teams in the PAC-10 conference.

EXAMPLE 5.10   Determine the number of different permutations of two of the five vowels a, e, i, o, and u, and list them all.

Solution   Since $m = 5$ and $n = 4$, there are $5 \cdot 4 = 20$ different permutations, and they are

ae	ai	ao	au	ei	eo	eu	io	iu	ou
ea	ia	oa	ua	ie	oe	ue	oi	ui	uo

∎

In general, it would be desirable to have a formula for the total number of permutations of r objects selected from a set of n distinct objects, such as the four universities chosen from among the ten universities with football teams in the PAC-10. To this end, observe that the first selection is made from the whole set of n objects, the second selection is made from the $n - 1$ objects that remain after the first selection has been made, the third selection is made from the $n - 2$ objects that remain after the first two selections have been made, ..., and the rth and final selection is made from the $n - (r - 1) = n - r + 1$ objects that remain after the first $r - 1$ selections have been made. Therefore, direct application of the generalized rule for the multiplication of choices yields the result that the total number of permutations of r objects selected from a set of n distinct objects, which we shall denote by $_nP_r$, is

$$n(n - 1)(n - 2) \cdots (n - r + 1)$$

Since products of consecutive integers arise in many problems relating to permutations and other kinds of special arrangements or selections, it is convenient to introduce here the **factorial notation.** In this notation, the product of all positive integers less than or equal to the positive integer n is called "n factorial" and denoted by $n!$. Thus,

$$1! = 1$$
$$2! = 2 \cdot 1 = 2$$
$$3! = 3 \cdot 2 \cdot 1 = 6$$
$$4! = 4 \cdot 3 \cdot 2 \cdot 1 = 24$$
$$5! = 5 \cdot 4 \cdot 3 \cdot 2 \cdot 1 = 120$$
$$6! = 6 \cdot 5 \cdot 4 \cdot 3 \cdot 2 \cdot 1 = 720$$

$$\cdot \quad \cdot \quad \cdot \quad \cdot \quad \cdot \quad \cdot$$

and in general

$$n! = n(n - 1)(n - 2) \cdots 3 \cdot 2 \cdot 1$$

Also, to make various formulas more generally applicable, we let $0! = 1$ by definition.

Since factorials grow very rapidly, it has been claimed that the exclamation mark reflects one's surprise. Indeed, the value of 10! exceeds three millions and 70! exceeds the memory limit of hand-held calculators.

To express the formula for $_nP_r$ more compactly in terms of factorials, we note, for example, that $12 \cdot 11 \cdot 10! = 12!$, that $9 \cdot 8 \cdot 7 \cdot 6! = 9!$, and that $37 \cdot 36 \cdot 35 \cdot 34 \cdot 33! = 37!$. Similarly,

$$_nP_r \cdot (n - r)! = n(n - 1)(n - 2) \cdots (n - r + 1) \cdot (n - r)!$$
$$= n!$$

so that $_nP_r = \dfrac{n!}{(n - r)!}$. To summarize,

The number of permutations of r objects selected from a set of n distinct objects is

$$_nP_r = n(n - 1)(n - 2) \cdots (n - r + 1)$$

or, in factorial notation,

$$_nP_r = \frac{n!}{(n-r)!}$$

where either formula can be used for $r = 1, 2, \ldots$, or n. (The second formula, but not the first, could be used also for $r = 0$, for which we would get the trivial result that there is

$$_nP_0 = \frac{n!}{(n-0)!} = 1$$

way of selecting none of n objects.) The first formula is generally easier to use because it requires fewer steps, but many students find the one in factorial notation easier to remember.

EXAMPLE 5.11   Find the number of permutations of $r = 4$ objects selected from a set of $n = 12$ distinct objects (say, the number of ways in which four of twelve new movies can be ranked first, second, third, and fourth by a panel of critics).

Solution   For $n = 12$ and $r = 4$, the first formula yields

$$_{12}P_4 = 12 \cdot 11 \cdot 10 \cdot 9 = 11,880$$

and the second formula yields

$$_{12}P_4 = \frac{12!}{(12-4)!} = \frac{12!}{8!} = \frac{12 \cdot 11 \cdot 10 \cdot 9 \cdot 8!}{8!} = 11,880$$

Essentially, the work is the same, but the second formula requires a few extra steps. ∎

To find the formula for the number of permutations of n distinct objects taken all together, we substitute $r = n$ into either formula for $_nP_r$ and get

Number of permutations of n objects taken all together

$$_nP_n = n!$$

EXAMPLE 5.12   In how many different ways can eight teaching assistants be assigned to eight sections of a course in economics?

Solution   Substituting $n = 8$, we get $_8P_8 = 8! = 40,320$. ∎

Throughout this discussion it has been assumed that the n objects are all distinct. When this is not the case, the formula for $_nP_n$ must be modified, and we shall illustrate how this is done for some special cases in Exercises 5.29 and 5.30. (When r is less than n, the corresponding modification of $_nP_r$ is complicated, and it will not be discussed in this book.)

5.3   COMBINATIONS

There are many problems in which we want to know the number of ways in which *r* objects can be selected from a set of *n* objects, but we do not care about the order in which the selection is made. For instance, we may want to know in how many ways a committee of four can be selected from among the 45 members of a college fraternity, or the number of ways in which the IRS can choose five of 36 tax returns for a special audit. To derive a formula that applies to problems like these, let us first examine the following 24 permutations of three of the first four letters of the alphabet:

abc   acb   bac   bca   cab   cba
abd   adb   bad   bda   dab   dba
acd   adc   cad   cda   dac   dca
bcd   bdc   cbd   cdb   dbc   dcb

If we do not care about the order in which the three letters are chosen from among the four letters *a*, *b*, *c*, and *d*, there are only four ways in which the selection can be made: *abc, abd, acd,* and *bcd*. Note that these are the groups of letters shown in the first column of the table, and that each row contains the $_3P_3 = 3! = 6$ permutations of the three letters in the first column.

In general, there are $_rP_r = r!$ permutations of *r* distinct objects, so that the $_nP_r$ permutations of *r* objects selected from among *n* distinct objects contain each group of *r* objects *r!* times. (In our example, the $_4P_3 = 4 \cdot 3 \cdot 2 = 24$ permutations of three letters selected from among the first four letters of the alphabet contain each group of three letters $_3P_3 = 3! = 6$ times.) Therefore, to get a formula for the number of ways in which *r* objects can be selected from a set of *n* distinct objects *without regard to their order,* we divide $_nP_r$ by *r!*. Referring to such a selection as a **combination** of *n* objects taken *r* at a time, we denote the number of combinations of *n* objects taken *r* at a time by $_nC_r$ or $\binom{n}{r}$, and write

*Number of
combinations of
n objects taken
r at a time*

> The number of ways in which r objects can be selected from a set of n distinct object is
>
> $$\binom{n}{r} = \frac{n(n-1)(n-2) \cdot \cdots \cdot (n-r+1)}{r!}$$
>
> or, in factorial notation,
>
> $$\binom{n}{r} = \frac{n!}{r!(n-r)!}$$

Like the two formulas for $_nP_r$, either formula can be used for $r = 1, 2, \ldots,$ or *n*, but the second one only for $r = 0$. Again, the first formula is generally easier to use because it requires fewer steps, but many students find the one in factorial notation easier to remember.

For $n = 0$ to $n = 20$, the values of $\binom{n}{r}$ may be read from Table XI, where these quantities are referred to as **binomial coefficients.** The reason for this is explained in Exercise 5.38.

EXAMPLE 5.13   On page 116 we asked for the number of ways in which a committee of four can be selected from among the 45 members of a college fraternity. Since the order in which the selections are made does not matter, we can now say that we want to find the value of $_{45}C_4$.

Solution   Substituting $n = 45$ and $r = 4$ into the first of the two formulas for $_nC_r$, we get

$$_{45}C_4 = \frac{45 \cdot 44 \cdot 43 \cdot 42}{4!} = 148,995.$$ ■

EXAMPLE 5.14   On page 116 we also asked for the number of ways in which the IRS can choose five of 36 tax returns for a special audit. Again, the order in which the choices are made does not matter, so we can now say that we want to find the value of $_{36}C_5$.

Solution   Substituting $n = 36$ and $r = 5$ into the first of the two formulas for $_nC_r$, we get

$$_{36}C_5 = \frac{36 \cdot 35 \cdot 34 \cdot 33 \cdot 32}{5!} = 376,992.$$ ■

EXAMPLE 5.15   In how many ways can a person choose three books from a list of ten best-sellers, assuming that the order in which the books are chosen is of no consequence?

Solution   Substituting $n = 10$ and $r = 3$ into the first of the two formulas for $_nC_r$, we get

$$\binom{10}{3} = \frac{10 \cdot 9 \cdot 8}{3!} = 120$$

Similarly, substitution into the second formula yields

$$\binom{10}{3} = \frac{10!}{3!7!} = \frac{10 \cdot 9 \cdot 8 \cdot 7!}{3!7!} = \frac{10 \cdot 9 \cdot 8}{3!} = 120$$

Essentially, the work is the same, but the first formula required fewer steps.   ■

EXAMPLE 5.16   In how many different ways can the director of a research laboratory choose two chemists from among seven applicants and three physicists from among nine applicants?

Solution   The two chemists can be selected in $\binom{7}{2}$ ways, the three physicists can be selected in $\binom{9}{3}$ ways, so that, by the multiplication of choices, all five of them can be selected in

$$\binom{7}{2} \cdot \binom{9}{3} = 21 \cdot 84 = 1,764$$

ways. The values of the two binomial coefficients were obtained from Table XI.

∎

In Section 5.2 we gave the special formula $_nP_n = n!$, but there is no need to do so here; substitution into either formula for $\binom{n}{r}$ yields $\binom{n}{n} = 1$. In other words, there is one and only one way in which we can select all n of the elements that constitute a set.

When r objects are selected from a set of n distinct objects, $n - r$ of the objects are left, and consequently, there are as many ways of leaving (or selecting) $n - r$ objects from a set of n distinct objects as there are ways of selecting r objects. Symbolically, we write

$$\binom{n}{r} = \binom{n}{n - r} \qquad \text{for } r = 0,1,2,\ldots,n$$

Sometimes this rule serves to simplify calculations and sometimes it is needed in connection with Table XI.

EXAMPLE 5.17   Determine the value of $\binom{75}{72}$.

Solution   Instead of having to write down the product $75 \cdot 74 \cdot 73 \cdots \cdot 4$ and then to cancel $72 \cdot 71 \cdot 70 \cdots \cdot 4$, we write directly

$$\binom{75}{72} = \binom{75}{3} = \frac{75 \cdot 74 \cdot 73}{3!} = 67,525$$

∎

EXAMPLE 5.18   Find the value of $\binom{19}{13}$.

Solution $\binom{19}{13}$ cannot be looked up directly in Table XI, but by making use of the fact that

$$\binom{19}{13} = \binom{19}{19 - 13} = \binom{19}{6},$$ we look up $\binom{19}{6}$ and obtain 27,132.

∎

For $r = 0$ we cannot substitute into the first of the two formulas for $\binom{n}{r}$, but if we substitute into the second formula, or if we write

$$\binom{n}{0} = \binom{n}{n-0} = \binom{n}{n}$$

we get $\binom{n}{0} = 1$. Evidently, there are as many ways of selecting none of the elements in a set as there are ways of choosing the n elements that are left.

■ Exercises

5.1   In the World Series of baseball the winner is the first team to win four games. Suppose that the American League champions lead the National League champions by three games to one. Construct a tree diagram showing the number of ways that these teams can continue to the completion of the series.

5.2   A student can study 1 or 2 hours for an astronomy test on any given night. Draw a tree diagram to find the number of ways in which the student can study altogether
(a) five hours on three consecutive nights;
(b) at least five hours on three consecutive nights.

5.3   A person with $2 in his pocket bets $1, even money, on the flip of a coin, and he continues to bet $1 so long as he has any money left. Draw a tree diagram to show the various things that can happen in the first three flips of the coin (provided, of course, that there will be a third flip). In how many of the cases will he be
(a) exactly $1 ahead;
(b) exactly $1 behind?

5.4   A food specialty shop stocks two cherry cheesecakes, reordering two more at the end of each day (for delivery early the next morning) if and only if both have been sold. Draw a tree diagram to show that if it receives two of these cheesecakes on a Monday morning, there are altogether eight different ways in which the store can make sales of the cheesecake on Monday and Tuesday.

5.5   With reference to Exercise 5.4, in how many ways can the store sell two or three of the cheesecakes on these two days?

5.6   On the faculty of a university there are three professors named Smith: Adam Smith, Brett Smith, and Craig Smith. Draw a tree diagram to show the various ways in which the payroll department can distribute their paychecks so that each of them receives a check made out to himself or to one of the other two Smiths. In how many of these possibilities will
(a) only one of them get his own check;
(b) at least one of them get his own check?

5.7   An artist has two paintings in an art exhibit that lasts two days. Draw tree diagrams to show in how many different ways he can make sales on the two days if
(a) we are interested only in how many of his paintings are sold on each day;
(b) we do care which painting is sold on what day.

5.8   In a union election, Mr. Brown, Ms. Green, and Ms. Jones are running for president. Mr. Adams, Ms. Roberts, and Mr. Smith are running for vice-president. Construct a tree diagram showing the nine possible outcomes, and use it to determine the number of ways in which the two unions officials will not be of the same sex.

5.9   In a political science survey voters are classified into six categories according to income and into four categories according to education. In how many different ways can a voter thus be classified?

5.10   In a traffic court, violators are classified according to whether they are properly licensed, whether their violations are major or minor, and whether or not they have committed any other violations during the preceding 12 months.
(a) Construct a tree diagram showing the various ways in which a violator can be classified by this court.
(b) If there are 20 violators in each of the eight categories obtained on part (a) and the judge gives each violator who is not properly licensed a stern lecture, how many of the violators will receive a stern lecture?
(c) If the judge gives an $80 fine to everyone who has committed a major violation and/or another violation in the preceding 12 months, how many of the violators will receive an $80 fine?
(d) How many of the violators will receive a stern lecture as well as an $80 fine?

5.11   A chain of convenience stores has four warehouses and 32 retail outlets. In how many different ways can it ship a carton of maple syrup from one of the warehouses to one of the retail outlets?

5.12   A purchasing agent places her orders by phone, by fax, by e-mail, by mail, or by an express carrier. She requests that her order be confirmed by phone or by fax. In how many different ways can one of her orders be placed and confirmed?

5.13   There are four different trails to the top of a mountain. In how many different ways can a person hike up and down the mountain if
(a) he must take the same trail both ways;
(b) he can, but need not, take the same trail both ways;
(c) he does not want to take the same trail both ways?

5.14   In an optics kit there are five concave lenses, five convex lenses, two prisms, and three mirrors. In how many different ways can a person choose one of each kind?

5.15   A cafeteria offers ten different soups or salads, eight entrees, and six desserts. In how many different ways can a customer choose a soup or salad, an entree, and a dessert?

5.16   A multiple-choice test consists of eight questions, each permitting a choice of three alternatives.
(a) In how many different ways can one choose an answer to each question?
(b) In how many different ways can one choose an answer to each question and get them all wrong? (It is assumed that each question has only one correct answer.)

5.17   A true-false test consists of 12 questions. In how many different ways can a person mark one answer to each question?

5.18   In how many different ways can a laboratory technician inject three of 15 mice with three different dosages of a serum?

5.19   Determine for each of the following whether it is true or false:
(a) $19! = 19 \cdot 18 \cdot 17 \cdot 16!$;
(d) $6! + 3! = 9!$;
(b) $\dfrac{12!}{3!} = 4!$;
(e) $\dfrac{9!}{7!2!} = 36$;
(c) $3! + 0! = 7$;
(f) $15! \cdot 2! = 17!$.

5.20   Determine for each of the following whether it is true or false:

(a) $\frac{1}{3!} + \frac{1}{4!} = \frac{5}{24}$;

(c) $5 \cdot 4! = 5!$;

(b) $0! \cdot 8! = 0$;

(d) $\frac{16!}{12!} = 16 \cdot 15 \cdot 14$.

5.21   In how many different ways can a television director schedule a sponsor's six different commercials during the telecast of the first half of a football game?

5.22   On her vacation, a person wants to visit three of the nation's 22 historical parks. If the order of the visits matters, in how many different ways can she plan her trip?

5.23   If the drama club of a college wants to present four of ten half-hour skits on one evening between 8 and 10 P.M., in how many different ways can it arrange its schedule?

5.24   In how many different ways can the curator of a museum arrange five of eight paintings horizontally on a wall?

5.25   In how many different ways can five graduate students choose one of ten research projects, if no two of them can choose the same project?

5.26   In how many different ways can the manager of a baseball team arrange the batting order of the nine players in the starting lineup?

5.27   Four married couples have bought eight seats in a row for a football game. In how many different ways can they be seated if
(a) each husband is to sit to the left of his wife;
(b) all the men are to sit together and all the women are to sit together?

5.28   A psychologist preparing three-letter nonsense words for use in a memory test chooses the first letter from among the consonants q, w, x, and z; the second letter from among the vowels e, i, and u; and the third letter from among the consonants c, f, p, and v.
(a) How many different three-letter nonsense words can he construct?
(b) How many of these nonsense words will begin with the letter w?
(c) How many of these nonsense words will end either with the letter f or the letter p?

5.29   If among n objects r are alike, and the others are all distinct, the number of permutations of these n objects taken all together is $\frac{n!}{r!}$.
(a) How many permutations are there of the letters in the word "class"?
(b) In how many ways (according to manufacturer only) can five cars place in a stock-car race if three of the cars are Fords, one is a Chevrolet, and one is a Dodge?
(c) In how many ways can the television director of Exercise 5.21 fill the six time slots allocated to commercials if she has four different commercials, of which a given one is to be shown three times, while each of the others is to be shown once?
(d) Present an argument to justify the formula given in this exercise.

5.30   If among n objects r_1 are identical, another r_2 are identical, and the rest (if any) are all distinct, the number of permutations of these n objects taken all together is $\frac{n!}{r_1! \cdot r_2!}$.
(a) How many permutations are there of the letters in the word "greater"?

(b) In how many ways can the television director of Exercise 5.21 fill the six time slots allocated to commercials if she has only two different commercials, each of which is to be shown three times?

(c) Generalize the formula so that it applies if among n objects r_1 are identical, another r_2 are identical, another r_3 are identical, and the rest (if any) are all distinct. In how many ways can the television director of Exercise 5.21 fill the six time slots allocated to commercials if she has three different commercials, each of which is to be shown twice?

5.31 Calculate the number of ways in which a chain of ice cream stores can choose two of 12 locations for new franchises.

5.32 A computer store carries 15 kinds of monitors. Calculate the number of ways in which a computer lab can purchase three different ones.

5.33 The dean of a college has to choose two chemists from among eight applicants, three physicists from among six applicants, and six mathematicians from among ten applicants. In how many different ways can he make his choice?

5.34 A student is required to report on three of eighteen books on a reading list. Calculate the number of ways in which a student can choose the three books, and verify the answer in Table XI.

5.35 A carton of 12 transistor batteries includes one that is defective. In how many different ways can an inspector choose three of the batteries and
(a) get the one that is defective;
(b) not get the one that is defective?

5.36 With reference to Exercise 5.35, suppose that two of the batteries are defective. In how many different ways can the inspector choose four of the batteries and get
(a) none of the defective batteries;
(b) one of the defective batteries;
(c) both of the defective batteries?

5.37 Counting the number of outcomes in games of chance has been popular for many centuries. Not only was gambling involved, but the outcomes were also taken as indications of divine intent. It was just about 1,000 years ago that a bishop in what is now Belgium determined that there are 56 different ways in which three dice can fall, provided one is interested only in the overall result and not in the outcomes of the individual dice. He assigned a virtue to each of these possibilities and each sinner had to concentrate for some time on the virtue that corresponded to his cast of the dice.
(a) Find the number of ways in which three dice can all come up with the same number of points.
(b) Find the number of ways in which two of the three dice can come up with the same number of points while the third die comes up with a different number of points.
(c) Find the number of ways in which all three of the dice can come up with a different number of points.
(d) Use parts (a), (b), and (c) to verify the bishop's calculation that there are altogether 56 possibilities.

5.38 The quantity $\binom{n}{r}$ is called a **binomial coefficient** because it is the coefficient of $a^{n-r}b^r$ in the binomial expansion of $(a + b)^n$.
(a) Verify this for $n = 2$ by expanding $(a + b)^2$.

(b) Verify this for $n = 3$ by expanding $(a + b)^3$.

(c) Verify this for $n = 4$ by expanding $(a + b)^4$.

5.39   Use Table XI to determine the following binomial coefficients:

(a) $\binom{16}{7}$;   (c) $\binom{19}{14}$;

(b) $\binom{13}{5}$;   (d) $\binom{15}{11}$.

5.40   A table of binomial coefficients is easy to construct by following the pattern shown here.

```
                    1
                1       1
            1       2       1
        1       3       3       1
    1       4       6       4       1
1       5      10      10       5       1
```
. . . .

Called **Pascal's triangle,** in this arrangement each row begins with a 1, ends with a 1, and each entry is the sum of the nearest two values from the row above. Construct the next three rows of Pascal's triangle, and verify from Table XI that they are, respectively, the binomial coefficients corresponding to

(a) $n = 6$;

(b) $n = 7$;

(c) $n = 8$.

5.41   Verify the identity $\binom{n + 1}{r} = \binom{n}{r} + \binom{n}{r - 1}$ by expressing each of the binomial coefficients in terms of factorials. Explain why this identity justifies the method used in the construction of Pascal's triangle in the preceding exercise.

5.42   Ben is one of six cameramen working for a TV station. If three of them are needed to cover a tennis match,

(a) in how many different ways can they be selected;

(b) in how many different ways can they be selected so that Ben will not be included;

(c) in how many different ways can they be selected so that Ben will be included? Also, verify that the results obtained for parts (b) and (c) add up to the result obtained for part (a). (Note that this is a special case of the identity of Exercise 5.41.)

5.4   PROBABILITY

So far in this chapter we have studied only what is possible in a given situation. In some instances we listed all the possibilities and in others we merely determined how many different possibilities there are. Now we shall go one step further and judge also what is probable and what is improbable.

The most common way of measuring the uncertainties connected with events (say, the outcome of a presidential election, the side effects of a new medication, the durability of an exterior paint, or the total number of points we may roll with a pair of dice) is to assign them **probabilities** or to specify the **odds** at which it would be fair to bet that the events will occur. In this section we shall learn how probabilities are interpreted and how their numerical values are determined; odds will be discussed in Section 6.3.

Historically, the oldest way of measuring uncertainties is the **classical probability concept.** It was developed originally in connection with games of chance, and it lends itself most readily to bridging the gap between possibilities and probabilities. The classical probability concept applies only when all possible outcomes are equally likely, in which case we say that

The classical probability concept

> *If there are n equally likely possibilities, of which one must occur and s are regarded as favorable, or as a "success," then the probability of a "success" is $\frac{s}{n}$.*

In the application of this rule, the terms "favorable" and "success" are used rather loosely—what is favorable to one player is unfavorable to his opponent, and what is a success from one point of view is a failure from another. Thus, the terms "favorable" and "success" can be applied to any particular kind of outcome, even if "favorable" means that a television set does not work, or "success" means that someone catches the flu. This usage dates back to the days when probabilities were quoted only in connection with games of chance.

EXAMPLE 5.19   What is the probability of drawing an ace from a well-shuffled deck of 52 playing cards?

Solution   By "well-shuffled" we mean that each card has the same chance of being drawn, so that the classical probability concept can be applied. Since there are $s = 4$ aces among the $n = 52$ cards, we find that the probability of drawing an ace is

$$\frac{s}{n} = \frac{4}{52} = \frac{1}{13}$$

∎

EXAMPLE 5.20   What is the probability of rolling a 3, a 4, or a 5 with a balanced die?

Solution   By "balanced" we mean that each face of the die has the same chance, so that the classical probability concept can be applied. Since $s = 3$ and $n = 6$, we find that the probability of rolling a 3, a 4, or a 5 is

$$\frac{s}{n} = \frac{3}{6} = \frac{1}{2}$$

∎

EXAMPLE 5.21   If H stands for heads and T for tails, the eight possible outcomes for three flips of a balanced coin are HHH, HHT, HTH, THH, HTT, THT, TTH, and TTT. What are the probabilities of getting two heads or three heads?

Solution   Again, by "balanced" we mean that all possible outcomes have the same chance, so that the classical probability concept can be applied. Counting the possibilities, we find that for two heads $s = 3$ and $n = 8$, and that for three heads $s = 1$ and $n = 8$. Thus, for two heads the probability is $\dfrac{s}{n} = \dfrac{3}{8}$ and for three heads it is $\dfrac{s}{n} = \dfrac{1}{8}$. ∎

Although equally likely possibilities are found mostly in games of chance, the classical probability concept applies also in a great variety of situations where gambling devices are used to make **random selections**—say, when offices are assigned to research assistants by lot, when laboratory animals are chosen for an experiment so that each one has the same chance of being selected (perhaps, by the method that is described in Section 10.1), when each family in a township has the same chance of being included in a survey, or when machine parts are chosen for inspection so that each part produced has the same chance of being selected.

EXAMPLE 5.22   If three of 20 weight lifters have used illegal steroids and four of them are randomly tested for the use of steroids, what is the probability that one of the three weight lifters who has used steroids will be included in the tests?

Solution   There are $n = \dbinom{20}{4} = \dfrac{20 \cdot 19 \cdot 18 \cdot 17}{4!} = 4{,}845$ ways of choosing the four weight lifters to be tested, and these possibilities may be regarded as equally likely by virtue of the random selection. The number of "favorable" outcomes is the number of ways in which one of the three weight lifters who have used steroids and three of the 17 weight lifters who have not used steroids can be selected, namely, $s = \dbinom{3}{1}\dbinom{17}{3} = 3 \cdot 680 = 2{,}040$, where the values of the binomial coefficients were obtained from Table XI. It follows that the probability that one and only one of the weight lifters who have used steroids will be caught is

$$\frac{s}{n} = \frac{2{,}040}{4{,}845} = \frac{8}{19}$$

or approximately 0.42. ∎

A major shortcoming of the classical probability concept is its limited applicability, because there are many situations in which the various possibilities cannot all be regarded as equally likely. This would be the case, for example, if we are concerned whether an experiment will support or refute a new theory; whether an expedition will be able to locate a ship wreck; whether a person's performance will justify a raise; or whether the Dow Jones Index will go up or down.

Among the various probability concepts, most widely held is the **frequency interpretation,** according to which probabilities are interpreted as follows:

The frequency interpretation of probability

> *The probability of an event (happening or outcome) is the proportion of the time that events of the same kind will occur in the long run.*

If we say that the probability is 0.78 that a jet from San Francisco to Phoenix will arrive on time, we mean that such flights arrive on time 78% of the time. Also, if the Weather Service predicts that there is a 40 percent chance for rain (that the probability is 0.40 that it will rain), they mean that under the same weather conditions it will rain 40% of the time. More generally, we say that an event has a probability of, say, 0.90, in the same sense in which we might say that our car will start in cold weather 90% of the time. We cannot guarantee what will happen on any particular occasion—the car may start and then it may not—but if we kept records over a long period of time, we should find that the proportion of "successes" is very close to 0.90.

In accordance with the frequency interpretation of probability, we estimate the probability of an event by observing what fraction of the time similar events have occurred in the past.

EXAMPLE 5.23   A sample survey conducted in 1990 showed that among 739 women in their twenties who remarried after being divorced, 114 divorced again. What is the probability that a woman in her twenties who remarried after being divorced will divorce again?

Solution   In the past this happened $\frac{114}{739} \cdot 100 \approx 15.4$ percent of the time, so we use 0.154 as an estimate of the desired probability.[†]   ∎

EXAMPLE 5.24   If records show that 34 of 956 persons who visited Central Africa came down with malaria, what is the probability that a person who visits Central Africa will not catch this disease?

Solution   Since $956 - 34 = 922$ of the persons who visited Central Africa did not catch malaria, we estimate the desired probability as $\frac{922}{956} \approx 0.96$.   ∎

[†]Usage of the symbol $\approx$ is discussed on page 55.

When probabilities are estimated in this way, it is only reasonable to ask whether the estimates are any good. In Chapter 14 we shall answer this question in some detail, but for now let us refer to an important theorem called the **law of large numbers.** Informally, this theorem may be stated as follows:

The Law of Large Numbers

> *If a situation, trial, or experiment is repeated again and again, the proportion of successes will tend to approach the probability that any one outcome will be a success.*

This theorem is known informally as the "law of averages." It is a statement about the long-run proportion of successes and it has little to say about any single trial.

In the first six editions of this book we illustrated the law of large numbers by repeatedly flipping a coin and recording the accumulated proportion of heads after each fifth flip. Since then we have used the **computer simulation** shown in Figure 5.3, where the 1's and 0's denote heads and tails.

```
MTB > Random 10 c21-c30;
SUBC>    Bernoulli 0.5.
MTB > Print c21-c30.
```

0	0	1	0	1	0	1	0	1	1
1	1	0	1	0	1	1	0	0	0
1	0	0	1	1	1	0	1	1	0
1	0	1	1	0	1	1	1	1	1
1	0	1	1	0	0	1	0	0	0
1	1	0	0	0	0	0	1	1	1
0	1	1	0	0	1	1	1	0	1
1	0	0	1	1	0	0	0	1	1
0	0	0	0	1	1	0	1	0	0
0	1	0	1	1	0	0	0	1	0

FIGURE 5.3
Computer simulation of 100 flips of a coin.

Reading across successive rows, we find that among the first five simulated flips there are 2 heads, among the first ten there are 5 heads, among the first fifteen there are 8 heads, among the first twenty there are 10 heads, among the first twenty-five there are 13 heads, ..., and among all hundred there are 51 heads. The corresponding proportions, plotted in Figure 5.4, are $\frac{2}{5} = 0.40$, $\frac{5}{10} = 0.50$, $\frac{8}{15} \approx 0.53$, $\frac{10}{20} = 0.50$, $\frac{13}{25} = 0.52$, ..., and $\frac{51}{100} = 0.51$. Observe that the proportion of heads fluctuates but comes closer and closer to 0.50, the probability of heads for each flip of the coin.

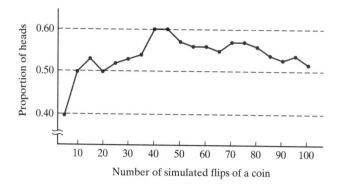

FIGURE 5.4
Graph illustrating law
of large numbers.

In the frequency interpretation, the probability of an event is defined in terms of what happens to similar events in the long run, so let us examine briefly whether it is at all meaningful to talk about the probability of an event that can occur only once. For instance, can we assign a probability to the event that Ms. Bertha Jones will be able to leave the hospital within four days after having an appendectomy, or to the event that a certain major-party candidate will win an upcoming gubernatorial election? If we put ourselves in the position of Ms. Jones's doctor, we might check medical records, discover that patients left the hospital within four days after an appendectomy in, say 78% of hundreds of cases, and apply this figure to Ms. Jones. This may not be of much comfort to Ms. Jones, but it does provide a meaning for a probability statement about her leaving the hospital within four days—the probability is 0.78.

This illustrates that when we make a probability statement about a specific (nonrepeatable) event, the frequency interpretation of probability leaves us no choice but to refer to a set of similar events. As can well be imagined, however, this can easily lead to complications, since the choice of similar events is generally neither obvious nor straightforward. With reference to Ms. Jones's appendectomy, we might consider as similar only cases in which the patients were of the same sex, only cases in which the patients were also of the same age as Ms. Jones, or only cases in which the patients were also of the same height and weight as Ms. Jones. Ultimately, the choice of similar events is a matter of personal judgment, and it is by no means contradictory that we can arrive at different probability estimates, all valid, concerning the same event.

With regard to the question whether a certain major-party candidate will win an upcoming gubernatorial election, suppose that we ask the persons who have conducted a poll how sure they are that the candidate will win. If they say they are "95 percent sure" (that is, if they assign a probability of 0.95 to the candidate's winning the election), this is not meant to imply that he would win 95% of the time if he ran for office a great number of times. Rather, it means that the pollsters' prediction is based on a method that works 95% of the time. It is in this way that we must interpret many of the probabilities attached to statistical results.

Finally, let us mention a third probability concept that is currently enjoying favor. According to this point of view, probabilities are interpreted as **personal** or **subjective** evaluations. They reflect one's belief with regard to the uncertainties that are involved, and they apply especially when there is little or no direct evidence, so that there really is no choice but to consider collateral (indirect) information, educated guesses, and perhaps intuition and other subjective factors. Subjective probabilities are sometimes determined by putting the issues in question on a money basis, as will be explained in Sections 6.3 and 7.1

Exercises

5.43 When one card is drawn from a well-shuffled deck of 52 playing cards, what are the probabilities of getting
(a) the queen of spades;
(b) a face card (jack, queen, or king) of any suit;
(c) a red card;
(d) a 2, a 3, or a 4;
(e) a club?

5.44 When two cards are drawn from a well-shuffled deck of 52 playing cards, what are the probabilities of getting
(a) two aces;
(b) two face cards (jack, queen, or king);
(c) two black cards;
(d) one red card and one black card;
(e) two spades?

5.45 When three cards are drawn from a well-shuffled deck of 52 playing cards, what are the probabilities of getting
(a) three hearts;
(b) two queens and a king;
(c) a jack, a queen, and a king?

5.46 If we roll a balanced die, what are the probabilities of getting
(a) a 5;
(b) an odd number;
(c) a number greater than 2?

5.47 If we roll a pair of balanced dice, one red and one green, list the 36 possible outcomes and determine the probabilities of getting a total of
(a) 4;
(b) 9;
(c) 7 or 11.

5.48 If H stands for heads and T for tails, the 16 possible outcomes for four flips of a coin are HHHH, HHHT, HHTH, HTHH, THHH, HHTT, HTHT, HTTH, THHT, THTH, TTHH, HTTT, THTT, TTHT, TTTH, and TTTT. Assuming that these outcomes are all equally likely, find the probabilities of getting 0, 1, 2, 3, or 4 heads.

5.49 A bowl contains 15 red beads, 30 white beads, 20 blue beads, and 7 black beads. If one of the beads is drawn at random, what are the probabilities that it will be
(a) red;   (c) black;
(b) white or blue;   (d) neither white nor black?

5.50 A bowl contains 25 red marbles, 18 yellow marbles, and 11 blue marbles. If one of the marbles is selected at random, find the probabilities that it will be
(a) red;   (c) blue;
(b) yellow or blue;   (d) neither red nor blue.

5.51 The balls used in selecting numbers for BINGO carry the numbers $1, 2, 3, \ldots, 75$. If one of the balls is selected at random, what are the probabilities that it will be
(a) an even number;
(b) a number that is 15 or below;
(c) a number that is 60 or above?

5.52 If a game has n equally likely outcomes, what is the probability of each individual outcome?

5.53 Among the 15 applicants for three positions at a newspaper, 10 are college graduates. If the selections are random, what are the probabilities that the positions will be filled with
(a) three applicants with college degrees;
(b) two applicants with college degrees and one without;
(c) three applicants without college degrees?

5.54 A carton of 24 light bulbs includes two that are defective. If two of the bulbs are chosen at random, what are the probabilities that
(a) neither bulb will be defective;
(b) one of the bulbs will be defective;
(c) both bulbs will be defective?

5.55 A person has four books, two fiction and two nonfiction, haphazardly arranged on a shelf. What is the probability that the two fiction books will be next to each other?

5.56 A hoard of medieval coins discovered in what is now Belgium included 20 struck in Antwerp and 16 struck in Brussels. If a person chooses five of these coins at random, what are the probabilities that she will get
(a) two coins struck in Antwerp and three struck in Brussels;
(b) four coins struck in Antwerp and one coin struck in Brussels?

5.57 The eight cities in the United States with the most violent crimes reported in a recent year were Atlanta, Miami, St. Louis, Newark, Tampa, Baton Rouge, Baltimore, and Washington, D.C. If a television news program randomly selects two of these cities for a special report, what are the probabilities that the selection will
(a) include Miami;
(b) consist of Baltimore and Washington, D.C.?

5.58 On a tray there are six pieces of apple pie and six pieces of cherry pie. If a waiter randomly picks two pieces of pie from the tray and gives them to persons who had ordered cherry pie, what is the probability that he is making a mistake?

5.59 In a poll conducted by a newspaper, 424 of 954 readers claimed that the coverage of local sports was inadequate. Estimate the probability that any one reader of the newspaper, selected at random, will support this claim.

5.60 If 816 of 4,800 shoplifters were not caught until they tried it for the fifth time, estimate the probability that a shoplifter will not get caught until his or her fifth try.

5.61 If 678 of 904 car owners passed a state-operated emission test on the first try, estimate the probability that any one car owner will pass this test on the first try.

5.62 If 1,558 of 2,050 persons visiting the Grand Canyon said that they expect to return within a few years, estimate the probability that any one person visiting the Grand Canyon expects to return within a few years.

5.63 Weather bureau statistics show that in a certain community it has rained 15 times in the last 48 years on the first Sunday in May, when a service club holds its annual picnic. Estimate the probability that it will not rain this year on the first Sunday in May.

5.64 Records show that it was overcast on 117 of the last 351 total eclipses of the sun. Estimate the probability that it will be overcast when there is the next total eclipse of the sun.

5.65 To get a "feeling" for the law of large numbers, flip a coin 200 times (or use a computer simulation) and record the accumulated proportion of heads after each tenth flip. Draw a diagram like the one shown in Figure 5.4.

5.66 With reference to Exercise 5.65, by at most how much do the proportions differ from 0.50
(a) after the 50th flip;
(b) after the 100th flip;
(c) after the 150th flip?

5.67 Record the last digit of the numbers on the license plates of 300 cars and plot the accumulated proportions of 3's after each ten cars. By how much do these proportions differ from 0.10
(a) after the first 100 cars;
(b) after the first 200 cars;
(c) after the 300th car?

(What we mean by "chance," "randomness," and "probability," is subject to all sorts of myths and misconceptions. Some of these are illustrated by the exercises that follow.)

5.68 Some philosophers have argued that if we have absolutely no information about the likelihood of the different possibilities, it is reasonable to regard them all as equally likely. This is sometimes referred to as the principle of equal ignorance. Discuss the argument that human life either does or does not exist elsewhere in the universe, and since we really have no information one way or the other, the probability that human life exists elsewhere in the universe is $\frac{1}{2}$.

5.69 The following illustrates how one's intuition can be misleading in connection with probabilities: A box contains 100 beads, some red and some white. One bead will be drawn, and you are asked to call beforehand whether it is going to be red or white. Would you be willing to bet even money (say, you will win $5 if you are right and lose $5 if you are wrong) if
(a) you have no idea how many of the beads are red and how many are white;
(b) you are told that 50 of the beads are red and 50 are white?
Strange as it may seem, most persons are more willing to gamble under condition (b) than under condition (a).

5.70 The following is a good example of the difficulties in which we may find ourselves if we use only "common sense," or intuition, in judgments concerning probabilities:

"Among three indistinguishable boxes one contains 2 pennies, one contains a penny and a dime, and one contains 2 dimes. Selecting one of these boxes at random (each box has a probability of 1/3), one coin is taken out at random (each coin has a probability of 1/2) without looking at the other. The coin that is taken out of the box is a penny, and without giving the matter too *much thought, we may well be inclined to say that there is a probability of 1/2 that the other coin in the box is also a penny. After all, the penny must have come either from the box with the penny and the dime or from the box with two pennies. In the first case the other coin is a dime, in the second case it is a penny, and it would seem reasonable to say that these two possibilities are equally likely."*

Actually, the correct value of the probability that the other coin is also a penny is 2/3, and it will be left to the reader to verify this result by mentally labeling the two pennies in the first box P_1 and P_2, the two dimes in the third box D_1 and D_2, and drawing a tree diagram showing the six possible (and equally likely) outcomes of the experiment.

5.71   Discuss the following assertion: If a meteorologist says that the probability for rain on the next day is 0.30, whatever happens on that day cannot prove him right or wrong.

5.72   Discuss the following assertion: Since probabilities are measures of uncertainty, the probability we assign to a future event will always increase as we get more information.

5.73   The probability that a patient will survive minor surgery is 0.98 for hospital A and 0.86 for hospital B; the probability that a patient will survive major surgery is 0.73 for hospital A and 0.66 for hospital B. Can we conclude that any kind of surgery at hospital A is a better risk than any kind of surgery at hospital B? (In case this looks familiar, take another look at Exercise R.22.)

5.74   No diagnostic tests are infallible, so imagine that the probability is 0.95 that a certain test will diagnose a diabetic correctly as being diabetic, and it is 0.05 that it will diagnose a person who is not diabetic as being diabetic. It is known that roughly 10% of the population is diabetic. Guess at the probability that a person diagnosed as being diabetic actually is diabetic. (This problem will be discussed further in Exercise 6.91.)

5.75   Some persons claim that if the probability of something happening is more than 0.50, it is confirmed if the event actually happens and refuted if the event does not happen. Correspondingly, if the probability of something happening is less than 0.50, it is confirmed if the event does not happen and it is refuted if the event happens. Comment on this method of confirming and refuting probabilities.

5.5 *Checklist of Key Terms* *(with page references to their definitions)*

Binomial coefficients, 117, 122
  Table, 590
Classical probability concept, 124
Combinations, 116
Computer simulation, 127
Factorial notation, 114
Frequency interpretation, 126
Law of large numbers, 127
Multiplication of choices, 110
  generalized, 112

Odds, 124
Pascal's triangle, 123
Permutations, 113
Personal probability, 129
Probability, 108, 124
Probability theory, 108
Random selection, 125
Subjective probability, 129
Tree diagram, 109

5.6 References

Informal introductions to probability, written primarily for the layperson, may be found in

GARVIN, A.D., *Probability in Your Life.* Portland, Maine: J. Weston Walch Publisher, 1978.

HUFF, D., and GEIS, I., *How to Take a Chance.* New York: W.W. Norton & Company, Inc., 1959.

KOTZ, S., and STROUP, D. E., *Educated Guessing: How to Cope in an Uncertain World.* New York: Marcel Dekker, Inc., 1983.

LEVINSON, H. C., *Chance, Luck, and Statistics.* New York: Dover Publications, Inc., 1963.

MOSTELLER, F., KRUSKAL, W. H., LINK, R. F., PIETERS, R. S., and RISING, G. R., *Statistics by Example: Weighing Chances.* Reading, Mass.: Addison-Wesley Publishing Company, Inc., 1973.

WEAVER, W., *Lady Luck: The Theory of Probability.* New York: Dover Publications, Inc., 1982.

For fascinating reading on the history of probability, see

DAVID, F. N., *Games, Gods and Gambling.* New York: Hafner Press, 1962.

and the first three chapters of

STIGLER, S. M., *The History of Statistics.* Cambridge, Mass.: Harvard University Press, 1986.

To supplement Exercises 5.68 through 5.75, further examples of myths and misconceptions about probability, including some fascinating paradoxes, may be found in

BENNETT, D. J., *Randomness.* Cambridge, Mass.: Harvard University Press, 1998.

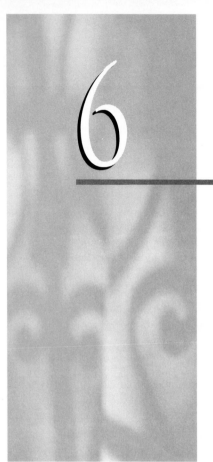

6 SOME RULES OF PROBABILITY

6.1   Sample Spaces and Events   135

6.2   The Postulates of Probability   145

6.3   Probabilities and Odds   148

6.4   Addition Rules   153

6.5   Conditional Probability   160

6.6   Multiplication Rules   163

*6.7   Bayes' Theorem   170

6.8   Checklist of Key Terms   175

6.9   References   175

*I*n the study of probability there are three fundamental kinds of questions:

1. What do we mean when we say that the probability of an event is, say, 0.50, 0.78, or 0.44?

2. How are the numbers we call probabilities determined, or measured in practice?

3. What are the mathematical rules that probabilities must obey?

For the most part, we have already studied the first two kinds of questions in Chapter 5. In the classical probability concept we are concerned with equally likely possibilities, count the ones that are favorable, and use the formula s/n. In the frequency interpretation we are concerned with proportions of "successes" in the long run and base our estimates on what has happened in the past. When it comes to subjective probabilities we are concerned with a measure of a person's belief, and in Section 6.3 (and later in Section 7.1) we shall see how such probabilities can actually be determined.

In this chapter, after some preliminaries in Section 6.1, we shall concentrate on the rules that probabilities must obey, namely, on the **theory of probability.** This includes the basic postulates, the relationship between probabilities and odds, the addition rules, the definition of conditional probability, the multiplication rules, and finally Bayes' theorem.

6.1   SAMPLE SPACES AND EVENTS

In statistics, the word "experiment" is used in a very broad and unconventional sense. For lack of a better term, it refers to any process of observation or measurement. Thus, an **experiment** may consist of counting how many times a mission into outer space had to be aborted; it may consist of the simple process of noting whether a certain switch is on or off or whether a person is single or married; or it may consist of the very complicated process of obtaining and evaluating data to predict trends in the economy, to find the source of social unrest, or to find the cause of a disease. The results one obtains from an experiment, whether they are instrument readings, counts, "yes or no" answers, or values obtained through extensive calculations, are called the **outcomes** of the experiment.

For each experiment, the set of all possible outcomes is called the **sample space** and it is usually denoted by the letter S. For instance, if a zoologist must choose three of 24 guinea pigs for an experiment, the sample space consists of the $\binom{24}{3} = 2{,}024$ different ways in which the selection can be made; or if the dean of a college must assign two of 84 members of the faculty as advisors to a political science club, the sample space consists of the $\binom{84}{2} = 3{,}486$ ways in which the selection can be made. Also, if we are concerned with what day in May the government will approve certain appropriations, the sample space is the set $S = \{1, 2, 3, 4, \ldots, 30, 31\}$.

When we study the outcomes of an experiment, we usually identify the various possibilities with numbers, points, or some other kinds of symbols, so that we can treat all questions about them mathematically, without having to go through long verbal descriptions of what has taken place, is taking place, or will take place. For instance, if there are eight candidates for a scholarship and we let a, b, c, d, e, f, g, and h denote that it is awarded to Ms. Adam, Mr. Bean, Miss Clark, and so on, then the sample space for this experiment is the set

$$S = \{a, b, c, d, e, f, g, h\}$$

The use of points rather than letters or numbers has the advantage that it makes it easier to visualize the various possibilities, and perhaps discover special features that several of the outcomes may have in common.

EXAMPLE   6.1   A used-car dealer has three 1998 Dodge Ram trucks on his lot and we are interested in how many of them each of his two salespersons will sell in a given week.

(a) Using two coordinates, so that $(1, 1)$, for example, denotes the outcome that each of the two salespersons will sell one of the three trucks, $(2, 0)$ denotes the outcome that the first salesperson will sell two of the trucks and the second salesperson will sell none, and so forth, list all possible outcomes of this "experiment."

(b) Draw a figure showing the corresponding points of the sample space.

Solution

(a) The ten possible outcomes are $(0, 0)$, $(0, 1)$, $(0, 2)$, $(0, 3)$, $(1, 0)$, $(1, 1)$, $(1, 2)$, $(2, 0)$, $(2, 1)$, and $(3, 0)$.

(b) The corresponding points are shown in Figure 6.1, from which it is apparent, for instance, that in three of the ten possibilities the first salesperson sells only one of the three trucks, and that in four of the ten possibilities the two salespersons, between them, sell all three of the trucks. ∎

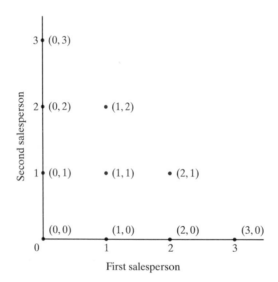

FIGURE 6.1
Sample space for two-salespersons example.

Usually, we classify sample spaces according to the number of elements, or points, that they contain. The ones we have mentioned so far in this section contained 2,024, 3,486, 31, 8, and 10 elements, and we refer to them all as **finite.** In this chapter we shall consider only sample spaces that are finite, but in later chapters we shall consider also samples spaces that are **infinite.** An infinite sample space arises, for example, when we deal with quantities such as temperatures, weights, or distances, which are measured on continuous scales. Even when we throw a dart at a target, there is a continuum of points we may hit.

In statistics, any **subset** of a sample space is referred to as an **event.** By subset we mean any part of a set, including the set as a whole and the **empty set,** denoted by ∅, which has no elements at all. For instance, for the example dealing with the days in May on which the government will approve certain appropriations, $F = \{25, 26, 27, 28, 29, 30, 31\}$ is the event that the government will approve the appropriations during the last week in May, and $G = \{1, 2, 3, 4\}$ is the event that the government will approve the appropriations during the first four days of May. As is the custom, we denoted these events by capital letters.

EXAMPLE   6.2   With reference to Example 6.1 and Figure 6.1, express in words what events are represented by

      (a) $A = \{(2, 0), (1, 1), (0, 2)\}$;
      (b) $B = \{(0, 0), (1, 0), (2, 0), (3, 0)\}$;
      (c) $C = \{(0, 2), (1, 2), (0, 3)\}$.

Solution

      (a) A is the event that, between them, the two salespersons will sell two of the trucks.
      (b) B is the event that the second salesperson will not sell any of the trucks.
      (c) C is the event that the second salesperson will sell at least two of the trucks.   ∎

    In Example 6.2, events B and C have no elements (outcomes) in common and they are referred to as **mutually exclusive;** that is, the occurrence of either one precludes the occurrence of the other. Clearly, if the second salesperson will not sell any of the trucks, he or she cannot very well sell at least two. Observe also that events A and B are not mutually exclusive, and neither are events A and C. The first pair shares the outcome $(2, 0)$ and the second pair shares the outcome $(0, 2)$.

    In many probability problems we are interested in events that can be expressed by forming **unions, intersections,** and/or **complements.** The reader is probably familiar with these elementary set operations, but if not, the union of two sets X and Y, denoted by $X \cup Y$, is the event that consists of all the elements (outcomes) contained in X, in Y, or in both. The intersection of two events X and Y, denoted by $X \cap Y$, is the event that consists of all the elements contained in both X and Y, and the complement of X, denoted by X', is the event that consists of all the elements of the sample space that are not contained in X. We usually read $\cup$ as "or," $\cap$ as "and," and X' as "not X."

EXAMPLE   6.3   With reference to Examples 6.1 and 6.2, list the outcomes comprising each of the following events, and express the events in words.

      (a) $B \cup C$;
      (b) $A \cap C$;
      (c) B'.

Solution

      (a) Since $B \cup C$ contains all the elements (outcomes) that are in B, in C, or in both, we find that

$$B \cup C = \{(0, 0), (1, 0), (2, 0), (3, 0), (0, 2), (1, 2), (0, 3)\}$$

and this is the event that the second salesperson will sell 0, 2, or 3 of the trucks. Namely, the event that the second salesperson will not sell just one of the trucks.

(b) As we already pointed out, events A and C share the outcome $(0, 2)$, and only $(0, 2)$, so we can write

$$A \cap C = \{(0, 2)\}$$

and this is the event that the first salesperson does not sell any of the trucks and the second salesperson sells two.

(c) Since B' contains all the elements not obtained in B, we can write

$$B' = \{(0, 1), (1, 1), (2, 1), (0, 2), (1, 2), (0, 3)\}$$

and this is the event that the second salesperson will sell at least one of the trucks.   ■

Sample spaces and events, particularly relationships among events, are often illustrated by **Venn diagrams** such as those of Figures 6.2 and 6.3. In each case, the sample space is represented by a rectangle, and events by circles or parts of circles within the rectangle. The tinted regions of the four Venn diagrams of Figure 6.2 represent event X, the complement of event X, the union of events X and Y, and the intersection of events X and Y.

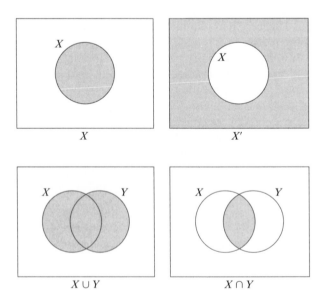

FIGURE 6.2
Venn diagrams.

EXAMPLE   6.4   If X is the event that Mr. Brown likes mathematics and Y is the event that he is a good chess player, what events are represented by the tinted regions of the four Venn diagrams of Figure 6.2?

Solution   The tinted region of the first diagram represents the event that Mr. Brown likes mathematics; the tinted region of the second diagram represents the event that he does not like mathematics; the tinted region of the third diagram represents the event that he likes mathematics and/or is a good chess player; and the tinted region of the fourth diagram represents the event that he likes mathematics and also is a good chess player.   ■

When we deal with three events, we draw circles as in Figure 6.3. In this diagram, the circles divide the sample space into eight regions, numbered 1 through 8, and it is easy to determine whether the corresponding events are in X or in X', in Y or in Y', and in Z or Z'.

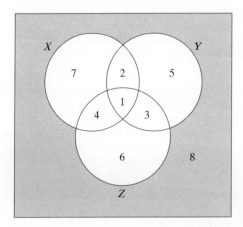

FIGURE 6.3
Venn diagram.

EXAMPLE   6.5   If X is the event that unemployment will go up, Y is the event that stock prices will go up, and Z is the event that interest rates will go up, express in words what events are represented by the following regions of the Venn diagram of Figure 6.3:

(a) region 4;
(b) regions 1 and 3 together;
(c) regions 3, 5, 6, and 8 together.

Solution

(a) Since this region is contained in X and Z but not in Y, it represents the event that unemployment and interest rates will go up, but stock prices will not go up.
(b) Since this is the region common to Y and Z, it represents the event that stock prices and interest rates will go up.
(c) Since this is the entire region outside X, it represents the event that unemployment will not go up.   ■

■ Exercises

6.1   With reference to the illustration on page 135, suppose that a, b, c, d, e, f, g, and h denote the events that the scholarship is awarded to Ms. Adams, Mr. Bean, Miss Clark, Mrs. Daly, Mr. Earl, Ms. Fuentes, Ms. Gardner, or Mr. Hall, and that $U = \{b, e, h\}$ and $V = \{e, f, g, h\}$. List the outcomes that comprise each of the following events and also express the events in words:
(a) U';
(b) $U \cap V$;
(c) $U \cup V'$.

6.2   With reference to Exercise 6.1, are the two events U and V mutually exclusive?

6.3   With reference to the sample space of Figure 6.1, list the sets of points that constitute the following events:
(a) One of the three trucks will remain unsold.
(b) The two salespersons will sell equally many trucks.
(c) Both salespersons will sell at least one truck.

6.4   At least one of two professors and two of five graduate assistants must be present when a chemistry lab is being used.
(a) Using two coordinates as in Example 6.1, so that $(1, 4)$, for example, denotes the presence of one of the professors and four of the graduate assistants, list the eight possibilities.
(b) Draw a diagram similar to that of Figure 6.1.

6.5   With reference to Exercise 6.4, list in words what events are represented by
(a) $K = \{(1, 2), (2, 3)\}$;
(b) $L = \{(1, 3), (2, 2)\}$;
(c) $M = \{(1, 2), (2, 2)\}$.
Also, determine which of the three pairs of events, K and L, K and M, and L and M, are mutually exclusive.

6.6   A literary critic has two days on which to take a look at some of the seven books that have recently been released. She wants to check out at least five of the books, but not more than four on either day.
(a) Using two coordinates so that $(2, 3)$, for example, represents the event that she will take a look at two books on the first day and three books on the second day, list the nine possibilities and draw a diagram of the sample space similar to that of Figure 6.1.
(b) List the points of the sample space that constitute event T that she will take a look at five of the books, event U that she will take a look at more books on the first day than on the second day, and event V that she will take a look at three books on the second day.

6.7   With reference to Exercise 6.6, list the elements of:
(a) $T \cap U$;
(b) $U \cap V$;
(c) $V \cap T'$.

6.8   A small marina has three fishing boats that are sometimes in dry dock for repairs.
(a) Using two coordinates, so that (2, 1), for example, represents the event that two
of the fishing boats are in dry dock and one is rented out for the day, and
(0, 2) represents the event that none of the boats is in dry dock and two are
rented out for the day, draw a diagram similar to that of Figure 6.1 showing the
10 points of the corresponding sample space.
(b) If K is the event that at least two of the boats are rented out for the day, L is
the event that more boats are in dry dock than are rented out for the day, and
M is the event that all the boats that are not in dry dock are rented out for the
day, list the outcomes that comprise each of these events.
(c) Which of the three pairs of events, K and L, K and M, and L and M, are mutually exclusive?

6.9   With reference to Exercise 6.8, list the outcomes that comprise each of the following events and express the events in words:
(a) K';   (b) $L \cap M$.

6.10   To construct sample spaces for experiments where we deal with categorical data,
we often code the various alternatives by assigning them numbers. For instance, if
individuals are asked whether their favorite color is red, yellow, blue, green, brown,
white, purple, or some other color, we might assign these alternatives the codes 1,
2, 3, 4, 5, 6, 7, and 8. If

$$A = \{3, 4\}, \quad B = \{1, 2, 3, 4, 5, 6, 7\}, \quad \text{and} \quad C = \{6, 7, 8\}$$

list the outcomes that comprise each of the following events and also express the
events in words:
(a) B';
(b) $A \cap B$;
(c) $B \cap C'$;
(d) $A \cup B'$.

6.11   Among six applicants for an executive job, A is a college graduate, foreign born,
and single; B is not a college graduate, foreign born, and married; C is a college
graduate, native born, and married; D is not a college graduate, native born, and
single; E is a college graduate, native born, and married; and F is not a college graduate, native born, and married. One of these applicants is to get the job, and the
event that the job is given to a college graduate, for example, is denoted $\{A, C, E\}$.
State in a similar manner the event that the job is given to
(a) a single person;
(b) a native-born college graduate;
(c) a married person who is foreign born.

6.12   Which of the following pairs of events are mutually exclusive? Explain your answers.
(a) A driver getting a ticket for speeding and a ticket for going through a red light.
(b) Being foreign-born and being President of the United States.
(c) A baseball player getting a walk and hitting a home run in the same at bat.
(d) A baseball player getting a walk and hitting a home run in the same game.

6.13   Which of the following pairs of events are mutually exclusive? Explain your answers.
(a) Having rain and sunshine on the 4th of July, 1999.
(b) A person wearing black shoes and green socks.
(c) A person leaving Los Angeles by jet at 11 P.M. and arriving in New York City
on the same day.
(d) A person having a degree from U.C.L.A. and a degree from the University of
Chicago.

6.14   With reference to Figure 6.4, F is the event that a graduate student can speak very good French and S is the event that he is studying at the Sourbonne. Explain in words what events are represented by regions 1, 2, 3, and 4.

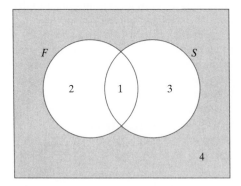

FIGURE 6.4
Venn diagram for
Exercise 6.14.

6.15   With reference to Exercise 6.14, what events are represented by
(a) regions 1 and 2 together;
(b) regions 2 and 4 together.

6.16   With reference to Figure 6.5, D is the event that a flight leaves Denmark on time and H is the event that it arrives in Holland on time. Explain in words what events are represented by regions 1, 2, 3, and 4.

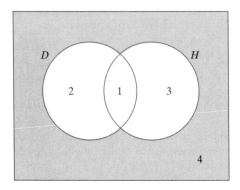

FIGURE 6.5
Venn diagram for
Exercise 6.16.

6.17   With reference to Exercise 6.16, what events are represented by
(a) regions 1 and 3 together;
(b) regions 3 and 4 together;
(c) regions 2, 3, and 4 together?

6.18   Venn diagrams are also useful in determining the numbers of possible outcomes associated with various events. Suppose that one of the 360 members of a golf club is to be chosen Player of the Year. If 224 of the members play at least once a week, 98 of them are lefthanded, and 50 of the lefthanded members play at least once a week, how many of the possible choices would be
(a) lefthanded members who do not play at least once a week;
(b) members who play at least once a week but are not lefthanded;
(c) members who do not play at least once a week and are not lefthanded?

6.19 One of the 200 business majors at a college is to be chosen for the student senate. If 77 of these students are enrolled in a course in accounting, 64 are enrolled in a course in business law, and 92 are not enrolled in either course, how many of the outcomes correspond to the choice of a business major who is enrolled in both courses?

6.20 With reference to Example 6.5 and Figure 6.3, what region or combination of regions represent the events that
(a) unemployment, stock prices, and interest rates will all go up;
(b) stock prices will go up, but unemployment and interest rates will not go up;
(c) unemployment or stock prices, but not interest rates, will go up?

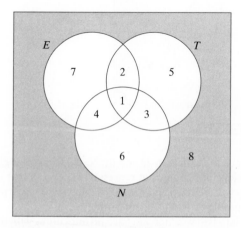

FIGURE 6.6
Venn diagram for
Exercise 6.21.

6.21 In Figure 6.6, E, T, and N are the events that a car brought to a garage needs an engine overhaul, transmission repairs, or new tires. Express in words what events are represented by
(a) region 1;
(b) region 3;
(c) region 7;
(d) regions 1 and 4 together;
(e) regions 2 and 5 together;
(f) regions 3, 5, 6, and 8 together.

6.22 With reference to Exercise 6.21 and Figure 6.6, list the regions or combinations of regions that represent the events that a car brought to the garage will need
(a) transmission repairs, but neither an engine overhaul nor new tires;
(b) an engine overhaul and transmission repairs;
(c) transmission repairs or new tires, but not an engine overhaul;
(d) new tires.

6.23 Suppose that a group of biologists plan a trip to study endangered species in the Lake Victoria region of Africa and that B is the event that they will run into bad weather, P is the event that they will have problems with local authorities, and E is the event that they will have difficulties with their photographic equipment. With reference to the Venn diagram of Figure 6.7, express in words what events are represented by the following regions:
(a) region 2;
(b) region 6;

(c) regions 1 and 2 together;
(d) region 8;
(e) regions 4 and 7 together;
(f) regions 4, 6, 7, and 8 together?

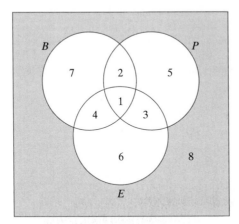

FIGURE 6.7
Venn diagram for
Exercise 6.23.

6.24   With reference to Exercise 6.23 and the Venn diagram of Figure 6.7, list the regions or combinations of regions that represent the events that they will
   (a) run into bad weather and have difficulties with their photographic equipment, but no problems with local authorities;
   (b) not run into bad weather and have no difficulties with their photographic equipment, but problems with local authorities;
   (c) run into bad weather, but have no difficulties with their photographic equipment;
   (d) have no problems with local authorities and no difficulties with their photographic equipment.

6.25   As we pointed out in Exercise 6.18, Venn diagrams are also useful in determining the numbers of outcomes associated with various circumstances. Among 60 houses advertised for sale there are 8 with swimming pools, three or more bedrooms, and wall-to-wall carpeting; 5 with swimming pools, three or more bedrooms, but no wall-to-wall carpeting; 3 with swimming pools, wall-to-wall carpeting, but fewer than three bedrooms; 8 with swimming pools but neither wall-to-wall carpeting nor three or more bedrooms; 24 with three or more bedrooms, but neither a swimming pool nor wall-to-wall carpeting; 2 with three or more bedrooms, wall-to-wall carpeting, but no swimming pool; 3 with wall-to-wall carpeting, but neither a swimming pool nor three or more bedrooms; and 7 without any of these features. If one of these houses is to be chosen for a television commercial, how many outcomes correspond to the choice of
   (a) a house with a swimming pool;
   (b) a house with wall-to-wall carpeting?

6.26   Venn diagrams are often used to verify relationships among sets, subsets, or events, without requiring formal proofs based on the algebra of sets. We simply check whether the expressions that are supposed to be equal are represented by the same region of a Venn diagram. Use Venn diagrams to show that
   (a) $A \cup (A \cap B) = A$;
   (b) $(A \cap B) \cup (A \cap B') = A$;

(c) $(A \cap B)' = A' \cup B'$ and also $(A \cup B)' = A' \cap B'$;

(d) $A \cup B = (A \cap B) \cup (A \cap B') \cup (A' \cap B)$;

(e) $A \cap (B \cup C) = (A \cap B) \cup (A \cap C)$.

6.2   THE POSTULATES OF PROBABILITY

Probabilities always pertain to the occurrence of events, and now that we have learned how to deal mathematically with events, let us turn to the rules that probabilities must obey. To formulate these rules, we continue the practice of denoting events by capital letters and write the probability of event A as $P(A)$, the probability of event B as $P(B)$, and so on. As before, we denote the set of all possible outcomes, the sample space, by the letter S.

Most basic among all the rules of probability are the three **postulates,** which, as we shall state them here, apply when the sample space S is finite. Beginning with the first two, we write

*First two
postulates of
probability*

> 1. *Probabilities are positive real numbers or zero; symbolically, $P(A) \geq 0$ for any event A.*
> 2. *Every sample space has probability 1; symbolically, $P(S) = 1$ for any sample space S.*

To justify these two postulates, as well as the third one that follows, let us show that they are in agreement with the classical probability concept as well as the frequency interpretation. In Section 6.3 we shall see to what extent the postulates are compatible also with subjective probabilities.

The first two postulates are in agreement with the classical probability concept because the fraction $\frac{s}{n}$ is always positive or zero, and for the entire sample space (which includes all n outcomes) the probability is $\frac{s}{n} = \frac{n}{n} = 1$. When it comes to the frequency interpretation, the proportion of the time that an event will occur cannot be a negative number, and one of the outcomes in the sample space must occur 100 percent of the time, that is, with probability 1.

Although a probability of 1 is thus identified with certainty, in actual practice we also assign a probability of 1 to events that are "practically certain" to occur. For instance, we would assign a probability of 1 to the event that at least one person will vote in the next presidential election, even though it is not logically certain. Similarly, we would assign a probability of 1 to the event that not every student entering college in the fall of 2000 will apply for admission to Princeton University.

The third postulate of probability is especially important, but it is not quite as obvious as the other two.

Third postulate of probability

> **3.** *If two events are mutually exclusive, the probability that one or the other will occur equals the sum of their probabilities. Symbolically,*
>
> $$P(A \cup B) = P(A) + P(B)$$
>
> *for any two mutually exclusive events A and B.*

For instance, if the probability that weather conditions will improve during a certain week is 0.62 and the probability that they will remain unchanged is 0.23, then the probability that they will either improve or remain unchanged is $0.62 + 0.23 = 0.85$. Similarly, if the probabilities that a student will get an A or a B in a course are 0.13 and 0.29, then the probability that he or she will get either an A or a B is $0.13 + 0.29 = 0.42$.

To show that the third postulate is also compatible with the classical probability concept, let s_1 and s_2 denote the number of equally likely possibilities that comprise events A and B. Since A and B are mutually exclusive, no two of these possibilities are alike and all $s_1 + s_2$ of them comprise event $A \cup B$. Thus,

$$P(A) = \frac{s_1}{n} \qquad P(B) = \frac{s_2}{n} \qquad P(A \cup B) = \frac{s_1 + s_2}{n}$$

and $P(A) + P(B) = P(A \cup B)$.

Insofar as the frequency interpretation is concerned, if one event occurs, say, 36% of the time, another event occurs 41% of the time, and they cannot both occur at the same time (that is, they are mutually exclusive), then one or the other will occur $36 + 41 = 77\%$ of the time. This is in agreement with the third postulate.

By using the three postulates of probability, we can derive many further rules according to which probabilities must "behave"—some of them are easy to prove and some are not, but they all have important applications. Among the immediate consequences of the three postulates we find that probabilities can never be greater than 1, that an event that cannot occur has probability 0, and that the probabilities that an event will occur and that it will not occur always add up to 1. Symbolically,

Further rules of probability

> $$P(A) \leq 1 \qquad \text{\textit{for any event A}}$$
>
> $$P(\varnothing) = 0$$
>
> $$P(A) + P(A') = 1 \qquad \text{\textit{for any event A}}$$

The first of these results simply expresses the fact that there cannot be more favorable outcomes than there are outcomes, or that an event cannot occur more than 100 percent of the time. The second result expresses the fact that when an event cannot occur there are $s = 0$ favorable outcomes, or that such an event occurs zero percent of the time. In actual practice, we also assign 0 probability to events that are so unlikely that we are "practically certain" they will not occur.

For instance, we would assign 0 probability to the event that a monkey set loose on a typewriter will by chance type Plato's *Republic* word for word without a single mistake. Or, when flipping a coin, we would assign 0 probability to the event that it will land on its edge.

The third result can also be derived from the postulates of probability, and it can easily be seen that it is compatible with the classical probability concept and the frequency interpretation. In the classical concept, if there are s "successes" there are $n - s$ "failures," the corresponding probabilities are $\frac{s}{n}$ and $\frac{n-s}{n}$, and their sum is

$$\frac{s}{n} + \frac{n-s}{n} = \frac{n}{n} = 1$$

In accordance with the frequency interpretation, we can say that if some given investments are successful 22% of the time, then they are not successful 78% of the time, the corresponding probabilities are 0.22 and 0.78, and their sum is 1.

The examples that follow show how the postulates and the rules we gave previously are put to use in actual practice.

EXAMPLE 6.6   If A and B are the events that a consumer magazine will rate a car stereo system good or poor, $P(A) = 0.24$ and $P(B) = 0.35$, determine the following probabilities:

(a) $P(A')$;
(b) $P(A \cup B)$;
(c) $P(A \cap B)$.

Solution

(a) Using the third of the "further rules," we find that $P(A')$, the probability that the consumer magazine will not rate the car stereo system as being good, is equal to $P(A') = 1 - P(A) = 1 - 0.24 = 0.76$. (Besides poor, this might also include very good and very poor.)

(b) Since A and B are mutually exclusive, we can use the third postulate and write $P(A \cup B) = P(A) + P(B) = 0.24 + 0.35 = 0.59$ for the probability that the car stereo system will be rated good or poor.

(c) Since A and B are mutually exclusive, they cannot both occur and, hence, $P(A \cap B) = P(\emptyset) = 0$.   ∎

In problems like this, it often helps to draw a Venn diagram, fill in the probabilities associated with the various regions, and then read the answers directly off the diagram.

EXAMPLE 6.7   If C and D are the events that our family doctor will be in his office at 8 A.M. or at the hospital, $P(C) = 0.48$ and $P(D) = 0.27$, find $P(C' \cap D')$, namely, the probability that he will be neither in his office nor at the hospital.

Solution   Drawing the Venn diagram as in Figure 6.8, we first put a 0 into region 1 since C and D are mutually exclusive events. It follows that the 0.48 probability of event C must go into region 2 and that the 0.27 probability of event D must go into region 3. Hence, the remaining probability, $1 - (0.48 + 0.27) = 0.25$, must go into region 4. Since $C' \cap D'$ is represented by the region outside both circles, namely, region 4, we find that the answer is $P(C' \cap D') = 0.25$.   ■

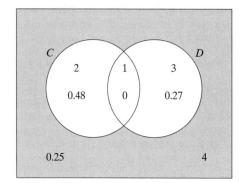

FIGURE 6.8
Venn diagram for Example 6.7.

6.3   PROBABILITIES AND ODDS

If an event is twice as likely to occur than not to occur, we say that the **odds** are 2 to 1 that it will occur; if an event is three times as likely to occur than not to occur, we say that the odds are 3 to 1 that it will occur; if an event is ten times as likely to occur than not to occur, we say that the odds are 10 to 1 that it will occur; and so forth. In general,

> **The odds that an event will occur are given by the ratio of the probability that it will occur to the probability that it will not occur.**

Symbolically,

Formula relating odds to probabilities

> *If the probability of an event is p, the odds for its occurrence are a to b, where a and b are positive values such that*
>
> $$\frac{a}{b} = \frac{p}{1-p}$$

It is customary to express odds as ratios of two positive integers having no common factors. Also, if an event is more likely not to occur than to occur, it is customary to quote the odds that it will not occur rather than the odds that it will occur.

EXAMPLE   6.8   What are the odds for the occurrence of an event if its probability is

(a) $\frac{5}{9}$;
(b) 0.85;
(c) 0.20?

Solution

(a) By definition, the odds are $\frac{5}{9}$ to $1 - \frac{5}{9} = \frac{4}{9}$, or 5 to 4.

(b) By definition, the odds are 0.85 to $1 - 0.85 = 0.15$, 85 to 15, or better, 17 to 3.

(c) By definition, the odds are 0.20 to $1 - 0.20 = 0.80$, 20 to 80, or 1 to 4, or better, the odds against the occurrence of the event are 4 to 1. In general, this is how we quote odds when an event is more likely not to occur than to occur.   ∎

In betting, the term "odds" is also used to denote the ratio of the wager of one party to that of another. For instance, if a gambler says that he will give 3 to 1 odds on the occurrence of an event, he means that he is willing to bet $3 against $1 (or perhaps $30 against $10 or $1,500 against $500) that the event will occur. If such **betting odds** actually equal the odds that the event will occur, we say that the betting odds are **fair.** If a gambler really believes that a bet is fair, then he is, at least in principle, willing to bet on either side. The gambler in this situation would also be willing to bet $1 against $3 (or $10 against $30 or $500 against $1,500) that the event will not occur.

EXAMPLE   6.9   Records show that $\frac{1}{12}$ of the trucks weighed at a certain check point in Nevada carry too heavy a load. If someone offers to bet $40 against $4 that the next truck weighed at this check point will not carry too heavy a load, are these betting odds fair?

Solution   Since the probability is $1 - \frac{1}{12} = \frac{11}{12}$ that the truck will not carry too heavy a load, the odds are 11 to 1, and the bet would be fair if the person offered to bet $44 against $4 that the next truck weighed at the check point will not carry too heavy a load. Thus, the $40 against $4 bet is not fair; it favors the person offering the bet.   ∎

This discussion of odds and betting odds provides the groundwork for a way of measuring **subjective probabilities.** If a businessman "feels" that the odds for the success of a new clothing store are 3 to 2, this means that he is willing to bet (or considers it fair to bet) $300 against $200, or perhaps $3,000 against $2,000, that the new store will be a success. In this way he is expressing his belief regarding the uncertainties connected with the success of the store, and to convert it into a probability we take the equation

$$\frac{a}{b} = \frac{p}{1 - p}$$

with $a = 3$ and $b = 2$ and solve it for p. In general, solving the equation $\frac{a}{b} = \frac{p}{1 - p}$, for p, we get the following result, which the reader will be asked to verify in Exercise 6.49.

Formula relating probabilities to odds

> If the odds are a to b that an event will occur, the probability of its occurrence is
>
> $$p = \frac{a}{a + b}$$

Now, if we substitute $a = 3$ and $b = 2$ into this formula for p, we find that, according to the businessman, the probability of the new clothing store's success is $\frac{3}{3 + 2} = \frac{3}{5}$ or 0.60.

EXAMPLE 6.10   If an applicant for a teaching position feels that the odds are 5 to 3 that she will get the job, what subjective probability is she thus assigning to her getting the job?

Solution   Substituting $a = 5$ and $b = 3$ into the preceding formula for p, we get $p = \frac{5}{5 + 3} = \frac{5}{8}$ or 0.625.   ∎

Let us now see whether subjective probabilities determined in this way "behave" in accordance with the postulates of probability. Since a and b are positive numbers, $\frac{a}{a + b}$ cannot be negative and this satisfies the first postulate. As for the second postulate, observe that the more certain we are that an event will occur, the "better" odds we should be willing to give—say, 99 to 1, 9,999 to 1, or perhaps even 999,999 to 1. The corresponding probabilities are 0.99, 0.9999, and 0.999999, and it can be seen that the more certain we are that an event will occur, the closer its probability will be to 1.

The third postulate does not necessarily apply to subjective probabilities, but proponents of the subjectivist point of view impose it as a **consistency criterion.** In other words, if a person's subjective probabilities "behave" in accordance with the third postulate, he or she is said to be consistent; otherwise, the person's probability judgments must be taken with a grain of salt.

EXAMPLE 6.11   A government economist feels that the odds are 2 to 1 that interest rates will go up before the end of the year, 1 to 5 that they will remain unchanged, and 8 to 3 that they will go up or remain unchanged. Are the corresponding probabilities consistent?

Solution   The corresponding probabilities that interest rates will go up before the end of the year, that they will remain unchanged, and that they will go up or remain unchanged are, respectively,

$$\frac{2}{2 + 1} = \frac{2}{3}, \quad \frac{1}{1 + 5} = \frac{1}{6}, \quad \text{and} \quad \frac{8}{8 + 3} = \frac{8}{11}$$

and since

$$\frac{2}{3} + \frac{1}{6} = \frac{5}{6} \quad \text{and not} \quad \frac{8}{11}$$

the probabilities are not consistent. Hence, the economist's judgment must be questioned.   ■

■ E x e r c i s e s

6.27 In a study of the future needs of a community, C is the event that there will be enough capital for expansion and T is the event that there will be adequate transportation. State in words that probabilities are expressed by
(a) $P(C')$;
(b) $P(T')$;
(c) $P(C \cup T)$;
(d) $P(C \cap T)$.

6.28 With reference to Exercise 6.27, express symbolically the probabilities that there will be
(a) adequate transportation but not enough capital for expansion;
(b) neither enough capital for expansion nor adequate transportation.

6.29 If T is the event that a space shuttle will be launched on schedule and U is the event that the whole mission will have to be scrapped, express symbolically the probabilities that
(a) the space shuttle will not be launched on schedule;
(b) the space shuttle will be launched on schedule or the whole mission will have to be scrapped;
(c) the space shuttle will not be launched on schedule but the whole mission will not have to be scrapped.

6.30 With reference to Exercise 6.29, state in words what probabilities are expressed by
(a) $P(U')$;
(b) $P(T \cap U')$;
(c) $P(T' \cap U)$.

6.31 Which of the postulates of probability are violated by the following statements?
(a) Since their car broke down, the probability that they will be late is -0.40.
(b) The probability that a mineral specimen will contain copper is 0.26 and the probability that it will not contain copper is 0.64.
(c) The probability that a lecture will be entertaining is 0.35, and the probability that it will not be entertaining is four times as large.
(d) The probabilities that a student will spend an evening studying or watching television are, respectively, 0.22 and 0.48, and the probability that it will be one or the other is 0.80.

6.32 Which of the "further" rules on page 146 are violated by the following statements?
(a) The probability that a chemistry experiment will succeed is 0.63 and the probability that it will fail is 0.47.
(b) The chances for a patient's recovery are excellent; in fact, the probability is 1.25.
(c) The probability that two mutually exclusive events will both occur is always equal to 1.

(d) The probability that an event will occur and that it will not occur is always equal to 1.

6.33 Use Venn diagrams to verify that
(a) $P(A \cup B) = P(A) + P(B \cap A')$;
(b) $P(A \cap B) = P(A) - P(A \cap B')$.

6.34 Explain in words why each of the following inequalities must be false:
(a) $P(A \cup B) < P(A)$;
(b) $P(A \cap B) > P(A)$.

6.35 Make up numerical examples in which two events A and B are mutually exclusive and events A' and B'
(a) are not mutually exclusive;
(b) are also mutually exclusive.

6.36 Given $P(L) = 0.33$ and $P(M) = 0.49$, where L and M are mutually exclusive, find
(a) $P(L')$;
(b) $P(M')$;
(c) $P(L \cup M)$;
(d) $P(L' \cap M')$.

6.37 When entering data into a computer, the probability that a student will make at most three mistakes per 1,000 keystrokes is 0.64, and the probability of making anywhere from 4 to 6 mistakes per 1,000 keystrokes is 0.21. Find the probabilities that in 1,000 keystrokes the student will make
(a) at least 4 mistakes;
(b) at most 6 mistakes;
(c) at least 7 mistakes.

6.38 The probabilities that a missile will explode during lift-off or have its guidance system fail in flight are, respectively, 0.0003 and 0.0005. Find the probabilities that the missile will
(a) not explode during lift-off;
(b) explode during lift-off or have its guidance system fail in flight;
(c) not explode during lift-off nor have its guidance system fail in flight.

6.39 Convert each of the following probabilities to odds:
(a) The probability of getting at least two heads in four flips of a balanced coin is $\frac{11}{16}$.
(b) The probability of rolling "7 or 11" with a pair of balanced dice is $\frac{2}{9}$.

6.40 If the probability is 0.35 that a certain stock index will top 10,000 by the end of the year, what are the odds that this will not occur?

6.41 Convert each of the following odds to probabilities:
(a) If three ceramic tiles are randomly chosen from a carton of twelve, of which three have minor blemishes, the odds are 34 to 21 that at least one of them will have minor blemishes.
(b) If we randomly arrange the letters in the word "nest," the odds are 5 to 1 that we will not get a meaningful word in the English language.

6.42 If the odds are better than 7 to 5 that the home team will win its next football game, what is the corresponding probability?

6.43 An entrepreneur claims that the odds are 5 to 1 that profits will increase and 3 to 1 that they will go down. Can both of these odds be right?

6.44 A television producer is willing to bet $1,200 against $1,000, but not $1,500 against $1,000 that a new game show will be a success. What does this tell us about the probability that the producer assigns to the show's success?

6.45 A soccer fan is offered a bet of $15 to his $5 that the United States will lose its first World Cup match. What does this tell us about the subjective probability he assigns to the United States winning this match if
(a) he considers the bet to be fair;
(b) he is unwilling to bet?

6.46 Asked about his political future, a party official replies that the odds are 2 to 1 that he will not run for the House of Representatives and 4 to 1 that he will not run for the Senate. Furthermore, he feels that the odds are 7 to 5 that he will run for one or the other. Are the corresponding probabilities consistent?

6.47 A high school principal feels that the odds are 7 to 5 against her getting a $1,000 raise and 11 to 1 against her getting a $2,000 raise. Furthermore, she feels that it is an even-money bet that she will get one of these raises or the other. Discuss the consistency of the corresponding subjective probabilities.

6.48 Some events are so unlikely that we choose to assign them probabilities of zero. Would you assign zero probabilities to the events that
(a) if you randomly strike the keys, your computer will correctly print Lincoln's Gettysburg Address;
(b) lightning will strike the same tree on four successive days;
(c) thirteen cards randomly dealt from an ordinary deck of 52 playing cards will all be spades?

6.49 Verify algebraically that the equation $\dfrac{a}{b} = \dfrac{p}{1-p}$, solved for p, yields $p = \dfrac{a}{a+b}$.

6.4 ADDITION RULES

The third postulate of probability applies only to two mutually exclusive events, but it can easily be generalized in two ways, so that it will apply to more than two mutually exclusive events and also to two events that need not be mutually exclusive. We say that k events are mutually exclusive if no two of them have any elements in common. In that case, we can repeatedly use the third postulate and, thus, show that

Generalization of Postulate 3

> *If k events are mutually exclusive, the probability that one of them will occur equals the sum of their individual probabilities; symbolically,*
>
> $$P(A_1 \cup A_2 \cup \cdots \cup A_k) = P(A_1) + P(A_2) + \cdots + P(A_k)$$
>
> *for any mutually exclusive events $A_1, A_2, \ldots,$ and A_k.*

Here again, we read $\cup$ as "or."

EXAMPLE 6.12 The probabilities that a person looking for a new car will end up buying a Chevrolet, a Ford, or a Honda are, respectively, 0.17, 0.22, and 0.08. Assuming that she will buy just one car, what is the probability that it will be one of the three kinds?

Solution Since the three possibilities are mutually exclusive, direct substitution into the formula yields $0.17 + 0.22 + 0.08 = 0.47$. ∎

EXAMPLE 6.13   The probabilities that a consumer testing service will rate a new VCR poor, fair, good, very good, or excellent are 0.07, 0.16, 0.34, 0.32, and 0.11. What are the probabilities that it will rate the new VCR

    (a) poor, fair, or good;

    (b) good, very good, or excellent?

Solution   Since the five possibilities are mutually exclusive, direct substitution into the formula yields

    (a) $0.07 + 0.16 + 0.34 = 0.57$;

    (b) $0.34 + 0.32 + 0.11 = 0.77$.   ∎

      The job of assigning probabilities to all possible events connected with a given situation can be very tedious. Indeed, it can be shown that if a sample space has 10 elements (points or outcomes) we can form more than 1,000 different events, and if a sample space has 20 elements we can form more than 1 million.[†] Fortunately, it is seldom necessary to assign probabilities to all possible events (that is, to all possible subsets of a sample space). The following rule, which is a direct application of the preceding generalization of the third postulate of probability, makes it easy to determine the probability of any event on the basis of the probabilities associated with the individual outcomes in a sample space:

Rule for calculating the probability of an event

> *The probability of any event A is given by the sum of the probabilities of the individual outcomes comprising A.*

EXAMPLE 6.14   Referring again to Example 6.1, which dealt with two car salespersons and three 1998 Dodge Ram trucks, suppose that the ten points of the sample space, shown originally in Figure 6.1, have the probabilities shown in Figure 6.9. Find the probabilities that

    (a) between them, the two salespersons will sell two of the trucks;

    (b) the second salesperson will not sell any of the trucks;

    (c) the second salesperson will sell at least two of the trucks.

Solution   **(a)** Adding the probabilities associated with the points $(2, 0)$, $(1, 1)$, and $(0, 2)$, we get $0.08 + 0.06 + 0.04 = 0.18$.

    (b) Adding the probabilities associated with the points $(0, 0)$, $(1, 0)$, $(2, 0)$, and $(3, 0)$, we get $0.44 + 0.10 + 0.08 + 0.05 = 0.67$.

    (c) Adding the probabilities associated with the points $(0, 2)$, $(1, 2)$, and $(0, 3)$, we get $0.04 + 0.02 + 0.09 = 0.15$.   ∎

[†]In general, if a sample space has n elements, we can form 2^n different events. Each element is either included or excluded for a given event, so by the multiplication of choices there are $2 \cdot 2 \cdot 2 \cdot \ldots \cdot 2 = 2^n$ possibilities. Note that $2^{10} = 1,024$ and $2^{20} = 1,048,576$.

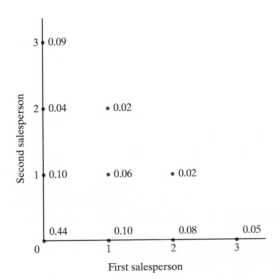

FIGURE 6.9
Sample space for
Example 6.14.

In the special case where the outcomes are all equiprobable, the preceding rule leads to the formula $P(A) = \dfrac{s}{n}$, which we used earlier in connection with the classical probability concept. Here, n is the total number of individual outcomes in the sample space and s is the number of "successes," namely, the number of outcomes comprising event A.

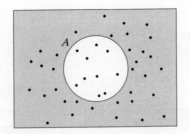

FIGURE 6.10
Sample space with 44
equiprobable
outcomes.

EXAMPLE 6.15   Given that the 44 points (outcomes) in the sample space of Figure 6.10 are all equiprobable, find $P(A)$.

Solution   Counting the number of outcomes in A—there are $s = 10$ among the $n = 44$ in the whole sample space—we find that

$$P(A) = \frac{10}{44} = \frac{5}{22}$$

or approximately 0.23.   ■

Since the third postulate and its generalization apply only to mutually exclusive events, they cannot be used, for example, to find the probability that at least one of two experiments will succeed; the probability that a bird-watcher will spot a roadrunner, a cactus wren, or a Gila woodpecker on a walk in the desert; or the probability that a customer will buy a shirt, a sweater, a belt, or a tie at a department store. Both experiments can succeed, a bird-watcher can spot more than one of these birds, and the customer of the department store can buy any number of these items.

To find a formula for $P(A \cup B)$ that holds regardless of whether events A and B are mutually exclusive, let us consider the Venn diagram of Figure 6.11. It concerns the job applications of a recent M.B.A. graduate, and the letters I and B denote her getting a job offer from an investment broker or from a bank. It follows from the probabilities shown in the Venn diagram that

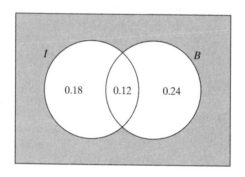

FIGURE 6.11
Venn diagram.

$$P(I) = 0.18 + 0.12 = 0.30$$
$$P(B) = 0.12 + 0.24 = 0.36$$

and

$$P(I \cup B) = 0.18 + 0.12 + 0.24 = 0.54$$

where we could add the respective probabilities because they pertain to mutually exclusive events (nonoverlapping regions of the Venn diagram).

Had we erroneously used the third postulate to calculate $P(I \cup B)$, we would have obtained $P(I) + P(B) = 0.30 + 0.36 = 0.66$, which exceeds the correct value by 0.12. This error results from adding $P(I \cap B) = 0.12$ in twice, once in $P(I) = 0.30$ and once in $P(B) = 0.36$, and we could correct for this by subtracting 0.12 from 0.66. Thus, we could write

$$P(I \cup B) = P(I) + P(B) - P(I \cap B)$$
$$= 0.30 + 0.36 - 0.12 = 0.54$$

and this agrees, as it should, with the result obtained before.

Since the argument used in this example holds for any two events A and B, we can now state the following **general addition rule,** which applies regardless of whether A and B are mutually exclusive events:

General
addition rule

$$P(A \cup B) = P(A) + P(B) - P(A \cap B)$$

When A and B are mutually exclusive, $P(A \cap B) = 0$ and the preceding formula reduces to that of the third postulate of probability. In this connection, the third postulate is also referred to as the **special addition rule.** To add to this terminology, the generalization of the third postulate on page 153 is sometimes referred to as the **generalized** (special) **addition rule.**

EXAMPLE 6.16   If the probabilities are, respectively, 0.27, 0.24, and 0.15 that it will rain in Tucson on a day in mid-August, that there will be a thunderstorm, or that it will rain and there will also be a thunderstorm, what is the probability that it will rain or there will be a thunderstorm on such a day?

Solution   If R denotes rain and T denotes a thunderstorm, we have $P(R) = 0.27$, $P(T) = 0.24$, and $P(R \cap T) = 0.15$. Substituting these values into the formula for the general addition rule, we get

$$P(R \cup T) = P(R) + P(T) - P(R \cap T)$$

$$= 0.27 + 0.24 - 0.15$$

$$= 0.36 \quad \blacksquare$$

EXAMPLE 6.17   In a sample survey conducted in a suburban community, the probabilities are 0.92, 0.53, and 0.48 that a family selected at random will own a family sedan, a recreational vehicle, or both. What is the probability that such a family will own a family sedan, a recreational vehicle, or both?

Solution   Substituting these values into the formula for the general addition rule, we get $0.92 + 0.53 - 0.48 = 0.97$. Note that if we had erroneously used the third postulate of probability in this example, we would have obtained the impossible result $0.92 + 0.53 = 1.45$. $\quad \blacksquare$

The general addition rule can be generalized further so that it will apply to more than two events that need not be mutually exclusive, but we shall not go into that in this book.

■ **E x e r c i s e s**

6.50   A city's police department needs new tires for its patrol cars. The probabilities that it will buy Firestone, Goodyear, Michelin, Goodrich, or Pirelli tires are, respectively, 0.19, 0.26, 0.25, 0.20, and 0.07. Find the probabilities that this police department will buy
   (a) Goodyear or Goodrich tires;
   (b) Firestone or Michelin tires;
   (c) Goodyear, Michelin, or Pirelli tires;

(d) Firestone, Goodyear, Goodrich, or Pirelli tires;

(e) some other kind of tires or no new tires at all.

6.51   The probabilities of rolling a total of 2, 3, 4, ... , 11, or 12 with a pair of balanced dice are, respectively, $\frac{1}{36}, \frac{2}{36}, \frac{3}{36}, \frac{4}{36}, \frac{5}{36}, \frac{6}{36}, \frac{5}{36}, \frac{4}{36}, \frac{3}{36}, \frac{2}{36}$, and $\frac{1}{36}$. What are the probabilities of rolling

(a) a total of 7 or 11;

(b) a total of 2, 3, or 12;

(c) a total that is an odd number, that is 3, 5, 7, 9, or 11?

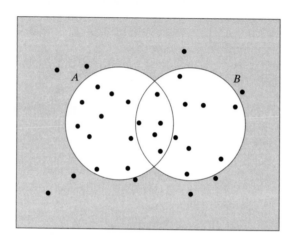

FIGURE 6.12
Sample space for Exercise 6.52.

6.52   If each point in the sample space of Figure 6.12 represents an outcome having the probability $\frac{1}{32}$, find

(a) $P(A)$;

(b) $P(B)$;

(c) $P(A \cap B)$;

(d) $P(A' \cap B')$.

6.53   Sometimes a laboratory assistant has lunch at the cafeteria where she works, sometimes she brings her own lunch, sometimes she has lunch at a nearby restaurant, sometimes she goes home for lunch, and sometimes she skips lunch to lose weight. If the corresponding probabilities are 0.23, 0.31, 0.15, 0.24, and 0.07, find the probabilities that she will

(a) have lunch at the cafeteria or the nearby restaurant;

(b) bring her own lunch, go home for lunch, or skip lunch altogether;

(c) have lunch at the cafeteria or go home for lunch;

(d) not skip lunch to lose weight.

6.54   With reference to Figure 6.10, suppose that each outcome in A is twice as likely as each outcome in A'. Find $P(A)$.

6.55   Figure 6.13 pertains to the number of persons who are invited to attend a conference and the number of persons who actually attend. If the 45 points of the sample space are all equiprobable, what are the probabilities that

(a) at most five persons attend the conference;

(b) at least six persons attend the conference;

(c) one invited person does not attend the conference?

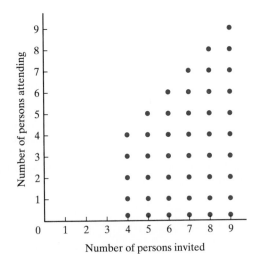

FIGURE 6.13
Sample space for
Exercise 6.55.

6.56   The probabilities that a television station will receive fewer than 10 complaints about a controversial program, fewer than 20 complaints, fewer than 30 complaints, fewer than 40 complaints, or 40 or more complaints are, respectively, 0.10, 0.35, 0.72, 0.95, and 0.05. What are the probabilities that the station will receive
(a) anywhere from 10 to 29 complaints;
(b) anywhere from 20 to 39 complaints;
(c) 30 or more complaints?

6.57   If H stands for heads and T for tails, the 32 possible outcomes for five flips of a coin are HHHHH, HHHHT, HHHTH, HHTHH, HTHHH, THHHH, HHHTT, HHTHT, HHTTH, HTHHT, HTHTH, HTTHH, THHHT, THHTH, THTHH, TTHHH, HHTTT, HTHTT, HTTHT, HTTTH, THHTT, THTHT, THTTH, TTHHT, TTHTH, TTTHH, HTTTT, THTTT, TTHTT, TTTHT, TTTTH, and TTTTT. If all these 32 possibilities are equally likely, what are the probabilities of getting 0, 1, 2, 3, 4, or 5 heads?

6.58   The probabilities that a person convicted of drunk driving will spend a night in jail, have his license revoked, or both are, respectively, 0.68, 0.51, and 0.22. What is the probability that a person convicted of drunk driving will spend a night in jail and/or have his license revoked?

6.59   An auction house has two appraisers of precious jewelry. The probability that the older of the two will not be available is 0.33, the probability that the other one will not be available is 0.27, and the probability that both of them will not be available is 0.19. What is the probability that either or both of them will not be available?

6.60   The probability that a certain movie will get an award for good acting is 0.16, the probability that it will get an award for good directing is 0.27, and the probability that it will get awards for both is 0.11. What is the probability that the movie will get
(a) either or both awards;
(b) only one of the two awards;
(c) neither award?

6.61 A professor feels that the odds are 3 to 2 against his getting a promotion, there is a fifty-fifty chance that he will get a raise, and the odds are 4 to 1 against his getting both. What are the odds that he will get a promotion and/or a raise?

6.62 For married couples living in a suburb, the probabilities that the husband, the wife, or both will vote in a gubernatorial election are, respectively, 0.39, 0.46, and 0.31. What is the probability that either or both will vote in the election?

6.63 Explain why there must be a mistake in each of the following statements:
   (a) The probabilities that the management of a professional basketball team will fire the coach, the general manager, or both are 0.85, 0.49, and 0.27.
   (b) The probabilities that a patient at a hospital will have a high temperature, high blood pressure, or both are 0.63, 0.29, and 0.45.

6.5   CONDITIONAL PROBABILITY

If we ask for the probability that a certain event will occur without specifying the sample space, we may well get different answers and they can all be correct. For instance, if we ask for the probability that a lawyer will make more than $200,000 a year within ten years after passing the bar, we may get one answer that applies to all persons practicing law in the United States, another one that applies to corporation lawyers, another one that applies to lawyers employed by the federal government, another one that applies to lawyers who specialize in divorce cases, and so forth. Since the choice of the sample space is by no means always self-evident, it helps to use the symbol $P(A|S)$ to denote the **conditional probability** of event A relative to the sample space S, or as we often call it "the probability of A given S." The symbol $P(A|S)$ makes it explicit that we are referring to a particular sample space S, and it is generally preferable to the abbreviated notation $P(A)$ unless the tacit choice of S is clearly understood. It is also preferable when we have to refer to different sample spaces in the same problem.

To elaborate on the idea of a conditional probability, suppose that a consumer research organization has studied the service under warranty provided by the 200 tire dealers in a large city, and that their findings are summarized in the following table:

	Good service under warranty	Poor service under warranty	Total
Name-brand tire dealers	64	16	80
Off-brand tire dealers	42	78	120
Total	106	94	200

If one of these tire dealers is randomly selected (that is, each one has the probability $\frac{1}{200}$ of being selected), we find that the probabilities of event N (choosing a name-brand dealer), event G (choosing a dealer who provides good service under warranty), and event $N \cap G$ (choosing a name-brand dealer who provides good service under warranty) are

$$P(N) = \frac{80}{200} = 0.40$$

$$P(G) = \frac{106}{200} = 0.53$$

and

$$P(N \cap G) = \frac{64}{200} = 0.32$$

All these probabilities were calculated by means of the formula $\dfrac{s}{n}$ for equally likely possibilities.

Since the second of these possibilities is particularly disconcerting—there is almost a fifty-fifty chance of choosing a dealer who does not provide good service under warranty—let us see what will happen if we limit the choice to name-brand dealers. This reduces the sample space to the 80 choices corresponding to the first row of the table, and we find that the probability of choosing a name-brand dealer who will provide good service under warranty is

$$P(G|N) = \frac{64}{80} = 0.80$$

This is quite an improvement over $P(G) = 0.53$, as might have been expected. Note that the conditional probability that we have obtained here,

$$P(G|N) = 0.80$$

can also be written as

$$P(G|N) = \frac{\frac{64}{200}}{\frac{80}{200}} = \frac{P(N \cap G)}{P(N)}$$

namely, as the ratio of the probability of choosing a name-brand dealer who provides good service under warranty to the probability of choosing a name-brand dealer.

Generalizing from this example, let us now make the following definition of conditional probability, which applies to any two events A and B belonging to a given sample space S:

Defintion of
conditional
probability

If $P(B)$ is not equal to zero, then the conditional probability of A relative to B, namely, the probability of A given B, is

$$P(A|B) = \frac{P(A \cap B)}{P(B)}$$

When $P(B)$ is equal to zero, the conditional probability of A relative to B is undefined.

EXAMPLE 6.18   With reference to the tire dealers of the preceding illustration, what is the probability that an off-brand dealer will provide good service under warranty?

Solution   As can be seen from the table,

$$P(G \cap N') = \frac{42}{200} = 0.21 \quad \text{and} \quad P(N') = \frac{120}{200} = 0.60$$

so that substitution into the formula yields

$$P(G|N') = \frac{P(G \cap N')}{P(N')} = \frac{0.21}{0.60} = 0.35$$

Of course, we could have obtained this result directly from the second row of the table on page 160 by writing

$$P(G|N') = \frac{42}{120} = 0.35$$ ∎

EXAMPLE 6.19   The probability that a research project will be well planned is 0.60 and the probability that it will be well planned and well executed is 0.54. What is the probability that a research project will be well executed given that it is well planned?

Solution   If A and B are, respectively, the events that an experiment will be well planned and well executed, we get

$$P(B|A) = \frac{P(B \cap A)}{P(A)} = \frac{0.54}{0.60} = 0.90$$

where we interchanged A and B in the formula that defines conditional probabilities and where we made use of the fact that $A \cap B = B \cap A$.   ∎

EXAMPLE 6.20   At a certain elementary school, the probability that a randomly selected student will come from a one-parent home is 0.36 and the probability that he or she will come from a one-parent home and be a low achiever (get mostly D's and F's) is 0.27. What is the probability that such a randomly selected student will be a low achiever given that he or she comes from a one-parent home?

Solution   If we let L denote a low achiever and O a student from a one-parent home, we have $P(O) = 0.36$ and $P(O \cap L) = 0.27$, and we get

$$P(L|O) = \frac{P(O \cap L)}{P(O)} = \frac{0.27}{0.36} = 0.75$$ ∎

The example that follows introduces another concept that is important in the study of probability.

EXAMPLE 6.21   The probability that Henry will like a new movie is 0.70 and the probability that Jean, his girlfriend, will like it is 0.60. If the probability is 0.28 that he will like it and she will dislike it, what is the probability that he will like it given that she is not going to like it?

Solution   If H and J are the events that Henry will like the new movie and that Jean will like it, we have $P(J') = 1 - 0.60 = 0.40$ and $P(H \cap J') = 0.28$, and we get

$$P(H|J') = \frac{P(H \cap J')}{P(J')} = \frac{0.28}{0.40} = 0.70$$ ∎

What is special and interesting about this result is that $P(H)$ and $P(H|J')$ both equal 0.70 and, as the reader will be asked to verify in Exercises 6.81, it follows from the given information that $P(H|J)$ is also equal to 0.70. Thus, the probability of event H is the same regardless of whether or not event J has occurred, occurs, or will occur, and we say that event H is independent of event J. In general, it can easily be verified that when one event is independent of another, the second event is also independent of the first, and we say that the two events are **independent.** When two events are not independent, we say that they are **dependent.**

When giving these definitions, we should have specified that neither H nor J can have zero probability, for in that case some of the conditional probabilities would not even have existed. This is why an alternative definition of independence, which we shall give on page 165, is often preferred.

6.6   MULTIPLICATION RULES

In Section 6.5 we used the formula $P(A|B) = \dfrac{P(A \cap B)}{P(B)}$ only to define and calculate conditional probabilities, but if we multiply by $P(B)$ on both sides of the equation, we get the following formula, which enables us to calculate the probability that two events will both occur:

*General
multiplication rule*

$$P(A \cap B) = P(B) \cdot P(A|B)$$

As we have indicated in the margin, this formula is called the **general multiplication rule,** and it states that the probability that two events will both occur is the product of the probability that one of the events will occur and the conditional probability that the other event will occur given that the first event has occurred, occurs, or will occur. As it does not matter which event is referred to as A and which is referred to as B, the formula can also be written as

General multiplication rule (alternative form)

$$P(A \cap B) = P(A) \cdot P(B \mid A)$$

EXAMPLE 6.22 A panel of jurors consists of 15 persons who have had no education beyond high school and 9 who have had some college education. If a lawyer randomly chooses two of them to ask some questions, what is the probability that neither of them will have had any college education?

Solution If A is the event that the first person selected has not had any college education, then $P(A) = \frac{15}{24}$. Also, if B is the event that the second person picked has not had any college education, it follows that $P(B \mid A) = \frac{14}{23}$, since there are only 14 persons without any college education among the 23 who are left after one person without any college education has been picked. Hence, the general multiplication rule yields

$$P(A \cap B) = P(A) \cdot P(B \mid A) = \frac{15}{24} \cdot \frac{14}{23} = \frac{105}{276}$$

or approximately 0.38. ∎

EXAMPLE 6.23 Suppose that the probability is 0.45 that a rare tropical disease is diagnosed correctly and, if diagnosed correctly, the probability is 0.60 that the patient will be cured. What is the probability that a person who has the disease will be diagnosed correctly and cured?

Solution Using the general multiplication rule, we get $(0.45)(0.60) = 0.27$. ∎

When A and B are independent events, we can substitute $P(A)$ for $P(A \mid B)$ in the first of the two formulas for $P(A \cap B)$, or $P(B)$ for $P(B \mid A)$ in the second, and we obtain

Special multiplication rule (independent events)

If A and B are independent events, then

$$P(A \cap B) = P(A) \cdot P(B)$$

In words, the probability that two independent events will both occur is simply the product of their respective probabilities.

As can easily be shown, it is also true that if $P(A \cap B) = P(A) \cdot P(B)$, then A and B are independent events. Dividing by $P(B)$, we get

$$\frac{P(A \cap B)}{P(B)} = P(A)$$

and then replacing $\dfrac{P(A \cap B)}{P(B)}$ with $P(A|B)$ in accordance with the definition of a conditional probability, we arrive at the result that $P(A|B) = P(A)$, namely, that A and B are independent. Therefore, we can use the special multiplication rule as a *definition* of independence that makes it very easy to check whether two events A and B are independent.

EXAMPLE 6.24   Check for each of the following pairs of events whether they are independent:

   (a) Events A and B for which $P(A) = 0.40$, $P(B) = 0.90$, and $P(A \cap B) = 0.36$.
   (b) Events C and D for which $P(C) = 0.75$, $P(D) = 0.80$, and $P(C \cap D') = 0.15$.
   (c) Events E and F for which $P(E) = 0.30$, $P(F) = 0.35$, and $P(E' \cap F') = 0.40$.

Solution

   (a) Since $(0.40)(0.90) = 0.36$, the two events are independent.
   (b) Since $P(D') = 1 - 0.80 = 0.20$ and $(0.75)(0.20) = 0.15$, the events C and D', and hence also the events C and D, are independent.
   (c) Since $P(E') = 1 - 0.30 = 0.70$, $P(F') = 1 - 0.35 = 0.65$, and $(0.70)(0.65) = 0.455$ and not 0.40, the events E' and F', and hence also the events E and F, are not independent. ∎

The special multiplication rule can easily be generalized so that it applies to the occurrence of three or more independent events—again, we multiply together all the individual probabilities.

EXAMPLE 6.25   If the probability is 0.70 that any one person interviewed at a shopping mall will be against an increase in the sales tax to finance a new football stadium, what is the probability that among four persons interviewed at the mall the first three will be against the increase in the sales tax, but the fourth will not be against it?

Solution   Assuming independence, we multiply the respective probabilities, getting $(0.70)(0.70)(0.70)(0.30) = 0.1029$. ∎

When three or more events are not independent, the multiplication rule becomes more complicated—we form the product of the probability that one of the events will occur, the probability that a second event will occur given that the first event has occurred, the probability that a third event will occur given that the first two events have occurred, and so on.

EXAMPLE 6.26   With reference to Example 6.22, what is the probability that if three of the members of the panel are randomly picked by the lawyer, none of them will have had any college education?

Solution   Since the probabilities are, respectively, $\frac{15}{24}$, $\frac{14}{23}$, and $\frac{13}{22}$ that the first person picked will not have had any college education, that the second person picked will not have had any college education given that the first person picked did not have any college education, and that the third person picked will not have had any college education given that the first two persons picked did not have any college education, it follows that the probability asked for is $\frac{15}{24} \cdot \frac{14}{23} \cdot \frac{13}{22} = \frac{455}{2,024}$, or approximately 0.225. ∎

■ Exercises

6.64   If A is the event that an astronaut is a member of the armed services, T is the event that he was once a test pilot, and W is the event that he is a well-trained scientist, express each of the following probabilities in symbolic form:
(a) the probability that an astronaut who was once a test pilot is a member of the armed services;
(b) the probability that an astronaut who is a member of the armed services is a well-trained scientist;
(c) the probability that an astronaut who is not a well-trained scientist was once a test pilot;
(d) the probability that an astronaut who is not a member of the armed services and was never a test pilot is a well-trained scientist.

6.65   With reference to Exercise 6.64, express in words the probabilities that are represented by
(a) $P(A|W)$;
(b) $P(A'|T')$;
(c) $P(A' \cap W|T)$;
(d) $P(A|W \cap T)$.

6.66   A guidance department tests students in various ways. If I is the event that a student scores high in intelligence, A is the event that a student rates high on a social adjustment scale, and N is the event that a student displays neurotic tendencies, express symbolically the probabilities that
(a) a student who scores high in intelligence will display neurotic tendencies;
(b) a student who does not rate high on the social adjustment scale will not score high in intelligence;
(c) a student who displays neurotic tendencies will neither score high in intelligence nor rate high on the social adjustment scale.

6.67   With reference to Exercise 6.66, express in words what probabilities are represented by
(a) $P(I|A)$;
(b) $P(A'|N')$;
(c) $P(I \cap N|A')$;
(d) $P(N'|I \cap A)$.

6.68   If H is the event that a job has a high starting salary and G is the event that it has a good future, state in words what probabilities are represented by
(a) $P(G|H)$;
(b) $P(H'|G)$;
(c) $P(H|G')$;
(d) $P(G'|H')$.

6.69   Among the 30 applicants for a position at a credit union, some are married and some are not, some have had experience in banking and some have not, with the exact breakdown being

	Married	Single
Some experience	6	3
No experience	12	9

If the branch manager randomly chooses the applicant to be interviewed first, M denotes the event that the first applicant to be interviewed is married, and E denotes the event that the first applicant to be interviewed has had some experience in banking, express in words and also evaluate the following probabilities:

(a) $P(M)$;
(b) $P(E')$;
(c) $P(M \cap E)$;
(d) $P(M' \cap E')$;
(e) $P(E|M)$;
(f) $P(M'|E')$.

6.70 Use the results obtained in Exercise 6.69 to verify that

(a) $P(E|M) = \dfrac{P(M \cap E)}{P(M)}$;

(b) $P(M'|E') = \dfrac{P(M' \cap E')}{P(E')}$.

6.71 Among the 400 inmates of a prison, some are first offenders, some are hardened criminals, some serve terms of less than five years, and some serve longer terms, with the exact breakdown being

	Terms of less than five years	Longer terms
First offenders	120	40
Hardened criminals	80	160

If one of the inmates is to be selected at random to be interviewed about prison conditions, H is the event that he is a hardened criminal, and L is the event that he is serving a longer term, determine each of the following probabilities directly from the entries and the row and column totals of the table:

(a) $P(H)$;
(b) $P(L)$;
(c) $P(L \cap H)$;
(d) $P(H' \cap L)$;
(e) $P(L|H)$;
(f) $P(H'|L)$.

6.72 Use the result obtained in Exercise 6.71 to verify that

(a) $P(L|H) = \dfrac{P(L \cap H)}{P(H)}$;

(b) $P(H'|L) = \dfrac{P(H' \cap L)}{P(L)}$.

6.73 The probability that a bus from Seattle to Vancouver will leave on time is 0.80, and the probability that it will leave on time and also arrive on time is 0.72.
(a) What is the probability that a bus that leaves on time will also arrive on time?
(b) If the probability that such a bus will arrive on time is 0.75, what is the probability that a bus that arrives on time also left on time?

6.74 A survey of women in executive positions showed that the probability is 0.80 that such a woman will enjoy making financial decisions, and 0.44 that such a woman will enjoy making financial decisions and also be willing to assume substantial risks. What is the probability that a woman in an executive position who enjoys making financial decisions will be willing to assume substantial risks?

6.75 The probability that a woman attending a junior college will buy a portable computer is 0.75; if she buys such a computer, the odds are 4 to 1 that her grades will go up. What is the probability that such a student will buy a portable computer and have her grades go up?

6.76 Among 40 pieces of luggage loaded on a bus from Heathrow airport to downtown London, 30 are destined for the Dorchester and 10 are destined for the Savoy. If two of them were stolen from the bus when it stopped at a red light, what is the probability that both of them should have gone to the Dorchester?

6.77 If two cards are drawn at random from an ordinary deck of 52 playing cards, what are the probabilities that they will both be hearts if
(a) the first card is replaced before the second card is drawn;
(b) the first card is not replaced before the second card is drawn?
The distinction between the two parts of this exercise is important in statistics. What we do in part (a) is called **sampling with replacement** and what we do in part (b) is called **sampling without replacement.**

6.78 Strange as it may seem, it is possible for an event A to be independent of two events B and C taken individually, but not when they are taken together. Verify that this is the case in the situation pictured in Figure 6.14 by showing that $P(A|B) = P(A)$, $P(A|C) = P(A)$, but $P(A|B \cup C) \neq P(A)$.

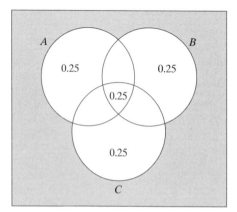

FIGURE 6.14
Venn diagram for
Exercise 6.78.

6.79 If $P(A) = 0.80$, $P(C) = 0.95$, and $P(A \cap C) = 0.76$, are events A and C independent?

6.80 If $P(M) = 0.15$, $P(N) = 0.82$, and $P(M \cap N) = 0.12$, are events M and N independent?

6.81 In Example 6.21 we gave $P(H \cap J') = 0.28$, and accordingly we wrote 0.28 in the region corresponding to $H \cap J'$ in the Venn diagram of Figure 6.15.

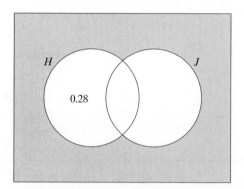

FIGURE 6.15
Venn diagram for
Exercise 6.81.

(a) Making use of the fact that we also gave $P(H) = 0.70$ and $P(J) = 0.60$, fill in the probabilities associated with the other three regions of the Venn diagram of Figure 6.15.

(b) Using the probabilities associated with the four regions of the Venn diagram of Figure 6.15, show that $P(H|J) = 0.70$, which verifies what we said on page 163 about event H being independent of event J.

(c) Using the probabilities associated with the four regions of the Venn diagram of Figure 6.15, show that

$$P(J) = P(J|H) = P(J|H') = 0.60$$

which verifies what we said on page 163 about event J also being independent of event H.

6.82 Using the generalization of the special multiplication rule mentioned on page 165, find

(a) the probability of getting six tails in a row with a balanced coin;

(b) the probability of rolling either a 5 or a 6 in four successive rolls of a die;

(c) the probability of drawing (with replacement) four hearts in a row from an ordinary deck of 52 playing cards;

(d) the probability of first getting four heads and then four tails with a balanced coin;

(e) the probability that a marksman will hit a target five times in a row, given that the probability of his hitting the target on any one try is 0.90 and that we can assume independence.

6.83 As part of an experiment, 80 patients received a new medication for the treatment of ulcers and 20 patients received a placebo. If three of the patients are later chosen for further tests, what is the probability that all three of them in this random sample will originally have received the medication?

6.84 In a Western community, the probability of passing the road test for a driver's license on the first try is 0.75. After that, the probability of passing becomes 0.60 regardless of how many times a person has failed. What is the probability of getting one's license on the fourth try?

6.85   During the month of September, the probability that a rainy day will be followed by another rainy day in a given city is 0.70, and the probability that a sunny day will be followed by a rainy day is 0.40. Assuming that each day is classified as being either rainy or sunny and that the weather on any one day depends only on what happened on the day before, what is the probability that a rainy day will be followed by two more rainy days, then two sunny days, and finally another rainy day?

*6.7 BAYES' THEOREM

Although the symbols $P(A|B)$ and $P(B|A)$ may look alike, there is a great difference between the probabilities that they represent. For instance, on page 161 we calculated the probability $P(G|N)$ that a name-brand tire dealer will provide good service under warranty, but what do we mean when we write $P(N|G)$? This is the probability that a tire dealer who provides good service under warranty is a name-brand dealer. To give another example, suppose that B represents the event that a person committed a burglary and G represents the event that he or she is found guilty of the crime. Then $P(G|B)$ is the probability that the person who committed the burglary will be found guilty of the crime, and $P(B|G)$ is the probability that the person who is found guilty of the burglary actually committed it. Thus, in both of these examples we turned things around—cause, so to speak, became effect and effect became cause.

Since there are many problems in statistics that involve such pairs of conditional probabilities, let us find a formula that expresses $P(B|A)$ in terms of $P(A|B)$ for any two events A and B. To this end we equate the expressions for $P(A \cap B)$ in the two forms of the general multiplication rule on pages 163 and 164 and we get

$$P(A) \cdot P(B|A) = P(B) \cdot P(A|B)$$

and, hence,

$$P(B|A) = \frac{P(B) \cdot P(A|B)}{P(A)}$$

after we divide by $P(A)$.

EXAMPLE 6.27   In a state where cars have to be tested for the emission of pollutants, 25% of all cars emit excessive amounts of pollutants. When tested, 99% of all cars that emit excessive amounts of pollutants will fail, but 17% of the cars that do not emit excessive amounts of pollutants will also fail. What is the probability that a car that fails the test actually emits excessive amounts of pollutants?

Solution   Letting A denote the event that a car fails the test and B the event that it emits excessive amounts of pollutants, we first translate the given percentages into probabilities and write $P(B) = 0.25$, $P(A|B) = 0.99$, and $P(A|B') = 0.17$. To calculate $P(B|A)$ by means of the formula

$$P(B|A) = \frac{P(B) \cdot P(A|B)}{P(A)}$$

we shall first have to determine $P(A)$, and to this end let us look at the tree diagram of Figure 6.16. Here A is reached either along the branch that passes through B or along the branch that passes through B', and the probabilities of this happening are, respectively, $(0.25)(0.99) = 0.2475$ and $(0.75)(0.17) = 0.1275$.

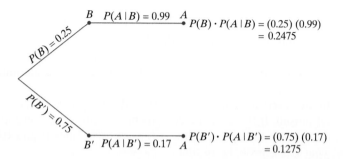

FIGURE 6.16
Tree diagram for emission-testing example.

Since the two alternatives represented by the two branches are mutually exclusive, we find that $P(A) = 0.2475 + 0.1275 = 0.3750$. Thus, substitution into the formula for $P(B|A)$ yields

$$P(B|A) = \frac{P(B) \cdot P(A|B)}{P(A)} = \frac{0.2475}{0.3750} = 0.66$$

This is the probability that a car that fails the test actually emits excessive amounts of pollutants. ∎

With reference to the tree diagram of Figure 6.16, we can say that $P(B|A)$ is the probability that event A was reached via the upper branch of the tree, and we showed that it equals the ratio of the probability associated with that branch of the tree to the sum of the probabilities associated with both branches. This argument can be generalized to the case where there are more than two possible "causes," namely, more than two branches leading to event A. With reference to Figure 6.17 we can say that $P(B_i|A)$ is the probability that event A was reached via the ith branch of the tree (for $i = 1, 2, 3, \ldots,$ or k), and it can be shown that it equals the ratio of the probability associated with the ith branch to the sum of the probabilities associated with all k branches leading to A. Formally, we write

FIGURE 6.17
Tree diagram for Bayes' theorem.

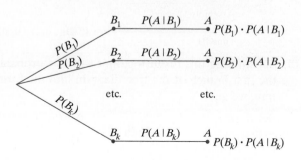

Bayes' theorem

If B_1, B_2, ..., and B_k are mutually *exclusive* events of which one must occur, then

$$P(B_i|A) = \frac{P(B_i) \cdot P(A|B_i)}{P(B_1) \cdot P(A|B_1) + P(B_2) \cdot P(A|B_2) + \cdots + P(B_k) \cdot P(A|B_k)}$$

for $i = 1, 2, ...,$ or k.

Note that the expression in the denominator actually equals $P(A)$. This formula for calculating $P(A)$ when A is reached via one of several intermediate steps is called the **rule of elimination** or the **rule of total probability.**

EXAMPLE 6.28   In a cannery, assembly lines I, II, and III account for 50, 30, and 20% of the total output. If 0.4% of the cans from assembly line I are improperly sealed, and the corresponding percentages for assembly lines II and III are 0.6% and 1.2%, what is the probability that

(a) a can produced by this cannery will be improperly sealed;

(b) an improperly sealed can (discovered at the final inspection of outgoing products) will have come from assembly line I?

Solution

(a) Letting A denote the event that a can is improperly sealed, and B_1, B_2, and B_3 denote the events that a can comes from assembly lines I, II, or III, we can translate the given percentages into probabilities and write $P(B_1) = 0.50$, $P(B_2) = 0.30$, $P(B_3) = 0.20$, $P(A|B_1) = 0.004$, $P(A|B_2) = 0.006$, and $P(A|B_3) = 0.012$. Thus, the probabilities associated with the three branches of the tree diagram of Figure 6.18 are $(0.50)(0.004) = 0.0020$, $(0.30)(0.006) = 0.0018$, and $(0.20)(0.012) = 0.0024$, and the rule of elimination yields $P(A) = 0.0020 + 0.0018 + 0.0024 = 0.0062$.

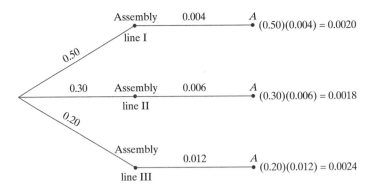

FIGURE 6.18
Tree diagram for
Example 6.28.

(b) Substituting this result together with the probability associated with the first branch of the tree diagram into the formula for Bayes' theorem, we get

$$P(B_1|A) = \frac{0.0020}{0.0062} \approx 0.32 \qquad \blacksquare$$

As can be seen from the two examples of this section, Bayes' formula is a relatively simple mathematical rule. There can be no question about its validity, but criticism has frequently been raised about its applicability. This is because it involves a "backward" or "inverse" sort of reasoning, namely, reasoning from effect to cause. For instance, in Example 6.27 we wondered whether a car's failing the test was brought about, or caused, by its emitting excessive amounts of pollutants. Similarly, in the example immediately preceding we wondered whether an improperly sealed can was produced, or caused, by assembly line I. It is precisely this aspect of Bayes' theorem that makes it play an important role in statistical inference, where our reasoning goes from sample data that are observed to the populations from which they came.

Exercises

*6.86   The probability is 0.60 that a famous Nigerian distance runner will enter the Boston marathon. If he does not enter, the probability that last year's winner will repeat is 0.66, but if he enters, the probability that last year's winner will repeat is only 0.18. What is the probability that last year's winner will repeat?

*6.87   With reference to Exercise 6.86, suppose that we have been out of touch, but hear on the radio that last year's winner won again. What is the probability that the Nigerian distance runner did not enter the race?

*6.88   In a T-maze, a rat is given food if it turns left and an electric shock if it turns right. On the first trial there is a fifty-fifty chance that a rat will turn either way; then, if it receives food on the first trial, the probability that it will turn left on the second trial is 0.68, and if it receives a shock on the first trial, the probability that it will turn left on the second trial is 0.84. What is the probability that a rat will turn left on the second trial?

*6.89   With reference to Exercise 6.88, what is the probability that a rat that turns left on the second trial will also have turned left on the first trial?

*6.90   At an electronics plant, it is known from past experience that the probability is 0.86 that a new worker who has attended the company's training program will meet his production quota, and that the corresponding probability is 0.35 for a new worker who has not attended the company's training program. If 80% of all new workers attend the training program, what is the probability that
(a) a new worker will not meet his production quota;
(b) a new worker who does not meet his production quota will not have attended the company's training program?

*6.91   In Exercise 5.74 we asked the reader to imagine that the probability is 0.95 that a certain test will correctly diagnose a person with diabetes as being diabetic, and that it is 0.05 that the test will incorrectly diagnose a person without diabetes as being diabetic. Given that roughly 10% of the population is diabetic, the reader was asked to guess at the probability that a person diagnosed as being diabetic actually has diabetes. Now use Bayes' theorem to answer this question correctly.

*6.92   With reference to Example 6.27, suppose that the state agency doing the testing has acquired new equipment. With this equipment, 99% of all cars that emit excessive amounts of pollutants will still fail, but only 3% of all cars that do not emit excessive amounts of pollutants will fail the test. If it can still be assumed that 25% of all cars emit excessive amounts of pollutants, what is the probability that a car that fails the emission test with the new equipment will actually emit excessive amounts of pollutants?

*6.93   (From Hans Reichenbach's *The Theory of Probability,* University of California Press, 1949.) Mr. Smith's gardener is not dependable; the probability that he will forget to water the rosebush during Smith's absence is $\frac{2}{3}$. The rosebush is in questionable condition anyhow; if watered, the probability of its withering is $\frac{1}{2}$, but if it is not watered, the probability of its withering is $\frac{3}{4}$. Upon returning, Smith finds that the rosebush has withered. What is the probability that the gardener did not water the rosebush?

*6.94   Two firms V and W consider bidding on a road-building job that may or may not be awarded depending on the amount of the bids. Firm V submits a bid and the probability is $\frac{3}{4}$ that it will get the job provided firm W does not bid. The odds are 3 to 1 that W will bid, and if it does, the probability that V will get the job is only $\frac{1}{3}$.
(a) What is the probability that V will get the job?
(b) If V gets the job, what is the probability that W did not bid?

*6.95   A computer software firm maintains a telephone hotline service for its customers. The firm finds that 48% of the calls involve questions about the application of the software, 38% involve issues of incompatibility with the hardware, and 14% involve the inability to install the software on the user's machine. These three categories of problems can be resolved with probabilities 0.90, 0.15, and 0.80, respectively.
(a) Find the probability that a call to the hotline involves a problem that cannot be resolved.
(b) If a call involves a problem that cannot be resolved, what is the probability that it concerned incompatibility with the hardware?

*6.96   An explosion in a liquefied natural gas tank undergoing repair could have occurred as the result of static electricity, malfunctioning electrical equipment, an open flame in contact with the liner, or purposeful action (industrial sabotage). Interviews with engineers who were analyzing the risks led to estimates that such an explosion would occur with probability 0.25 as a result of static electricity, 0.20 as a result of malfunctioning electric equipment, 0.40 as a result of an open flame, and 0.75 as a result of purposeful action. These interviews also yielded subjective estimates of the probabilities of the four causes of 0.30, 0.40, 0.15, and 0.15, respectively. Based on all this information, what is the most likely cause of the explosion? (From I. Miller, and J. E. Freund, *Probability and Statistics for Engineers,* 3rd ed. Upper Saddle River, NJ: Prentice Hall, Inc., 1985.)

*6.97   To get answers to sensitive questions, we sometimes use a method called the **randomized response technique.** Suppose, for instance, that we want to determine what percentage of the students at a large university smoke marijuana. We construct 20 flash cards, write "I smoke marijuana at least once a week" on 12 of the cards, where 12 is an arbitrary choice, and "I do not smoke marijuana at least once a week" on the others. Then we let each student (in the sample interviewed) select one of the cards at random, and respond "yes" or "no" without divulging the question.
(a) Establish a relationship between $P(Y)$, the probability that a student will give a "yes" response, and $P(M)$, the probability that a student randomly selected at that university smokes marijuana at least once a week.

(b) If 106 of 250 students answered "yes" under these conditions, use the result of part (a) and $\frac{106}{250}$ as an estimate of $P(Y)$ to estimate $P(M)$.

*6.98 The following story illustrates what can happen when we use "common sense," or intuition, in connection with probabilities:

"Among three indistinguishable boxes one contains two pennies, one contains a penny and a dime, and one contains two dimes. We randomly choose one of the three boxes, and randomly pick one of the two coins in that box without looking at the other. If the coin we pick is a penny, what is the probability that the other coin in the box is also a penny? Without giving the matter too much thought, we might argue that there is a fifty-fifty chance that the other coin is also a penny. After all, the coin we picked must have come from the box with the penny and the dime or from the box with the two pennies. In the first case the other coin is a dime, in the second case the other coin is a penny, and it would seem reasonable to say that these two possibilities are equally likely."

Use Bayes' theorem to show that the probability is actually $\frac{2}{3}$ that the other coin is also a penny.

6.8 Checklist of Key Terms *(with page references to their definitions)*

Addition rules, 156, 157
*Bayes' theorem, 172
Betting odds, 149
Complement, 137
Conditional probability, 160, 162
Consistency criterion, 150
Dependent events, 163
Empty set, 136
Event, 136
Experiment, 135
Fair odds, 149
Finite sample space, 136
General addition rule, 156
General multiplication rule, 163, 164
Generalized addition rule, 157
Independent events, 163
Infinite sample space, 136
Intersection, 137

Multiplication rules, 163, 164
Mutually exclusive events, 137
Odds, 148, 149
Outcome, 135
Postulates of probability, 145, 146
*Randomized response technique, 174
*Rule of elimination, 172
*Rule of total probability, 172
Sample space, 135
Sampling with replacement, 168
Sampling without replacement, 168
Special addition rule, 157
Special multiplication rule, 164
Subjective probability, 149
Subset, 136
Theory of probability, 135
Union, 137
Venn diagram, 138

6.9 References

More detailed, though still elementary, treatments of probability may be found in

BARR, D. R., and ZEHNA, P. W., *Probability: Modeling Uncertainty*. Reading, Mass.: Addison-Wesley Publishing Company, Inc., 1983.

DRAPER, N. R., and LAWRENCE, W. E., *Probability: An Introductory Course*. Chicago: Markham Publishing Co., 1970.

FREUND, J. E., *Introduction to Probability*. New York: Dover Publications, Inc., 1993 Reprint.

GOLDBERG, S., *Probability—An Introduction.* Englewood Cliffs, N.J.: Prentice Hall, 1960.

HODGES, J. L., and LEHMANN, E. L., *Elements of Finite Probability,* 2nd ed. San Francisco: Holden-Day, Inc., 1970.

MOSTELLER, F., ROURKE, R. E. K., and THOMAS, G. B., *Probability with Statistical Applications,* 2nd ed. Reading, Mass.: Addison-Wesley Publishing Company, Inc., 1970.

SCHEAFFER, R. L. and MENDENHALL, W., *Introduction to Probability: Theory and Applications.* North Scituate, Mass.: Duxbury Press, 1975.

The following is an introduction to the mathematics of gambling and various games of chance:

PACKEL, E. W., *The Mathematics of Games and Gambling.* Washington, D.C.: Mathematics Association of America, 1981.

7

EXPECTATIONS AND DECISIONS†

7.1   Mathematical Expectation 178

*7.2   Decision Making 184

*7.3   Statistical Decision Problems 188

7.4   Checklist of Key Terms 190

7.5   References 191

*W*hen decisions are made in the face of uncertainty, they are rarely based on probabilities alone; in most cases we must also know something about the potential consequences (profits, losses, penalties, or rewards). If we must decide whether to buy a new car, knowing that our old car will soon require repairs is not enough—to make an intelligent decision we must know, among other things, the cost of the repairs and the trade-in value of the old car. Also, suppose that a building contractor has to decide whether to bid on a construction job that promises a profit of $120,000 with probability 0.20 or a loss of $27,000 (perhaps, due to a strike) with probability 0.80. The probability that the contractor will make a profit is not very high, but on the other hand, the amount he stands to gain is much greater than the amount he stands to lose. Both of these examples demonstrate the need for a

---

†Except for the discussion of mathematical expectation in Section 7.1, the material in this chapter is rarely covered in general introductory courses in statistics. It is included here because of the basic role that decision theory plays in the foundations of statistics. Since there is very little chance that the reader will be exposed to this material elsewhere (except, perhaps, in more advanced work in business statistics), it is hoped that it will not be omitted altogether: Actually, it could be omitted without loss of continuity.

method of combining probabilities and consequences, and this is why we introduce the concept of a **mathematical expection** in Section 7.1.

In Chapters 11 through 18 we deal with many different problems of inference. We estimate unknown quantities, test hypotheses (assumptions or claims), and make predictions, and in problems like these it is essential that, directly or indirectly, we pay attention to the consequences of what we do. After all, if there is nothing at stake, no penalties or rewards, and nobody cares, why not estimate the average weight of gray squirrels as 315.2 pounds, why not accept the claim that by adding water to gasoline we can get 150 miles per gallon with the old family car, and why not predict that by the year 2050 the average person will live to be 200 years old? **WHY NOT, if there is nothing at stake, no penalties or rewards, and nobody cares?** So, in Section 7.2 we give some examples that show how mathematical expectations based on penalties, rewards, and other kinds of payoffs can be used in making decisions, and in Section 7.3 we show how such factors may have to be considered in choosing appropriate statistical techniques.

7.1 MATHEMATICAL EXPECTATION

If a mortality table tells us that in the United States a 50-year-old woman can expect to live 31 more years, this does not mean that anyone really expects a 50-year-old woman to live until her 81st birthday and then die the next day. Similarly, if we read that in the United States a person can expect to eat 104.4 pounds of beef and drink 39.6 gallons of soft drinks per year, or that a child in the age group from 6 to 16 can expect to visit a dentist 2.2 times a year, it must be apparent that the word "expect" is not being used in its colloquial sense. A child cannot go to the dentist 2.2 times, and it would be surprising, indeed, if we found somebody who actually ate 104.4 pounds of beef and drank 39.6 gallons of soft drinks in any given year. So far as 50-year-old women are concerned, some will live another 12 years, some will live another 20 years, some will live another 33 years, . . . , and the life expectancy of "31 more years" will have to be interpreted as an average, or as we call it here, a **mathematical expectation.**

Originally, the concept of a mathematical expectation arose in connection with games of chance, and in its simplest form it is the product of the amount that a player stands to win and the probability that the player will win.

EXAMPLE   7.1   What is the mathematical expectation if a player will win $100 if and only if a balanced coin comes up tails?

Solution   If the coin is balanced and randomly tossed, namely, if the probability of tails is $\frac{1}{2}$, the mathematical expectation is $100 \cdot \frac{1}{2} = \$50$.   ∎

EXAMPLE   7.2   What is the mathematical expectation if someone buys one of 2,000 raffle tickets for a trip to the Grand Canyon worth $640?

Solution   Since the probability of winning the trip is $\dfrac{1}{2,000} = 0.0005$, the mathematical expectation is $640(0.0005) = \$0.32$. Thus, financially speaking, it would be foolish to pay more than 32 cents for one of these tickets, unless the proceeds of the raffle go to a worthy cause or unless one derives a certain pleasure from placing the bet.   ∎

In each of these two examples there is but one prize, yet two possible payoffs—$100 or nothing in Example 7.1 and the trip worth $640 or nothing in Example 7.2. Indeed, in Example 7.2 we could argue that one of the tickets will pay the equivalent of $640, each of the other 1,999 tickets will not pay anything at all, so that, altogether, the 2,000 tickets will pay the equivalent of $640 or on the average $0.32 per ticket. This average is the mathematical expectation.

To generalize the concept of a mathematical expectation, consider the following change in the raffle of Example 7.2:

EXAMPLE   7.3   What is the mathematical expectation if the raffle also awards a second prize consisting of dinner for two at a famous restaurant worth $100 and a third prize of two movie tickets worth $12?

Solution   Now we can argue that one of the tickets will pay the equivalent of $640, another ticket will pay the equivalent of $100, and a third ticket will pay the equivalent of $12, while the other 1,997 tickets will not pay anything at all. Altogether, the 2,000 tickets will thus pay the equivalent of $640 + 100 + 12 = \$752$, or on the average the equivalent of $\dfrac{752}{2,000} = \$0.376$ per ticket. Again, this average is the mathematical expectation.   ∎

Looking at Example 7.3 in a different way, we could argue that if the raffle were repeated many times, a person holding one of the tickets would get nothing $\dfrac{1,997}{2,000} \cdot 100 = 99.85\%$ of the time (or with probability 0.9985) and each of the three prizes $\dfrac{1}{2,000} \cdot 100 = 0.05\%$ of the time (or with probability 0.0005). On the average, a person holding one of the tickets would thus win the equivalent of

$$0(0.9985) + 640(0.0005) + 100(0.0005) + 12(0.0005) \;=\; \$0.376$$

which is the sum of the products obtained by multiplying each payoff by the corresponding proportion or probability. Generalizing from this example leads to the following definition:

Mathematical expectation

> If the probabilities of obtaining the amounts $a_1, a_2, \ldots,$ or a_k are $p_1, p_2, \ldots,$ and p_k, where $p_1 + p_2 + \cdots + p_k = 1$, then the mathematical expectation is
> $$E = a_1 p_1 + a_2 p_2 + \cdots + a_k p_k$$

Each amount is multiplied by the corresponding probability, and the mathematical expectation, E, is given by the sum of all these products. In the Σ notation, $E = \Sigma\, a \cdot p$.

Insofar as the a's are concerned, it is important to keep in mind that they are positive when they represent profits, winnings, or gains (namely, amounts that we receive), and that they are negative when they represent losses, penalties, or deficits (namely, amounts that we have to pay).

EXAMPLE   7.4   What is our mathematical expectation if we win \$25 when a die comes up 1 or 6, and lose \$12.50 when it comes up 2, 3, 4, or 5?

Solution   The amounts are $a_1 = 25$ and $a_2 = -12.5$, and the probabilities are $p_1 = \frac{2}{6} = \frac{1}{3}$ and $p_2 = \frac{4}{6} = \frac{2}{3}$ (if the die is balanced and randomly tossed). Thus, our mathematical expectation is

$$E = 25 \cdot \frac{1}{3} + (-12.5) \cdot \frac{2}{3} = 0 \qquad\blacksquare$$

EXAMPLE   7.5   The probabilities are 0.22, 0.36, 0.28, and 0.14 that an investor will be able to sell a piece of property at a profit of \$2,500, at a profit of \$1,500, at a profit of \$500, or at a loss of \$500. What is the investor's expected profit?

Solution   Substituting $a_1 = 2,500$, $a_2 = 1,500$, $a_3 = 500$, $a_4 = -500$, $p_1 = 0.22$, $p_2 = 0.36$, $p_3 = 0.28$, and $p_4 = 0.14$ into the formula for E, we get

$$E = 2,500(0.22) + 1,500(0.36) + 500(0.28) - 500(0.14)$$
$$= \$\,1,160 \qquad\blacksquare$$

The first of these two examples illustrates what we mean by an **equitable** or **fair game.** It is a game that does not favor either player; namely, a game in which each player's mathematical expectation is zero.

Although we referred to the quantities $a_1, a_2, \ldots,$ and a_k as "amounts," they need not be amounts of money. On page 178 we said that a child in the age group from 6 to 16 can expect to visit a dentist 2.2 times a year. This value is a mathematical expectation, namely, the sum of the products obtained by multiplying $0, 1, 2, 3, 4, \ldots,$ by the corresponding probabilities that a child in that age group will visit a dentist that many times a year.

EXAMPLE   7.6   If the probabilities are 0.06, 0.21, 0.24, 0.18, 0.14, 0.10, 0.04, 0.02, and 0.01 that an airline office at a certain airport will receive 0, 1, 2, 3, 4, 5, 6, 7, or 8 complaints per day about its luggage handling, how many such complaints can it expect per day?

Solution   Substituting into the formula for a mathematical expectation, we get

$$E = 0(0.06) + 1(0.21) + 2(0.24) + 3(0.18) + 4(0.14)$$
$$+ 5(0.10) + 6(0.04) + 7(0.02) + 8(0.01)$$
$$= 2.75$$ ■

In all of the examples of this section we were given the values of a and p (or the values of the a's and p's) and calculated E. Now let us consider an example in which we are given values of a and E to arrive at some result about p, and also an example in which we are given values of p and E to arrive at some result about a.

EXAMPLE 7.7   To defend a client in a liability suit resulting from a car accident, a lawyer must decide whether to charge a straight fee of $7,500 or a contingent fee, which she will get only if her client wins. How does she feel about her client's chances if

(a) she prefers the straight fee of $7,500 to a contingent fee of $25,000;
(b) she prefers a contingent fee of $60,000 to the straight fee of $7,500?

Solution   **(a)** If she feels that the probability is p that her client will win, the lawyer associates a mathematical expectation of $25,000p$ with the contingent fee of $25,000. Since she feels that $7,500 is preferable to this expectation, we can write $7,500 > 25,000p$ and, hence,

$$p < \frac{7,500}{25,000} = 0.30$$

(b) Now the mathematical expectation associated with the contingent fee is $60,000p$, and since she feels that this is preferable to $7,500, we can write $60,000p > 7,500$ and, hence,

$$p > \frac{7,500}{60,000} = 0.125$$ ■

Combining the results of parts (a) and (b) of Example 7.7, we have shown here that $0.125 < p < 0.30$, where p is the lawyer's subjective probability about her client's success. To narrow it down further, we might vary the contingent fee as in Exercises 7.13 and 7.14.

EXAMPLE 7.8   A friend says that he would "give his right arm" for our two tickets to an NBA playoff game. To put this on a cash basis, we propose that he pay us $220 (the actual price of the two tickets), but he will get the tickets only if he draws a jack, queen, king, or ace from an ordinary deck of 52 playing cards; otherwise, we keep the tickets and his $220. What are the two tickets worth to our friend if he feels that this arrangement is fair?

Solution Since there are four jacks, four queens, four kings, and four aces, the probability that our friend will get the two tickets is $\frac{16}{52}$. Hence, the probability that he will not get the tickets is $1 - \frac{16}{52} = \frac{36}{52}$, and the mathematical expectation associated with the gamble is

$$E = a \cdot \frac{16}{52} + 0 \cdot \frac{36}{52} = a \cdot \frac{16}{52}$$

where a is the amount, he feels, the tickets are worth. Putting this mathematical expectation equal to $220, which he considers a fair price to pay for taking the risk, we get

$$a \cdot \frac{16}{52} = 220 \quad \text{and} \quad a = \frac{52 \cdot 220}{16} = \$715$$

This is what the two tickets are worth to our friend. ■

■ E x e r c i s e s

7.1 If a service club has printed and sold 2,000 raffle tickets for a painting worth $500, what is the mathematical expectation of a person who bought one of these tickets?

7.2 With reference to Exercise 7.1, what would have been the mathematical expectation if the service club had sold only 1,250 of the tickets?

7.3 As part of a promotional scheme, a soap manufacturer offers a first prize of $300,000 and a second prize of $100,000 to persons willing to try a new product (distributed without charge) and send in their names on the label. The winners will be drawn at random in front of a television audience.

(a) What would be each entrant's mathematical expectation if 1,500,000 persons were to send in their names?

(b) Would that have made it worth the cost of a first-class postage stamp to send in an entry?

7.4 A jeweler wants to unload 45 men's watches that cost her $12 each. She wraps these 45 watches and also five men's watches that cost her $600 each in identically shaped unmarked boxes and lets each customer take his or her pick.

(a) Find each customer's mathematical expectation.

(b) What is the jeweler's expected profit per customer if she charges $100 for the privilege of taking a pick?

7.5 At the end of a golf tournament paying the winner $300,000 and the runner-up $120,000, two golfers are tied for first place. In the play-off, what are the two golfers' mathematical expectations if

(a) they are evenly matched;

(b) the younger one is favored by odds of 3 to 2?

7.6 A student's parents promise her a gift of $100 if she gets an A in a course in statistics, $50 if she gets a B, and otherwise no reward. What is the student's mathematical expectation if the probabilities of her getting an A or a B are, respectively, 0.32 and 0.40?

7.7   If it is very cold in the Midwest, a guest ranch in Arizona will have 120 guests during the holiday season. If it is cold (but not very cold) they will have 104 guests, and if the weather is moderate they will have only 75 guests. How many guests can they expect if the probabilities for very cold, cold, or moderate weather in the Midwest are, respectively, 0.36, 0.54, and 0.10?

7.8   If the two teams in a "best of seven" play-off are evenly matched, the probabilities that the series will last 4, 5, 6, or 7 games are, respectively, $1/8$, $1/4$, $5/16$, and $5/16$. Under these conditions, how many games can such a series be expected to last?

7.9   An importer pays $12,000 for a shipment of bananas, and the probabilities that she will be able to sell them for $16,000, $13,000, $12,000, or only $10,000 are, respectively, 0.25, 0.46, 0.19, and 0.10. What is her expected gross profit?

7.10   A security service knows from experience that the probabilities of 2, 3, 4, 5, or 6 false alarms on any given evening are 0.12, 0.26, 0.37, 0.18, and 0.07. How many false alarms can they expect on any given evening?

7.11   The probabilities that a person entering a certain department store will make 0, 1, 2, 3, 4, 5, or 6 purchases are, respectively, 0.15, 0.32, 0.27, 0.11, 0.09, 0.04, and 0.02. How many purchases can a person entering the store be expected to make?

7.12   The wage negotiator of a labor union feels that the odds are 3 to 1 that the members of the union will get a $1.00 raise in their hourly wage, 17 to 3 that they will not get a $1.40 raise in their hourly wage, and 9 to 1 that they will not get a $2.00 raise in their hourly wage. What is the corresponding expected raise in the hourly wage?

7.13   With reference to Example 7.7, suppose that the lawyer prefers a straight fee of $7,500 to a contingent fee of $30,000. How does she feel about the chances that her client will win?

7.14   With reference to Example 7.7, suppose that the lawyer prefers a contingent fee of $37,500 to a straight fee of $7,500. How does she feel about the chances that her client will win?

7.15   Mr. Williams is negotiating with his teenage son, Tommy, regarding payment for mowing the lawn during the summer months. Mr. Williams offers his son a choice: $100 for the season or $10 for each time he mows the lawn, as determined by the height of the grass. What does Tommy believe about the anticipated number of mowings if he prefers the second choice?

7.16   One contractor offers to do a road repair job for $45,000, while another contractor offers to do the job for $50,000 with a penalty of $12,500 if the job is not finished on time. If the person who lets out the contract for the job prefers the second offer, what does this tell us about her assessment of the probability that the second contractor will not finish the job on time?

7.17   Mr. Smith feels that it is just about a toss-up whether to accept a cash prize of $26 or to gamble on two flips of a coin, where he is to receive an electric drill if the coin comes up heads both times, while otherwise he is to receive $5. What cash value does he attach to owning the drill?

7.18   Mr. Jones would like to beat Mr. Brown in an upcoming golf tournament, but his chances are nil unless he takes $400 worth of lessons, which (according to the pro at his club) will give him a fifty-fifty chance. If Mr. Jones can expect to break even if he takes the lessons and bets Mr. Brown $1,000 against x dollars that he will win, find x.

*7.2 DECISION MAKING

In the face of uncertainty, mathematical expectations can often be used to great advantage in making decisions. In general, if we must choose between two or more alternatives, it is considered rational to select the one with the "most promising" mathematical expectation—the one that maximizes expected profits, minimizes expected costs, maximizes expected tax advantages, minimizes expected losses, and so on. This approach to decision making has intuitive appeal, but it is not without complications. In many problems it is difficult, if not impossible, to assign numerical values to all of the *a*'s (amounts) and *p*'s (probabilities) in the formula for *E*. Some of these will be illustrated in the examples that follow.

EXAMPLE   7.9   The research division of a pharmaceutical company has already spent $400,000 to determine whether a new medication for the prevention of seasickness is effective. Now the director of the division must decide whether to spend an additional $200,000 to complete the tests, knowing that the probability of success is only $\frac{1}{3}$. He also knows that if the tests are continued and the medication proves to be effective, this would result in a profit of $1,500,000 to his company. Of course, if the tests are continued and the medication turns out to be ineffective, this would mean $600,000 "down the drain." He also knows that if the tests are not continued and the medication is successfully produced by a competitor, this would entail an additional loss of $100,000 due to his company's being put at a competitive disadvantage. What should he decide to do in order to maximize his company's expected profit?

Solution   In problems like this it usually helps to present the given information in a table such as the following:

	Continue tests	*Discontinue tests*
Medication is effective	1,500,000	−500,000
Medication is not effective	−600,000	−400,000

Using all this information and the $\frac{1}{3}$ and $\frac{2}{3}$ probabilities for the effectiveness and ineffectiveness of the medication, the director of the research division can argue that the expected profit is

$$1,500,000 \cdot \frac{1}{3} + (-600,000) \cdot \frac{2}{3} = \$100,000$$

if the tests are continued, and

$$(-500,000) \cdot \frac{1}{3} + (-400,000)\frac{2}{3} \approx -\$433,333$$

if the tests are discontinued. Since an expected profit of $100,000 is preferable to an expected loss of $433,333, the director's decision is to continue the tests. ∎

The way in which we have studied this problem is called a **Bayesian analysis.** In this kind of analysis, probabilities are assigned to the alternatives about which uncertainties exist (the **states of nature,** which in our example were the effectiveness and the ineffectiveness of the medication); then we choose whichever alternative promises the greatest expected profit or the smallest expected loss. As we have said, this approach to decision making is not without complications. If mathematical expectations are to be used for making decisions, it is essential that our appraisals of all relevant probabilities and payoffs are fairly close.

EXAMPLE 7.10   With reference to Example 7.9, suppose that the director of the research division has an assistant who strongly feels that he greatly overestimated the probability of success; namely, that the probability of the medication's effectiveness should be $\frac{1}{15}$ instead of $\frac{1}{3}$. How does this change in the probability of success affect the result?

Solution   With this change, the expected profit becomes

$$1,500,000 \cdot \frac{1}{15} + (-600,000) \cdot \frac{14}{15} = -\$460,000$$

if the tests are continued, and

$$(-500,000) \cdot \frac{1}{15} + (-400,000) \cdot \frac{14}{15} \approx -\$406,667$$

if the tests are discontinued. Neither of these alternatives looks very promising, but since an expected loss of $406,667 is preferable to an expected loss of $460,000, the decision reached in Example 7.9 should be reversed. ∎

EXAMPLE 7.11   Now suppose that the same assistant tells the director of the research division that the anticipated profit of $1,500,000 was also wrong; he feels that it should have been $2,300,000. How will this change affect the result?

Solution   With this change and the $\frac{1}{15}$ probability of success as in Example 7.10, the expected profit becomes

$$2,300,000 \cdot \frac{1}{15} + (-600,000) \cdot \frac{14}{15} = -\$406,667$$

if the tests are continued, and it is also $-\$406,667$ if the tests are discontinued, exactly as in Example 7.10. Again, the result has changed; now it seems that the decision might be left to the toss of a coin. ∎

■ Exercises

*7.19   A grab-bag contains 5 packages worth $1 apiece, 5 packages worth $3 apiece, and 10 packages worth $5 apiece. Is it rational to pay $4 for the privilege of selecting one of these packages at random?

*7.20   A contractor must choose between two jobs. The first promises a profit of $120,000 with a probability of $\frac{3}{4}$ or a loss of $30,000 (due to strikes and other delays) with a probability of $\frac{1}{4}$; the second job promises a profit of $180,000 with a probability of $\frac{1}{2}$ or a loss of $45,000 with a probability of $\frac{1}{2}$. Which job should the contractor choose so as to maximize his expected profit?

*7.21   A landscape architect must decide whether to bid on the landscaping of a public building. What should she do if she figures that the job promises a profit of $10,800 with probability of 0.40 or a loss of $7,000 (due to a lack of rain or perhaps an early frost) with probability 0.60, and it is not worth her time unless the expected profit is at least $1,000?

*7.22   A truck driver has to deliver a load of building materials to one of two construction sites, a barn that is 18 miles from the lumberyard or a shopping center that is 22 miles from the lumberyard. He has misplaced the order indicating where the load should go. Also, he must return to the lumberyard after the delivery. The barn and shopping center are 8 miles apart. To complicate matters, the telephone at the lumberyard is not working. If the driver feels that the probability is $\frac{1}{6}$ that the load should go to the barn and $\frac{5}{6}$ that the load should go to the shopping center, where should he go first so as to minimize the distance that he will have to drive?

*7.23   With reference to Exercise 7.22, where should the driver go first so as to minimize the expected driving distance if instead of $\frac{1}{6}$ and $\frac{5}{6}$ the probabilities are
(a) $\frac{1}{3}$ for the barn and $\frac{2}{3}$ for the shopping center;
(b) $\frac{1}{4}$ for the barn and $\frac{3}{4}$ for the shopping center?

*7.24   The management of a mining company must decide whether to continue an operation at a certain location. If they continue and are successful, they will make a profit of $4,500,000; if they continue and are not successful, they will lose $2,700,000; if they do not continue but would have been successful if they had continued, they will lose $1,800,000 (for competitive reasons); and if they do not continue and would not have been successful if they had continued, they will make a profit of $450,000 (because funds allocated to the operation remain unspent). What decision would maximize the company's expected profit if it is felt that there is a fifty-fifty chance for success?

*7.25   With reference to Exercise 7.24, show that it does not matter what they decide to do if it is felt that the probabilities for and against success are $\frac{1}{3}$ and $\frac{2}{3}$.

*7.26   A group of investors must decide whether to arrange for the financing to build a new arena or to continue holding its sports promotions at the gymnasium of a community college. They figure that if the new arena is built and they can get a professional basketball franchise, there will be a profit of $2,050,000 over the next five years; if the new arena is built and they cannot get a professional basketball franchise, there will be a deficit of $500,000; if the new arena is not built and they get a professional basketball franchise, there will be a profit of $1,000,000; and if the new arena is not built and they cannot get a professional basketball franchise, there will be a profit of only $100,000 from their other promotions.
(a) Present all this information in a table like that on page 184.

(b) If the investors believe an official of the professional basketball league who tells them that the odds are 2 to 1 against their getting the franchise, what should they decide to do so as to maximize the expected profit over the next five years?

(c) If the investors believe the sports editor of a local newspaper who tells them that the odds are really only 3 to 2 against their getting the franchise, what should they decide to do so as to maximize the expected profit over the next five years?

*7.27 In the absence of any information about relevant probabilities, a pessimist may well try to minimize the maximum loss or maximize the minimum profit, that is, use the **minimax** or **maximin** criterion.

(a) With reference to Example 7.9, suppose that the director of the research division of the pharmaceutical company has no idea about the probability that the medication will be effective. What decision would minimize the maximum loss?

(b) With reference to Exercise 7.22, suppose that the truck driver has no idea about the chances that the building materials should go to either of the two construction sites. Where should he go first so as to minimize the maximum distance he has to drive?

*7.28 In the absence of any information about relevant probabilities, an optimist may well try to minimize the minimum loss or maximize the maximum profit, that is, use the **minimin** or **maximax** criterion.

(a) With reference to Example 7.10, suppose that the assistant of the director of research also has no idea about the probability that the medication will be effective. What should he recommend so as to maximize the maximum profit?

(b) With reference to Exercise 7.22, suppose that the truck driver has no idea about the probabilities that the building material should go to one of the construction sites or the other. Where should he go first so as to minimize the minimum distance he has to drive?

*7.29 With reference to Exercise 7.24, suppose that there is no information about the potential success of the mining operation. What should the CEO recommend to the board of directors, if he is always

(a) very optimistic;

(b) very pessimistic?

*7.30 With reference to Exercise 7.26, suppose that there is no information about the group's chances of getting the franchise. Would one of the investors vote for or against building the new arena if she is

(a) a confirmed optimist;

(b) a confirmed pessimist?

*7.31 There are situations where the various criteria we have discussed are outweighed by special considerations.

(a) With reference to Example 7.9, what should the director of the research division decide to do if he knew that the pharmaceutical company would go out of business unless it can make a profit of at least $1,000,000 on the new medication for the prevention of seasickness?

(b) With reference to Exercise 7.26, how should the investors vote on the construction of the new arena if they knew that the group cannot afford a deficit of more than $300,000?

*7.3 STATISTICAL DECISION PROBLEMS

Modern statistics, with its emphasis on inference, may be looked upon as the art, or science, of decision making under uncertainty. This approach to statistics, called **decision theory,** dates back only to the middle of this century and the publication of John von Neumann and Oscar Morgenstern's *Theory of Games and Economic Behavior* in 1944 and Abraham Wald's *Statistical Decision Functions* in 1950. Since the study of decision theory is quite complicated mathematically, we limit our discussion here to an example in which the method of Section 7.2 is applied to a problem that is of a statistical nature.

EXAMPLE 7.12   On the five teams appointed by the government to study gender discrimination in business, 1, 2, 5, 1, and 6 of the members are women. The teams are randomly assigned to various cities, and the mayor of one city hires a consultant to predict how many of the members on the team sent to her city will be women. If the consultant is paid \$300 plus a bonus of \$600, which he will receive only if his prediction is correct (that is, exactly on target), what prediction maximizes the amount of money he can expect to get?

Solution   If the consultant's prediction is 1, which is the mode of the five numbers, he will make \$300 with probability $\frac{3}{5}$ and \$900 with probability $\frac{2}{5}$. So he can expect to make

$$300 \cdot \frac{3}{5} + 900 \cdot \frac{2}{5} = \$540$$

As can easily be verified, that is the best he can do. If his prediction is 2, 5, or 6, he can expect to make

$$300 \cdot \frac{4}{5} + 900 \cdot \frac{1}{5} = \$420$$

and for any other prediction his expectation is only \$300.   ∎

This example illustrates the (perhaps obvious) fact that if one has to pick the exact value on the nose and there is no reward for being close, the best prediction is the mode. To illustrate further how the consequences of one's decisions may dictate the choice of a statistical method of decision or prediction, let us consider the following variation of Example 7.12.

EXAMPLE 7.13   Suppose that the consultant is paid \$600 minus an amount of money equal in dollars to 40 times the magnitude of the error. What prediction will maximize the amount of money he can expect to get?

Solution   Now it is the median that yields the best predictions. If the consultant's prediction is 2, that is, the median of 1, 1, 2, 5, and 6, the magnitude of the error will be 1, 0, 3, or 4, depending on whether 1, 2, 5, or 6 of the members of the team sent to the city will be women. Consequently, he will get $560, $600, $480, or $440 with probabilities $\frac{2}{5}$, $\frac{1}{5}$, $\frac{1}{5}$, and $\frac{1}{5}$, and he can expect to make

$$560 \cdot \frac{2}{5} + 600 \cdot \frac{1}{5} + 480 \cdot \frac{1}{5} + 440 \cdot \frac{1}{5} = \$528$$

It can be shown that the consultant's expectation is less than $528 for any value other than 2, but we shall verify this only for the mean of the five numbers, which is 3. In that case, the magnitude of the error will be 2, 1, 2, or 3, depending on whether 1, 2, 5, or 6 of the members of the team sent to her city will be women. Thus, the consultant will get $520, $560, $520, or $480 with probabilities $\frac{2}{5}$, $\frac{1}{5}$, $\frac{1}{5}$, and $\frac{1}{5}$, and he can expect to make

$$520 \cdot \frac{2}{5} + 560 \cdot \frac{1}{5} + 520 \cdot \frac{1}{5} + 480 \cdot \frac{1}{5} = \$520 \qquad \blacksquare$$

The mean comes into its own right when the penalty, the amount subtracted, increases more rapidly with the size of the error; namely, when it is proportional to its square.

EXAMPLE 7.14   Suppose that the consultant is paid $600 minus an amount of money equal in dollars to 20 times the square of the error. What prediction will maximize the amount of money that he can expect to get?

Solution   If the consultant's prediction is $\dfrac{1 + 2 + 1 + 5 + 6}{5} = 3$, the squares of the errors will be 4, 1, 4, or 9 depending on whether 1, 2, 5, or 6 of the members of the team sent to the city will be women. Correspondingly, the consultant will get $520, $580, $520, or $420 with probabilities $\frac{2}{5}$, $\frac{1}{5}$, $\frac{1}{5}$, and $\frac{1}{5}$, and he can expect to make

$$520 \cdot \frac{2}{5} + 580 \cdot \frac{1}{5} + 520 \cdot \frac{1}{5} + 420 \cdot \frac{1}{5} = \$512$$

As can be verified, the consultant's expectation is less than $512 for any other prediction (see Exercise 7.32). $\qquad \blacksquare$

This third case is of special importance in statistics, as it ties in closely with the *method of least squares*. We shall study this method in Chapter 16, where it is used in fitting curves to observed data, but besides this it has other important applications in the theory of statistics. The idea of working with the squares of the errors is justified on the grounds that in actual practice the seriousness of an error often increases rapidly with the size of the error, more rapidly than the magnitude of the error itself.

The greatest difficulty in applying the methods of this chapter to realistic problems in statistics is that we seldom know the exact values of all the risks that are involved; that is, we seldom know the exact values of the "payoffs" corresponding to the various eventualities. For instance, if the Food and Drug Administration must decide whether or not to release a new drug for general use, how can it put a cash value on the damage that might be done by not waiting for a more thorough analysis of possible side effects or on the lives that might be lost by not making the drug available to the public right away? Similarly, if a faculty committee must decide which of several applicants should be admitted to a medical school or, perhaps, receive a scholarship, how can they possibly foresee all the consequences that might be involved?

The fact that we seldom have adequate information about relevant probabilities also provides obstacles to finding suitable decision criteria; without them, is it reasonable to base decisions, say, on optimism or pessimism, as we did in Exercises 7.28 and 7.27? Questions like this are difficult to answer, but their analysis serves the important purpose of revealing the logic that underlies statistical thinking.

Exercises

*7.32 With reference to Example 7.14, where the consultant is paid $600 minus an amount of money equal in dollars to 20 times the square of the error, what can the consultant expect to make if
(a) his prediction is 1, the mode;
(b) his prediction is 2, the median?

(In general, it can be shown that for any set of numbers $x_1, x_2, \ldots,$ and x_n, the quantity $\Sigma(x - k)^2$ is smallest when $k = \bar{x}$. In this case, the amount subtracted from $600 is smallest when the prediction is $\bar{x} = 3$.)

*7.33 The ages of the seven entries in an essay contest are 17, 17, 17, 18, 20, 21, and 23, and their chances of winning are all equal. If we want to predict the age of the winner and there is a reward for being right, but none for being close, what prediction maximizes the expected reward?

*7.34 With reference to Exercise 7.33, what prediction maximizes the expected reward if
(a) there is a penalty proportional to the size of the error;
(b) there is a penalty proportional to the square of the error?

*7.35 Some of the used cars on a lot are priced at $1,895, some are priced at $2,395, some are priced at $2,795, and some are priced at $3,495. If we want to predict the price of the car that will be sold first, what prediction would minimize the maximum size of the error? What is the name of this statistic, which is mentioned in one of the exercises in Chapter 3?

7.4 Checklist of Key Terms (with page references to their definitions)

*Bayesian analysis, 185
*Decision theory, 188
Equitable game, 180
Fair game, 180
Mathematical expectation, 178, 180

*Maximax criterion, 187
*Maximin criterion, 187
*Minimax criterion, 187
*Minimin criterion, 187
*States of nature, 185

7.5 References

More detailed treatments of the subject matter of this chapter may be found in

BROSS, I. D. J., *Design for Decision*. New York: Macmillan Publishing Co., Inc., 1953.

JEFFREY, R. C., *The Logic of Decision*. New York: McGraw-Hill Book Company, 1965.

and in some textbooks on business statistics. Some fairly elementary material on decision theory can be found in

CHERNOFF, H., and MOSES, L. E. *Elementary Decision Theory*. New York: Dover Publications, Inc., 1987 reprint.

"The Dowry Problem" and "A Tie Is Like Kissing Your Sister" are two amusing examples of decision making given in

HOLLANDER, M., and PROSCHAN, F., *The Statistical Exorcist: Dispelling Statistics Anxiety*. New York: Marcel Dekker, Inc., 1984.

REVIEW EXERCISES for Chapters 5, 6, and 7

R.48 With reference to Figure R.1, express symbolically the events that are represented by regions 1, 2, 3, and 4 of the Venn diagram.

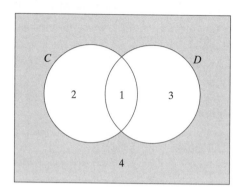

FIGURE R.1
Venn diagram for
Exercises R.48, R.49,
and R.50.

R.49 With reference to Figure R.1, assign the probabilities 0.48, 0.12, 0.32, and 0.08 to the events represented by regions 1, 2, 3, and 4 of the Venn diagram. Find
(a) $P(C)$;
(b) $P(D')$;
(c) $P(C \cup D)$;
(d) $P(C \cap D')$.

R.50 With reference to Exercise R.49, use the results to check whether events C and D' are independent.

R.51 The probabilities that a newspaper will receive $0, 1, 2, \ldots, 7$, or at least 8 letters to the editor about an unpopular decision of the school board are $0.01, 0.02, 0.05, 0.14,$ $0.16, 0.20, 0.18, 0.15,$ and 0.09. What are the probabilities that the newspaper will receive

(a) at most 4 letters to the editor about the school board decision;
(b) at least 6;
(c) from 3 to 5?

R.52 Determine whether each of the following is true or false:
(a) $\dfrac{1}{4!} + \dfrac{1}{6!} = \dfrac{31}{6!}$;

(b) $\dfrac{20!}{17!} = 20 \cdot 19 \cdot 18 \cdot 17$;

(c) $5! + 6! = 7 \cdot 5!$;

(d) $3! + 2! + 1! = 6$.

R.53 A small real estate office has five part-time salespersons. Using two coordinates, so that $(3, 1)$, for example, represents the event that three of the salespersons are at work and one of them is busy with a customer, and $(2, 0)$ represents the event that two of the salespersons are at work but neither of them is busy with a customer, draw a diagram similar to that of Figure 6.1, showing the 21 points of the corresponding sample space.

R.54 With reference to Exercise R.53, assume that each of the 21 points of the sample space has the probability $\frac{1}{21}$. Find the probabilities that
(a) at least three salespersons are at work;
(b) at least three salespersons are busy with customers;

(c) none of the salespersons are busy with customers;

(d) Only one salesperson at work is not busy with customers.

R.55 One substitute quarterback is willing to give odds of 3 to 1, but not odds of 4 to 1, that he will be able to beat another substitute quarterback in the 40-yard dash. What does this tell us about the probability he assigns to his being faster than the other substitute quarterback?

*R.56 The manufacturer of a new battery additive has to decide whether to sell his product for $1.00 a can or for $1.25 with a "double-your-money-back-if-not-satisfied guarantee." How does he feel about the chances that a person will actually ask for double his or her money back if

(a) he decides to sell the product for $1.00;

(b) he decides to sell the product for $1.25 with the guarantee;

(c) he cannot make up his mind?

R.57 In how many different ways can a person buy half a pound each of four of the 15 kinds of coffee carried by a gourmet food shop?

R.58 Suppose that someone flips a coin 100 times and gets 34 heads, which is far short of the number of heads he might expect. Then he flips the coin another 100 times and gets 46 heads, which is again short of the number of heads he might expect. Comment on his claim that the law of large numbers is letting him down.

R.59 If 1,134 of the 1,800 students attending a small college are residents of the community in which the college is located, estimate the probability that any one student attending the college, chosen at random, will be a resident of the community in which the college is located.

R.60 On her low-cholesterol diet, a person is allowed to eat four eggs in three weeks, with no more than two eggs in any one week. Draw a tree diagram to show the various ways in which he or she can plan to distribute the four eggs among three weeks.

R.61 As part of a promotional scheme in Arizona and New Mexico, a company distributing frozen foods will award a grand prize of $100,000 to some person sending in his or her name on an entry blank, with the option of including a label from one of the company's products. A breakdown of the 225,000 entries received is shown in the following table:

	With label	*Without label*
Arizona	120,000	42,000
New Mexico	30,000	33,000

If the winner of the grand prize is chosen by lot, but the drawing is rigged so that by including a label the probability of winning the grand prize is tripled, what are the probabilities that the grand prize will be won by someone who

(a) included a label;

(b) is from New Mexico?

*R.62 The mortgage manager for a bank figures that if an applicant for a $150,000 home mortgage is a good risk and the bank accepts him, the bank's profits will be $8,000. If the applicant is a bad risk and the bank accepts him, the bank will lose $20,000. If the manager turns down the applicant, there will be no profit or loss either way. What should the manager do if

(a) he wishes to maximize the expected profit and feels that the probability is 0.10 that the applicant is a bad risk;

(b) he wishes to maximize the expected profit and feels that the probability is 0.30 that the applicant is a bad risk;

(c) he wishes to minimize the maximum loss and has no idea about the probability that the applicant is a bad risk?

R.63 A test consists of eight true-false questions and four multiple-choice questions, with each having four different answers. In how many different ways can a student check off one answer to each question?

R.64 Among 1,200 women interviewed, 972 said that they prefer being called "Mrs. or Miss" rather than "Ms." Estimate the probability that a woman prefers being called "Mrs. or Miss" rather than "Ms."

R.65 A donut shop has 12 jelly donuts on hand and 15 donuts with chocolate icing. If the first customer buys one donut and the second customer buys one of each kind, how many choices does the second customer have if

(a) the first customer buys a jelly donut;

(b) the first customer buys a donut with chocolate icing?

R.66 A hotel gets cars for its guests from three rental agencies, 20% from agency X, 40% from agency Y, and 40% from agency Z. If 14% of the cars from X, 4% from Y, and 8% from Z need tune-ups, what is the probability that

(a) a car needing a tune-up will be delivered to one of the guests;

(b) if a car needing a tune-up is delivered to one of the guests, it came from agency X?

R.67 If the probability is 0.24 that any one sportswriter will give the University of Nebraska's football team a preseason ranking of No. 1, what is the probability that three sportswriters, chosen at random, will all rank the University of Nebraska football team as No. 1?

R.68 If Q is the event that a person is qualified for a job and G is the event that he or she will get the job, express in words what probabilities are represented by

(a) $P(Q')$;   (c) $P(G \mid Q)$;

(b) $P(G' \mid Q')$;   (d) $P(Q' \mid G)$.

R.69 Convert each of the following odds to probabilities.

(a) The odds are 21 to 3 that a certain driver will not win the Indiannapolis 500.

(b) If four cards are drawn with replacement from an ordinary deck of 52 playing cards, the odds are 11 to 5 that at most two of them will be black.

R.70 The probabilities that a towing service will receive 0, 1, 2, 3, 4, 5, or 6 calls for help during the evening rush hour are 0.05, 0.12, 0.31, 0.34, 0.12, 0.05, and 0.01. How many calls for help can the towing service expect during the evening rush hour?

R.71 If $P(A) = 0.37$, $P(B) = 0.25$, and $P(A \cup B) = 0.62$, are events A and B
(a) mutually exclusive;
(b) independent?

R.72 Explain why there must be a mistake in each of the following:
(a) $P(A) = 0.53$ and $P(A \cap B) = 0.59$.
(b) $P(C) = 0.83$ and $P(C') = 0.27$.
(c) For the independent events E and F, $P(E) = 0.60$, $P(F) = 0.15$, and
$P(E \cap F) = 0.075$.
(d) If $P(G) = 0.40$ and $P(G \cap H) = 0.30$, then $P(G \mid H) = 0.75$.

R.73 Mrs. Jones feels that it is a toss-up whether to accept $30 in cash or to gamble on drawing a bead from an urn containing 15 red beads and 45 blue beads, with the provision that she is to receive $3 if she draws a red bead or a bottle of fancy perfume if she draws a blue bead. What value, or utility, does she assign to the bottle of perfume?

R.74 An artist feels that she can get $5,000 for one of her paintings on display at an art show if it wins a prize, but only $2,000 if it does not win a prize. How does she rate her chances of its winning a prize if she decides to sell the painting for $3,000 before the prize winners are announced?

*R.75 The kinds of cars considered for the fleet of a cab company average 32, 30, 30, 33, and 30 miles per gallon. Assuming that each of the five kinds of cars has an equal chance of being selected, the cab company's accountant wants to prepare the company's budget. What figure would he use for the cars' mileage per gallon if
(a) it is much more important to him to be right rather than to be close;
(b) he wants to minimize the square of his error?

R.76 The storage room of a medical laboratory has five humidity settings and eight temperature settings. In how many different ways can these two variables be set?

R.77 How many different permutations are there of the letters in the word
(a) cheap;
(b) helmet;
(c) little?

R.78 The faculty of an industrial design department of an engineering school includes three with Ph.D.'s, two with M.A.'s, and five with M.F.A.'s. If three of them are randomly selected to serve on a curriculum committee, find the probabilities that the committee will include
(a) one with each kind of degree;
(b) only those with M.F.A.'s;
(c) one with a Ph.D. and two with M.F.A.'s.

R.79 A movie producer feels that the odds are 5 to 1 that his new picture will not get an X rating, 8 to 1 that it will not get an R rating, and 2 to 1 that it will get neither an X rating nor an R rating. Are the corresponding probabilities consistent?

R.80 In Figure R.2, B is the event that a person traveling in the South Pacific will visit Bora Bora, M is the event that he will visit Moorea, and T is the event that he will visit Tahiti. Explain in words what events are represented by the following regions or combinations of regions of the Venn diagram:

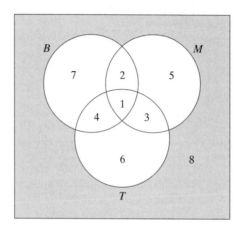

FIGURE R.2 Venn diagram for Exercise R.80.

(a) region 4;

(b) regions 1 and 3;

(c) regions 3 and 6;

(d) regions 2, 5, 7, and 8.

R.81 Lie detectors have been used during wartime to uncover security risks. As is well known, lie detectors are not infallible. Let us suppose that the probability is 0.10 that the lie detector will fail to detect a person who is a security risk and that the probability is 0.08 that the lie detector will incorrectly label a person who is not a security risk. If 2% of the persons who are given the test are actually security risks, what is the probability that

(a) a person labeled a security risk by a lie detector is in fact a security risk;

(b) a person cleared by a lie detector is in fact not a security risk?

R.82 One of two partners in a salvage operation feels that the odds are at most 3 to 1 that they have located the right shipwreck, and the other partner feels that the odds are at least 13 to 7. Find betting odds that would be agreeable to both partners. (Note that the answer is not unique.)

R.83 Sometimes we may prefer an option that has an inferior mathematical expectation. Suppose that you have the choice of investing $5,000 in a federally insured certificate of deposit paying 4.5% or in a mining stock that pays no dividend but has averaged a growth rate of 6.2%. Why may a person conceivably prefer the certificate of deposit?

R.84 The probabilities that a research worker's success will lead to a raise in salary, a promotion, or both, are, respectively, 0.33, 0.40, and 0.25. What is the probability that it will lead to either or both kinds of rewards?

R.85 In Figure R.3 the 20 outcomes are all equiprobable, but those constituting event A are twice as likely as those constituting A'. What is the probability of event A?

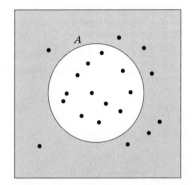

FIGURE R.3 Venn diagram for Exercise R.85.

R.86 Draw a tree diagram to determine the number of ways in which we can get a total of 6 in three rolls of a die.

R.87 Suppose that we answer a true-false test consisting of 15 questions by flipping a coin. What is the probability of getting 9 right and 6 wrong?

R.88 Four persons are getting ready for a game of bridge.

(a) In how many different ways can they choose partners?

(b) In how many different ways can they be seated at the table if it matters only who sits to whose left and who sits to whose right.

R.89 In how many different ways can the members of a family put their four cars in

(a) a four-car garage;

(b) in four of five parking spaces?

R.90 Two friends are betting on repeated flips of a balanced coin. One has $7 at the start and the other has $3, and after each flip the loser pays the winner $1. If p is the probability that the one who starts with $7 will win his friend's $3 before he loses his own $7, explain why $3p - 7(1 - p)$ should equal 0, and then solve the equation

$$3p - 7(1 - p) = 0$$

for p. Generalize this result to the case where the two players start with a dollars and b dollars, respectively.

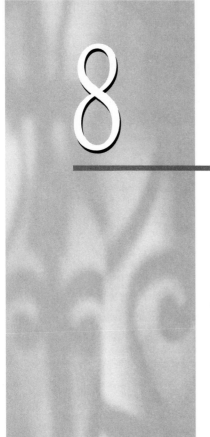

PROBABILITY DISTRIBUTIONS[†]

8.1   Random Variables 199

8.2   Probability Distributions 200

8.3   The Binomial Distribution 202

8.4   The Hypergeometric Distribution 211

8.5   The Poisson Distribution 215

*8.6   The Multinomial Distribution 219

8.7   The Mean of a Probability Distribution 223

8.8   The Standard Deviation of a Probability Distribution 225

8.9   Checklist of Key Terms 232

8.10   References 232

*I*n most problems of statistics we are interested only in one aspect, or at most in a few aspects, of the outcomes of experiments. For instance, a student taking a true–false test may be interested only in the number of questions he answers correctly and not which ones; a geologist may be interested only in the age of a rock sample and not in its hardness; and a sociologist may be interested only in the socioeconomic status of a person interviewed in a survey and not in her age or weight. Also, an agronomist may be interested in determining not only the yield per acre of a new variety of corn but also the temperature at which it will germinate; and an automotive engineer may be interested in the brightness and

[†]Since there are quite a few computer printouts and reproductions from the display screen of a graphing calculator in this chapter and in subsequent chapters, let us repeat from the Preface and the footnote on page 12 that the purpose of the printouts and the graphing calculator reproductions is to make the reader aware of the existence of these technologies for work in statistics. Let us make it clear, however, that neither computers nor graphing calculators are required for the use of our text. Indeed, the book can be used effectively by readers who do not possess or have easy access to computers and statistical software, or to graphing calculators. Some of the exercises are labeled with special icons for the use of a computer or a graphing calculator, but these exercises are optional.

the durability of the headlights proposed for a new model car and also in their projected cost.

In these five examples, the student, the geologist, the sociologist, the agronomist, and the automotive engineer are all interested in numbers that are associated with the outcomes of situations involving an element of chance, or more specifically, in values of **random variables.** Since random variables are neither random nor variables, why do they go by this name? This is hard to say, but a mathematics professor with a good sense of humor likened them to alligator pears, or avocados, which are neither alligators nor pears.

In the study of random variables we are usually interested in the probabilities with which they take on the various values within their range, namely, in their **probability distributions.** The general introduction of random variables and probability distributions in Sections 8.1 and 8.2 will be followed by the discussion of some of the most important probability distributions in Sections 8.3 through 8.6. Then, we discuss some ways of describing the most relevant features of probability distributions in Sections 8.7 and 8.8.

8.1   RANDOM VARIABLES

To be more explicit about the concept of a random variable, let us consider Figure 8.1, which, like Figure 6.9, pictures the sample space for the example dealing with the two car salesmen who hope to sell three 1998 Dodge Rams. The point $(2, 1)$, for example, represents the outcome where the first salesperson will sell two of the trucks and the second salesperson will sell the other one. Note that we added another number to each point—0 to the point $(0, 0)$; 1 to the points $(1, 0)$ and $(0, 1)$; 2 to the points $(2, 0)$, $(1, 1)$, and $(0, 2)$; and 3 to the points $(3, 0)$, $(2, 1)$, $(1, 2)$, and $(0, 3)$. In this way we have associated with each point of the sample space the number of trucks that, between them, the two salespersons will sell.

FIGURE 8.1 Sample space with values of random variable.

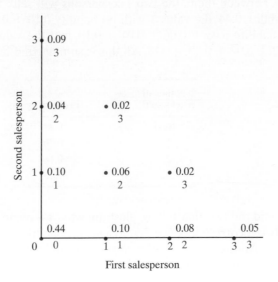

Since associating numbers with the points of a sample space is just a way of defining a function over the points of the sample space, random variables are really functions and not variables. Conceptually, though, most beginners find it easier to think of random variables simply as quantities that can take on different values depending on chance. For instance, the number of speeding tickets issued each day on the freeway between Indio and Blythe in California is a random variable, and so is the annual production of coffee in Brazil, the number of persons visiting Disneyland each week, the wind velocity at Kennedy airport, the size of the audience at a baseball game, and the number of mistakes a person makes when typing a report.

It is customary to classify random variables according to the number of values that they can assume, and in this chapter we shall consider only random variables that are **discrete;** namely, random variables that can take on a finite number of values or a countable infinity (as many values as there are whole numbers). For instance, the number of trucks that, between them, the two salespersons will sell is a discrete random variable that can take on a finite set of values, the four numbers 0, 1, 2, and 3. In contrast, the roll on which a die comes up 6 for the first time is a discrete random variable that can take on the countable infinity of values 1, 2, 3, 4, ... It is possible, though highly unlikely, that it will take a thousand rolls of the die, a million rolls, or even more, until we finally get a 6. There are also continuous random variables that arise when we deal with quantities measured on a continuous scale, say, time, weight, or distance. These will be taken up in Chapter 9.

8.2   PROBABILITY DISTRIBUTIONS

In Figure 8.1, the probabilities associated with the points of the sample space are shown in black. For instance, the 0.06 attached to the point (1, 1) indicates that the probability is 0.06 that each of the two salespersons will sell one of the three trucks. Thus, we find from Figure 8.1 that the random variable "the number of trucks that, between them, the two salespersons will sell" takes on the value 0 with probability 0.44, the value 1 with probability $0.10 + 0.10 = 0.20$, the value 2 with probability $0.08 + 0.06 + 0.04 = 0.18$, and the value 3 with probability $0.05 + 0.02 + 0.02 + 0.09 = 0.18$. All this is summarized in the following table:

Number of trucks sold	Probability
0	0.44
1	0.20
2	0.18
3	0.18

This table, and the two that follow, illustrate what we mean by a **probability distribution.** It is a correspondence that assigns probabilities to the values of a random variable. Another example of such a correspondence is given by the following table, which pertains to the number of points we roll with a balanced die:

Number of points we roll with a die	Probability
1	$\frac{1}{6}$
2	$\frac{1}{6}$
3	$\frac{1}{6}$
4	$\frac{1}{6}$
5	$\frac{1}{6}$
6	$\frac{1}{6}$

Finally, for four flips of a balanced coin there are the sixteen equally likely possibilities HHHH, HHHT, HHTH, HTHH, THHH, HHTT, HTHT, HTTH, THHT, THTH, TTHH, HTTT, THTT, TTHT, TTTH, and TTTT, where H stands for heads and T for tails. Counting the number of heads in each case and using the formula $\frac{s}{n}$ for equiprobable outcomes, we get the following probability distribution for the total number of heads:

Number of heads	Probability
0	$\frac{1}{16}$
1	$\frac{4}{16}$
2	$\frac{6}{16}$
3	$\frac{4}{16}$
4	$\frac{1}{16}$

When possible, we try to express probability distributions by means of formulas that enable us to calculate the probabilities associated with the various values of a random variable. For instance, for the number of points we roll with a balanced die we can write

$$f(x) = \frac{1}{6} \qquad \text{for } x = 1,2,3,4,5, \text{ and } 6$$

where $f(1)$ denotes the probability of rolling a 1, $f(2)$ denotes the probability of rolling a 2, and so on, in the usual functional notation. Here we wrote the probability that the random variable will take on the value x as $f(x)$, but we could just as well write it as $g(x)$, $h(x)$, $m(x)$, etc.

EXAMPLE   8.1   Verify that for the number of heads obtained in four flips of a balanced coin the probability distribution is given by

$$f(x) = \frac{\binom{4}{x}}{16} \qquad \text{for } x = 0,1,2,3, \text{ and } 4$$

Solution   By direct calculation, or by using Table XI at the end of the book, we find that $\binom{4}{0} = 1$, $\binom{4}{1} = 4$, $\binom{4}{2} = 6$, $\binom{4}{3} = 4$, and $\binom{4}{4} = 1$. Thus, the probabilities for $x = 0, 1, 2, 3,$ and 4 are $\frac{1}{16}, \frac{4}{16}, \frac{6}{16}, \frac{4}{16},$ and $\frac{1}{16}$, which agrees with the values given in the table on page 201.   ∎

Since the values of probability distributions are probabilities, and since random variables have to take on one of their values, we have the following two rules that apply to any probability distribution:

The values of a probability distribution must be numbers on the interval from 0 to 1.

The sum of all the values of a probability distribution must be equal to 1.

These rules enable us to determine whether or not a function (given by an equation or by a table) can serve as the probability distribution of some random variable.

EXAMPLE 8.2   Check whether the correspondence given by

$$f(x) = \frac{x + 3}{15} \qquad \text{for } x = 1, 2, \text{ and } 3$$

can serve as the probability distribution of some random variable.

Solution   Substituting $x = 1, 2,$ and 3 into $\frac{x + 3}{15}$, we get $f(1) = \frac{4}{15}$, $f(2) = \frac{5}{15}$, and $f(3) = \frac{6}{15}$.

Since none of these values is negative or greater than 1, and since their sum is

$$\frac{4}{15} + \frac{5}{15} + \frac{6}{15} = 1$$

the given function can serve as the probability distribution of some random variable.   ∎

8.3   THE BINOMIAL DISTRIBUTION

There are many applied problems in which we are interested in the probability that an event will occur x times out of n. For instance, we may be interested in the probability of getting 45 responses to 400 questionnaires sent out as part of a sociological study, the probability that 5 of 12 mice will survive for a given length of time after the injection of a cancer-inducing substance, the probability that 45 of 300 drivers stopped at a road block will be wearing their seat belts, or the probability that 66 of 200 television viewers (interviewed by a rating service) will recall what products were advertised on a given program. To borrow from the language of games of chance, we could say that in each of these examples we are interested in the probability of getting "x successes in n trials," or in other words, "x successes and $n - x$ failures in n attempts."

In the problems we shall study in this section, we always make the following assumptions:

There is a fixed number of trials.

The probability of a success is the same for each trial.

The trials are all independent.

Thus, the theory we develop does not apply, for example, if we are interested in the number of dresses that a woman may try on before she buys one (where the number of trials is not fixed), if we check every hour whether traffic is congested at a certain intersection (where the probability of "success" is not constant), or if we are interested in the number of times that a person voted for the Republican candidate in the last five presidential elections (where the trials are not independent).

In what follows, we will be able to obtain a formula to solve problems that meet the conditions listed in the preceding paragraph. If p and $1 - p$ are the probabilities of a success and a failure on any given trial, then the probability of getting x successes and $n - x$ failures *in some specific order* is $p^x(1 - p)^{n-x}$; clearly, in this product of p's and $(1 - p)$'s there is one factor p for each success, one factor $1 - p$ for each failure, and the x factors p and $n - x$ factors $1 - p$ are all multiplied together by virtue of the generalization of the special multiplication rule for more than two independent events. Since this probability applies to any point of the sample space that represents x successes and $n - x$ failures (in some specific order), we have only to count how many points of this kind there are, and then multiply $p^x(1 - p)^{n-x}$ by this number. Clearly, the number of ways in which we can choose the x trials on which the successes are to occur is $\binom{n}{x}$, and we have thus arrived at the following result:

Binomial distribution

The probability of getting x successes in n independent trials is

$$f(x) = \binom{n}{x} p^x(1 - p)^{n-x} \qquad \text{for } x = 0, 1, 2, \ldots, \text{ or } n$$

where p is the constant probability of a success for each trial.

It is customary to say here that the number of successes in n trials is a random variable having the **binomial probability distribution,** or simply the **binomial distribution.** The binomial distribution is called by this name because for $x = 0, 1, 2, \ldots$, and n, the values of the probabilities are the successive terms of the binomial expansion of $[(1 - p) + p]^n$.

EXAMPLE   8.3   Verify that the formula we gave in Example 8.1 for the probability of getting x heads in four flips of a balanced coin is, in fact, the one for the binomial distribution with $n = 4$ and $p = \frac{1}{2}$.

Solution   Substituting $n = 4$ and $p = \frac{1}{2}$ into the formula for the binomial distribution, we get

$$f(x) = \binom{4}{x}\left(\frac{1}{2}\right)^x\left(1 - \frac{1}{2}\right)^{4-x} = \binom{4}{x}\left(\frac{1}{2}\right)^4 = \frac{\binom{4}{x}}{16}$$

for $x = 0, 1, 2, 3,$ and 4. This is exactly the formula given in Example 8.1.   ∎

EXAMPLE 8.4   If the probability is 0.70 that any one registered voter (randomly selected from official rolls) will vote in a given election, what is the probability that two of five registered voters will vote in the election?

Solution   Substituting $x = 2, n = 5, p = 0.70,$ and $\binom{5}{2} = 10$ into the formula for the binomial distribution, we get

$$f(2) = \binom{5}{2}(0.70)^2(1 - 0.70)^{5-2} = 10(0.70)^2(0.30)^3 \approx 0.132$$ ∎

Following is an example where we calculate all the probabilities of a binomial distribution.

EXAMPLE 8.5   The probability is 0.30 that a person shopping at a certain supermarket will take advantage of its special promotion of ice cream. Find the probabilities that among six persons shopping at this market there will be 0, 1, 2, 3, 4, 5, or 6 who will take advantage of the promotion. Also, draw a histogram of this probability distribution.

Solution   Assuming that the selection is random, we substitute $n = 6, p = 0.30,$ and, respectively, $x = 0, 1, 2, 3, 4, 5,$ and 6 into the formula for the binomial distribution, and we have

$$f(0) = \binom{6}{0}(0.30)^0(0.70)^6 \approx 0.118$$

$$f(1) = \binom{6}{1}(0.30)^1(0.70)^5 \approx 0.303$$

$$f(2) = \binom{6}{2}(0.30)^2(0.70)^4 \approx 0.324$$

$$f(3) = \binom{6}{3}(0.30)^3(0.70)^3 \approx 0.185$$

$$f(4) = \binom{6}{4}(0.30)^4(0.70)^2 \approx 0.060$$

$$f(5) = \binom{6}{5}(0.30)^5(0.70)^1 \approx 0.010$$

$$f(6) = \binom{6}{6}(0.30)^6(0.70)^0 \approx 0.001$$

A histogram of this distribution is shown in Figure 8.2.   ∎

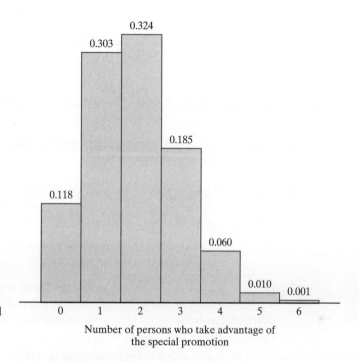

FIGURE 8.2
Histogram of binomial distribution with $n = 6$ and $p = 0.30$.

In case the reader does not care much for ice cream and is not particularly interested in personal buying habits, let us stress the importance of the binomial distribution as a **statistical model.** The results of the preceding example apply also if the probability is 0.30 that the energy cell of a watch will last two years under normal usage, and we want to know the probabilities that, among six of these cells, 0, 1, 2, 3, 4, 5, or 6 will last two years under normal usage; if the probability is 0.30 that an embezzler will be caught and brought to trial, and we want to know the probabilities that, among six embezzlers, 0, 1, 2, 3, 4, 5, or 6 will be caught and brought to trial; if the probability is 0.30 that the head of a household owns at least one life insurance policy, and we want to know the probabilities that, among six heads of households, 0, 1, 2, 3, 4, 5, or 6 will own at least one life insurance policy; or if the probability that a person having a certain disease will live for another ten years is 0.30, and we want to know the probabilities that, among six persons having the disease, 0, 1, 2, 3, 4, 5, or 6 will live another ten years. The argument we have presented here is precisely like the one we used in Section 1.2, where we tried to impress upon the reader the generality of statistical techniques.

In actual practice, binomial probabilities are seldom found by direct substitution into the formula. Sometimes we use approximations such as those discussed later in this chapter and in Chapter 9, and sometimes we use special tables such as Table V at the end of this book. Computer software is now the most common source for binomial probabilities. At the end of this chapter are listed some references for detailed binomial tables in book form (although they are falling into disuse).

Table V is limited to the binomial probabilities for $n = 2$ to $n = 20$ and $p = 0.05, 0.1, 0.2, 0.3, 0.4, 0.5, 0.6, 0.7, 0.8, 0.9$, and 0.95, all rounded to three decimals. Where values are omitted in this table, they are 0.0005 or less and, hence, rounded to 0.000 to three decimals.

EXAMPLE 8.6   Suppose that the probability is 0.60 that a lunar eclipse will be obscured by clouds. Use Table V to find the probabilities that

(a) at most three of ten lunar eclipses will be obscured by clouds;
(b) at least seven of ten lunar eclipses will be obscured by clouds.

Solution   (a) For $n = 10$ and $p = 0.60$, the entries in Table V corresponding to $x = 0, 1, 2$, and 3 are 0.000, 0.002, 0.011, and 0.042. Thus, the probability that at most three of ten lunar eclipses will be obscured by clouds is

$$0.000 + 0.002 + 0.011 + 0.042 = 0.055$$

(b) For $n = 10$ and $p = 0.60$, the entries in Table V corresponding to $x = 7, 8, 9$, and 10 are 0.215, 0.121, 0.040, and 0.006. Thus, the probability that at least seven of ten lunar eclipses will be obscured by clouds is

$$0.215 + 0.121 + 0.040 + 0.006 = 0.382 \qquad \blacksquare$$

EXAMPLE 8.7   If the probability is 0.05 that the baby anteater will be visible during a visit to the Phoenix Zoo, what is the probability that the baby anteater will be visible on at least three of 20 visits to this zoo?

Solution   For $n = 20$ and $p = 0.05$, the entries in Table V corresponding to $x = 3, 4$, and 5 are 0.060, 0.013, and 0.002, while those corresponding to $x = 6, 7, 8, \ldots$, and 20 are all 0.000. Thus, the probability that the baby anteater will be visible on at least three of twenty visits to this zoo is

$$0.060 + 0.013 + 0.002 = 0.075 \qquad \blacksquare$$

In Example 8.6 we could not have used Table V if the probability that a lunar eclipse will be obscured by clouds had been, say, 0.63 instead of 0.60. In general, if n is greater than 20 or p takes on a value other than $0.05, 0.1, 0.2, \ldots, 0.9$, or 0.95, we will have to use one of the more detailed tables referred to on page 232, use a computer printout as in the example that follows, or, as a last resort, refer to the formula for the binomial distribution.

EXAMPLE   8.8   Rework Example 8.6, using 0.63 instead of 0.60 as the probability that a lunar eclipse will be obscured by clouds.

olution

    (a) Referring to the computer printout of Figure 8.3, we find that the probabilities corresponding to $x = 0, 1, 2,$ and 3 are 0.0000, 0.0008, 0.0063, and 0.0285. Thus, the probability that at most three of ten lunar eclipses will be obscured by clouds is

$$0.0000 + 0.0008 + 0.0063 + 0.0285 = 0.0356$$

    Note that this answer is also shown in the lower part of the printout under the $P(X <= x)$ caption corresponding to 3.00.

FIGURE 8.3
Computer printout for the binomial distribution with $n = 10$ and $p = 0.63$.

```
MTB > PDF c1;
SUBC>    Binomial 10 0.63.


Binomial with n = 10 and p = 0.630000

            x        P( X = x)
          0.00         0.0000
          1.00         0.0008
          2.00         0.0063
          3.00         0.0285
          4.00         0.0849
          5.00         0.1734
          6.00         0.2461
          7.00         0.2394
          8.00         0.1529
          9.00         0.0578
         10.00         0.0098

MTB > CDF c1;
SUBC>    Binomial 10 0.63.


Binomial with n = 10 and p = 0.630000

            x        P( X <= x)
          0.00         0.0000
          1.00         0.0009
          2.00         0.0071
          3.00         0.0356
          4.00         0.1205
          5.00         0.2939
          6.00         0.5400
          7.00         0.7794
          8.00         0.9323
          9.00         0.9902
         10.00         1.0000
```

(b) Also using the printout of Figure 8.3, we find that the probabilities corresponding to $x = 7, 8, 9$, and 10 are 0.2394, 0.1529, 0.0578, and 0.0098. Thus, the probability that at least seven of the ten lunar eclipses will be obscured by clouds is

$$0.2394 + 0.1529 + 0.0578 + 0.0098 = 0.4599$$

Since the probability of "at least 7" is 1 minus the probability of "at most 6," we can get the answer also by subtracting from 1 the entry under the $P(X <= x)$ caption corresponding to 6.00, namely, $1 - 0.5400 = 0.4600$. The difference between the two results, 0.4599 and 0.4600, is due to rounding.   ■

When we observe a value of a random variable having the binomial distribution—for instance, when we observe the number of heads in 25 flips of a coin, the number of seeds (in a package of 24 seeds) that germinate, the number of students (among 200 interviewed) who are opposed to a change in student activity fees, or the number of automobile accidents (among 20 investigated) that are due to drunk driving—we say that we are **sampling a binomial population.** This terminology is widely used in statistics.

■ E x e r c i s e s

8.1   In each case determine whether the given values can serve as the values of the probability distribution of some random variable that can take on the values 1, 2, and 3, and explain your answers:
(a) $f(1) = 0.52, f(2) = 0.26$. and $f(3) = 0.32$;
(b) $f(1) = 0.18, f(2) = 0.02$, and $f(3) = 1.00$;
(c) $f(1) = \frac{10}{33}, f(2) = \frac{1}{3}$, and $f(3) = \frac{12}{33}$.

8.2   In each case determine whether the given values can serve as the values of the probability distribution of some random variable that can take on the values 1, 2, 3, and 4, and explain your answers:
(a) $f(1) = 0.20, f(2) = 0.80, f(3) = 0.20$, and $f(4) = -0.20$;
(b) $f(1) = 0.25, f(2) = 0.17, f(3) = 0.39$, and $f(4) = 0.19$;
(c) $f(1) = \frac{1}{17}, f(2) = \frac{7}{17}, f(3) = \frac{6}{17}$, and $f(4) = \frac{2}{17}$.

8.3   For each of the following, determine whether it can serve as the probability distribution of some random variable:
(a) $f(x) = \frac{1}{7}$ for $x = 1, 2, 3, 4, 5, 6, 7$;
(b) $g(y) = \frac{1}{9}$ for $y = 0, 1, 2, 3, 4, 5, 6, 7, 8, 9$;
(c) $f(x) = \dfrac{x + 2}{18}$ for $x = 1, 2, 3, 4$.

8.4   For each of the following, determine whether it can serve as the probability distribution of some random variable:
(a) $f(x) = \dfrac{x - 1}{10}$ for $x = 0, 1, 2, 3, 4, 5$;

(b) $h(z) = \dfrac{z^2}{30}$ for $z = 0, 1, 2, 3, 4$;

(c) $f(y) = \dfrac{y + 4}{y - 4}$ for $y = 1, 2, 3, 4, 5$.

8.5 In a certain city, medical expenses are given as the reason for 75% of all personal bankruptcies. Use the formula for the binomial distribution to calculate the probability that medical expenses will be given as the reason for two of the next three personal bankruptcies filed in that city.

8.6 Use the formula for the binomial distribution to calculate the probability that four of six hibiscus plants will fail to survive a frost if the probability is 0.30 that any such plant will survive a frost. Also, check the answer in Table V.

8.7 A doctor knows from experience that 10% of the patients to whom he prescribes a certain blood pressure medication will have undesirable side effects. Use the formula for the binomial distribution to calculate the probability that none of four patients to whom he prescribes the medication will have undesirable side effects. Also, check the answer in Table V.

8.8 With reference to Exercise 8.7, use the formula for the binomial distribution to calculate the probability that at most two of four patients to whom the doctor prescribes the medication will have undesirable side effects. Also, use Table V to check the answer.

8.9 If 40% of the mice used in an experiment will become very aggressive within two minutes after having been administered an experimental drug, find the probability that exactly four of nine mice that have been administered the drug will become very aggressive within two minutes, using
(a) the formula for the binomial distribution;
(b) Table V.

8.10 If it is correct to say that 80% of all industrial accidents can be prevented by paying strict attention to safety regulations, find the probability that four of six industrial accidents can thus be prevented, using
(a) the formula for the binomial distribution;
(b) Table V.

8.11 Experience has shown that 30% of the rocket launchings at a NASA base have to be delayed due to weather conditions. Use Table V to determine the probabilities that among ten rocket launchings at that base
(a) at most three will have to be delayed due to weather conditions;
(b) at least six will have to be delayed due to weather conditions.

8.12 A study shows that 60% of the divorce cases filed in a certain county give incompatibility as the legal reason. Find the probabilities that among 15 divorce cases filed in that county
(a) at most five will give incompatibility as the legal reason;
(b) anywhere from eight to eleven will give incompatibility as the legal reason;
(c) at least eleven will give incompatibility as the legal reason.

8.13 An agricultural cooperative claims that 95% of the watermelons shipped out are ripe and ready to eat. Find the probabilities that among 20 watermelons shipped out
(a) at least 17 are ripe and ready to eat;
(b) all of them are ripe and ready to eat;
(c) anywhere from 16 to 19 are ripe and ready to eat.

8.14   A frozen food distributor claims that 80% of its frozen chicken dinners contain at least three ounces of chicken. To check on this claim, a consumer testing service decides to check ten of these frozen chicken dinners and reject the claim unless at least seven of them contain at least three ounces of chicken. Find the probabilities that the testing service will make the error of
(a) rejecting the claim even though it is true;
(b) not rejecting the claim when in reality only 70% of the frozen chicken dinners contain at least three ounces of chicken.

8.15   A quality control engineer wants to check whether, in accordance with specifications, 90% of the products shipped are in perfect working condition. To this end, she randomly selects 12 items from each lot ready to be shipped and passes the lot only if all 12 are in perfect working condition. If one or more items are not in perfect working condition, she holds the lot for a complete inspection. Find the probabilities that she will commit the error of
(a) holding a lot for a complete inspection even though 90% of the items are in perfect working condition;
(b) letting a lot pass through even though only 80% of the items are in perfect working condition;
(c) letting a lot pass through even though only 70% of the items are in perfect working condition.

8.16   A study shows that 70% of all patients coming to a certain medical clinic have to wait at least 15 minutes to see their doctor. Find the probabilities that among ten patients coming to this clinic 0, 1, 2, 3, ..., or 10 have to wait at least 15 minutes to see their doctor, and draw a histogram of this probability distribution.

8.17   The records of a computer retail store indicate that 80% of all persons buying a new computer will request an internal modem. Find the probabilities that among 15 persons buying a new computer, 0, 1, 2, 3, ..., or 15 will request an internal modem. Also draw a histogram of this probability distribution.

 8.18   With reference to Exercise 8.17, suppose that the percentage had been 78% instead of 80%. Use appropriate computer software or a graphing calculator to rework the exercise.

8.19   Use Figure 8.3 to determine the probability that a random variable having the binomial distribution with $n = 10$ and $p = 0.63$ will take on a value less than five using
(a) the binomial probabilities;
(b) the cumulative probabilities.

8.20   Use Figure 8.3 to determine the probability that a random variable having the binomial distribution with $n = 10$ and $p = 0.63$ will take on a value greater than eight using
(a) the binomial probabilities;
(b) the cumulative probabilities.

 8.21   With reference to Exercise 8.11, suppose that the percentage had been 28% instead of 30%. Use appropriate computer software or a graphing calculator to rework that exercise.

 8.22   With reference to Exercise 8.12, suppose that the percentage had been 54% instead of 60%. Use appropriate computer software or a graphing calculator to rework that exercise.

8.23   In some situations where otherwise the binomial distribution applies, we are interested in the probability that the first success will occur on a given trial. For this to happen on the xth trial, it must be preceded by $x - 1$ failures for which the probability is $(1 - p)^{x-1}$, and it follows that the probability that the first success will occur on the xth trial is

$$f(x) = p(1 - p)^{x-1} \qquad \text{for } x = 1, 2, 3, 4, \ldots$$

This distribution is called the **geometric distribution** (because its successive values constitute a geometric progression) and it should be observed that there is a countable infinity of possibilities.[†] Using the formula, we find, for example, that for repeated rolls of a balanced die the probability that the first 6 will occur on the fifth roll is

$$\frac{1}{6}\left(\frac{5}{6}\right)^{5-1} = \frac{625}{7,776} \approx 0.080$$

(a) When taping a television commercial, the probability that a child actor will get his lines straight on any one take is 0.40. What is the probability that this child actor will finally get his lines straight on the fourth take?

(b) Suppose the probability is 0.25 that any given person will believe a rumor about the private life of a certain politician. What is the probability that the fifth person to hear the rumor will be the first one to believe it?

(c) The probability is 0.70 that a child exposed to a certain contagious disease will catch it. What is the probability that the third child exposed to the disease will be the first one to catch it?

8.4   THE HYPERGEOMETRIC DISTRIBUTION

In Exercise 6.77 we introduced the terms "sampling with replacement" and "sampling without replacement" in connection with drawings from a deck of cards. To carry this distinction further, let us point out that the binomial distribution applies when we sample with replacement and the trials are all independent, but not when we sample without replacement. To introduce a probability distribution that applies when we sample without replacement, let us consider the following example: A factory ships audio tape players in lots of 24, and when they arrive at their destination, an inspector randomly selects three from each lot. If they are all in good working condition, the entire lot is accepted; otherwise, the entire lot is inspected. Since a lot can be accepted without further inspection even though quite a few of the tape players are defective, there is a considerable risk. So let us find the probability that a lot will be accepted without further inspection even though, say, six of the 24 tape players are defective. This means that we must find the probability of three successes (three nondefective tape players) in three trials (among the three tape players inspected), and we might

[†]As formulated in Chapter 6, the postulates of probability apply only when the sample space is finite. When the sample space is countably infinite, as is the case here, the third postulate must be modified accordingly. This will be explained on page 216.

be tempted to argue that since 18 of the 24 tape players are not defective, the probability of a success is $\frac{18}{24} = \frac{3}{4}$, and hence the desired probability is

$$f(3) = \binom{3}{3}\left(\frac{3}{4}\right)^3\left(1 - \frac{3}{4}\right)^{3-3} \approx 0.42$$

This result, obtained with the formula for the binomial distribution, would be correct if sampling is with replacement, but that is not what we do in realistic problems of sampling inspection. To get the correct answer for our problem when sampling is without replacement, we might argue as follows: There are altogether $\binom{24}{3} = 2{,}024$ ways of choosing three of the tape recorders, and they are all equiprobable by virtue of the assumption that the selection is random. Among these, there are $\binom{18}{3} = 816$ ways of selecting three of the nondefective tape recorders, and it follows by the special formula $\frac{s}{n}$ for equiprobable outcomes that the desired probability is $\dfrac{816}{2{,}024} \approx 0.40$.

To generalize the method we used here, suppose that n objects are to be chosen from a set of a objects of one kind (successes) and b objects of another kind (failures), the selection is without replacement, and we are interested in the probability of getting x successes and $n - x$ failures. Arguing as before, we find that the n objects can be chosen from the whole set of $a + b$ objects in $\binom{a + b}{n}$ ways, and that x of the a successes and $n - x$ of the b failures can be chosen in $\binom{a}{x}\cdot\binom{b}{n - x}$ ways. It follows that for sampling without replacement the probability of "x successes in n trials" is

Hypergeometric distribution

$$f(x) = \frac{\binom{a}{x}\cdot\binom{b}{n - x}}{\binom{a + b}{n}} \qquad for\ x = 0, 1, 2, \ldots, or\ n$$

where x cannot exceed a and $n - x$ cannot exceed b. This is the formula for the **hypergeometric distribution.**

The following are two examples where we sample without replacement and, hence, can use the formula for the hypergeometric distribution:

EXAMPLE   8.9   A mailroom clerk is supposed to send 6 of 15 packages to Europe by airmail, but he gets them all mixed up and randomly puts airmail postage on 6 of the packages. What is the probability that only three of the packages that are sup-posed to go by airmail will go by airmail?

Solution   Substituting $a = 6, b = 9$, and $x = 3$ into the formula for the hypergeometric dis-tribution, we get

$$f(3) = \frac{\binom{6}{3} \cdot \binom{9}{6-3}}{\binom{15}{6}} = \frac{20 \cdot 84}{5{,}005} \approx 0.336 \qquad \blacksquare$$

EXAMPLE 8.10   Among an ambulance service's 16 ambulances, five emit excessive amounts of pollutants. If eight of the ambulances are randomly picked for inspection, what is the probability that this sample will include at least three of the ambulances that emit excessive amounts of pollutants?

Solution   The probability we must find is $f(3) + f(4) + f(5)$, where each term in this sum is a value of the hypergeometric distribution with $a = 5, b = 11$, and $n = 8$. Sub-stituting these quantities together with $x = 3, 4$, and 5 into the formula for the hypergeometric distribution, we get

$$f(3) = \frac{\binom{5}{3} \cdot \binom{11}{5}}{\binom{16}{8}} = \frac{10 \cdot 462}{12{,}870} \approx 0.359$$

$$f(4) = \frac{\binom{5}{4} \cdot \binom{11}{4}}{\binom{16}{8}} = \frac{5 \cdot 330}{12{,}870} \approx 0.128$$

$$f(5) = \frac{\binom{5}{5} \cdot \binom{11}{3}}{\binom{16}{8}} = \frac{1 \cdot 165}{12{,}870} \approx 0.013$$

and the probability that the sample will include at least three of the ambulances that emit excessive amounts of pollutants is

$$0.359 + 0.128 + 0.013 = 0.500$$

This result suggests that the inspection should, perhaps, have included more than eight of the ambulances, and it will be left to the reader to show in Exercise 8.24 that the probability of catching at least three of the ambulances that emit ex-cessive amounts of pollutants would have been 0.76 if the inspection had included ten of the ambulances. $\qquad \blacksquare$

In the beginning of this section we gave an example where we erroneously used the binomial distribution instead of the hypergeometric distribution. The error was quite small, however—we got 0.42 instead of 0.40—and in actual practice the binomial distribution is often used to approximate the hypergeometric distribution. It is generally agreed that this approximation is satisfactory if n does not exceed 5 percent of $a + b$, namely, if

$$n \leq (0.05)(a + b)$$

The main advantages of the approximation are that the binomial distribution has been tabulated much more extensively than the hypergeometric distribution, and that, between them, the one for the binomial distribution is easier to use; that is, the binomial calculations are usually less complicated. Observe also that the binomial distribution is described by two parameters (n and p), while the hypergeometric distribution requires three (a, b, and n).

EXAMPLE 8.11   In a federal prison, 120 of the 300 inmates are serving time for drug-related offenses. If eight of them are to be chosen at random to appear before a legislative committee, what is the probability that three of the eight will be serving time for drug-related offenses?

Solution   Since $n = 8$ and $a + b = 300$ and 8 is less than $0.05(300) = 15$, we can use the binomial approximation to the hypergeometric distribution. From Table V we find that for $n = 8$, $p = \frac{120}{300} = 0.40$, and $x = 3$, the probability asked for is 0.279. Fairly extensive calculations would show that the error of this approximation is only 0.003.   ∎

■ Exercises

8.24   In Example 8.10 we indicated that if ten of the ambulances had been inspected, the probability of including at least three of them that emit excessive amounts of pollutants would have been 0.76. Verify this probability.

8.25   Among the 12 scientists working on a research project, nine have Ph.D.'s. If four of them are chosen at random to attend a convention, what is the probability that three of them will have Ph.D.'s?

8.26   Among the 20 solar collectors on display at a trade show, 12 are flat-plate collectors and the others are concentrating collectors. If a person visiting the show randomly selects six of the solar collectors to check out, what is the probability that three of them will be flat-plate collectors?

8.27   What is the probability that an IRS auditor will catch only two income tax returns with illegitimate deductions if she randomly chooses six returns from among eighteen returns of which eight contain illegitimate deductions for a special audit?

8.28   Among the 12 male applicants for a job with the postal service, nine have working wives. If two of the applicants are randomly chosen for further consideration, what are the probabilities that
(a) neither has a working wife;
(b) only one has a working wife;
(c) both have working wives?

8.29 A customs inspector decides to inspect 3 of 16 shipments that arrive from Caracas by plane. If the selection is random and five of the shipments contain contraband, find the probabilities that the customs inspector will catch
(a) none of the shipments with contraband;
(b) only one of the shipments with contraband;
(c) two of the shipments with contraband;
(d) three of the shipments with contraband.

8.30 A collection of eight Spanish gold doubloons contains five counterfeits. If two of these coins are randomly selected to be sold at auction, what are the probabilities that
(a) neither coin is a counterfeit;
(b) either or both coins are counterfeits?

8.31 To pass a quality control inspection, two batteries are chosen from each lot of 12 car batteries, and the lot is passed only if neither battery has any defects; otherwise, each of the batteries in the lot is checked. If the selection of the batteries is random, find the probabilities that a lot will
(a) pass the inspection when one of the 12 batteries is defective;
(b) fail the inspection when three of the batteries are defective;
(c) fail the inspection when six of the batteries are defective.

8.32 Check in each case whether the condition for the binomial approximation to the hypergeometric distribution is satisfied:
(a) $a = 140, b = 60,$ and $n = 12$;
(b) $a = 220, b = 280,$ and $n = 20$;
(c) $a = 250, b = 390,$ and $n = 30$;
(d) $a = 220, b = 220,$ and $n = 25$.

8.33 A shipment of 300 burglar alarms contains six that are defective. Use the binomial approximation to the hypergeometric distribution to find the probability that if five of these burglar alarms are randomly selected to be shipped to a customer, they will include one that is defective.

8.34 With reference to Exercise 8.33, find the error of the approximation.

8.35 Among the 200 employees of a company, 120 are union members while the others are not. If six of the employees are to be chosen by lot to serve on a committee that administers the pension fund, find the probability that three of them will be union members while the others are not, using
(a) the formula for the hypergeometric distribution;
(b) the binomial distribution with $p = \frac{120}{200} = 0.60$ and $n = 6$ as an approximation.

8.36 With reference to Exercise 8.35, what is the percentage error of the approximation?

8.5 THE POISSON DISTRIBUTION

When n is large and p is small, binomial probabilities are often approximated by means of the formula

Poisson approximation to binomial distribution

$$f(x) = \frac{(np)^x \cdot e^{-np}}{x!} \quad \text{for } x = 0, 1, 2, 3, \ldots$$

which is a special form of the **Poisson distribution,** named after the French mathematician and physicist S. D. Poisson (1781–1840); the more general form is given on page 219. In this formula, the irrational number $e = 2.71828\ldots$ is the base of the system of natural logarithms, and the necessary values of e^{-np} may be obtained from Table XII at the end of the book. Note also that, as in Exercise 8.23, we are faced here with a random variable that can take on a countable infinity of values (namely, as many values as there are whole numbers). Correspondingly, the third postulate of probability must be modified so that for any sequence of mutually exclusive events $A_1, A_2, A_3, \ldots$, the probability that one of them will occur is

$$P(A_1 \cup A_2 \cup A_3 \cup \cdots) = P(A_1) + P(A_2) + P(A_3) + \cdots$$

It is difficult to give precise conditions under which the Poisson approximation to the binomial distribution may be used; that is, explain precisely what we mean here by "when n is large and p is small." Although other books may give less stringent rules of thumb, we shall play it relatively safe and use the Poisson approximation to the binomial distribution only when

$$n \geq 100 \quad \text{and} \quad np < 10$$

To get some idea about the closeness of the Poisson approximation to the binomial distribution, consider the computer printouts of Figure 8.4, which show, one next to the other, the binomial probabilities with $n = 150$ and $p = 0.05$ and the Poisson probabilities with $np = 150(0.05) = 7.5$. [Comparing the probabilities in the columns headed $P(X = x)$, we find that the greatest difference, corresponding to $x = 8$, is $0.1410 - 0.1373 = 0.0037$.]

FIGURE 8.4
Computer printout of the binomial distribution with $n = 150$ and $p = 0.05$ and the Poisson distribution with $np = 7.5$.

```
MTB > PDF c1;                             MTB > PDF c1;
SUBC>    Binomial 150 0.05.               SUBC>    Poisson 7.5.

Binomial with n = 150 and p = 0.0500000   Poisson with mu = 7.50000

         x      P( X = x)                         x      P( X = x)
      0.00        0.0005                        0.00        0.0006
      1.00        0.0036                        1.00        0.0041
      2.00        0.0141                        2.00        0.0156
      3.00        0.0366                        3.00        0.0389
      4.00        0.0708                        4.00        0.0729
      5.00        0.1088                        5.00        0.1094
      6.00        0.1384                        6.00        0.1367
      7.00        0.1499                        7.00        0.1465
      8.00        0.1410                        8.00        0.1373
      9.00        0.1171                        9.00        0.1144
     10.00        0.0869                       10.00        0.0858
     11.00        0.0582                       11.00        0.0585
     12.00        0.0355                       12.00        0.0366
     13.00        0.0198                       13.00        0.0211
     14.00        0.0102                       14.00        0.0113
     15.00        0.0049                       15.00        0.0057
     16.00        0.0022                       16.00        0.0026
     17.00        0.0009                       17.00        0.0012
     18.00        0.0003                       18.00        0.0005
     19.00        0.0001                       19.00        0.0002
     20.00        0.0000                       20.00        0.0001
```

EXAMPLE 8.12   It is known that 2% of the books bound at a certain bindery have defective bindings. Use the Poisson approximation to the binomial distribution to find the probability that 5 of 400 books bound at this bindery will have defective bindings.

Solution   Substituting $x = 5$, $np = 400(0.02) = 8$, and $e^{-8} = 0.00033546$ (from Table XII) into the formula for the Poisson distribution, we get

$$f(5) = \frac{8^5 \cdot e^{-8}}{5!} = \frac{(32,768)(0.00033546)}{120} \approx 0.0916$$ ∎

EXAMPLE 8.13   Records show that the probability is 0.00006 that a car will have a flat tire while being driven through a certain tunnel. Use the Poisson approximation to the binomial distribution to find the probability that at least two of 10,000 cars will have a flat tire while being driven through that tunnel.

Solution   Rather than add the probabilities for $x = 2, 3, 4, \ldots$, we shall subtract from 1 the sum of the probabilities for $x = 0$ and $x = 1$. Thus, substituting $np = 10,000(0.00006) = 0.6$, $e^{-0.6} = 0.5488$, and respectively, $x = 0$ and $x = 1$ into the formula for the Poisson distribution, we get

$$f(0) = \frac{(0.6)^0 \cdot e^{-0.6}}{0!} = \frac{1(0.5488)}{1} = 0.5488$$

$$f(1) = \frac{(0.6)^1 \cdot e^{-0.6}}{1!} = \frac{(0.6)(0.5488)}{1} \approx 0.3293$$

and, finally, $1 - (0.5488 + 0.3293) = 0.1219$.   ∎

In actual practice, Poisson probabilities are seldom obtained by direct substitution into the formula or by the use of statistical tables. These tasks are greatly simplified by utilizing appropriate computer software. For instance, had we used the computer printout of Figure 8.5 in Example 8.13, we would have obtained the result directly as $1 - 0.8781 = 0.1219$, where 0.8781 is the value corresponding to $x = 1$ in the column headed $P(X <= x)$.

Since in some cases the hypergeometric distribution can be approximated by a binomial distribution, and the binomial distribution can in some cases be approximated by a Poisson distribution, there are situations in which the hypergeometric distribution can be approximated by a Poisson distribution. Consider, for instance, the following example:

```
MTB > PDF c1;                          MTB > CDF c1;
SUBC>    Poisson 0.6.                   SUBC>    Poisson 0.6.


Poisson with mu = 0.600000            Poisson with mu = 0.600000

          x      P( X = x)                      x      P( X <= x)
        0.00       0.5488                      0.00       0.5488
        1.00       0.3293                      1.00       0.8781
        2.00       0.0988                      2.00       0.9769
        3.00       0.0198                      3.00       0.9966
        4.00       0.0030                      4.00       0.9996
        5.00       0.0004                      5.00       1.0000
        6.00       0.0000                      6.00       1.0000

                                       MTB >
```

FIGURE 8.5
Computer printout of the Poisson distribution with $np = 0.60$.

EXAMPLE 8.14 Suppose that 28 of 4,000 sales invoices contain errors. If a CPA randomly chooses 150 of them for an audit, what is the probability that exactly two of them will contain errors?

Solution This calls for the hypergeometric probability with $x = 2$, $a = 28$, $b = 3,972$, and $n = 150$, but since

$$150 \leqslant 0.05(4,000) = 200$$

we can use the binomial approximation to the hypergeometric distribution. Furthermore, since this is the binomial distribution with $n = 150$ and $p = \dfrac{a}{a+b} = \dfrac{28}{4,000} = 0.007$, for which

$$n \geqslant 100 \text{ and } np = 150(0.007) = 1.05 < 10$$

we can approximate it with the Poisson distribution with $np = 1.05$. Thus, we can approximate the original hypergeometric probability with the Poisson probability with $x = 2$ and $np = 1.05$. This is

$$f(2) = \frac{1.05^2 \cdot e^{-1.05}}{2!} = \frac{1.1025(0.349938)}{2} = 0.1929$$

where the value of $e^{-1.05}$ was obtained with a statistical calculator. ∎

Since auditors deal frequently with situations with large values of n and even larger values of $a + b$, they find the Poisson distribution to be quite useful. Observe that, aside from x, the hypergeometric distribution involves three parameters, $(a, b, \text{and } n)$, whereas the binomial distribution involves two, $(n \text{ and } p)$, and the Poisson distribution involves only one, (np).

The Poisson distribution has many important applications that have no direct connection with the binomial distribution. In that case np is replaced by the parameter λ (Greek lowercase *lambda*) and we calculate the probability of getting x successes by means of the formula

Poisson distribution

$$f(x) = \frac{\lambda^x \cdot e^{-\lambda}}{x!} \qquad for\ x = 0, 1, 2, 3, \ldots$$

where λ is interpreted as the expected, or average, number of successes, as is explained in the last paragraph of Section 8.7.

This formula applies to many situations where we can expect a fixed number of "successes" per unit time (or for some other kind of unit), say, when a bank can expect to receive six bad checks per day, when 1.6 accidents can be expected per day at a busy intersection, when eight small pieces of meat can be expected in a frozen meat pie, when 5.6 imperfections can be expected per roll of cloth, when 0.03 complaint per passenger can be expected by an airline, and so on.

EXAMPLE 8.15   If a bank receives on the average $\lambda = 6$ bad checks per day, what is the probability that it will receive four bad checks on any given day?

Solution   Substituting $x = 4$ and $\lambda = 6$ into the preceding formula for the Poisson distribution, we get

$$f(4) = \frac{6^4 \cdot e^{-6}}{4!} = \frac{1{,}296\,(0.002479)}{24} \approx 0.1339$$

where the value of e^{-6} was obtained from Table XII.   ∎

EXAMPLE 8.16   If $\lambda = 5.6$ imperfections can be expected per roll of a certain kind of cloth, what is the probability that a roll will have three imperfections?

Solution   Substituting $x = 3$ and $\lambda = 5.6$ into the preceding formula for the Poisson distribution, we get

$$f(3) = \frac{5.6^3 \cdot e^{-5.6}}{3!} = \frac{175.616\,(0.003698)}{6} \approx 0.1082$$

where the value of $e^{-5.6}$ was obtained from Table XII.   ∎

*8.6   THE MULTINOMIAL DISTRIBUTION

An important generalization of the binomial distribution arises when there are more than two possible outcomes for each trial, the probabilities of the various outcomes remain the same for each trial, and the trials are all independent. This is the case, for example, when we repeatedly roll a die, where each trial has six possible outcomes; when students are asked whether they like a certain new recording, dislike it, or don't care; or when a U.S. Department of Agriculture inspector grades beef as prime, choice, good, commercial, or utility.

If there are k possible outcomes for each trial and their probabilities are $p_1, p_2, \ldots,$ and p_k, it can be shown that the probability of x_1 outcomes of the first kind, x_2 outcomes of the second kind, ..., and x_k outcomes of the kth kind in n trials is given by

Multinomial distribution

$$\frac{n!}{x_1! x_2! \cdots x_k!} \, p_1^{x_1} \cdot p_2^{x_2} \cdots p_k^{x_k}$$

This distribution is called the **multinomial distribution.**

EXAMPLE 8.17   In a large city, network TV has 30% of the viewing audience on Friday nights, a local channel has 20%, cable TV has 40%, and 10% of the viewing audience is watching videocassettes. What is the probability that among seven television viewers randomly selected in that city on a Friday night three will be viewing network TV, one will be watching the local channel, two will be watching cable TV, and one will be watching a videocassette?

Solution   Substituting $n = 7$, $x_1 = 3$, $x_2 = 1$, $x_3 = 2$, $x_4 = 1$, $p_1 = 0.30$, $p_2 = 0.20$, $p_3 = 0.40$, and $p_4 = 0.10$ into the formula for the multinomial distribution, we get

$$\frac{7!}{3! \cdot 1! \cdot 2! \cdot 1!} \cdot (0.30)^3 \, (0.20)^1 \, (0.40)^2 \, (0.10)^1 \approx 0.036 \qquad \blacksquare$$

Strictly speaking, the multinomial distribution does not apply when we sample without replacement, but this does not matter in Example 8.17 since the city is large and the sample is very small. Otherwise, when n objects are chosen without replacement from a set of objects consisting of a_1 objects of one kind, a_2 objects of a second kind, ..., and a_k objects of a kth kind, the probability of getting x_1 objects of the first kind, x_2 objects of the second kind, ..., and x_k objects of the kth kind is given by

Multivariate hypergeometric distribution

$$f(x_1, x_2, \ldots, x_k) = \frac{\binom{a_1}{x_1}\binom{a_2}{x_2}\cdots\binom{a_k}{a_k}}{\binom{a_1 + a_2 + \cdots + a_k}{n}}$$

where $n = x_1 + x_2 + \cdots + x_k$. This distribution is called the **multivariate hypergeometric distribution.**

EXAMPLE 8.18   A panel of prospective jurors consists of six married men, three single men, seven married women, and four single women. If the selection is random, what is the probability that a jury (including alternates) will consist of four married men, one single man, five married women, and two single women?

Solution   Substituting $x_1 = 4$, $x_2 = 1$, $x_3 = 5$, $x_4 = 2$, $a_1 = 6$, $a_2 = 3$, $a_3 = 7$, $a_4 = 4$, and, hence, $n = 12$, into the formula for the multivariate hypergeometric distribution, we get

$$f(x_1, x_2, x_3, x_4) = \frac{\binom{6}{4}\binom{3}{1}\binom{7}{5}\binom{4}{2}}{\binom{20}{12}} \approx 0.0450 \qquad \blacksquare$$

■ Exercises

8.37   Check in each case whether the values of n and p satisfy the rule of thumb that we gave on page 216 for using the Poisson approximation to the binomial distribution:
(a) $n = 250$ and $p = \frac{1}{20}$;
(b) $n = 400$ and $p = \frac{1}{50}$;
(c) $n = 90$ and $p = \frac{1}{10}$.

8.38   Check in each case whether the values of n and p satisfy the rule of thumb that we gave on page 216 for using the Poisson approximation to the binomial distribution:
(a) $n = 300$ and $p = 0.01$;
(b) $n = 600$ and $p = 0.02$;
(c) $n = 75$ and $p = 0.1$.

8.39   It is known from experience that 8% of the calls one receives around dinner time are solicitations for funds. Use the Poisson approximation to the binomial distribution to determine the probability that among 120 such calls only four will be solicitations for funds.

8.40   If 0.8% of the fuses delivered to an arsenal are defective, use the Poisson approximation to the binomial distribution to determine the probability that in a random sample of 500 fuses delivered to this arsenal, five will be defective.

8.41   Suppose that in a certain city, 5% of all licensed drivers will be involved in at least one car accident in any given year. Use the formula for the Poisson distribution to approximate the probability that among 150 licensed drivers in that city, at most two will be involved in at least one car accident in a given year.

8.42   Use the computer printouts of Figure 8.4 to determine the error of the approximation of Exercise 8.41.

8.43   With reference to Example 8.13, where we had $n = 10,000$ and $p = 0.00006$, use the formula for the Poisson distribution to approximate the probability that three or four of 10,000 cars will have a flat tire while being driven through the tunnel.

8.44   Referring to the computer printout of Figure 8.5, rework Exercise 8.43 using
(a) the values in the $P(X = x)$ column;
(b) the values in the $P(X <= x)$ column.

8.45   If we use the same rules as in Example 8.14, can the hypergeometric distribution with $n = 120$, $a = 50$, and $b = 3,150$ be approximated with a Poisson distribution?

8.46   The number of complaints that a dry cleaning establishment receives per day is a random variable having the Poisson distribution with $\lambda = 3.5$. Find the probabilities that on any given day it will receive
(a) only two complaints;
(b) at most two complaints.

8.47   The number of emergency calls that an ambulance service gets per day is a random variable having the Poisson distribution with $\lambda = 4.8$. What is the probability that on a given day it will receive only four emergency calls?

8.48   The number of monthly breakdowns of the kind of computer used by an office is a random variable having the Poisson distribution with $\lambda = 1.6$. Find the probabilities that this kind of computer will function for a month
(a) without a breakdown;
(b) with one breakdown;
(c) with two breakdowns.

8.49   With reference to Exercise 8.48, suppose that the office has four of the computers. Find the probability that all four of them will function for a month without a breakdown.

*8.50 With reference to Exercise 8.48, suppose that the office has five of the computers. Find the probability that in a given month two of them will function without a breakdown, two of them will have one breakdown, and the other one will have two breakdowns.

*8.51 For a car being tested at a state inspection station, the probability that it will pass on the first try is 0.70, the probability that it will pass on the second try is 0.20, and the probability that it will pass on the third try is 0.10. What is the probability that among ten cars being tested, six will pass on the first try, three will pass on the second try, and one will pass on the third try?

*8.52 According to the Mendelian theory of heredity, if plants with round yellow seeds are crossbred with plants with wrinkled green seeds, the probabilities of getting a plant that produces round yellow seeds, wrinkled yellow seeds, round green seeds, or wrinkled green seeds are, respectively, $\frac{9}{16}, \frac{3}{16}, \frac{3}{16},$ and $\frac{1}{16}$. What is the probability that among nine plants thus obtained there will be four that produce round yellow seeds, two that produce wrinkled yellow seeds, three that produce round green seeds, and none that produce wrinkled green seeds?

*8.53 The probabilities are 0.60, 0.20, 0.10, and 0.10 that a state income tax form will be filled out correctly, that it will contain only errors favoring the taxpayer, that it will contain only errors favoring the government, and that it will contain both kinds of errors. What is the probability that among ten such tax forms (randomly selected for audit) seven will be filled out correctly, one will contain only errors favoring the taxpayer, one will contain only errors favoring the government, and one will contain both kinds of errors?

*8.54 If 18 defective glass bricks include 10 that have cracks but no discoloration, five that have discoloration but no cracks, and three that have cracks and discoloration, what is the probability that among six of the bricks (chosen at random for further checks) three will have cracks but no discoloration, one will have discoloration but no cracks, and two will have cracks and discoloration?

*8.55 Among 25 silver dollars struck in 1903 there are 15 from the Philadelphia mint, seven from the New Orleans mint, and three from the San Francisco mint. If five of these silver dollars are picked at random, find the probabilities of getting
(a) four from the Philadelphia mint and one from the New Orleans mint;
(b) three from the Philadelphia mint and one from each of the other two mints.

8.7   THE MEAN OF A PROBABILITY DISTRIBUTION

When we showed on page 181 that an airline office at a certain airport can expect 2.75 complaints per day about its luggage handling, we arrived at this result by using the formula for a mathematical expectation, namely, by adding the products obtained by multiplying 0, 1, 2, 3, . . . by the corresponding probabilities that the office will receive 0, 1, 2, 3, . . . complaints about its luggage handling on any given day. Here the number of complaints is a random variable and 2.75 is its **expected value.**

If we apply the same argument to the first example of Section 8.2, we find that, between them, the two salespersons can expect to sell

$$0(0.44) + 1(0.20) + 2(0.18) + 3(0.18) = 1.10$$

1.1 of the trucks. In this case, the number of trucks is a random variable and 1.1 is its expected value.

As we explained in Chapter 7, mathematical expectations must be interpreted as averages, or means, and it is customary to refer to the expected value of a random variable as its **mean,** or as the **mean of its probability distribution.** In general, if a random variable takes on the values $x_1, x_2, x_3, \ldots,$ or x_k, with the probabilities $f(x_1), f(x_2), f(x_3), \ldots,$ and $f(x_k)$, its expected value is

$$x_1 \cdot f(x_1) + x_2 \cdot f(x_2) + x_3 \cdot f(x_3) + \cdots + x_k \cdot f(x_k)$$

and in the Σ notation we write

Mean of a probability distribution

$$\mu = \sum x \cdot f(x)$$

Like the mean of a population, the mean of a probability distribution is denoted by the Greek lowercase μ (*mu*). The notation is the same, for as we pointed out in connection with the binomial distribution, when we observe a value of a random variable, we refer to its distribution as the population we are sampling. For instance, the histogram of Figure 8.2 on page 205 may be looked upon as the population we are sampling when we observe a value of a random variable having the binomial distribution with $n = 6$ and $p = 0.30$.

EXAMPLE 8.19   Find the mean of the second probability distribution of Section 8.2: the one that pertained to the number of points we roll with a balanced die.

Solution   Since the probabilities of rolling a 1, 2, 3, 4, 5, or 6 are all $\frac{1}{6}$, we get

$$\mu = 1 \cdot \frac{1}{6} + 2 \cdot \frac{1}{6} + 3 \cdot \frac{1}{6} + 4 \cdot \frac{1}{6} + 5 \cdot \frac{1}{6} + 6 \cdot \frac{1}{6} = 3\frac{1}{2}$$

∎

EXAMPLE 8.20   With reference to Example 8.5, find the mean number of persons, among six shopping at the supermarket, who will take advantage of the special promotion.

Solution   Substituting $x = 0, 1, 2, 3, 4, 5,$ and 6, and the probabilities on pages 204 and 205 into the formula for μ, we get

$$\mu = 0(0.118) + 1(0.303) + 2(0.324) + 3(0.185)$$
$$+ 4(0.060) + 5(0.010) + 6(0.001)$$
$$= 1.802 \qquad \blacksquare$$

When a random variable can take on many different values, the calculation of μ may become very laborious. For instance, if we want to know how many persons can be expected to contribute to a charity, when 2,000 are solicited for funds and the probability is 0.40 that any one of them will make a contribution, we might consider calculating the 2,001 probabilities corresponding to $0, 1, 2, 3, \ldots, 1,999,$ or $2,000$ of them making a contribution. Not seriously, though, and we might argue instead that in the long run 40% of the persons will make a contribution, 40% of 2,000 is 800, and hence we can expect that 800 of the 2,000 persons will make a contribution. Similarly, if a balanced coin is flipped 1,000 times, we might argue that in the long run heads will come up 50% of the time, and hence that we can expect $1,000(0.50) = 500$ heads. These two results are correct; both problems deal with random variables having binomial distributions, and it can be shown that in general

Mean of a binomial distribution

$$\mu = n \cdot p$$

In words, the mean of a binomial distribution is simply the product of the number of trials and the probability of success on an individual trial.

EXAMPLE 8.21   With reference to Examples 8.5 and 8.20, use this formula to determine the mean number of persons, among six shopping at the supermarket, who can be expected to take advantage of the special promotion.

Solution   Since we are dealing with the binomial distribution having $n = 6$ and $p = 0.30$, we get $\mu = 6(0.30) = 1.80$. The small difference between the values obtained here and in Example 8.20 is due to rounding the probabilities used in Example 8.20 to three decimals. $\qquad \blacksquare$

It is important to remember that the formula $\mu = n \cdot p$ applies only to binomial distributions. There are other formulas for other distributions; for instance, for the hypergeometric distribution the formula for the mean is

Mean of a
hypergeometric
distribution

$$\mu = \frac{n \cdot a}{a + b}$$

EXAMPLE 8.22   Among twelve school buses, five have worn brakes. If six of these buses are randomly picked for inspection, how many of them can be expected to have worn brakes?

Solution   Since we are sampling without replacement, we have here a hypergeometric situation with $a = 5, b = 7$, and $n = 6$. Substituting these values into the special formula for the mean of a hypergeometric distribution, we get

$$\mu = \frac{6 \cdot 5}{5 + 7} = 2.5$$

This should not come as a surprise—half of the school buses are chosen for inspection and, hence, half of the ones with faulty brakes can be expected to be included in the sample.   ∎

Also, the mean of the Poisson distribution with the parameter λ is $\mu = \lambda$, and this agrees with what we suggested earlier, namely, that λ is to be interpreted as an average. Derivations of all these special formulas may be found in textbooks on mathematical statistics.

8.8   THE STANDARD DEVIATION OF A PROBABILITY DISTRIBUTION

In Chapter 4 we saw that the most widely used measures of variation are the variance and its square root, the standard deviation, which measure variability by averaging the squared deviations from the mean. For probability distributions we measure variability in nearly the same way, but instead of averaging the squared deviations from the mean, we find their expected value. In general, if a random variable takes on the values $x_1, x_2, x_3, \ldots$, or x_k, with the probabilities $f(x_1), f(x_2), f(x_3), \ldots$, and $f(x_k)$, and the mean of this probability distribution is μ, then the deviations from the mean are $x_1 - \mu, x_2 - \mu, x_3 - \mu, \ldots$, and $x_k - \mu$, and the expected value of their squares is

$$(x_1 - \mu)^2 \cdot f(x_1) + (x_2 - \mu)^2 \cdot f(x_2) + \cdots + (x_k - \mu)^2 \cdot f(x_k)$$

Thus, in the Σ notation we write

Variance of a
probability
distribution

$$\sigma^2 = \sum (x - \mu)^2 \cdot f(x)$$

which we refer to as the **variance of the random variable** or **the variance of its probability distribution.** As in the preceding section, and for the same reason, we denote this description of a probability distribution with the same symbol as the corresponding description of a population.

The square root of the variance defines the **standard deviation of a probability distribution** and we write

Standard deviation of a probability distribution

$$\sigma = \sqrt{\sum (x - \mu)^2 \cdot f(x)}$$

EXAMPLE 8.23   With reference to Example 8.5, find the standard deviation of the number of persons, among six shopping at the supermarket, who will take advantage of the special promotion.

Solution   As we have seen in Example 8.21, $\mu = 6(0.30) = 1.80$ for this random variable. Thus, we can arrange the calculation of the variance as follows:

Number of persons	Probability	Deviation from mean	Squared deviation from mean	$(x - \mu)^2 f(x)$
0	0.118	−1.8	3.24	0.38232
1	0.303	−0.8	0.64	0.19392
2	0.324	0.2	0.04	0.01296
3	0.185	1.2	1.44	0.26640
4	0.060	2.2	4.84	0.29040
5	0.010	3.2	10.24	0.10240
6	0.001	4.2	17.64	0.01764

$$\sigma^2 = 1.26604$$

The values in the right-hand column were obtained by multiplying the squared deviations from the mean by their probabilities, and the total of this column is the variance of the distribution. Thus, the standard deviation is

$$\sigma = \sqrt{1.26604} \approx 1.13$$

The calculations were easy in this example because the deviations from the mean were small numbers given to one decimal. If the deviations from the mean are large numbers, or if they are given to several decimals, it is usually worthwhile to simplify the calculations by using the following computing formula, which does not require that we work with the deviations from the mean:

Computing formula for the variance of a probability distribution

$$\sigma^2 = \sum x^2 \cdot f(x) - \mu^2$$

EXAMPLE 8.24   In Example 7.6 we showed that if the probabilities are 0.06, 0.21, 0.24, 0.18, 0.14, 0.10, 0.04, 0.02, and 0.01 that an airline office at a certain airport will receive 0, 1, 2, 3, 4, 5, 6, 7, or 8 complaints per day about its luggage handling, the mean of this probability distribution is $\mu = 2.75$. Use the preceding computing formula to find its standard deviation.

Solution First calculating

$$\sum x^2 \cdot f(x) = 0^2(0.06) + 1^2(0.21) + 2^2(0.24) + 3^2(0.18)$$
$$+ 4^2(0.14) + 5^2(0.10) + 6^2(0.04)$$
$$+ 7^2(0.02) + 8^2(0.01)$$
$$= 10.59$$

we get

$$\sigma^2 = 10.59 - (2.75)^2 \approx 3.03$$

and, hence, $\sigma = \sqrt{3.03} \approx 1.74$. The calculations would have been much more tedious in this case if we had not used the computing formula. ∎

As in the case of the mean, the calculation of the variance or the standard deviation can generally be simplified when we deal with special kinds of probability distributions. For instance, for the binomial distribution we have the formula

Standard deviation of a binomial distribution

$$\sigma = \sqrt{np(1 - p)}$$

EXAMPLE 8.25 Use this formula to verify the result obtained in Example 8.23 that dealt with the number of persons, among six shopping at the supermarket, who will take advantage of the special promotion.

Solution Since we are dealing with a binomial distribution with $n = 6$ and $p = 0.30$, the formula yields $\sigma = \sqrt{6(0.30)(0.70)} \approx 1.12$. Thus, the difference, due to rounding, is only $1.13 - 1.12 = 0.01$. ∎

There also exist special formulas for the standard deviation of other probability distributions. In Exercise 8.74 the reader will find the one for the hypergeometric distribution, and in Exercise 8.78 the one for the Poisson distribution.

Intuitively speaking, the standard deviation of a probability distribution measures the expected size of the chance fluctuations of a corresponding random variable. When σ is small, there is a high probability that we will get a value close to the mean; when σ is large, we are more likely to get a value far away from the mean. This important idea is expressed formally by Chebyshev's theorem, which we introduced in Section 4.3, as it pertains to numerical data. For probability distributions, Chebyshev's theorem may be stated as follows:

Chebyshev's theorem

The probability that a random variable will take on a value within k standard deviations of the mean is at least $1 - \dfrac{1}{k^2}$.

Thus, the probability of getting a value within two standard deviations of the mean (a value between $\mu - 2\sigma$ and $\mu + 2\sigma$) is at least $1 - \dfrac{1}{2^2} = \dfrac{3}{4}$, the probability of getting a value within five standard deviations of the mean (a value between $\mu - 5\sigma$ and $\mu + 5\sigma$) is at least $1 - \dfrac{1}{5^2} = \dfrac{24}{25}$, and so forth. Note that in the formulation of this theorem, the phrase "within k standard deviations of the mean" does not include the endpoints $\mu - k\sigma$ and $\mu + k\sigma$.

EXAMPLE 8.26   The number of telephone calls that an answering service receives between 9 A.M. and 10 A.M. is a random variable whose distribution has the mean $\mu = 27.5$ and the standard deviation $\sigma = 3.2$. What does Chebyshev's theorem with $k = 3$ tell us about the number of telephone calls that the answering service may receive between 9 A.M. and 10 A.M.?

Solution   Since $\mu - 3\sigma = 27.5 - 3(3.2) = 17.9$ and $\mu + 3\sigma = 27.5 + 3(3.2) = 37.1$, we can assert with a probability of at least $1 - \dfrac{1}{3^2} = \dfrac{8}{9}$, or approximately 0.89, that the answering service will receive between 17.9 and 37.1 calls, namely, anywhere from 18 to 37 calls. ∎

EXAMPLE 8.27   What does Chebyshev's theorem with $k = 5$ tell us about the number of heads, and hence the proportion of heads, we might get in 400 flips of a balanced coin?

Solution   Here we are dealing with a random variable having the binomial distribution with $n = 400$ and $p = 0.50$, so that $\mu = 400(0.50) = 200$ and $\sigma = \sqrt{400(0.50)(0.50)} = 10$. Since $\mu - 5\sigma = 200 - 5 \cdot 10 = 150$ and $\mu + 5\sigma = 200 + 5 \cdot 10 = 250$, we can assert with a probability of at least $1 - \dfrac{1}{5^2} = \dfrac{24}{25} = 0.96$ that we will get between 150 and 250 heads or that the proportion of heads will be between $\dfrac{150}{400} = 0.375$ and $\dfrac{250}{400} = 0.625$. ∎

To continue with this example, the reader will be asked to show in Exercise 8.82 that for $n = 10{,}000$ flips of a balanced coin the probability is at least 0.96 that the proportion of heads will be between 0.475 and 0.525, and that for 1,000,000 flips of a balanced coin the probability is at least 0.96 that the proportion of heads will be between 0.4975 and 0.5025. All this provides support for the law of large numbers, introduced in Chapter 5 in connection with the frequency interpretation of probability.

Exercises

8.56 Suppose that the probabilities are 0.4, 0.3, 0.2, and 0.1 that 1, 2, 3, or 4 new anti-inflammatory drugs will be approved by the FDA in the year 2001.
   (a) Use the formula that defines the mean of a probability distribution to find the mean of this distribution.
   (b) Use the formula that defines the variance of a probability distribution to find the variance of this distribution.
   (c) Use the computing formula for the variance of a probability distribution to find the variance of this distribution.

8.57 Under ordinary circumstances, the probabilities are 0.2, 0.6, 0.1, and 0.1 that 0, 1, 2, or 3 hurricanes will reach the U.S. coast in any given year.
   (a) Use the formula that defines μ to find the mean of this probability distribution.
   (b) Use the formula that defines σ^2 to find the variance and the standard deviation of this probability distribution.
   (c) Use the computing formula for σ^2 to find the variance and the standard deviation of this probability distribution.

8.58 The following table gives the probabilities that a probation officer will be informed of 0, 1, 2, 3, 4, 5, or 6 probation violations on any given day:

Number of violations	0	1	2	3	4	5	6
Probability	0.15	0.22	0.31	0.18	0.09	0.04	0.01

Use the formulas that define μ and σ^2 to find
   (a) the mean of this probability distribution;
   (b) the standard deviation of this probability distribution.

8.59 Use the computing formula for σ^2 to rework part (b) of Exercise 8.58.

8.60 In Section 8.2 we showed that the probabilities of getting 0, 1, 2, 3, or 4 heads in four flips of a balanced coin are, respectively, $\frac{1}{16}, \frac{4}{16}, \frac{6}{16}, \frac{4}{16}$, and $\frac{1}{16}$. Use the formulas that define μ and σ^2 to find the mean and the standard deviation of this probability distribution.

8.61 Rework Exercise 8.60, using the special formulas for the mean and the standard deviation of a binomial distribution.

8.62 A study shows that 60 percent of all first-class letters between two cities are delivered within 48 hours. Find the mean and the variance of the number of first-class letters between the two cities, among eight randomly selected, which will be delivered within 48 hours, using
   (a) Table V, the formula that defines μ, and the computing formula for σ^2;
   (b) the special formulas for the mean and the standard deviation of a binomial distribution.

8.63 If 80 percent of certain videocasette recorders will function successfully through the 90-day warranty period, find the mean and standard deviation of the number of these videocasette recorders, among 10 randomly selected, which will function successfully through the 90-day warranty period, using
   (a) Table V, the formula that defines μ, and the computing formula for σ^2;
   (b) the special formulas for the mean and the standard deviation of the binomial distribution.

8.64   A study shows that 27 percent of all patients coming to a certain medical clinic have to wait at least half an hour to see their doctor. Use the probabilities in Figure 8.6 to calculate μ and σ for the number of patients, among 18 coming to this clinic, who have to wait at least half an hour to see their doctor.

```
MTB > PDF c1;
SUBC>    Binomial 18 0.27


Binomial with n = 18 and p = 0.270000

          x           P( X = x)
        0.00            0.0035
        1.00            0.0231
        2.00            0.0725
        3.00            0.1431
        4.00            0.1985
        5.00            0.2055
        6.00            0.1647
        7.00            0.1044
        8.00            0.0531
        9.00            0.0218
       10.00            0.0073
       11.00            0.0020
       12.00            0.0004
```

FIGURE 8.6
Binomial distribution
with $n = 18$ and
$p = 0.27$.

8.65   Use the special formulas for the mean and the standard deviation of a binomial distribution to verify the results obtained in Exercise 8.64.

8.66   Find the mean and the standard deviation of each of the following binomial random variables:
(a) the number of heads obtained in 484 flips of a balanced coin;
(b) the number of 3's obtained in 720 rolls of a balanced die;
(c) the number of persons, among 600 invited, who will attend the opening of a new branch bank, when the probability is 0.30 that any one of them will attend;
(d) the number of defectives in a sample of 600 parts made by a machine, when the probability is 0.04 that any one of the parts is defective;
(e) the number of students, among 800 interviewed, who do not like the food served at the university cafeteria, when the probability is 0.65 that any one of them does not like the food.

8.67   Use the results of Exercise 8.28 to find the mean number of applicants, among the two randomly selected, who have working wives.

8.68   Use the special formula for the mean of a hypergeometric distribution to verify the result of Exercise 8.67.

8.69   Use the results of Exercise 8.29 to find the mean number of shipments containing contraband that the customs inspector will catch.

8.70   Use the special formula for the mean of a hypergeometric distribution to verify the result of Exercise 8.69.

8.71   In Example 8.10, which dealt with a random variable having a hypergeometric distribution with $a = 5, b = 11$, and $n = 8$, we showed that the probabilities are, respectively, 0.359, 0.128, and 0.013 that it will take on the values 3, 4, or 5. As can easily be verified, the corresponding probabilities for 0, 1, or 2 "successes" are 0.013,

0.128, and 0.359. Use all these probabilities to calculate the mean of this hyperge-ometric distribution. Also, use the special formula for the mean of a hypergeomet-ric distribution to verify the result.

8.72 Among eight faculty members considered for promotions, four have Ph.D. degrees and four do not.
  (a) If four of them are chosen at random, find the probabilities that 0, 1, 2, 3, or all 4 of them have Ph.D.'s.
  (b) Use the results of part (a) to determine the mean of this probability distribu-tion.
  (c) Use the special formula for the mean of a hypergeometric distribution to ver-ify the result of part (b).

8.73 Use parts (a) and (c) of Exercise 8.72 to find the standard deviation of the number of faculty members, among the four chosen at random, who have Ph.D.'s.

*8.74 For the variance of a hypergeometric distribution with the parameters a, b, and n, there is the special formula

$$\sigma^2 = \frac{nab(a + b - n)}{(a + b)^2(a + b - 1)}$$

Use this formula to verify the result of the preceding exercise.

*8.75 Use the formula of Exercise 8.74 to find the variance of the random variable of
  (a) Example 8.10, where $a = 5$, $b = 11$, and $n = 8$;
  (b) Exercise 8.67, where $a = 9$, $b = 3$, and $n = 2$;
  (c) Exercise 8.69, where $a = 5$, $b = 11$, and $n = 3$.

8.76 The probabilities that there will be 0, 1, 2, 3, 4, or 5 fires caused by lightning dur-ing a summer storm in the Flagstaff area are, respectively, 0.449, 0.360, 0.144, 0.038, 0.008, and 0.001. Calculate the mean of this Poisson distribution with $\lambda = 0.8$ and use it to verify the special formula $\mu = \lambda$ mentioned on page 225.

8.77 If the number of gamma rays emitted per second by a certain radioactive substance is a random variable having the Poisson distribution with $\lambda = 2.5$, the probabilities that it will emit 0, 1, 2, 3, 4, 5, 6, 7, 8, or 9 gamma rays in any one second are, re-spectively, 0.082, 0.205, 0.256, 0.214, 0.134, 0.067, 0.028, 0.010, 0.003, and 0.001. Cal-culate the mean and use the result to verify the special formula $\mu = \lambda$ for a random variable having the Poisson distribution with the parameter λ.

*8.78 For the standard deviation of a Poisson distribution there is the special formula $\sigma = \sqrt{\lambda}$. Calculate the standard deviation of the probability distribution of Exer-cise 8.76 and use the result to verify this special formula.

*8.79 Calculate the standard deviation of the probability distribution of Exercise 8.77 and use the result to verify the special formula $\sigma = \sqrt{\lambda}$ for the standard deviation of a Poisson distribution.

8.80 If a student answers the 144 questions of a true–false test by flipping a balanced coin—heads is "true" and tails is "false"—what does Chebyshev's theorem with $k = 4$ tell us about the number of correct answers he will get?

8.81 In a certain Southwestern city, the annual number of days with above 100 degree temperature is a random variable with $\mu = 138$ and $\sigma = 9$.
  (a) What does Chebyshev's theorem with $k = 12$ tell us about the number of days with above 100 degree temperature there will be in this city in any one year?

(b) According to Chebyshev's theorem, with what probability can we assert that there will be between 108 and 168 days with above 100 degree temperature in this city in any one year?

8.82 Use Chebyshev's theorem to show that the probability is at least 0.96 that
(a) for 10,000 flips of a balanced coin the proportion of heads will be between 0.475 and 0.525;
(b) for 1,000,000 flips of a balanced coin the proportion of heads will be between 0.4975 and 0.5025.

8.9 Checklist of Key Terms (with page references to their definitions)

Binomial distribution, 203
Binomial population, 208
Discrete random variable, 200
Expected value of a random variable, 223
Geometric distribution, 211
Hypergeometric distribution, 212
Mean of a probability distribution, 223
*Multinomial distribution, 220
*Multivariate hypergeometric distribution, 220

Poisson distribution, 215, 216
Probability distribution, 199, 200
Random variable, 199
Standard deviation of probability distribution, 226
Statistical model, 205
Variance of a probability distribution, 225
Variance of a random variable, 225

8.10 References

A great deal of information about various probability distributions may be found in

HASTINGS, N. A. J., and PEACOCK, J. B., *Statistical Distributions*. London: Butterworth & Company (Publishers) Ltd., 1975.

More detailed tables of binomial probabilities may be found in

ROMIG, H. G., *50–100 Binomial Tables*. New York: John Wiley & Sons, Inc., 1953.

Tables of the Binomial Probability Distribution, National Bureau of Standards Applied Mathematics Series No. 6. Washington, D.C.: U.S. Government Printing Office, 1950.

and a detailed table of Poisson probabilities is given in

MOLINA, E. C., *Poisson's Exponential Binomial Limit*. Princeton, N.J.: D. Van Nostrand Company, Inc., 1947.

The wide availability of computer programs for binomial and Poisson probabilities makes it unlikely that any of the aforementioned tables will be extended or updated.

THE NORMAL
DISTRIBUTION

9.1   Continuous Distributions 234

9.2   The Normal Distribution 237

*9.3   A Check for Normality 247

9.4   Applications of the Normal Distribution 250

9.5   The Normal Approximation to the Binomial Distribution 253

9.6   Checklist of Key Terms 259

9.7   References 260

*C*ontinuous sample spaces and **continuous random variables** arise when we deal with quantities that are measured on a continuous scale—for instance, when we measure wind velocity or the altitude of a plane, the amount of alcohol in a person's blood, the net weight of a package of frozen chicken livers, or the amount of tar in a cigarette. In situations like these there is always a continuum of possibilities, but in practice we have no choice but to round measurements to the nearest unit or to a few decimals. Thus, if we say that on a given day the maximum temperature in Palm Springs was 109 degrees, rounded to the nearest degree Fahrenheit, this means that it could have been anywhere between 108.5 and 109.5. Similarly, if we say that a plane is flying at 33,000 feet, rounded to the nearest 100 feet, this means that its altitude could have been anywhere between 32,950 feet and 33,050 feet. We must keep this in mind when we ask for probabilities relating to continuous random variables. In other words, we associate probabilities with intervals or regions of a sample space, and not with individual points. For instance, we may want to know the probability that at a given moment a car is moving between 50 and 55 miles per hour, not that it is moving at exactly

$16\pi = 50.26548246\ldots$ miles per hour. Similarly, we may want to know whether a package of frozen food weighs at least 5.95 ounces, not that it weighs exactly $\sqrt{35.9} = 5.99166087\ldots$ ounces.

In this chapter we shall learn how to determine, and work with, probabilities relating to continuous random variables. The place of histograms will be taken by continuous curves, as in Figure 9.1, picturing them mentally as being approximated by histograms with narrower and narrower classes. After a general introduction to **continuous distributions** in Section 9.1, we shall devote the remainder of this chapter to the **normal distribution,** which is basic to most of the "bread and butter" techniques of modern statistics. Various applications of the normal distribution will be discussed in Sections 9.4 and 9.5, following the optional material in Section 9.3 that concerns a method of deciding whether observed data follow the general pattern of a normal distribution.

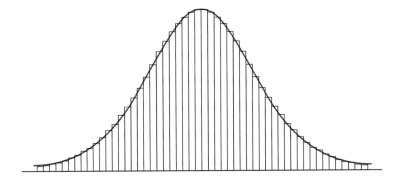

FIGURE 9.1
Continuous
distribution curve.

9.1   CONTINUOUS DISTRIBUTIONS

In the histograms we have seen thus far, the frequencies, percentages, proportions, or probabilities were represented by the heights of the rectangles, or by their areas. In the continuous case, we also represent probabilities by areas—not by areas of rectangles, but by areas under continuous curves. This is illustrated by Figure 9.2, where the diagram on the left shows a histogram of the probability distribution of a discrete random variable that takes on only the values $0, 1, 2, \ldots$ and 10. The probability that it will take on the value 3, for example, is given by the area of the lightly tinted rectangle, and the probability that it will take on a value greater than or equal to 8 is given by the sum of the areas of the

FIGURE 9.2
Histogram of a
probability
distribution and graph
of a continuous
distribution.

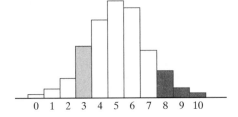
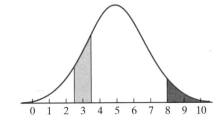

three darker tinted rectangles. The diagram on the right pertains to a continuous random variable that can take on any value on the interval from 0 to 10. The probability that it will take on a value on the interval from 2.5 to 3.5, for example, is given by the area of the lightly tinted region under the curve, and the probability that it will take on a value greater than or equal to 8 is given by the area of the darker tinted region under the curve.

Continuous curves such as the one shown on the right in Figure 9.2 are the graphs of functions called **probability densities,** or informally, **continuous distributions.** The term "probability density" comes from physics, where the terms "weight" and "density" are used in just about the same way in which we use the terms "probability" and "probability density" in statistics. As is illustrated by Figure 9.3, probability densities are characterized by the fact that

> **The area under the curve between any two values a and b gives the probability that a random variable having this continuous distribution will take on a value on the interval from a to b.**

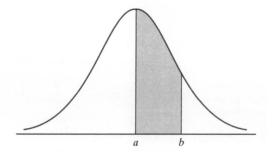

FIGURE 9.3
Continuous
distribution.

Observe from Figure 9.1 that when a and b are very close together, the height of the curve and the height of the rectangle with the interval from a to b as its base are nearly equal. Thus, the area under the curve from a to b nearly equals that of the corresponding rectangle.

It follows that the values of a continuous distribution must be nonnegative, and that the total area under the curve, representing the certainty that a random variable must take on one of its values, is always equal to 1. In contrast with the two rules on page 202, there is no requirement that the values of a continuous distribution be less than or equal to 1.

EXAMPLE   9.1   Verify that $f(x) = \dfrac{x}{8}$ can serve as the probability density of a continuous random variable that is defined over the interval from $x = 0$ to $x = 4$.

$\mathbf{S}$olution   The first condition is satisfied since $\dfrac{x}{8}$ is nonnegative (positive or zero) for all values of x on the interval from 0 to 4. Insofar as the second condition is concerned, it can be seen from Figure 9.4 that the total area under the curve from $x = 0$ to $x = 4$ is that of a triangle, whose base is 4 and whose height is $\frac{4}{8} = \frac{1}{2}$. The usual formula for the area of a triangle yields the required $\frac{1}{2} \cdot 4 \cdot \frac{1}{2} = 1$. ∎

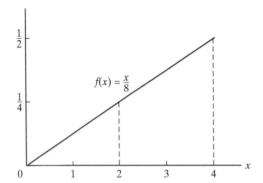

FIGURE 9.4
Diagram for Example 9.1.

EXAMPLE   9.2   With reference to Example 9.1, find the probabilities that a random variable having the given probability density will take on a value

(a) less than 2;
(b) less than or equal to 2.

Solution

(a) The probability is given by the smaller triangle of Figure 9.4, which is bounded on the right by the dashed line at $x = 2$. Its base is 2, its height is $\frac{2}{8} = \frac{1}{4}$, and its area is $\frac{1}{2} \cdot 2 \cdot \frac{1}{4} = \frac{1}{4}$.
(b) The probability is the same as that of part (a), namely, $\frac{1}{4}$. ∎

This example illustrates the important fact that, in the continuous case, the probability is zero that a random variable will take on any particular value. In our example, the probability is zero that the random variable will take on the value 2, and by 2 we mean *exactly* 2, and we do not include nearby values such as 1.9999998 or 2.0000001.

A consequence of measuring (rather than counting) is that we must assign probability zero to any particular outcome. We assert that the probability is zero that an individual will have a weight of *exactly* 145.27 pounds or that a horse will run a race in *exactly* 58.442 seconds. Observe, however, that even though every particular outcome has probability zero, the process will still produce a value (whether or not we can measure it with extra-fine precision); thus, events of probability zero not only can occur, but must occur when we deal with measured (continuous) random variables.

Statistical descriptions of continuous distributions are as important as descriptions of probability distributions or distributions of observed data, but most of them, including the mean and the standard deviation, cannot be defined without using calculus. Informally, though, we can always picture continuous distributions as being approximated by histograms of probability distributions (see Figure 9.1), whose mean and standard deviation we can calculate. Then, if we choose histograms with narrower and narrower classes, the means and the standard deviations of the corresponding probability distributions will approach the mean and the standard deviation of the continuous distribution. Actually, the mean and the standard deviation of a continuous distribution measure the same properties as the mean and the standard deviation of a probability distribution—the expected value of a random variable having the given distribution, and the square root of the expected value of its squared deviations from the mean. More intuitively, the mean μ of a continuous distribution is a measure of its center, or middle, and the standard deviation σ of a continuous distribution is a measure of its dispersion, or spread.

9.2   THE NORMAL DISTRIBUTION

Among many different continuous distributions used in statistics, the most important is the **normal distribution**, whose study dates back to eighteenth-century investigations concerning the nature of errors of measurement. It was observed that discrepancies among repeated measurement of the same physical quantity displayed a surprising degree of regularity. The distribution of the discrepancies could be closely approximated by a certain continuous curve, referred to as the "normal curve of errors" and attributed to the laws of chance. The mathematical equation for this type of curve is

$$f(x) = \frac{1}{\sigma\sqrt{2\pi}}\, e^{-\frac{1}{2}\left(\frac{x-\mu}{\sigma}\right)^2}$$

for $-\infty < x < \infty$, where e is the irrational number 2.71828 . . . , which we met on page 216 in connection with the Poisson distribution. We have given this equation only to point out some of the key features of normal distributions; it is not used in any of our calculations.

The graph of a normal distribution is a bell-shaped curve that extends indefinitely in both directions. Although this may not be apparent from a small drawing like that of Figure 9.5, the curve comes closer and closer to the horizontal axis without ever reaching it, no matter how far we go in either direction away from the mean. Fortunately, it is seldom necessary to extend the tails of a normal distribution very far because the area under the curve more than four or five standard deviations away from the mean is negligible for most practical purposes (see Exercise 9.15).

FIGURE 9.5
Normal distribution curve.

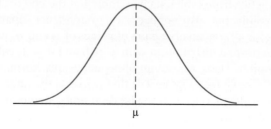

μ

An important feature of normal distributions, apparent from the preceding equation, is that they depend only on the two quantities μ and σ, which are, indeed, the mean and the standard deviation. In other words, there is one and only one normal distribution with a given mean μ and a given standard deviation σ. The fact that we will get different curves depending on the values of μ and σ is illustrated by Figure 9.6. At the top there are two normal curves with unequal means but equal standard deviations; the curve to the right has the higher mean. In the middle are two normal curves with equal means but unequal standard deviations; the lower and flatter curve has the higher standard deviation. At the bottom are two normal curves with unequal means and unequal standard deviations.

FIGURE 9.6

Three pairs of normal distributions.

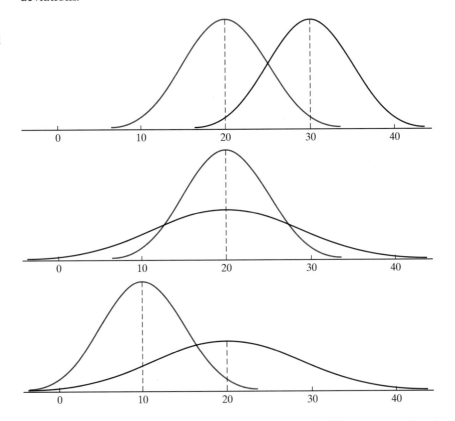

In all our work with normal distributions, we shall be concerned only with areas under their curves—so-called **normal-curve areas**—and such areas are found in practice from tables such as Table I at the end of the book. As it is physically impossible, but also unnecessary, to construct separate tables of normal-curve areas for all conceivable pairs of values of μ and σ, we tabulate these areas only for the normal distribution with $\mu = 0$ and $\sigma = 1$, called the **standard normal distribution.** Then, we obtain areas under any normal curve by performing the change of scale (see Figure 9.7) that converts the units of measurement from the original scale, or x-scale, into **standard units, standard scores,** or **z-scores,** by means of the formula

Standard units

$$z = \frac{x - \mu}{\sigma}$$

In this new scale, the *z*-scale, a value of *z* simply tells us how many standard deviations the corresponding value of *x* lies above or below the mean of its distribution.

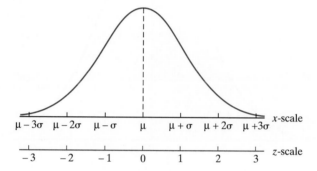

FIGURE 9.7
Change of scale to standard units.

The entries in Table I are the areas under the standard normal curve between the mean $z = 0$ and $z = 0.00, 0.01, 0.02, \ldots, 3.08,$ and 3.09, and also $z = 4.00$, $z = 5.00$, and $z = 6.00$. In other words, the entries in Table I are normal-curve areas like that of the tinted region of Figure 9.8.

Table I has no entries corresponding to negative values of *z*, for these are not needed by virtue of the symmetry of any normal curve about its mean. This follows from the equation on page 237, which remains unchanged when we substitute $-(x - \mu)$ for $x - \mu$. Specifically, $f(\mu - a) = f(\mu + a)$, meaning that we get the same value for $f(x)$ when we go the distance *a* to the left or to the right of μ.

FIGURE 9.8
Tabulated normal-curve area.

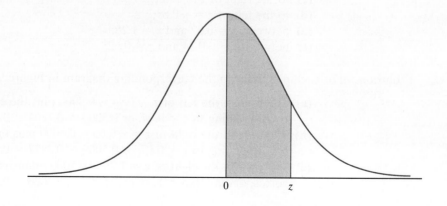

EXAMPLE   9.3   Find the standard normal-curve area between $z = -1.20$ and $z = 0$.

Solution   As can be seen from Figure 9.9, the area under the curve between $z = -1.20$ and $z = 0$ equals the area under the curve between $z = 0$ and $z = 1.20$. So we look up the entry corresponding to $z = 1.20$ in Table I and we get 0.3849.   ∎

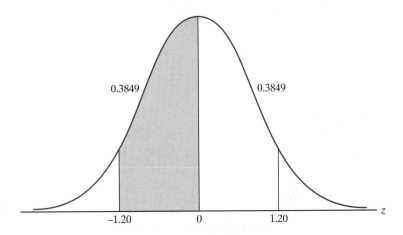

0.3849   0.3849

−1.20   0   1.20   z

FIGURE 9.9
Diagram for Example
9.3.

Questions concerning areas under normal distributions arise in various ways, and the ability to find any desired area quickly can be a big help. Although the table gives only areas between $z = 0$ and selected positive values of z, we often have to find areas to the left or to the right of given positive or negative values of z, or areas between two given values of z. This is easy, provided that we remember exactly what areas are represented by the entries in Table I, and also that the standard normal distribution is symmetrical about $z = 0$, so that the area under the curve to the left of $z = 0$ and that to the right of $z = 0$ are both equal to 0.5000.

EXAMPLE   9.4   Find the standard normal-curve area

    (a) to the left of $z = 0.94$;
    (b) to the right of $z = -0.65$;
    (c) to the right of $z = 1.76$;
    (d) to the left of $z = -0.85$;
    (e) between $z = 0.87$ and $z = 1.28$;
    (f) between $z = -0.34$ and $z = 0.62$.

Solution   For each part refer to the corresponding diagram in Figure 9.10.

    (a) The area to the left of $z = 0.94$ is 0.5000 plus the entry in Table I corresponding to $z = 0.94$, or $0.5000 + 0.3264 = 0.8264$.
    (b) The area to the right of $z = -0.65$ is 0.5000 plus the entry in Table I corresponding to $z = 0.65$, or $0.5000 + 0.2422 = 0.7422$.
    (c) The area to the right of $z = 1.76$ is 0.5000 minus the entry in Table I corresponding to $z = 1.76$, or $0.5000 - 0.4608 = 0.0392$.

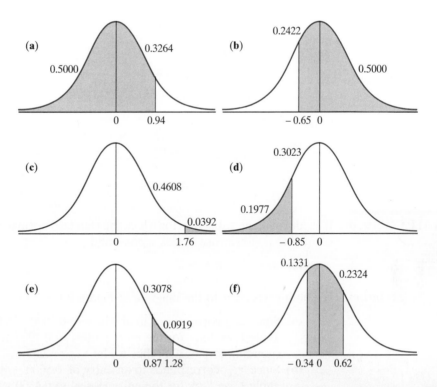

FIGURE 9.10
Diagram for Example 9.4.

(d) The area to the left of $z = -0.85$ is 0.5000 minus the entry in Table I corresponding to $z = 0.85$, or $0.5000 - 0.3023 = 0.1977$.

(e) The area between $z = 0.87$ and $z = 1.28$ is the difference between the entries in Table I corresponding to $z = 0.87$ and $z = 1.28$, or $0.3997 - 0.3078 = 0.0919$.

(f) The area between $z = -0.34$ and $z = 0.62$ is the sum of the entries in Table I corresponding to $z = 0.34$ and $z = 0.62$, or $0.1331 + 0.2324 = 0.3655$. ∎

In both of the preceding examples we dealt directly with the standard normal distribution. Now let us consider an example where μ and σ are not 0 and 1, so that we must first convert to standard units.

EXAMPLE 9.5   If a random variable has the normal distribution with $\mu = 10$ and $\sigma = 5$, what is the probability that it will take on a value on the interval from 12 to 15?

Solution   The probability is given by the area of the tinted region of Figure 9.11. Converting $x = 12$ and $x = 15$ to standard units, we get

$$z = \frac{12 - 10}{5} = 0.40 \quad \text{and} \quad z = \frac{15 - 10}{5} = 1.00$$

and since the corresponding entries in Table I are 0.1554 and 0.3413, we find that the probability asked for in this example is $0.3413 - 0.1554 = 0.1859$. ∎

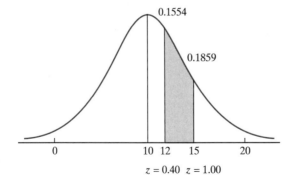

FIGURE 9.11
Diagram for Example
9.5.

EXAMPLE   9.6   If z_α denotes the value of z for which the standard normal-curve area to its right is equal to α (lowercase Greek *alpha*), find

      (a) $z_{0.01}$;   **(b)** $z_{0.05}$.

Solution   For both parts refer to the diagram of Figure 9.12.

    (a) Since $z_{0.01}$ corresponds to an entry of $0.5000 - 0.0100 = 0.4900$ in Table I, we look for the entry closest to 0.4900 and find 0.4901 corresponding to $z = 2.33$; thus we let $z_{0.01} = 2.33$.

    (b) Since $z_{0.05}$ corresponds to an entry of $0.5000 - 0.0500 = 0.4500$ in Table I, we look for the entry closest to 0.4500 and find 0.4495 and 0.4505 corresponding to $z = 1.64$ and $z = 1.65$; thus, we let $z_{0.05} = 1.645$.   ■

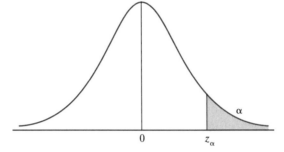

FIGURE 9.12
Diagram for Example
9.6.

      Table I also enables us to verify the remark on page 86 that for frequency distributions having the general shape of the cross section of a bell, about 68 percent of the values will lie within one standard deviation of the mean, about 95 percent will lie within two standard deviations of the mean, and about 99.7% will lie within three standard deviations of the mean. These percentages apply to frequency distributions having the general shape of a normal distribution, and the reader will be asked to verify them in the first three parts of Exercise 9.15. The other two parts of that exercise show that, although the "tails" extend indefinitely in both directions, the standard-normal-curve area to the right of $z = 4$ or $z = 5$, or to the left of $z = -4$ or $z = -5$, is negligible.

Although this chapter is devoted to the normal distribution and its importance cannot be denied, it would be a serious mistake to think that the normal distribution is the only continuous distribution that matters in the study of statistics. In Chapter 11 and in subsequent chapters we shall meet other continuous distributions that play important roles in problems of statistical inference.

■ Exercises

9.1 Explain in each case why the given equation cannot serve as the probability density of a random variable that takes on values on the interval from 1 to 4:
(a) $f(x) = 0.25$;
(b) $f(x) = \frac{1}{9}(4x - 7)$.

9.2 Explain in each case why the given equation cannot serve as the probability density of a random variable that takes on values on the interval from 0 to 5:
(a) $f(x) = \frac{1}{10}(x - 4)$;
(b) $f(x) = \frac{1}{50}(x + 1)$.

9.3 Figure 9.13 shows the graph of the **uniform distribution** of a random variable that takes on values on the interval from 2 to 10. Find the probabilities that this random variable will take on
(a) a value less than 8;
(b) the value 8;
(c) a value between 3.4 and 9.8.

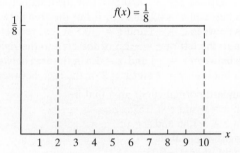

FIGURE 9.13
Diagram for Exercise 9.3.

9.4 Suppose that a random variable has the uniform distribution (see Exercise 9.3) on the interval from 20 to 50. Find the probabilities that this random variable will take on a value
(a) from 30 to 50;
(b) less than 20;
(c) greater than 45;
(d) between 25 and 35.

9.5 Figure 9.14 shows the graph of the distribution of a continuous random variable that takes on values on the interval from -1 to 1. Find the probabilities that this random variable will take on a value
(a) between $\frac{1}{3}$ and 1;

(b) between $-\frac{1}{2}$ and $\frac{1}{2}$;

(c) greater than $-\frac{1}{5}$.

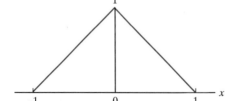

FIGURE 9.14
Diagram for Exercise 9.5.

9.6   For each case involving areas under the standard normal curve, decide whether the first area is bigger, the second area is bigger, or the two areas are equal:

(a) the area to the right of $z = 1.5$ or the area to the right of $z = 2$;

(b) the area to the left of $z = -1.5$ or the area to the left of $z = -2$;

(c) the area to the right of $z = 1$ or the area to the left of $z = -1.5$;

(d) the area to the right of $z = 2$ or the area to the left of $z = -2$;

(e) the area to the right of $z = -2.5$ or the area to the right of $z = -1.5$;

(f) the area to the left of $z = 0$ or the area to the right of $z = -0.1$;

(g) the area to the right of $z = 0$ or the area to the left of $z = 0$;

(h) the area to the right of $z = -1.4$ or the area to the left of $z = -1.4$.

9.7   For each case involving areas under the standard normal curve, decide whether the first area is bigger, the second area is bigger, or the two areas are equal:

(a) the area between $z = 0$ and $z = 1.3$ or the area between $z = 0$ and $z = 1$;

(b) the area between $z = -0.2$ and $z = 0.2$ or the area between $z = -0.4$ and $z = 0.4$;

(c) the area between $z = -1$ and $z = 1$ or the area between $z = 0$ and $z = 2$;

(d) the area between $z = 0.1$ and $z = 0.2$ or the area between $z = 1.1$ and $z = 1.2$;

(e) the area between $z = -1$ and $z = -0.5$ or the area between $z = 0.5$ and $z = 1$;

(f) the area to the left of $z = -1.5$ or the area to the right of $z = -0.5$;

(g) the area between $z = -1$ and $z = 1.5$ or the area between $z = -1.5$ and $z = 1$;

(h) the area between $z = 1$ and $z = 2$ or the area between $z = 2$ and $z = 3$.

9.8   Find the standard normal-curve area that lies

(a) between $z = 0$ and $z = 0.87$;

(b) between $z = -1.66$ and $z = 0$;

(c) to the right of $z = 0.48$;

(d) to the right of $z = -0.27$;

(e) to the left of $z = 1.30$;

(f) to the left of $z = -0.79$.

9.9   Find the area under the standard normal curve that lies

(a) between $z = 0.45$ and $z = 1.23$;

(b) between $z = -1.15$ and $z = 1.85$;

(c) between $z = -1.35$ and $z = 0.48$.

9.10   Find the standard normal-curve area that lies

(a) between -0.77 and $z = 0.77$;

(b) to the right of $z = -1.39$;

(c) to the left of $z = 0.27$;

(d) between $z = 1.69$ and $z = 2.33$.

9.11   For each case involving areas under a normal curve with $\mu = 100$, decide whether the first area is bigger, the second area is bigger, or the two areas are equal:
  (a) the area to the right of 120 or the area to the left of 70;
  (b) the area between 88 and 112 or the area between 100 and 124;
  (c) the area to the right of 90 or the area to the left of 110.

9.12   For each case involving random variables with normal distributions, decide whether the first probability is bigger, the second probability is bigger, or the two probabilities are equal:
  (a) the probability that a random variable having the normal distribution with $\mu = 50$ and $\sigma = 10$ takes on a value less than 60 or the probability that a random variable having the normal distribution with $\mu = 500$ and $\sigma = 100$ takes on a value less than 600;
  (b) the probability that a random variable having the normal distribution with $\mu = 40$ and $\sigma = 5$ takes on a value greater than 40 or the probability that a random variable having the normal distribution with $\mu = 50$ and $\sigma = 5$ takes on a value greater than 40;
  (c) the probability that a random variable having the normal distribution with $\mu = 50$ and $\sigma = 10$ takes on a value less than 60 or the probability that a random variable having the normal distribution with $\mu = 50$ and $\sigma = 20$ takes on a value less than 60;
  (d) the probability that a random variable having the normal distribution with $\mu = 100$ and $\sigma = 5$ takes on a value greater than 110 or the probability that a random variable having the normal distribution with $\mu = 108$ and $\sigma = 5$ takes on a value greater than 110.

9.13   Find z if the standard normal-curve area
  (a) between 0 and z is 0.4726;
  (b) to the left of z is 0.9868;
  (c) to the left of z is 0.3085;
  (d) between $-z$ and z is 0.8502.

9.14   Find z if the standard normal-curve area
  (a) between 0 and z is 0.1443;
  (b) to the right of z is 0.7389;
  (c) to the right of z is 0.3409;
  (d) between $-z$ and z is 0.9282.

9.15   Find the area under the standard normal curve between $-z$ and z if
  (a) $z = 1$;
  (b) $z = 2$;
  (c) $z = 3$;
  (d) $z = 4$;
  (e) $z = 5$.

9.16   With z_α defined as in Example 9.6, verify that
  (a) $z_{0.025} = 1.96$;
  (b) $z_{0.005} = 2.575$.

9.17   If a random variable has the normal distribution with $\mu = 82.0$ and $\sigma = 4.8$, find the probabilities that it will take on a value
  (a) less than 89.2;
  (b) greater than 78.4;
  (c) between 83.2 and 88.0;
  (d) between 73.6 and 90.4.

9.18 If a random variable has the normal distribution with $\mu = 77.5$ and $\sigma = 12.4$, find the probabilities that it will take on a value
(a) less than 55.1;
(b) greater than 84.3;
(c) between 80.0 and 90.0;
(d) between 72.4 and 82.6;

9.19 A normal distribution has the mean $\mu = 62.4$. Find its standard deviation if 20% of the area under the curve lies to the right of 79.2.

9.20 A random variable has a normal distribution with $\sigma = 10$. If the probability that the random variable will take on a value less than 82.5 is 0.8212, what is the probability that it will take on a value greater than 58.3?

*9.21 Another continuous distribution, called the **exponential distribution,** has many important applications. If a random variable has an exponential distribution with the mean μ, the probability that it will take on a value between 0 and any given non-negative value of x is $1 - e^{-x/\mu}$ (see Figure 9.15). Here e is the irrational number that also appears in the formula for the normal distribution. Many calculators have keys for computing expressions of the form $e^{-x/\mu}$, and selected values may be obtained from Table XII. Find the probability that a random variable having the exponential distribution with $\mu = 10$ will take on a value

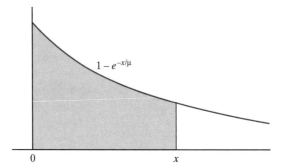

FIGURE 9.15
Exponential
distribution.

(a) less than 4;
(b) between 5 and 9;
(c) greater than 16.

*9.22 The lifetime of a certain electronic component is a random variable that has the exponential distribution with the mean $\mu = 2,000$ hours. Use the formula of Exercise 9.21 to find the probabilities that such a component will last
(a) at most 2,400 hours;
(b) at least 1,600 hours;
(c) between 1,800 and 2,200 hours.

*9.23 According to medical research, the time between successive reports of a rare tropical disease is a random variable having the exponential distribution with the mean $\mu = 120$ days. Find the probabilities that the time between successive reports of the disease will
(a) exceed 240 days;
(b) exceed 360 days;
(c) be less than 60 days.

*9.24   In the region around a geological fault line, the time between aftershocks that follow a major earthquake is a random variable with the exponential distribution with the mean $\mu = 36$ hours. Find the probabilities that the time between successive aftershocks will
(a) be less than 18 hours;
(b) exceed 72 hours;
(c) be between 36 and 108 hours.

*9.3   A CHECK FOR NORMALITY

In the introduction to this chapter we said that the normal distribution is basic to most of the "bread and butter" techniques of modern statistics. Indeed, in many of the procedures we shall discuss in subsequent chapters it will be assumed that our data come from normal populations, namely, that they are values of random variables having normal distributions. For many years, this assumption was checked with the use of a special kind of graph paper, called **normal probability paper** or **arithmetic probability paper,** that could be obtained at most college or university bookstores, stationary stores, and businesses offering technical supplies. Today it has become a curio, difficult to find, since the job of checking for normality has been taken over by computers and other technology. As we have pointed out repeatedly, statistical software and graphing calculators can be very helpful, but they are not required for the use of this text. This is why we marked this section as being optional.

Actually, computer-generated plots like the one shown in Figure 9.16 share the characteristic scales of normal probability paper. The horizontal axis, used for the values of the random variable (measurements or observations), is an ordinary scale with equal subdivisions. On the other hand, the vertical axis is scaled in such a way that the graph of any normal distribution becomes a straight line, and this is precisely how we judge the normality of our data. *If the pattern we get for our data is nearly that of a straight line, we are justified in assuming that they come from a normal population.*

Figure 9.16 shows an example of a **normal probability plot** obtained with the use of a computer and MINITAB software. It pertains to the waiting times between eruptions of Old Faithful, introduced first in Example 2.4, and it plots each of the original values (measured along the horizontal axis) against the probability that a normal curve fit to the data will yield a value smaller than that. In other words, we are plotting the "less than" probabilities. Sometimes the probabilities are given as percentages (as in Figure 9.16) and sometimes they are given as proportions.

Except for the four smallest values, the points in Figure 9.16 are all very close to the straight line, and if we disregard these four points as possible outliers that can be attributed to an assignable cause (possibly a seismological phenomenon), we conclude that the waiting times are values of a random variable having a near-normal distribution. Actually, a closer inspection may reveal a pattern that can be described as having the shape of an elongated letter S. This is

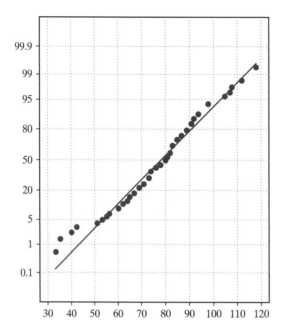

FIGURE 9.16
Normal probability
plot of the waiting-
time data.

indicative of a bell-shaped distribution with a moderate skewness, which is precisely what we inferred earlier from Figure 4.5. In Example 4.8 we actually showed that the Pearsonian coefficient of skewness for the waiting-time data is -0.41.

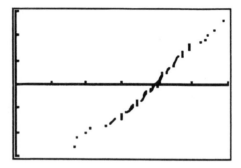

FIGURE 9.17
Normal probability
plot reproduced from
the display screen of a
TI-83 graphing
calculator.

Graphing calculators generate probability plots in a way that is somewhat different from the MINITAB-generated plot of Figure 9.16. The one shown in Figure 9.17 also pertains to the waiting times between eruptions of Old Faithful, but it plots the original values (again measured on the horizontal scale) against the corresponding fractiles, z-values, of the standard normal distribution. Again, the linearity of the plot is a sign of normality; that is, it indicates that the data are values of a random variable having a normal or a near-normal distribution.

■ E x e r c i s e s

 *9.25 Use appropriate computer software or a graphing calculator to produce a normal probability plot for the 60 lengths of sea trout given in the first paragraph of Section 2.1. Can these data be looked upon as values of a random variable having a normal distribution?

 *9.26 Use appropriate computer software or a graphing calculator to produce a normal probability plot for the room occupancy data given in the third paragraph of Section 2.2. Can these data be looked upon as values of a random variable having a normal distribution?

 *9.27 Use appropriate computer software or a graphing calculator to produce a normal probability plot for the data of Exercise 2.22, pertaining to the numbers of operations performed at a hospital during 80 weeks. Can these data be looked upon as values of a random variable having a near-normal distribution?

 *9.28 Use appropriate computer software or a graphing calculator to produce a normal probability plot for the data of Exercise 2.36, pertaining to the percent shrinkages on drying of 40 plastic clay specimens. Can these data be looked upon as values of a random variable having a normal distribution?

 *9.29 Use appropriate computer software or a graphing calculator to produce a normal probability plot for the data of Exercise 2.42, pertaining to the lengths of root penetrations of 120 crested wheatgrass seedlings. Can these data be looked upon as values of a random variable having a normal distribution?

*9.30 To judge the normality of grouped data (that is, the shape of a frequency distribution) it was a common practice to use normal probability paper, plotting the cumulative "less than" percentages at the corresponding class boundaries. Convert the cumulative "less than" frequency distribution obtained for the Old Faithful waiting-time data on page 26 into a cumulative "less than" percentage distribution, and use normal probability paper to judge the normality of the data.

*9.31 Following the suggestion of Exercise 9.30, convert the distribution obtained in Exercise 2.42 into a cumulative "less than" percentage distribution, and use normal probability paper to judge the normality of the data.

*9.32 Following is the distribution of the numbers of inquiries that a realty firm received with regard to 500 pieces of property:

Number of inquiries	Pieces of property
3–6	55
7–10	227
11–14	170
15–18	37
19–22	9
23–26	2

Convert this distribution into a cumulative "less than" percentage distribution, and following the suggestion of Exercise 9.30 use normal probability paper to judge the normality of the data. Does it confirm the impression one gets just by looking at the distribution?

*9.33 Normal probability paper or a normal probability plot like the one shown in Figure 9.16 can also be used to get crude estimates of the mean and the standard deviation of grouped or ungrouped data judged to be near-normal. For the mean, we read off the value on the horizontal scale that, according to the line we fit to the data, corresponds to the 50 percent (or 0.50) mark on the vertical scale. Since the normal-curve areas to the left of $z = -1$ and $z = 1$ are approximately 0.16 and 0.84, we read off the values on the horizontal scale that, according to the line we fit to the data, correspond to the 16% (or 0.16) and 84% (or 0.84) marks on the vertical scale. Their difference divided by 2 provides an estimate of the standard deviation. Use this method to estimate the mean and the standard deviation of the Old Faithful waiting-time data from Figure 9.16. Also, compare the results with the values obtained in Examples 3.21 and 4.7.

9.4   APPLICATIONS OF THE NORMAL DISTRIBUTION

Let us now consider some applications where it is assumed in each case that the distribution of the data, or the distribution of the random variable under consideration, can be approximated closely with a normal distribution.

EXAMPLE   9.7   If the amount of cosmic radiation to which a person is exposed while flying by jet across the United States is a random variable having a normal distribution with $\mu = 4.35$ mrem and $\sigma = 0.59$ mrem, find the probabilities that a person on such a flight will be exposed to

(a) more than 5.00 mrem of cosmic radiation;
(b) anywhere from 3.00 to 4.00 mrem of cosmic radiation.

Solution   (a) This probability is given by the area of the tinted region under the curve of the diagram at the top of Figure 9.18, namely, the area under the curve to the right of

$$z = \frac{5.00 - 4.35}{0.59} \approx 1.10$$

Since the entry in Table I corresponding to $z = 1.10$ is 0.3643, we find that the probability is $0.5000 - 0.3643 = 0.1357$, or approximately 0.14, that a person will be exposed to more than 5.00 mrem of cosmic radiation on such a flight.

(b) This probability is given by the area of the tinted region under the curve of the diagram at the bottom of Figure 9.18, namely, the area under the curve between

$$z = \frac{3.00 - 4.35}{0.59} \approx -2.29 \quad \text{and} \quad z = \frac{4.00 - 4.35}{0.59} \approx -0.59$$

Since the entries in Table I corresponding to $z = 2.29$ and $z = 0.59$ are, respectively, 0.4890 and 0.2224, we find that the probability is $0.4890 - 0.2224 = 0.2666$, or approximately 0.27, that a person will be exposed to anywhere from 3.00 to 4.00 mrem of cosmic radiation on such a flight. ∎

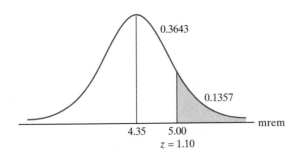

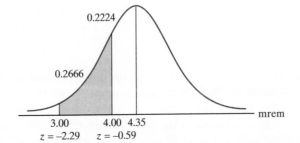

FIGURE 9.18
Diagram for Example
9.7.

EXAMPLE   9.8   The actual amount of instant coffee that a filling machine deposits into "6-ounce" jars varies from jar to jar and may be looked upon as a random variable having a normal distribution with a standard deviation of 0.04 ounce. If only 2 percent of the jars are to contain less than 6 ounces of coffee, what must be the mean fill of these jars?

Solution   Here we are given $\sigma = 0.04$, $x = 6.00$, a normal-curve area (that of the tinted region under the curve in Figure 9.19), and we are asked to find μ. The value of z for which the entry in Table I is closest to $0.5000 - 0.0200 = 0.4800$ is $z = 2.05$ corresponding to 0.4798, so that

$$-2.05 = \frac{6.00 - \mu}{0.04}$$

Then, solving for μ, we get $6.00 - \mu = -2.05(0.04) = -0.082$ and $\mu = 6.00 + 0.082 = 6.082$, or approximately 6.08. The mean fill must be 6.08 ounces.   ∎

FIGURE 9.19
Diagram for Example
9.8.

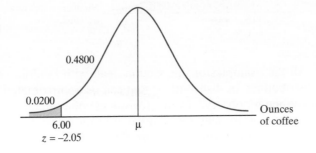

Although the normal distribution is a continuous distribution that applies to continuous random variables, it is often used to approximate distributions of discrete random variables, which can take on only a finite number of values or as many values as there are positive integers. To do this, we must use the **continuity correction** illustrated in the following example.

EXAMPLE 9.9 In a study of aggressive behavior, male white mice, returned to the group in which they live after four weeks of isolation, averaged 18.6 fights in the first five minutes with a standard deviation of 3.3 fights. If it can be assumed that the distribution of this random variable (the number of fights into which such a mouse gets under the stated conditions) can be approximated closely with a normal distribution, what is the probability that such a mouse will get into at least 15 fights in the first five minutes?

Solution The answer is given by the area of the tinted region under the curve in Figure 9.20; namely, by the area under the curve to the right of 14.5, and not 15. The reason for this is that the number of fights in which such a mouse gets involved is a whole number. Hence, if we want to approximate the distribution of this random variable with a normal distribution, we must "spread" its values over a continuous scale, and we do this by representing each whole number k by the interval from $k - \frac{1}{2}$ to $k + \frac{1}{2}$. For instance, 5 is represented by the interval from 4.5 to 5.5, 10 is represented by the interval from 9.5 to 10.5, 20 is represented by the interval from 19.5 to 20.5, and the probability of 15 or more is given by the area under the curve to the right of 14.5. Accordingly, we get

$$z = \frac{14.5 - 18.6}{3.3} \approx -1.24$$

and it follows from Table I that the area of the tinted region under the curve in Figure 9.20, the probability that such a mouse will get into at least 15 fights in the first five minutes, is $0.5000 + 0.3925 = 0.8925$, or approximately 0.89. ∎

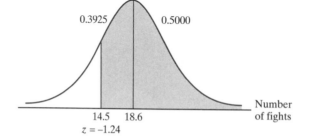

FIGURE 9.20
Diagram for Example 9.9.

All the examples of this section dealt with random variables having normal distributions, or distributions that can be approximated closely with normal curves. When we observe a value (or values) of a random variable having a normal distribution, we may say that we are sampling a normal population; this is consistent with the terminology introduced at the end of Section 8.3.

9.5   THE NORMAL APPROXIMATION TO THE BINOMIAL DISTRIBUTION

The normal distribution provides a close approximation to the binomial distribution when n, the number of trials, is large and p, the probability of a success on an individual trial, is close to $\frac{1}{2}$. Figure 9.21 shows the histograms of binomial distributions with $p = \frac{1}{2}$ and $n = 2, 5, 10,$ and 25, and it can be seen that with increasing n these distributions approach the symmetrical bell-shaped pattern of

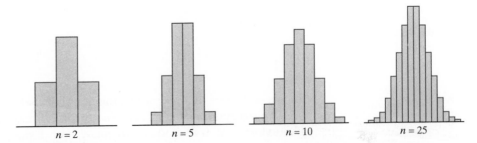

FIGURE 9.21
Binomial distributions with $p = \frac{1}{2}$.

the normal distribution. In fact, normal distributions with the mean $\mu = np$ and the standard deviation $\sigma = \sqrt{np(1 - p)}$ can often be used to approximate binomial probabilities when n is not all that large and p differs quite a bit from $\frac{1}{2}$. Since "not all that large" and "differs quite a bit" are not very precise terms, let us state the following rule of thumb:

> **It is considered sound practice to use the normal approximation to the binomial distribution only when np and $n(1 - p)$ are both greater than 5; symbolically, when**
>
> $$np > 5 \quad \text{and} \quad n(1 - p) > 5$$

EXAMPLE 9.10   Use the normal distribution to approximate the probability of getting 6 heads and 10 tails in 16 flips of a balanced coin, and compare the result with the corresponding value in Table V.

Solution   Since $np = 16 \cdot \frac{1}{2} = 8$ and $n(1 - p) = 16 \cdot (1 - \frac{1}{2}) = 8$ are both greater than 5, it is reasonable to use the normal approximation to the binomial distribution. In accordance with the continuity correction explained on page 252, we represent 6 heads by the interval from 5.5 to 6.5; that is, we must determine the area of the tinted region under the curve in Figure 9.22. Since $\mu = 16 \cdot \frac{1}{2} = 8$ and $\sigma = \sqrt{16 \cdot \frac{1}{2} \cdot \frac{1}{2}} = 2$, we get

$$z = \frac{5.5 - 8}{2} = -1.25 \quad \text{and} \quad z = \frac{6.5 - 8}{2} = -0.75$$

in standard units for $x = 5.5$ and $x = 6.5$. The entries corresponding to $z = 1.25$ and $z = 0.75$ in Table I are 0.3944 and 0.2734, and hence we get $0.3944 - 0.2734 = 0.1210$ for the normal approximation to the binomial probability of getting 6 heads and 10 tails in 16 flips of a balanced coin. Since the corresponding entry in Table V is 0.122, we find that the error of this approximation is only 0.122 − 0.121 = 0.001. ∎

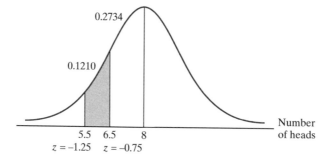

FIGURE 9.22
Diagram for Example 9.10.

To assess the size of an error like this, it usually helps to determine the corresponding **percentage error,** namely, the error expressed as a percentage of the quantity we are trying to approximate. For Example 9.10 we thus get $\frac{0.001}{0.122} \cdot 100 \approx 0.82\%$, which would seem to indicate that the approximation is quite good. Actually, this is hard to judge, since the assessment of any approximation depends to a large extent on what we intend to do with it.

EXAMPLE 9.11   A study shows that 44% of all self-employed authors in a certain income bracket have tax-sheltered retirement accounts. Use the normal distribution to approximate the probability that among 20 such authors, randomly selected from IRS files, 10 will have tax-sheltered retirement accounts. Also compare the result with the corresponding binomial probability.

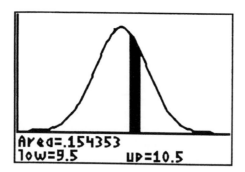

FIGURE 9.23
Reproduction from the display screen of a TI-83 graphing calculator for Example 9.11.

Solution   Since $np = 20(0.44) = 8.8$ and $n(1 - p) = 20(1 - 0.44) = 11.2$ are both greater than 5, it is again reasonable to use the normal approximation to the binomial distribution, and we could proceed exactly as in Example 9.10. Instead, let us use a graphing calculator to find the area between 9.5 and 10.5 under the normal curve with $\mu = np = 20(0.44) = 8.8$ and $\sigma = \sqrt{20(0.44)(0.56)} \approx 2.22$. As can be seen from Figure 9.23, the answer is 0.154353. To determine the error of this approximation, we cannot use Table V, but the *National Bureau of Standards Table,* listed among the references on page 232, yields the binomial probability of 0.152407, rounded like our approximation to six decimals. Thus the error of the approximation is 0.154353 − 0.152407 = 0.001946, or approximately 0.002. Also,

the corresponding percentage error is $\dfrac{0.001946}{0.152407} \cdot 100 \approx 1.3\%$.   ∎

The normal approximation to the binomial distribution is especially useful in problems that would otherwise require that we use the formula for the binomial distribution repeatedly to calculate the values of many different terms. This is illustrated by the example that follows.

EXAMPLE 9.12   Suppose that 5% of the adobe bricks shipped by a manufacturer have minor blemishes. Use the normal approximation to the binomial distribution to approximate the probability that among 150 adobe bricks shipped by the manufacturer at least nine will have minor blemishes. Also use the computer printout of Figure 8.4 on page 216 to calculate the percentage error of this approximation.

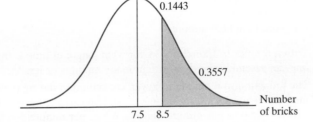

FIGURE 9.24
Diagram for Example 9.12.

Solution   Since 150(0.05) = 7.5 and 150(1 − 0.05) = 142.5 are both greater than 5, the rule of thumb for using the normal approximation to the binomial distribution is satisfied. In accordance with the continuity correction explained on page 252, we represent 9 of the bricks by the interval from 8.5 to 9.5; that is, we shall have to determine the tinted region under the curve in Figure 9.24. Since $\mu = 150(0.05) = 7.5$ and $\sigma = \sqrt{150(0.05)(0.95)} \approx 2.67$, we get

$$z = \frac{8.5 - 7.5}{2.67} \approx 0.37$$

in standard units for $x = 8.5$. The corresponding entry in Table I is 0.1443, and hence we get 0.5000 − 0.1443 = 0.3557 for the normal approximation to the probability that at least 9 of the adobe bricks will have minor blemishes. Since the printout of Figure 8.4 shows that the corresponding binomial probability is 0.1171 + 0.0869 + 0.0582 + · · · + 0.0001 = 0.3361, we find that the error of our approximation is 0.3557 − 0.3361 = 0.0196, or approximately 0.02. The corresponding percentage error is $\dfrac{0.0196}{0.3361} \cdot 100 \approx 5.8\%$.   ∎

The percentage error obtained in this example is fairly substantial, and it serves to illustrate that satisfying the rule of thumb on page 253 does not necessarily guarantee that we will get a good approximation. For instance, when p is quite small and we want to approximate the probability of a value that is not too close to the mean, the approximation may be quite poor even though the rule of thumb is satisfied. As the reader will be asked to verify in Exercise 9.55, if we had wanted to approximate the probability that there will be only one brick with minor blemishes among the 150 shipped by the manufacturer, the percentage error would have been more than 100%. Since we do not expect the reader to become an instant expert at using the normal approximation to the binomial distribution, let us add that we have presented it here mainly because it will be needed in later chapters for large-sample inferences concerning proportions.

Exercises

9.34   If the time to assemble an "easy to assemble" computer desk from a kit is a random variable having the normal distribution with $\mu = 55.8$ minutes and $\sigma = 12.2$ minutes, what are the probabilities that this kind of desk can be assembled in
(a) less than 49.7 minutes;
(b) anywhere from 61.9 to 74.1 minutes;
(c) more than 86.3 minutes?

9.35   With reference to Exercise 9.34, for what length of time is the probability 0.90 that one can assemble the desk in that many minutes or less?

9.36   The reduction of a person's oxygen consumption during periods of transcendental meditation may be looked upon as a random variable having the normal distribution with $\mu = 38.6$ cc per minute and $\sigma = 6.5$ cc per minute. Find the probabilities that during a period of transcendental meditation a person's oxygen consumption will be reduced by
(a) at least 33.4 cc per minute;
(b) at most 34.7 cc per minute.

9.37   The grapefruits grown in a large orchard have a mean weight of 18.2 ounces with a standard deviation of 1.2 ounces. Assuming that the distribution of the weights of these grapefruits has roughly the shape of a normal distribution, what percentage of the grapefruits weigh
(a) less than 16.1 ounces;
(b) more than 17.3 ounces;
(c) anywhere from 16.7 to 18.8 ounces?

9.38   With reference to Exercise 9.37, find
(a) the weight above which we will find the heaviest 15 percent of the grapefruits;
(b) the weight above which we will find the heaviest 75 percent of the grapefruits.

9.39   A manufacturer needs coil springs that can stand a load of at least 20.0 pounds. Among two suppliers, Supplier A can supply coil springs that, on the average, can stand a load of 24.5 pounds with a standard deviation of 2.1 pounds, and Supplier B can supply coil springs that, on the average, can stand a load of 23.3 pounds with a standard deviation of 1.6 pounds. If we can assume that the distributions of these loads can be approximated with normal distributions, determine which of the two suppliers can provide the manufacturer with the smaller percentage of unsatisfactory coil springs.

9.40 The lengths of full-grown scorpions of a certain variety have a mean of 1.96 inches and a standard deviation of 0.08 inch. Assuming that the distribution of these lengths has roughly the shape of a normal distribution, find what percentage of these scorpions have a length of
(a) 2.20 inches or more;
(b) at least 1.80 inches.

9.41 With reference to Exercise 9.40, above what value would we find the longest 6 percent of these scorpions?

9.42 The yearly number of major earthquakes, the world over, is a random variable having approximately a normal distribution with $\mu = 20.8$ and $\sigma = 4.5$. Find the probabilities that in any given year there will be
(a) exactly 19 major earthquakes;
(b) at least 19 major earthquakes.

9.43 In 1945, after World War II, all servicemen were given point scores based on length of service, number of purple hearts, number of other decorations, number of campaigns, etc. Assuming that the distribution of these point scores can be approximated closely with a normal distribution with $\mu = 63$ and $\sigma = 20$, how many men from an army of 8,000,000 would be discharged if the army discharged all men with more than 79.0 points? (Courtesy Department of Mathematics, U.S. Military Academy.)

9.44 The distribution of the IQ's of the 4,000 employees of a large company has a mean of 104.5, a standard deviation of 13.9, and its shape is roughly that of a normal distribution. Given that a certain job requires a minimum IQ of 95 and bores those with an IQ over 110, how many of the company's employees are suitable for this job on the basis of IQ alone? (Use the continuity correction.)

9.45 The head of the complaints department of a department store knows from experience that the number of complaints she receives per day is a random variable having roughly the shape of a normal distribution with $\mu = 33.4$ and $\sigma = 5.5$. Find the probabilities that on any given day she will receive
(a) more than 40 complaints;
(b) at least 40 complaints.

9.46 A student decides to check off the answer to each of the 16 questions in a true–false test by flipping a balanced coin. Find the probability that he will get at least 9 correct answers, using
(a) the normal approximation to the binomial distribution with $n = 16$ and $p = 0.50$;
(b) Table V.

9.47 With reference to Exercise 4.46, find the percentage error of the approximation.

9.48 Check in each case whether the conditions for the normal approximation to the binomial distribution are satisfied:
(a) $n = 20$ and $p = \frac{1}{5}$;
(b) $n = 65$ and $p = 0.10$;
(c) $n = 40$ and $p = 0.95$.

9.49   Check in each case whether the conditions for the normal approximation to the binomial distribution are satisfied:
(a) $n = 200$ and $p = 0.05$;
(b) $n = 150$ and $p = 0.97$;
(c) $n = 100$ and $p = \frac{1}{8}$.

9.50   Records show that 60% of the customers of a service station pay with a credit card. Use the normal approximation to find the probability that among 500 customers at most 175 will pay cash.

9.51   The probability is 0.72 that any one cloud seeded with silver iodine will show spectacular growth. Use the normal approximation to find the probabilities that among 26 clouds seeded with silver iodine
(a) exactly 20 will show spectacular growth;
(b) at least 20 will show spectacular growth.

9.52   With reference to Exercise 9.51, use the printout shown in Figure 9.25 to find the percentage errors of both parts of that exercise.

```
MTB > PDF c1;
SUBC>    Binomial 26 0.72.


Binomial with n = 26 and p = 0.720000

            x          P( X = x)
          9.00          0.0001
         10.00          0.0003
         11.00          0.0011
         12.00          0.0034
         13.00          0.0095
         14.00          0.0226
         15.00          0.0464
         16.00          0.0821
         17.00          0.1241
         18.00          0.1596
         19.00          0.1728
         20.00          0.1555
         21.00          0.1143
         22.00          0.0668
         23.00          0.0299
         24.00          0.0096
         25.00          0.0020
         26.00          0.0002
```

FIGURE 9.25
Binomial distribution
with $n = 26$ and
$p = 0.72$.

9.53   Studies have shown that 22% of all patients taking a certain medication will get a headache. Use the normal approximation to find the probabilities that among 30 patients taking this medication
(a) exactly 3 will get a headache;
(b) exactly 9 will get a headache;
(c) at most 6 will get a headache.

9.54   With reference to Exercise 9.53, use the printout shown in Figure 9.26 to find the percentage errors of all three parts of that exercise.

```
MTB > PDF c1;
SUBC>   Binomial 30 0.22.


Binomial with n = 30 and p = 0.220000

          x         P( X = x)
        0.00          0.0006
        1.00          0.0049
        2.00          0.0200
        3.00          0.0528
        4.00          0.1005
        5.00          0.1473
        6.00          0.1732
        7.00          0.1674
        8.00          0.1358
        9.00          0.0936
       10.00          0.0554
       11.00          0.0284
       12.00          0.0127
       13.00          0.0050
       14.00          0.0017
       15.00          0.0005
       16.00          0.0001
```

FIGURE 9.26
Binomial distribution
with $n = 30$ and
$p = 0.22$.

9.55   Use the printout of Figure 8.4 to show that if we use the normal approximation to the binomial probability with $x = 1, n = 150$, and $p = 0.05$, the percentage error is about 117%.

9.56   Use the normal distribution to approximate the probability that at least 55 of 90 persons flying across the Atlantic Ocean will feel the effect of the time difference for at least 24 hours if the probability is 0.70 that any one person flying across the Atlantic Ocean will feel the effect of the time difference for at least 24 hours.

 9.57   Use a graphing calculator or a computer printout of the binomial distribution with $n = 90$ and $p = 0.70$ to find the percentage error of the approximation of Exercise 9.56.

9.6 *Checklist of Key Terms* (with page references to their definitions)

*Arithmetic probability paper, 247
Continuity correction, 252
Continuous distribution, 234, 235
Continuous random variable, 233
Exponential distribution, 246
Normal approximation to binomial
  distribution, 253
Normal distribution, 234, 237
Normal population, 247
*Normal probability paper, 247

*Normal probability plot, 247
Normal-curve area, 238
Percentage error, 254
Probability density, 235
Standard normal distribution, 238
Standard scores, 238
Standard units, 238
Uniform distribution, 243
z-scores, 238

9.7 References

Detailed information about various continuous distributions may be found in

HASTINGS, N. A. J., and PEACOCK, J. B., *Statistical Distributions*. London: Butterworth & Company (Publishers) Ltd., 1975.

and further information about the normal approximation to the binomial distribution in

GREEN, J., and ROUND-TURNER, J., "The Error in Approximating Cumulative Binomial and Poisson Probabilities," *Teaching Statistics*, May 1986.

More extensive tables of normal-curve areas, as well as tables for other continuous distributions, may be found in

FISHER, R. A., and YATES, F., *Statistical Tables for Biological, Agricultural and Medical Research*. Cambridge: The University Press, 1954.

PEARSON, E. S., and HARTLEY, H. O., *Biometrika Tables for Statisticians*, Vol. I. New York: John Wiley & Sons, Inc., 1968.

10

SAMPLING AND SAMPLING DISTRIBUTIONS

10.1   Random Sampling 262

*10.2   Sample Designs 269

*10.3   Systematic Sampling 269

*10.4   Stratified Sampling 270

*10.5   Cluster Sampling 272

10.6   Sampling Distributions 276

10.7   The Standard Error of the Mean 279

10.8   The Central Limit Theorem 282

10.9   Some Further Considerations 285

*10.10 Technical Note (Simulation) 290

10.11   Checklist of Key Terms 293

10.12   References 293

*T*he main objective of most statistical studies, analyses, or investigations is to make sound generalizations on the basis of samples about the populations from which the samples came. Note the word "sound," because the question of when and under what conditions samples permit such generalizations is not easily answered. For instance, if we want to estimate the average amount of money a person spends on a vacation, would we take as our sample the amounts spent by passengers on a deluxe Concorde flight around the world; would we attempt to estimate the average price of all farm products on the basis of the price of fresh asparagus alone; or would we try to predict December temperatures in Florida on the basis of December temperatures reported in previous years in Siberia? Obviously not, but just what vacationers, what farm products, and December temperatures in what places we should include in our samples, and how many of them, is neither intuitively clear nor self-evident.

In most of the methods we shall study in the remainder of this book, it will be assumed that we are dealing with so-called **random samples.** We pay this much attention to random samples, which are defined and discussed in Section 10.1, because they permit valid, or logical, generalizations. As we shall see, however, random sampling is not always feasible, or even desirable, and some alternative sampling procedures are mentioned in the optional Sections 10.2 through 10.5.

In Section 10.6 we introduce the related concept of a **sampling distribution,** which tells us how quantities determined from samples may vary from sample to sample. Then, in Sections 10.7 through 10.9 we learn how such chance variations can be measured, predicted, and perhaps even controlled.

10.1   RANDOM SAMPLING

In Section 3.1 we distinguished between populations and samples, stating that a population consists of all conceivably possible (or hypothetically possible) observations of a given phenomenon, while a sample is simply part of a population. In what follows, we shall distinguish further between two kinds of populations— **finite populations** and **infinite populations.**

A population is finite if it consists of a finite, or fixed, number of elements, measurements, or observations. Examples of finite populations are the net weights of the 3,000 cans of paint in a certain production lot, the SAT scores of all the freshmen admitted to a given college in the fall of the year 2001, the 52 different cards in an ordinary deck of playing cards, and the daily high temperatures recorded at a weather station during the years 1995 through 1999.

In contrast, a population is infinite if, hypothetically at least, it contains infinitely many elements. This would be the case, for example, when we observe values of a continuous random variable, say, when we repeatedly measure the boiling point of a silicon compound. It would also be the case when we observe the totals obtained in repeated rolls of a pair of dice and when we sample with replacement from a finite population. There is no limit to the number of times that we can measure the boiling point of the silicon compound, no limit to the number of times that we can roll a pair of dice, and no limit to the number of times that we can draw an element from a finite population and replace it before the next one is drawn.

To present the idea of **random sampling from a finite population,** let us first see how many different samples of size n can be drawn from a finite population of size N. Referring to the rule for the number of combinations of n objects taken r at a time on page 116, we find that, with a change of letters, the answer is $\binom{N}{n}$.

EXAMPLE 10.1   How many different samples of size n can be drawn from a finite population of size N, when

(a) $n = 2$ and $N = 14$;
(b) $n = 3$ and $N = 100$?

Solution   (a) There are $\binom{14}{2} = \dfrac{14 \cdot 13}{2!} = 91$ different samples;

(b) there are $\binom{100}{3} = \dfrac{100 \cdot 99 \cdot 98}{3!} = 161{,}700$ different samples.   ∎

Based on the result that there are $\binom{N}{n}$ different samples of size n from a finite population of size N, let us now give the following definition of a **random sample** (sometimes referred to also as a **simple random sample**) from a finite population:

> **A sample of size n from a finite population of size N is random if it is chosen in such a way that each of the $\binom{N}{n}$ possible samples has the same probability, $\dfrac{1}{\binom{N}{n}}$, of being selected.**

For instance, if a population consists of the $N = 5$ elements a, b, c, d, and e (which might be the annual incomes of five persons, the weights of five cows, or five different models of airplanes), there are $\binom{5}{3} = 10$ possible samples of size $n = 3$. They consist of the elements $abc, abd, abe, acd, ace, ade, bcd, bce, bde$, and cde. If we choose one of these samples in such a way that each one has the probability $\frac{1}{10}$ of being selected, we call this sample a random sample.

Next comes the question of how random samples are drawn in actual practice. In a simple situation like the one described immediately above, we could write each of the ten samples on a slip of paper, put them in a hat, shuffle them thoroughly, and then draw one without looking. Obviously, though, this would be impractical in a more realistically complex situation where n and N, or only N, are large. For instance, for $n = 4$ and $N = 100$ we would have to label and draw one of $\binom{100}{4} = 3{,}921{,}225$ slips of paper.

Fortunately, we can draw a random sample from a finite population without listing all possible samples, which we mentioned here only to stress the point that the selection of a random sample must depend entirely on chance. Instead of listing all possible samples, we can write each of the N elements of the finite population on a slip of paper, and draw n of them one at a time without replacement, making sure in each of the successive drawings that all of the remaining elements of the population have the same chance of being selected. It is easy to show mathematically that this also leads to the probability $\dfrac{1}{\binom{N}{n}}$ for each possible sample. For instance, to take a random sample of $n = 12$ of $N = 138$ archaeological sites for supplementary funding of excavations, we could write the numbers $1, 2, 3, \ldots, 137$, and 138 on 138 slips of paper, mix them up thoroughly (say, in a proverbial hat), and then draw 12, one at a time without replacement, without looking.

Even an easy procedure like this can be simplified further. Nowadays, the easiest way of taking a random sample from a finite population is to base the selection on **random numbers** that are generated by means of statistical calculators or computers. For instance, the HEWLETT PACKARD 21S STAT/MATH calculator yields four-digit random numbers so that each four-digit number from 0000 to 9999 has the probability 0.0001. In order to avoid the possibility of getting the same random numbers for different applications, we can "seed" the calculator by entering an arbitrary four-digit number with which to start the operation.

EXAMPLE 10.2   With reference to the aforementioned archaeological sites, which presumably have been numbered 001 to 138, use a statistical calculator to generate a random sample of $n = 12$ of the sites.

Solution   Using only the first three digits of the four-digit random numbers generated, omitting 000 and those exceeding 138, and also omitting numbers that have already been recorded, we get

041	021	079	084	016	108
029	003	100	046	136	075

The archaeological sites associated with these numbers constitute the sample. ■

If we wanted to use a computer instead of a statistical calculator in Example 10.2, we could have used MINITAB (or some other statistical software) to produce a sample like the one shown in Figure 10.1. This sample consists of the archaeological sites associated with the numbers 90, 45, 83, 1, 110, 26, 72, 91, 76, 129, 38, and 71.

FIGURE 10.1
Computer-generated sample for Example 10.2.

```
MTB > Random 2 c1-c6;
SUBC>    Integer 1 138.
MTB > Print c1-c6.

        90    45    83     1   110    26
        72    91    76   129    38    71
```

Only a few decades ago, random samples were generated almost exclusively with the use of published tables of random numbers. Such tables consist of page after page of nothing but the digits 0, 1, 2, 3, . . . , 8, and 9 arranged in rows and columns. Such tables were constructed with the use of census data, decimal expansions of irrational numbers, tables of logarithms, Selective Service lotteries, and assorted electronic gadgetry. There were even some arithmetical procedures, such as the one where we begin with a three- or four-digit number, square it, and use the middle three or four digits to begin the sequence of random digits. Then we square this number and repeat this process over and over again. Eventually, these methods were replaced by computer-generated tables of random numbers, and even these have been replaced by the methods that we used in Example 10.2 to generate our own random numbers.

Nevertheless, since the necessary technology may not always be available, we shall illustrate here the use of a published table of random numbers. For this purpose, we shall refer to Figure 10.2, which consists of a page reproduced

FIGURE 10.2
Sample page of
random digits.

94620	27963	96478	21559	19246	88097	44926
60947	60775	73181	43264	56895	04232	59604
27499	53523	63110	57106	20865	91683	80688
01603	23156	89223	43429	95353	44662	59433
00815	01552	06392	31437	70385	45863	75971
83844	90942	74857	52419	68723	47830	63010
06626	10042	93629	37609	57215	08409	81906
56760	63348	24949	11859	29793	37457	59377
64416	29934	00755	09418	14230	62887	92683
63569	17906	38076	32135	19096	96970	75917
22693	35089	72994	04252	23791	60249	83010
43413	59744	01275	71326	91382	45114	20245
09224	78530	50566	49965	04851	18280	14039
67625	34683	03142	74733	63558	09665	22610
86874	12549	98699	54952	91579	26023	81076
54548	49505	62515	63903	13193	33905	66936
73236	66167	49728	03581	40699	10396	81827
15220	66319	13543	14071	59148	95154	72852
16151	08029	36954	03891	38313	34016	18671
43635	84249	88984	80993	55431	90793	62603
30193	42776	85611	57635	51362	79907	77364
37430	45246	11400	20986	43996	73122	88474
88312	93047	12088	86937	70794	01041	74867
98995	58159	04700	90443	13168	31553	67891
51734	20849	70198	67906	00880	82899	66065
88698	41755	56216	66852	17748	04963	54859
51865	09836	73966	65711	41699	11732	17173
40300	08852	27528	84648	79589	95295	72895
02760	28625	70476	76410	32988	10194	94917
78450	26245	91763	73117	33047	03577	62599
50252	56911	62693	73817	98693	18728	94741
07929	66728	47761	81472	44806	15592	71357
09030	39605	87507	85446	51257	89555	75520
56670	88445	85799	76200	21795	38894	58070
48140	13583	94911	13318	64741	64336	95103
36764	86132	12463	28385	94242	32063	45233
14351	71381	28133	68269	65145	28152	39087
81276	00835	63835	87174	42446	08882	27067
55524	86088	00069	59254	24654	77371	26409
78852	65889	32719	13758	23937	90740	16866
11861	69032	51915	23510	32050	52052	24004
67699	01009	07050	73324	06732	27510	33761
50064	39500	17450	18030	63124	48061	59412
93126	17700	94400	76075	08317	27324	72723
01657	92602	41043	05686	15650	29970	95877
13800	76690	75133	60456	28491	03845	11507
98135	42870	48578	29036	69876	86563	61729
08313	99293	00990	13595	77457	79969	11339
90974	83965	62732	85161	54330	22406	86253
33273	61993	88407	69399	17301	70975	99129

from *Tables of 105,000 Random Decimal Digits,* published by the Interstate Commerce Commission, Bureau of Transport Economics and Statistics, Washington D.C., in 1949.

EXAMPLE 10.3   Repeat Example 10.2, reading three-digit numbers off Figure 10.2, with the same restrictions as in the solution of Example 10.2. Use the 11th, 12th, and 13th columns, starting with the 6th row and going down the page. If necessary, continue with the 16th, 17th, and 18th columns, also starting with the 6th row and going down the page.

Solution   Again omitting 000 and numbers exceeding 138, and making sure that no value occurs more than once, it can easily be verified that the random numbers we get for the sample are

007	012	031	135	114	120
047	124	070	009	118	094

Thus, the sample consists of the archaeological sites associated with the numbers 7, 12, 31, 135, 114, 120, 47, 124, 70, 9, 118, and 94.   ■

When lists are available so that items can readily be numbered, it is easy to draw random samples with the use of calculators, computers, or random number tables. Unfortunately, however, there are many situations where it is impossible to proceed in the ways we have just described. For instance, if we want to use a sample to estimate the mean outside diameter of thousands of ball bearings packed in a large crate, or if we want to estimate the mean height of the trees in a forest, it would be impossible to number the ball bearings or the trees, choose random numbers, and then locate and measure the corresponding ball bearings or trees. In these and in many similar situations, all we can do is proceed according to the dictionary definition of the word "random," namely, "haphazardly, without aim or purpose." That is, we must not select or reject any element of a population because of its seeming typicalness or lack of it, nor must we favor or ignore any part of a population because of its accessibility or lack of it, and so forth. With some reservations, such samples can often be treated as if they were, in fact, random samples.

So far we have discussed random sampling only in connection with finite populations. For infinite populations we say that

A sample of size n from an infinite population is random if it consists of values of independent random variables having the same distribution.

As we pointed out in connection with the binomial and normal distributions, it is this "same" distribution that we refer to as the population being sampled. Also, by "independent" we mean that the probabilities relating to any one of the random variables are the same regardless of what values may have been observed for the other random variables.

For instance, if we get 2, 5, 1, 3, 6, 4, 4, 5, 2, 4, 1, and 2 in twelve rolls of a die, these numbers constitute a random sample if they are values of independent random variables having the same probability distribution.

$$f(x) = \frac{1}{6} \qquad \text{for } x = 1, 2, 3, 4, 5, \text{ or } 6$$

To give another example of a random sample from an infinite population, suppose that eight students obtained the following measurements of the boiling point of a silicon compound: 136, 153, 170, 148, 157, 152, 143, and 150 degrees Celsius. According to the definition, these values constitute a random sample if they are values of independent random variables having the same distribution, say, the normal distribution with $\mu = 152$ and $\sigma = 10$. To judge whether this is actually the case, we would have to ascertain, among other things, that the eight students' measuring techniques are equally precise (so that σ is the same for each of the random variables) and that there was no collaboration (which might make the random variables dependent). In practice, it is not an easy task to judge whether a set of data may be looked upon as a random sample, and we shall go into this further in Chapter 18.

■ Exercises

10.1 How many different samples of size $n = 2$ can be selected from a finite population of size
   (a) $N = 3$;
   (b) $N = 16$;
   (c) $N = 25$;
   (d) $N = 100$?

10.2 How many different samples of size $n = 3$ can be drawn from a finite population of size
   (a) $N = 5$;
   (b) $N = 50$;
   (c) $N = 35$;
   (d) $N = 200$?

10.3 What is the probability of each possible sample if a random sample of size $n = 4$ is drawn from a finite population of size
   (a) $N = 16$;
   (b) $N = 40$?

10.4 What is the probability of each possible sample if a random sample of size $n = 5$ is drawn from a finite population of size
   (a) $N = 14$;
   (b) $N = 45$?

10.5 List the $\binom{6}{3} = 20$ possible samples of size $n = 3$ that can be drawn from a finite population whose elements are denoted by $u, v, w, x, y,$ and z.

10.6 With reference to Exercise 10.5, what is the probability that one of the random samples will include the element denoted by u?

10.7 With reference to Exercise 10.5, what is the probability that one of the random samples will include the elements denoted by u and v?

10.8 A bookstore near a college campus specializes in books on art, history, literature, music, philosophy, poetry, and religion. If the owner wants to give a 10% discount on books in three of these categories, what is the probability of each possible random selection? Also, what is the probability that the random selection will include books on religion?

10.9 A mathematics department offers senior electives in Group Theory, Number Theory, Analysis, Noneuclidean Geometry, and Differential Equations. If a student randomly selects three of these courses, what are the probabilities
(a) of each possible selection;
(b) that the selection will include Differential Equations;
(c) that the selection will include both Number Theory and Noneuclidean Geometry?

10.10 A research organization wants to include 6 of the 50 states of the United States in a marketing survey. If the states are numbered 01, 02, 03, ..., 49, and 50 in alphabetic order and the research organization uses the 6th and 7th columns of the table in Figure 10.2, going down the page starting with the third row, which states will be included in the survey. A list that associates numbers with the elements of a population for obtaining a sample is called a **sampling frame.** For this exercise, such a list may be obtained from a listing of area codes in a telephone directory.

 10.11 Repeat Exercise 10.10, making the selection with a statistical calculator or a computer.

 10.12 Random samples from finite populations can also be obtained with the use of graphing calculators. Using the **MATH PRB** menu, the appropriate command for Exercise 10.10 would be **randInt (1, 50, 6).**
(a) Use a graphing calculator to repeat Exercise 10.10.
(b) Use a graphing calculator to rework Example 10.2.

Since the display screen of the TI-83 graphing calculator is small, getting all the sample values will require some scrolling with the ▣ key.

10.13 A hematologist wants to recheck a sample of $n = 10$ of the 653 blood specimens analyzed by her laboratory in a given month. In her records, these blood specimens are numbered 3250, 3251, 3252, ..., 3901, and 3902. Which specimens would she select if she used columns 21, 22, and 23 of the table in Figure 10.2, going down the page starting with row 16? (Since all the numbers begin with a 3, this digit can be omitted in the selection of the sample.)

 10.14 Use a statistical calculator, a graphing calculator, or a computer to rework Exercise 10.13.

10.15 Three hundred sales have been recorded in the shoe department of a department store over a period of two weeks, with the corresponding invoices numbered from 251 through 550. If an auditor wants to verify $n = 12$ of the invoices selected at random, which ones would he check if he used the 18th, 19th, and 20th columns of the table in Figure 10.2, going down the page starting with the top row?

 10.16 Use a statistical calculator, a graphing calculator, or a computer to rework Exercise 10.15.

 10.17 A county assessor wants to reassess a random sample of 50 of the county's 7,964 one-family homes, and he asks you to help him with the selection of the sample. First you create a sampling frame by assigning these homes the numbers 0001, 0002, 0003, ..., 7963, and 7964, and then you use the first four columns of the table in

Figure 10.2, going down the page starting with the first row, and continuing with the next four columns, also going down the page starting with the first row. By numbers, which of the one-family homes would thus be selected?

 10.18 Use a computer to rework Exercise 10.17.

10.19 On page 263 we said that a random sample can be drawn from a finite population without listing all possible samples; instead, we merely number (or label) the N elements of the finite population, and then draw n of them one at a time without replacement, making sure in each of the successive drawings that all of the remaining elements have the same probability of being selected. Verify this for the example on page 263, which dealt with random samples of size $n = 3$ drawn from the finite population that consists of the elements a, b, c, d, and e, by showing that the probability of any particular sample drawn by this method (say, bce) is again $\frac{1}{10}$.

10.20 Use the same kind of argument as in Exercise 10.19 to verify that each possible random sample of size $n = 3$, drawn one at a time from a finite population of size $N = 100$, has the probability $1/\binom{100}{3} = \frac{1}{161,700}$.

10.21 Use the same kind of argument as in Exercise 10.19 to verify that each possible random sample of size n, drawn one at a time from a finite population of size N, has the probability $1/\binom{N}{n}$.

*10.2 SAMPLE DESIGNS

The only kind of samples we have discussed so far are random samples, and we did not even consider the possibility that under certain conditions there may be samples that are better (say, easier to obtain, cheaper, or more informative) than random samples, and we did not go into any details about the question of what might be done when random sampling is impossible. Indeed, there are many other ways of selecting a sample from a population, and there is an extensive literature devoted to the subject of designing sampling procedures.

In statistics, a **sample design** is a definite plan, determined completely before any data are actually collected, for obtaining a sample from a given population. Thus, the plan to take a simple random sample of 12 of a city's 247 drugstores by using a table of random numbers in a prescribed way constitutes a sample design. In the next three sections we discuss briefly some of the most widely used kinds of sample designs.

*10.3 SYSTEMATIC SAMPLING

In some instances, the most practical way of sampling is to select, say, every 20th name on a list, every 12th house on one side of a street, every 50th piece coming off an assembly line, and so on. This is called **systematic sampling,** and an element of randomness can be introduced into this kind of sampling by using random numbers to pick the unit with which to start. Although a systematic sample may not be a random sample in accordance with the definition, it is often reasonable to treat systematic samples as if they were random samples; indeed, in some instances, systematic samples actually provide an improvement over simple random samples inasmuch as the samples are spread more evenly over the entire populations.

The real danger in systematic sampling lies in the possible presence of hidden periodicities. For instance, if we inspect every 40th piece made by a particular machine, the results would be very misleading if, because of a regularly recurring failure, every 10th piece produced by the machine has blemishes. Also, a systematic sample might yield biased results if we interview the residents of every 12th house along a certain route and it so happens that each 12th house along the route is a corner house on a double lot.

*10.4 STRATIFIED SAMPLING

If we have information about the makeup of a population (that is, its composition) and this is of relevance to our investigation, we may be able to improve on random sampling by **stratification.** This is a procedure that consists of stratifying (or dividing) the population into a number of non-overlapping subpopulations, or **strata,** and then taking a sample from each stratum. If the items selected from each stratum constitute simple random samples, the entire procedure—first stratification and then random sampling—is called **stratified (simple) random sampling.**

Suppose, for instance, that we want to estimate the mean weight of four persons on the basis of a sample of size 2, and that the (unknown) weights of the four persons are 115, 135, 185, and 205 pounds. Thus, the mean weight we want to estimate is

$$\mu = \frac{115 + 135 + 185 + 205}{4} = 160 \text{ pounds}$$

If we take an ordinary random sample of size 2 from this population, the $\binom{4}{2} = 6$ possible samples are 115 and 135, 115 and 185, 115 and 205, 135 and 185, 135 and 205, and 185 and 205, and the corresponding means are 125, 150, 160, 160, 170, and 195. Observe that since each of these samples has the probability $\frac{1}{6}$, the probabilities are $\frac{1}{3}, \frac{1}{3},$ and $\frac{1}{3}$ that our error (the difference between the sample mean and $\mu = 160$) will be 0, 10, or 35.

Now suppose that we know that two of these persons are men and two are women, and suppose that the (unknown) weights of the men are 185 and 205 pounds, while the (unknown) weights of the women are 115 and 135 pounds. Stratifying our sample (by sex) and randomly choosing one of the two men and one of the two women, we find that there are only the four stratified samples 115 and 185, 115 and 205, 135 and 185, and 135 and 205. The means of these samples are 150, 160, 160, and 170, and now the probabilities are $\frac{1}{2}$ and $\frac{1}{2}$ that our error will be 0 or 10. Clearly, stratification has greatly improved our chances of getting a good (close) estimate of the weight of the four persons. See, however, Exercise 10.27.

Essentially, the goal of stratification is to form strata in such a way that there is some relationship between being in a particular stratum and the answer sought in the statistical study, and that within the separate strata there is as much homogeneity (uniformity) as possible. In our example there is such a connection between sex and weight and there is much less variability in weight within each of the two groups than there is within the entire population.

In the preceding example, we used **proportional allocation,** which means that the sizes of the samples from the different strata are proportional to the sizes of the strata. In general, if we divide a population of size N into k strata of size $N_1, N_2, \ldots,$ and N_k, and take a sample of size n_1 from the first stratum, a sample of size n_2 from the second stratum, $\ldots,$ and a sample of size n_k from the kth stratum, we say that the allocation is proportional if

$$\frac{n_1}{N_1} = \frac{n_2}{N_2} = \cdots = \frac{n_k}{N_k}$$

or if these ratios are as nearly equal as possible. In the example dealing with the weights we had $N_1 = 2$, $N_2 = 2$, $n_1 = 1$, and $n_2 = 1$, so that

$$\frac{n_1}{N_1} = \frac{n_2}{N_2} = \frac{1}{2}$$

and the allocation was, indeed, proportional.

As the reader will be asked to verify in Exercise 10.30, allocation is proportional if

Sample sizes for proportional allocation

$$n_i = \frac{N_i}{N} \cdot n \qquad for\ i = 1, 2, \ldots, and\ k$$

where $n = n_1 + n_2 + \cdots + n_k$ is the total size of the sample. When necessary, we use the integers closest to the values given by this formula.

EXAMPLE 10.4   A stratified sample of size $n = 60$ is to be taken from a population of size $N = 4,000$, which consists of three strata of size $N_1 = 2,000$, $N_2 = 1,200$, and $N_3 = 800$. If the allocation is to be proportional, how large a sample must be taken from each stratum?

Solution   Substituting into the formula, we get

$$n_1 = \frac{2,000}{4,000} \cdot 60 = 30 \qquad n_2 = \frac{1,200}{4,000} \cdot 60 = 18$$

and

$$n_3 = \frac{800}{4,000} \cdot 60 = 12$$

∎

This example illustrates proportional allocation, but we should add that there exist other ways of allocating portions of a sample to the different strata. One of these, called **optimum allocation,** is described in Exercise 10.33. It accounts not only for the size of the strata, as in proportional allocation, but also for the variability (of whatever characteristic is of concern) within the strata.

Also, stratification is not limited to a single variable of classification, or characteristic, and populations are often stratified according to several characteristics. For instance, in a systemwide survey designed to determine the attitude of its students, say, toward a new tuition plan, a state college system with 17 colleges might stratify its sample not only with respect to the colleges, but also with respect to students' class standing, sex, and major. So, part of the sample would be allocated to junior women in college A majoring in engineering, another part to sophomore men in college L majoring in English, and so on. Up to a point, stratification like this, called **cross stratification,** will increase the precision (reliability) of estimates and other generalizations, and it is widely used, particularly in opinion sampling and market research.

In stratified sampling, the cost of taking random samples from the individual strata is often so high that interviewers are simply given quotas to be filled from the different strata, with few (if any) restrictions on how they are to be filled. For instance, in determining voters' attitudes toward increased medical coverage for elderly persons, an interviewer working a certain area might be told to interview 6 male self-employed homeowners under 30 years of age, 10 female wage earners in the 45–60 age bracket who live in apartments, 3 retired males over 60 who live in trailers, and so on, with the actual selection of the individuals being left to the interviewer's discretion. This is called **quota sampling,** and it is a convenient, relatively inexpensive, and sometimes necessary procedure, but as it is often executed, the resulting samples do not have the essential features of random samples. In the absence of any controls on their choice, interviewers naturally tend to select individuals who are most readily available—persons who work in the same building, shop in the same store, or perhaps reside in the same general area. Quota samples are thus essentially **judgment samples,** and inferences based on such samples generally do not lend themselves to any sort of formal statistical evaluation.

*10.5 CLUSTER SAMPLING

To illustrate another important kind of sampling, suppose that a large foundation wants to study the changing patterns of family expenditures in the San Diego area. In attempting to complete schedules for 1,200 families, the foundation finds that simple random sampling is practically impossible, since suitable lists are not available and the cost of contacting families scattered over a wide area (with possibly two or three callbacks for the not-at-homes) is very high. One way in which a sample can be taken in this situation is to divide the total area of interest into a number of smaller, nonoverlapping areas, say, city blocks. A number of these blocks are then randomly selected, with the ultimate sample consisting of all (or samples of) the families residing in these blocks.

In this kind of sampling, called **cluster sampling,** the total population is divided into a number of relatively small subdivisions, and some of these subdivisions, or clusters, are randomly selected for inclusion in the overall sample. If the clusters are geographic subdivisions, as in the preceding example, this kind of sampling is also called **area sampling.** To give another example of cluster sampling, suppose that the dean of students of a university wants to know how fraternity men at the school feel about a certain new regulation. He can take a cluster sample by interviewing some or all of the members of several randomly selected fraternities.

Although estimates based on cluster samples are usually not as reliable as estimates based on simple random samples of the same size (see Exercise 10.32), they are often more reliable per unit cost. Referring again to the survey of family expenditures in the San Diego area, it is easy to see that it may well be possible to take a cluster sample several times the size of a simple random sample for the same cost. It is much cheaper to visit and interview families living close together in clusters than families selected at random over a wide area.

In practice, several of the methods of sampling we have discussed may well be used in the same study. For instance, if government statisticians want to study the attitude of American elementary school teachers toward certain federal programs, they might first stratify the country by states or some other geographic subdivisions. To take a sample from each stratum, they might then use cluster sampling, subdividing each stratum into a number of smaller geographic subdivisions (say, school districts), and finally, they might use simple random sampling or systematic sampling to select a sample of elementary school teachers within each cluster.

■ Exercises

*10.22 Following are the percentages of persons 25 years old and over with some college education, but no degree, in the 50 states, listed in alphabetic order, as reported by the 1990 census:

16.8	27.6	25.4	16.6	22.6	24.0	15.9	16.9	19.4	17.0
20.1	24.2	19.4	16.6	17.0	21.9	15.2	17.2	16.1	18.6
15.8	20.4	19.0	16.9	18.4	22.1	21.1	25.8	18.0	15.5
20.9	15.7	16.8	20.5	17.0	21.3	25.0	12.9	15.0	15.8
18.8	16.9	21.1	27.9	14.7	18.5	25.0	13.2	16.7	24.2

List the ten possible systematic samples of size $n = 5$ that can be taken from this list by starting with one of the first ten numbers and then taking each tenth number.

*10.23 With reference to Exercise 10.22, list the five possible systematic samples of size $n = 10$ that can be taken from this list by starting with one of the first five numbers and then taking each fifth number.

*10.24 The following are consecutive monthly figures on the volume of mail (in millions of ton-miles) carried by domestic air operations over a four-year period.

67	62	75	67	70	68	64	70	66	73	73	97
76	73	80	78	78	72	75	75	73	83	76	108
84	78	86	85	81	78	78	75	78	86	76	111
79	77	87	84	82	77	79	77	80	84	78	117

List the six possible systematic samples of size $n = 8$ that can be taken from this list by starting with one of the first six numbers and then taking each sixth number.

*10.25 If one of the six systematic samples of Exercise 10.24 is randomly chosen to estimate the average monthly volume of mail, explain why there is a serious risk of getting a very misleading result.

*10.26 The rule for systematic sampling needs to be modified if n is not an exact divisor of N. If N/n is not an integer, let k be the next integer; for example, if $N = 73$ and $n = 10$, then $N/n = 7.3$ and we use $k = 8$. Now select one of the N items from the population, using probability $1/N$ for each item, Beginning at the selected item, select each kth item thereafter, continuing until n items have been selected. If you reach the end of the list, continue counting at the beginning of the list.
(a) Consider a population consisting of 11 items, identified as the letters a through k. List all possible systematic samples of size $n = 3$ obtained by applying this sampling rule.
(b) Consider a population consisting of 14 items, identified as the numbers 01 through 14. List all possible systematic samples of size $n = 4$ obtained by applying this sampling rule.

*10.27 To generalize the example on page 270, suppose that we want to estimate the mean weight of six persons, whose (unknown) weights are 115, 125, 135, 185, 195, and 205 pounds.
(a) List all possible random samples of size 2 that can be taken from this population, calculate their means, and determine the probability that the mean of such a sample will differ by more than 5 from 160, the actual mean weight of the six persons.
(b) Suppose that the first three weights are those of women and the other three are those of men. List all possible stratified samples of size 2 that can be taken by randomly choosing one of the three women and one of the three men, calculate their means, and determine the probability that the mean of such a sample will differ by more than 5 from 160, the actual mean weight of the six persons.
(c) Suppose that three of the persons, those with weights 125, 135, and 185 pounds, are under 25 years of age, while those remaining are over 25 years of age. List all possible stratified samples of size 2 that can be taken by randomly choosing one of the three younger persons and one of the three older persons, calculate their means, and determine the probability that the mean of such a sample will differ by more than 5 from 160, the actual mean weight of the six persons.
(d) Compare the results of parts (a), (b), and (c).

*10.28 Based on their volume of sales, 9 of the 12 new car dealers in a city are classified as being small, and the other three are classified as being large. How many different stratified samples of four of these new car dealers can we choose if
(a) half of the sample is to be allocated to each of the strata;
(b) the allocation is to be proportional?

*10.29 Among 36 persons nominated for a city council, 18 are lawyers, 12 are business executives, and 6 are teachers. How many different stratified samples of six of these persons can we choose if
(a) a third of the sample is to be allocated to each of the strata;
(b) the allocation is to be proportional?

*10.30 Verify that if the formula

$$n_i = \frac{N_i}{N} \cdot n \qquad \text{for } i = 1, 2, \ldots, \text{and } k$$

is used to determine the sample sizes allocated to the k strata, then
(a) the allocation is proportional; that is, the ratios n_i/N_i all equal the same constant;
(b) the sum of the n_i is equal to n.

*10.31 A stratified sample of size $n = 40$ is to be taken from a population of size $N = 1,000$, which consists of four strata of size $N_1 = 250$, $N_2 = 600$, $N_3 = 100$, and $N_4 = 50$. If the allocation is to be proportional, how large a sample must be taken from each of the four strata?

*10.32 With reference to Exercise 10.27, list all possible cluster samples of size $n = 2$ that can be taken by randomly choosing either two of the three women or two of the three men, calculate their means, and determine the probability that the mean of such a sample will differ by more than 5 from 160, the actual mean weight of the six persons. Compare this probability with those obtained in parts (a) and (b) of Exercise 10.27. What does this show about the relative merits of simple random sampling, stratified sampling, and cluster sampling in the given situation?

*10.33 In stratified sampling with proportional allocation, the importance of differences in stratum size is accounted for by letting the larger strata contribute relatively more items to the sample. However, strata differ not only in size but also in variability, and it would seem reasonable to take larger samples from the more variable strata and smaller samples from the less variable strata. If we let $\sigma_1, \sigma_2, \ldots,$ and σ_k denote the standard deviations of the k strata, we can account for both differences in stratum size and differences in stratum variability, but requiring that

$$\frac{n_1}{N_1\sigma_1} = \frac{n_2}{N_2\sigma_2} = \cdots = \frac{n_k}{N_k\sigma_k}$$

The sample sizes for this kind of allocation, called **optimum allocation,** are given by the formula

$$n_i = \frac{n \cdot N_i\sigma_i}{N_1\sigma_1 + N_2\sigma_2 + \cdots + N_k\sigma_k}$$

for $i = 1, 2, \ldots,$ and k, where, if necessary, we round to the nearest integer. Verify that
(a) with this formula, the quantities $\dfrac{n_i}{N_i\sigma_i}$ all equal the same constant;
(b) the sum of the n_i is equal to n.

*10.34 A sample of size $n = 100$ is to be taken from a population consisting of two strata for which $N_1 = 10,000$, $N_2 = 30,000$, $\sigma_1 = 45$, and $\sigma_2 = 60$. To attain optimum allocation, how large a sample must be taken from each of the two strata?

*10.35 A sample of size $n = 84$ is to be taken from a population consisting of three strata for which $N_1 = 5,000$, $N_2 = 2,000$, $N_3 = 3,000$, $\sigma_1 = 15$, $\sigma_2 = 18$, and $\sigma_3 = 5$. To attain optimum allocation, how large a sample must be taken from each of the three strata?

*10.36 To estimate the mean of a population on the basis of a stratified sample, we calculate the weighted mean of the means $\bar{x}_1, \bar{x}_2, \ldots$, and $\bar{x}_k$ obtained for the k strata, using as weights the strata sizes N_i. Verify that for proportional allocation, the weighted mean thus obtained equals the mean of the values obtained for all the strata.

*10.37 The records of a casualty insurance company show that among 3,800 claims filed against the company over a period of time 2,600 were minor claims (under $200), while the other 1,200 were major claims ($200 or more). To estimate the average size of these claims, the company takes a 1 percent sample, proportionally allocated to the two strata, with the following results (rounded to the nearest dollar):

Minor claims:	42	115	63	78	45	148	195
	66	18	73	55	89	170	41
	92	103	22	138	49	62	88
	113	29	71	58	83		
Major claims:	246	355	872	649	253	338	
	491	860	755	502	488	311	

(a) Find the means of these two samples and then determine their weighted mean, using as weights the two strata sizes $N_1 = 2,600$ and $N_2 = 1,200$.

(b) Verify that the result of part (a) equals the mean of the combined sample data. This being the case, proportional allocation is said to be **self-weighting**.

10.6 SAMPLING DISTRIBUTIONS

The sample mean, the sample median, and the sample standard deviation are examples of random variables, whose values will vary from sample to sample. Their distributions, which reflect such chance variations, play a fundamental role in statistical inference, and they are referred to as **sampling distributions**. In this chapter, we shall concentrate primarily on the sample mean and its sampling distribution, but in some of the exercises on page 287 and in later chapters we shall also consider the sampling distributions of other statistics.

To give an example of a sampling distribution, let us construct the one for the mean of a random sample of size $n = 2$ drawn without replacement from the finite population of size $N = 5$, whose elements are the numbers 3, 5, 7, 9, and 11. The mean of this population is

$$\mu = \frac{3 + 5 + 7 + 9 + 11}{5} = 7$$

and its standard deviation is

$$\sigma = \sqrt{\frac{(3-7)^2 + (5-7)^2 + (7-7)^2 + (9-7)^2 + (11-7)^2}{5}} = \sqrt{8}$$

Now, if we take a random sample of size $n = 2$ from this population, there are the $\binom{5}{2} = 10$ possibilities 3 and 5, 3 and 7, 3 and 9, 3 and 11, 5 and 7, 5 and 9, 5 and 11, 7 and 9, 7 and 11, and 9 and 11. Their means are 4, 5, 6, 7, 6, 7, 8, 8, 9, and 10, and since sampling is random, each of these ten values has the probability $\frac{1}{10}$. Thus, we arrive at the following sampling distribution of the mean:

$\bar{x}$	Probability
4	$\frac{1}{10}$
5	$\frac{1}{10}$
6	$\frac{2}{10}$
7	$\frac{2}{10}$
8	$\frac{2}{10}$
9	$\frac{1}{10}$
10	$\frac{1}{10}$

A histogram of this probability distribution is shown in Figure 10.3.

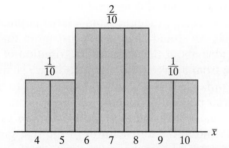

FIGURE 10.3
Sampling distribution
of the mean.

An examination of this sampling distribution reveals some pertinent information about the chance variations of the mean of a random sample of size $n = 2$ from the given finite population. For instance, we find that the probability is $\frac{6}{10}$ that a sample mean will differ from the population mean $\mu = 7$ by 1 or less, and that the probability is $\frac{8}{10}$ that a sample mean will differ from the population mean $\mu = 7$ by 2 or less. The first case corresponds to $\bar{x} = 6, 7$, or 8, and the second case corresponds to $\bar{x} = 5, 6, 7, 8$, or 9. So, if we did not know the mean of the given population and wanted to estimate it with a random sample of size $n = 2$, this would give us some idea about the potential size of our error.

Further useful information about this sampling distribution of the mean can be obtained by calculating its mean and its standard deviation, denoted, respectively, by $\mu_{\bar{x}}$ and $\sigma_{\bar{x}}$. Here, the subscript $\bar{x}$ is used to distinguish between the parameters of this sampling distribution and those of the original population. Using again the definitions of the mean and the standard deviation of a probability distribution, we get

$$\mu_{\bar{x}} = 4 \cdot \frac{1}{10} + 5 \cdot \frac{1}{10} + 6 \cdot \frac{2}{10} + 7 \cdot \frac{2}{10} + 8 \cdot \frac{2}{10} + 9 \cdot \frac{1}{10} + 10 \cdot \frac{1}{10}$$

$$= 7$$

and

$$\sigma_{\bar{x}}^2 = (4-7)^2 \cdot \frac{1}{10} + (5-7)^2 \cdot \frac{1}{10} + (6-7)^2 \cdot \frac{2}{10} + (7-7)^2 \cdot \frac{2}{10}$$

$$+ (8-7)^2 \cdot \frac{2}{10} + (9-7)^2 \cdot \frac{1}{10} + (10-7)^2 \cdot \frac{1}{10}$$

$$= 3$$

so that $\sigma_{\bar{x}} = \sqrt{3}$. Observe that, at least for this example,

$\mu_{\bar{x}}$, the mean of the sampling distribution of $\bar{x}$, equals μ, the mean of the population;

$\sigma_{\bar{x}}$, the standard deviation of the sampling distribution of $\bar{x}$, is smaller than σ, the standard deviation of the population.

These relationships are of fundamental importance, and we shall return to them in Section 10.7.

To illustrate the concept of a sampling distribution, we took a very small sample of size $n = 2$ from a very small finite population of size $N = 5$, but it would be difficult to duplicate this method to construct a sampling distribution of the mean of a large random sample from a large population. For instance, for $n = 10$ and $N = 100$, we would have had to list more than 17 trillion samples.

To get some idea about the sampling distribution of the mean of a somewhat larger sample from a large finite population, we shall use a **computer simulation.** In other words, we shall leave it to a computer to take repeated random samples from a given population, determine their means, and describe the distribution of these means in various ways. This will give us some idea about the overall shape and some of the key features of the real sampling distribution of the mean for random samples of the given size from the given population.

Without a computer, we can picture the simulation as follows: First, the numbers from 1 to 1,000 are written on 1,000 slips of paper (poker chips, small balls, or whatever may lend itself to drawing random samples). Then, a random sample of size $n = 15$ is drawn without replacement from this population and its values are recorded. We replace the sample before the next one is drawn, and we repeat this process until 100 random samples have been obtained.

Actually using an appropriate computer package, MINITAB in this case, we obtained the printout shown in Figure 10.4. Here, the first two lines are instructions to the computer to generate the population, the integers from 1 to 1,000, and we enter these numbers in column 1. In the next step, this population is described as having the mean 500.50 and the standard deviation 288.82. After that, the 100 random samples are generated and their means are put into column 3. The printout following the instructions MEAN C3 and STDEV C3 tells us that the mean of the 100 sample means is 495.75 and that their standard deviation is 75.748. Finally, following the instruction HIST C3 225.5 50, the 100 sample means are grouped into a distribution with a class interval of 50 and the first class mark at 225.5, which is shown in the printout as a histogram having the same format as the one shown in Figure 2.4. (The class marks were chosen so that the corresponding class boundaries are "impossible" values; namely, values that cannot be taken on by a mean of 15 positive integers.)

```
MTB > SET C1
DATA> 1:1000
DATA> END
MTB > MEAN C1
   MEAN    =    500.50
MTB > STDEV C1
   ST.DEV. =    288.82
MTB > SET C3
DATA> *
DATA> END
MTB > STORE 'E'
STOR> SAMPLE 15 C1 C2
STOR> LET K1 = AVER (C2)
STOR> STACK C3 K1 PUT INTO C3
STOR> ERASE C2
STOR> END
STOR> NOECHO
MTB > EXECUTE 'E' 100 TIMES
MTB > MEAN C3
   MEAN    =    495.75
MTB > STDEV C3
   ST.DEV =    75.748
MTB > HIST C3 225.5 50

Histogram of C3  N = 100  N* = 1

Midpoint Count
   225.5      0
   275.5      0
   325.5      2 **
   375.5     12 ************
   425.5     16 ****************
   475.5     20 ********************
   525.5     26 **************************
   575.5     15 ***************
   625.5      8 ********
   675.5      1 *
```

FIGURE 10.4
Computer simulation
of a sampling
distribution of the
mean.

As can be seen from Figure 10.4, the distribution of the 100 sample means is fairly symmetrical and bell shaped. In fact, the overall pattern seems to follow quite closely that of a normal curve. Note also that the simulation supports the two points made on page 278. The mean of the 100 sample means, shown to be 495.75, nearly equals the population mean of 500.50, and the standard deviation of the 100 sample means, shown to be 75.748, is smaller than the population standard deviation of 288.82.

10.7 THE STANDARD ERROR OF THE MEAN

In most practical situations we cannot proceed as in the two examples of Section 10.6. That is, we cannot enumerate all possible samples or simulate a sampling distribution in order to judge how close a sample mean might be to the mean of the population from which the sample came. Fortunately, though, we can usually get the information we need from two theorems, which express essential facts about sampling distributions of the mean. One of these is discussed in this section and the other in Section 10.8.

The first of these two theorems expresses formally what we discovered from both of the examples of the preceding section—the mean of the sampling distribution of $\bar{x}$ equals the mean of the population sampled, and the standard deviation of the sampling distribution of $\bar{x}$ is smaller than the standard deviation of the population sampled. It may be phrased as follows: For random samples of size n taken from a population with the mean μ and the standard deviation σ, the sampling distribution of $\bar{x}$ has the mean

Mean of sampling distribution of $\bar{x}$

$$\mu_{\bar{x}} = \mu$$

and the standard deviation

Standard error of the mean

$$\sigma_{\bar{x}} = \frac{\sigma}{\sqrt{n}} \quad or \quad \sigma_{\bar{x}} = \frac{\sigma}{\sqrt{n}} \cdot \sqrt{\frac{N-n}{N-1}}$$

depending on whether the population is infinite or finite of size N.

It is customary to refer to $\sigma_{\bar{x}}$ as the **standard error of the mean,** where "standard" is used in the sense of an average, as in "standard deviation." Its role in statistics is fundamental, as it measures the extent to which sample means can be expected to fluctuate, or vary, due to chance. If $\sigma_{\bar{x}}$ is small, the chances are good that the mean of a sample will be close to the mean of the population; if $\sigma_{\bar{x}}$ is large, we are more likely to get a sample mean which differs considerably from the mean of the population.

What determines the size of $\sigma_{\bar{x}}$ can be seen from the two preceding formulas. Both formulas (for infinite and finite populations) show that $\sigma_{\bar{x}}$ increases as the variability of the population increases, and that it decreases as the sample size increases. In fact, it is directly proportional to σ and inversely proportional to the square root of n. (For finite populations it decreases even faster due to the n appearing in $\sqrt{\dfrac{N-n}{N-1}}$.)

EXAMPLE 10.5   When we take a random sample from an infinite population, what happens to the standard error of the mean, and hence to the error we might expect when we use the mean of the sample to estimate the mean of the population, if the sample size is

(a) increased from 50 to 200;

(b) decreased from 360 to 40?

Solution   **(a)** The ratio of the two standard errors is

$$\frac{\frac{\sigma}{\sqrt{200}}}{\frac{\sigma}{\sqrt{50}}} = \frac{\sqrt{50}}{\sqrt{200}} = \sqrt{\frac{50}{200}} = \sqrt{\frac{1}{4}} = \frac{1}{2}$$

and where n is quadrupled, the standard error of the mean is reduced, but only divided by 2.

(b) The ratio of the two standard errors is

$$\frac{\frac{\sigma}{\sqrt{40}}}{\frac{\sigma}{\sqrt{360}}} = \frac{\sqrt{360}}{\sqrt{40}} = \sqrt{9} = 3$$

and where n is divided by 9, the standard error of the mean is increased, but only multiplied by 3.   ∎

The factor $\sqrt{\dfrac{N-n}{N-1}}$ in the second formula for $\sigma_{\bar{x}}$ is called the **finite population correction factor,** for without it the two formulas for $\sigma_{\bar{x}}$ (for infinite and finite populations) are the same. In practice, it is omitted unless the sample constitutes at least 5 percent of the population, for otherwise it is so close to 1 that it has little effect on the value of $\sigma_{\bar{x}}$.

EXAMPLE 10.6   Find the value of the finite population correction factor for $N = 10,000$ and $n = 100$.

Solution   Substituting $N = 10,000$ and $n = 100$, we get

$$\sqrt{\frac{N-n}{N-1}} = \sqrt{\frac{10,000 - 100}{10,000 - 1}} = 0.995$$

and this is so close to 1 that the correction factor can be omitted for all practical purposes.   ∎

Since we stated the formulas for the standard error of the mean without proof, let us verify the one for finite populations by referring to the results of the two illustrations of Section 10.6.

EXAMPLE 10.7   With reference to the illustration on page 276, where we had $N = 5, n = 2$, and $\sigma = \sqrt{8}$, verify that the second formula for $\sigma_{\bar{x}}$ will yield $\sqrt{3}$, namely, the value that we obtained on page 278.

Solution   Substituting $N = 5, n = 2,$ and $\sigma = \sqrt{8}$ into the second of the two formulas for $\sigma_{\bar{x}}$, we get

$$\sigma_{\bar{x}} = \frac{\sqrt{8}}{\sqrt{2}} \cdot \sqrt{\frac{5-2}{5-1}} = \frac{\sqrt{8}}{\sqrt{2}} \cdot \sqrt{\frac{3}{4}} = \sqrt{\frac{8}{2} \cdot \frac{3}{4}} = \sqrt{3} \qquad \blacksquare$$

EXAMPLE 10.8   With reference to the computer simulation of Figure 10.4, where we had $N = 1,000, n = 15,$ and $\sigma = 288.82$, what value could we have expected for the standard deviation of the 100 sample means?

Solution   Substituting $N = 1,000, n = 15,$ and $\sigma = 288.82$ into the second of the two formulas for $\sigma_{\bar{x}}$, we get

$$\sigma_{\bar{x}} = \frac{288.82}{\sqrt{15}} \cdot \sqrt{\frac{1,000 - 15}{1,000 - 1}} = 74.05$$

and this is quite close to 75.748, the value actually obtained in the computer simulation of Figure 10.4. $\qquad \blacksquare$

10.8   THE CENTRAL LIMIT THEOREM

When a sample mean is used to estimate the mean of a population, the uncertainties about its potential error can be expressed in various ways. If we knew the exact sampling distribution of the mean, which, of course, we never do, we could proceed as in the first illustration of Section 10.6 and calculate the probabilities associated with errors of various size. Something else we rarely, if ever, do, is to use Chebyshev's theorem and assert with a probability of at least $1 - \frac{1}{k^2}$ that the mean of a random sample will differ from the mean of the population sampled by less than $k \cdot \sigma_{\bar{x}}$.

EXAMPLE 10.9   Based on Chebyshev's theorem with $k = 2$, what can we say about the potential size of our error if we are going to use a random sample of size $n = 64$ to estimate the mean of an infinite population with $\sigma = 20$?

Solution   Substituting $n = 64$ and $\sigma = 20$ into the appropriate formula for the standard error of the mean, we get

$$\sigma_{\bar{x}} = \frac{20}{\sqrt{64}} = 2.5$$

and it follows that we can assert with a probability of at least $1 - \frac{1}{2^2} = 0.75$ that the error will be less than $k \cdot \sigma_{\bar{x}} = 2(2.5) = 5$. $\qquad \blacksquare$

The importance of this example is that it shows *how* we can make exact probability statements about the potential error when we estimate the mean of a population. The trouble with the use of Chebyshev's theorem is that "at least 0.75" does not tell us enough when in reality that probability may be, say, 0.998 or even 0.999. Whereas Chebyshev's theorem provides a logical link between errors and the probabilities that they may be committed, there exists another mathematical theorem that, in many instances, enables us to make much stronger probability statements about such potential errors.

This theorem, which is the second of the two theorems mentioned on page 279, is called the **central limit theorem** and, informally, it states that for large samples the sampling distribution of the mean can be approximated closely with a normal distribution. Recalling from Section 10.7 that

$$\mu_{\bar{x}} = \mu \quad \text{and} \quad \sigma_{\bar{x}} = \frac{\sigma}{\sqrt{n}}$$

for random samples from infinite populations, we can now say formally that

Central limit theorem

> *If $\bar{x}$ is the mean of a random sample of size n from an infinite population with the mean μ and the standard deviation σ and n is large, then*
>
> $$z = \frac{\bar{x} - \mu}{\sigma/\sqrt{n}}$$
>
> *has approximately the standard normal distribution.*

This theorem is of fundamental importance in statistics, as it justifies the use of normal-curve methods in a wide range of problems; it applies to infinite populations, and also to finite populations when n, though large, constitutes but a small portion of the population. We cannot say precisely how large n must be so that the central limit theorem can be applied, but unless the population distribution has a very unusual shape, $n = 30$ is usually regarded as sufficiently large. When the population we are sampling has, itself, roughly the shape of a normal curve, the sampling distribution of the mean can be approximated closely with a normal distribution regardless of the size of n.

The central limit theorem can also be used for finite populations, but a precise description of the situations under which this can be done is rather complicated. The most common proper use is the case in which n is large while $\frac{n}{N}$ is small.

Let us now see what probability will take the place of "at least 0.75" if we use the central limit theorem instead of Chebyshev's theorem in Example 10.9.

EXAMPLE 10.10   Based on the central limit theorem, what is the probability that the error will be less than 5, when the mean of a random sample of size $n = 64$ is used to estimate the mean of an infinite population with $\sigma = 20$.

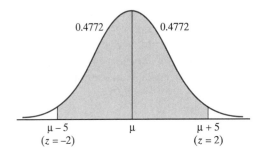

FIGURE 10.5
Diagram for Example
10.10.

$\mathbf{S}$olution   The probability is given by the area of the tinted region under the curve in Figure 10.5, namely, by the standard-normal-curve area between

$$z = \frac{-5}{20/\sqrt{64}} = -2 \quad \text{and} \quad z = \frac{5}{20/\sqrt{64}} = 2$$

Since the entry in Table I corresponding to $z = 2.00$ is 0.4772, the probability asked for is $0.4772 + 0.4772 = 0.9544$. Thus, the statement that the probability is "at least 0.75" is replaced by the much stronger statement that the probability is about 0.95 (that the mean of a random sample of size $n = 64$ from the given population will differ from the mean of the population by less than 5).   ■

Since we presented the central limit theorem without proof, let us consider the following example:

EXAMPLE 10.11   In the simulation of Figure 10.4, we created a population with the mean $\mu = 500.5$ and the standard deviation $\sigma = 288.82$. Then we generated 100 random samples of size $n = 15$, and their distribution shows that 97 of them, all but three, fell between 350.5 (the upper boundary of the third class) and 650.5 (the lower boundary of the class at the bottom of the distribution). According to the central limit theorem, what is the probability that any one of the sample means should have fallen between 350.5 and 650.5?

FIGURE 10.6
Diagram for Example
10.11.

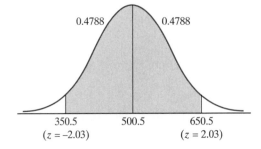

Solution   The probability asked for is given by the area of the tinted region under the curve in Figure 10.6. Making use of the fact that $\sigma_{\bar{x}} = 74.05$ according to Example 10.8, we find that this area is the standard-normal-curve area between

$$z = \frac{350.5 - 500.5}{74.05} \approx -2.03$$

and

$$z = \frac{650.5 - 500.5}{74.05} \approx 2.03$$

Since the entry in Table I corresponding to $z = 2.03$ is 0.4788, the probability asked for is $0.4788 + 0.4788 = 0.9576$, or approximately 0.96. In other words, we could have expected 96 of the 100 means to fall between 350.5 and 650.5, and this is very close to the 97 that did. Thus, the simulation provides strong support for the central limit theorem.   ∎

10.9   SOME FURTHER CONSIDERATIONS

In Sections 10.6 through 10.8, our main goal was to introduce the concept of a sampling distribution, and the one that we chose as an illustration was the sampling distribution of the mean. It should be clear, though, that instead of the mean we could have studied the median, the standard deviation, or some other statistic, and investigated its chance fluctuations. So far as the corresponding theory is concerned, this would have required a different formula for the standard error and theory analogous to, yet different from, the central limit theorem.

For instance, for large samples from continuous populations, the **standard error of the median** is approximately

$$\sigma_{\tilde{x}} = 1.25 \cdot \frac{\sigma}{\sqrt{n}}$$

where n is the size of the sample and σ is the population standard deviation. Note that comparison of the two formulas

$$\sigma_{\bar{x}} = \frac{\sigma}{\sqrt{n}} \quad \text{and} \quad \sigma_{\tilde{x}} = 1.25 \cdot \frac{\sigma}{\sqrt{n}}$$

reflects the fact that the mean is generally more reliable than the median (that is, it tends to expose us to smaller errors) when we estimate the mean of a symmetrical population. For symmetrical populations, the means of the sampling distributions of $\bar{x}$ and $\tilde{x}$ are both equal to the population mean μ.

EXAMPLE 10.12   How large a random sample do we need so that its mean is as reliable an estimate of the mean of a symmetrical continuous population as the median of a random sample of size $n = 200$?

Solution   Equating the two standard error formulas and substituting $n = 200$ into the one for the standard error of the median, we get

$$\frac{\sigma}{\sqrt{n}} = 1.25 \cdot \frac{\sigma}{\sqrt{200}}$$

which, solved for n, yields $n = 128$. Thus, for the stated purpose, the mean of a random sample of size $n = 128$ is as "good" as the median of a random sample of size $n = 200$. ∎

Also, a point worth emphasizing is that the illustrations of Section 10.6 were used as teaching aids, designed to convey the idea of a sampling distribution, but they do not reflect what we do in actual practice. In practice, we can seldom list all possible samples, and ordinarily we base an inference on one sample and not on 100 samples. In Chapter 11 and in subsequent chapters we shall go further into the problem of translating theory about sampling distributions into methods of evaluating the merits and shortcomings of statistical procedures.

Another point worth repeating concerns the $\sqrt{n}$ that appears in the denominator of the formulas for the standard error of the mean. It makes sense that when n becomes larger and larger, our generalizations will be subject to smaller errors, but the $\sqrt{n}$ in the formulas for the standard error of the mean tells us that the gain in reliability is not proportional to the increase in the size of the sample. As we saw, quadrupling the size of the sample will only double the reliability of a sample mean as an estimate of the mean of a population. Indeed, to quadruple the reliability, we would have to multiply the sample size by 16. This relationship between reliability and sample size indicates that there are, to use a phrase from economics, diminishing returns to increasing the size of a sample. It seldom pays to take samples that are massively large.

■ E x e r c i s e s[†]

10.38 Suppose that in the illustration on page 276, where random samples of size $n = 2$ were drawn without replacement from the finite population that consists of the numbers 3, 5, 7, 9, and 11, sampling had been with replacement.

(a) List the 25 ordered samples that can be drawn with replacement from the given population and calculate their means. (By "ordered" we mean that 3 and 7, for example, is a different sample than 7 and 3.)

(b) Assuming that sampling is random, namely, that each of the ordered samples of part (a) has the probability $\frac{1}{25}$, construct the sampling distribution of the mean for random samples of size $n = 2$ drawn with replacement from the given population.

10.39 With reference to Exercise 10.38, find the probabilities that the mean of a random sample of size $n = 2$ drawn with replacement from the given population will differ from $\mu = 7$ by

(a) 1 or less;   (b) at most 2.

[†]Exercises 10.54 through 10.57 are marked optional as they pertain to the optional material on sample designs in Section 10.2 through 10.5.

10.40 Calculate the standard deviation of the sampling distribution obtained in part (b) of Exercise 10.38, and verify the result by substituting $n = 2$ and $\sigma = \sqrt{8}$ into the first of the two standard error formulas on page 280.

10.41 A finite population consists of the $N = 8$ elements 12, 12, 12, 12, 12, 14, 20, and 42. As can easily be verified, the mean of this population is $\mu = 17$ and $\sigma = \sqrt{96} = 4\sqrt{6}$.

(a) List the $\binom{8}{2}$ possible samples of size $n = 2$ that can be drawn without replacement from this finite population. [*Hint*: In $\binom{5}{2} = 10$ of these samples, both values are equal to 12.]

(b) Calculate the mean of each of the samples obtained in part (a).

(c) Assigning to each of the samples obtained in part (a) the probability $\frac{1}{28}$, construct the sampling distribution of the mean for random samples of size $n = 2$ drawn without replacement from the given population.

(d) Find the mean and the standard deviation of the sampling distribution of the mean obtained in part (c).

10.42 Use the second of the two standard error formulas on page 280 to verify the result obtained in part (d) of Exercise 10.41.

10.43 A finite population consists of the $N = 6$ numbers 4, 5, 6, 7, 8, and 9. As can easily be verified, the mean of this population is $\mu = 6.5$ and its standard deviation is

$$\sigma = \frac{1}{2}\sqrt{\frac{35}{3}}.$$

(a) List the $\binom{6}{3} = 20$ possible samples of size $n = 3$ that can be drawn without replacement from the given population.

(b) Calculate the means of the 20 samples obtained in part (a).

(c) Assigning the probability $\frac{1}{20}$ to each of the samples obtained in part (a), construct the sampling distribution of the mean for random samples of size $n = 3$ drawn without replacement from the given population.

(d) What is the probability that the mean of a random sample of size $n = 3$ drawn without replacement from the given population will differ from $\mu = 6.5$ by less than 1?

10.44 With reference to Exercise 10.43, calculate the mean and the standard deviation of the sampling distribution obtained in part (c) and verify the results using the theory of Section 10.7.

10.45 With reference to Exercise 10.43, determine the medians of the 20 samples obtained in part (a), construct the sampling distribution of the median for random samples of size $n = 3$ drawn without replacement from the given population, and find the probability that the median of such a random sample will differ from $\mu = 6.5$ by less than 1. Also, compare this probability with the one obtained in part (d) of Exercise 10.43.

10.46 With reference to Exercise 10.43, determine the ranges of the 20 samples obtained in part (a) and construct the sampling distribution of the range for random samples of size $n = 3$ drawn without replacement from the given population.

10.47 When we sample from an infinite population, what happens to the standard error of the mean when the sample size is
(a) increased from 25 to 225;

(b) decreased from 180 to 45;

(c) increased from 200 to 450?

10.48 When we sample from an infinite population, what happens to the standard error of the mean when the sample size is

(a) decreased from 1,000 to 10;

(b) increased from 20 to 500;

(c) decreased from 500 to 80?

10.49 When we sample from a finite population of size $N = 82$, what happens to the standard error of the mean when the sample size is

(a) increased from 18 to 33;

(b) increased from 57 to 81?

10.50 For what positive value of n are the two formulas for the standard error of the mean on page 280 identical?

10.51 What is the value of the finite population correction factor when

(a) $N = 200$ and $n = 5$;

(b) $N = 300$ and $n = 10$;

(c) $N = 4,000$ and $n = 100$?

10.52 For what value of n is the finite population correction factor equal to zero?

*10.53 Show that if the mean of a random sample of size n is used to estimate the mean of an infinite population with the standard deviation σ and n is large, there is a fifty-fifty chance that the magnitude of the error will be less than

$$0.6745 \cdot \frac{\sigma}{\sqrt{n}}$$

It has been the custom to refer to this quantity as the **probable error of the mean;** nowadays, it is used mainly in military applications.

(a) If a random sample of size $n = 64$ is drawn from an infinite population with $\sigma = 24.8$, what is the probable error of the mean?

(b) If a random sample of size $n = 144$ is drawn from a very large population (consisting, say, of the fines paid for various traffic violations in a certain county in 1999) with $\sigma = \$219.12$, what is the probable error of the mean? Explain its significance.

*10.54 With reference to Exercise 10.24, assign each of the six systematic samples the probability $\frac{1}{6}$ and calculate the standard deviation of the corresponding sampling distribution of the mean. Compare the result with $\sigma_{\bar{x}} = 3.5$, the standard error of the mean for ordinary random samples of size $n = 8$ from the given finite population.

*10.55 In the illustration on page 270, we compared stratified samples from a population consisting of four weights with ordinary random samples of the same size.

(a) Assigning each of the random samples on page 270 the probability $\frac{1}{6}$, show that the mean and the standard deviation of this sampling distribution of the mean are $\mu_{\bar{x}} = 160$ and $\sigma_{\bar{x}} = 21.0$.

(b) Assigning each of the stratified samples on page 270 the probability $\frac{1}{4}$, show that the mean and the standard deviation of this sampling distribution of the mean are $\mu_{\bar{x}} = 160$ and $\sigma_{\bar{x}} = 7.1$.

Compare the results of parts (a) and (b).

*10.56 If $\bar{x}$ is the mean of a stratified random sample of size n obtained by proportional allocation from a finite population of size N, which consists of k strata of size N_1, $N_2, \ldots,$ and N_k, then

$$\sigma_{\bar{x}}^2 = \sum_{i=t}^{k} \frac{(N - n)N_i^2}{nN^2(N_i - 1)} \cdot \sigma_i^2$$

where $\sigma_1^2, \sigma_2^2, \ldots,$ and σ_k^2 are the corresponding variances for the individual strata. Use this formula to verify the result of part (b) of Exercise 10.55.

*10.57 In the illustration on page 270, the two cluster samples that consist of choosing and weighing either the two women or the two men have the means 125 and 195. Assigning the probability $\frac{1}{2}$ to each of these means, calculate the standard deviation of this sampling distribution of the mean, and compare it with the corresponding values for ordinary random sampling and stratified sampling obtained in the two parts of Exercise 10.55.

10.58 The mean of a random sample of size $n = 36$ is used to estimate the mean of an infinite population with the standard deviation $\sigma = 9$. What can we assert about the probability that the error of this estimate will be less than 4.5 if we use
(a) Chebyshev's theorem;
(b) the central limit theorem?

10.59 The mean of a random sample of size $n = 25$ is used to estimate the mean attention span of persons over 65. Given that the standard deviation of the population sampled is $\sigma = 2.4$ minutes, what can we assert about the probability that the error of the estimate is less than 1.2 minutes if we use
(a) Chebyshev's theorem;
(b) the central limit theorem?

10.60 The mean of a random sample of size $n = 100$ is going to be used to estimate the mean daily milk production of a very large herd of dairy cows. Given that the standard deviation of the population to be sampled is $\sigma = 3.6$ quarts, what can we assert about the probabilities that the error of this estimate will be
(a) more than 0.72 quart;
(b) less than 0.45 quart?

10.61 If measurements of the specific gravity of a metal can be looked upon as a random sample from a normal population with the standard deviation $\sigma = 0.025$ ounce, what is the probability that the mean of a random sample of size $n = 16$ will be off by at most 0.01 ounce?

10.62 If the distribution of the weights of all men traveling by air between Dallas and El Paso has a mean of 163 pounds and a standard deviation of 18 pounds, what is the probability that the combined weight of 36 men traveling on such a flight is more than 6,012 pounds?

10.63 Verify that the mean of a random sample of size $n = 256$ is as reliable an estimate of the mean of a symmetrical continuous population as the median of a random sample of size $n = 400$.

10.64 How large a random sample do we have to take so that its median is as reliable an estimate of the mean of a symmetrical continuous population as the mean of a random sample of size $n = 144$?

*10.65 In elementary algebra it can be shown by mathematical induction that the sum of the first n positive integers is $\dfrac{n(n+1)}{2}$ and that the sum of their squares is $\dfrac{n(n+1)(2n+1)}{6}$. Using these two formulas, it can be shown that the mean and the standard deviation of the finite population that consists of the first n positive integers are

$$\mu = \frac{n+1}{2} \quad \text{and} \quad \sigma = \sqrt{\frac{n^2-1}{12}}$$

Verify that these formulas yield the population mean and standard deviation shown in the printout of Figure 10.4.

*10.10 TECHNICAL NOTE (Simulation)

Simulation provides one of the most effective ways of illustrating, and thus teaching, some of the basic concepts of statistics. As we shall see in the chapters that follow, the evaluation and interpretation of statistical techniques will often require that we imagine what would happen if experiments are repeated over and over again. Since, most of the time, such repetitions are neither practical nor feasible, we can resort instead to simulations, preferable with the use of computers. Simulation also plays a role in the development of statistical theory, for there are situations where simulation is easier than a detailed mathematical analysis.

Simulations of random samples can also be made with the use of a table of random numbers, but for use in the exercises that follow, we present in Figure 10.7 forty computer-simulated random samples, each consisting of $n = 5$ values of a random variable having the Poisson distribution with $\lambda = 16$, and hence $\mu = 16$ and $\sigma = 4$. (The reader may picture these figures as data on the number of emergency calls that an ambulance service receives in an afternoon, the number of calls that a switchboard receives during a ten-minute interval, or the number of pieces of junk mail that a person receives in Monday's mail.)

Exercises

*10.66 In Figure 10.7, each row constitutes a random sample of size $n = 5$ from a Poisson population with $\lambda = 16$, and hence with $\mu = 16$ and $\sigma = 4$.
(a) Calculate the means of the 40 samples in Figure 10.7.
(b) Calculate the mean and the standard deviation of the 40 means obtained in part (a), and compare the results with the corresponding values expected in accordance with the theory of Section 10.7.

```
MTB > Random 40 cl-c5;
SUBC>    Poisson 16.
MTB > Print cl-c5.


Row

   1     9    15     6    19    11
   2    16    15    19    14    14
   3    14    20    11    22    19
   4    14    20    17    22    14
   5    21    11    13    18    14
   6    13    11    13    15    12
   7    14    12    14    17    10
   8    17    13    25    17    20
   9    21    16    16    18    21
  10    15    12    16    11    14
  11    21    12    19    14    14
  12    20    22    16    19    17
  13    25    15     8    16    21
  14    15    19    18    12    18
  15    17    23    20    11    13
  16    18    16    16    21    22
  17    20    19    21    17     9
  18    19    17    11    14    19
  19    12    18    16    10    14
  20    11    14    11    12    26
  21    17    16    11    11     9
  22    15    16    16    19    18
  23    16    12    18    16    15
  24    20    19    23    14    14
  25    19    18    16    24    13
  26    13    18    14    17    25
  27    16    17    18    14    22
  28    15    17    11    15    13
  29    23    12    13    13    16
  30    16    28    11    14    11
  31    14    15    18     7    16
  32    19    17    11    16    13
  33    13    14    16    12    17
  34    25    14     8    15    16
  35    12    17    12    12    13
  36    12    16    17    15    25
  37    20    14    14    16    17
  38    13    19    19    16    17
  39    12    14    11    19    14
  40    14     9    17    24    19
```

FIGURE 10.7
Computer simulation
of 40 random samples.

*10.67 Determine the medians of the 40 samples shown in Figure 10.7, calculate their standard deviation, and compare the result with the corresponding value expected in accordance with the theory of Section 10.9.

*10.68 In Exercise 11.42 we shall present a way of estimating the population standard deviation in terms of the range (largest value minus the smallest). For this purpose, the range is divided by a constant depending on the size of the sample; for instance, by 2.33 for $n = 5$. Another way of saying this is that for samples of size $n = 5$, the mean of the sampling distribution of the range is 2.33σ. To verify this, determine the ranges of the 40 samples shown in Figure 10.7 and then calculate their mean. This is an estimate of the mean of the sampling distribution of the range, and since the sample size is $n = 5$, divided by 2.33 it provides an estimate of σ, which is known to equal 4. Find the percentage error of this estimate.

*10.69 On page 83 we explained that we divide by $n - 1$ in the formulas of the sample standard deviation and the sample variance, to make s^2 an unbiased estimator of σ^2—namely, to make the mean of the sampling distribution of s^2 equal to σ^2. To verify this, here are the variances of the 40 samples shown in Figure 10.7:

26.0	4.3	20.7	12.8	16.3	2.2	6.8	19.8
6.3	4.3	14.5	5.7	41.5	8.3	24.2	7.8
23.2	12.0	10.0	40.7	12.2	2.7	4.8	15.5
16.5	22.3	8.8	5.2	20.3	49.5	17.5	10.2
4.3	37.3	4.7	23.5	6.2	6.2	9.5	31.3

Find the mean of these 40 sample variances, which, according to what we said previously, provides an estimate of the population variance. Since σ^2 is known to equal 16, calculate the percentage error of this estimate.

*10.70 Following are 20 simulated random samples of size $n = 4$ (with their values rounded to two decimals) from a normal population with $\mu = 5$ and $\sigma = 1$.

6.33	4.83	5.08	5.48
6.08	6.18	4.40	4.60
5.16	6.84	6.42	4.09
5.29	3.65	4.72	4.78
5.77	5.87	5.93	5.65
5.22	5.95	6.48	4.16
5.31	4.31	5.39	5.61
5.02	6.41	7.28	5.53
5.01	4.97	6.24	4.97
6.49	4.78	4.90	3.53
2.91	5.21	5.68	4.50
5.30	4.20	6.68	3.90
6.94	5.67	3.41	5.75
6.73	5.73	2.55	4.06
4.22	6.32	5.06	3.92
3.27	3.55	3.49	3.72
4.49	4.42	3.44	3.93
7.60	4.38	4.77	4.67
4.87	5.04	6.67	6.74
5.35	4.21	5.94	3.19

(a) Calculate the means of these 20 samples.
(b) Calculate the standard deviation of the 20 sample means obtained in part (a).
(c) Compare the result of part (b) with the corresponding standard error calculated in accordance with the theory of Section 10.7.

*10.71 Determine the medians of the 20 samples given in Exercise 10.70, calculate their standard deviation, and compare the result with the corresponding value expected in accordance with the theory of Section 10.9.

*10.72 Use appropriate computer software to simulate 30 random samples of size $n = 3$ from a normal population with $\mu = 12$ and $\sigma = 2$. Round these figures to two decimals and calculate the means of the 30 samples. Then calculate the standard deviation of these sample means and compare the result with the corresponding standard error expected in accordance with the theory of Section 10.7.

*10.73 Determine the medians of the 30 random samples obtained in Exercise 10.72, with its values rounded to two decimals. Then calculate the standard deviation of these sample medians and compare the result with the corresponding standard error expected in accordance with the theory of Section 10.9.

10.11 Checklist of Key Terms *(with page references to their definitions)*

*Area sampling, 273
Central limit theorem, 283
*Cluster sampling, 273
Computer simulation, 278, 279, 291
*Cross stratification, 272
Finite population, 262
Finite population correction factor, 281
Infinite population, 262, 266
*Judgment sample, 272
*Optimum allocation, 275
*Probable error of the mean, 288
Proportional allocation, 271
*Quota sampling, 272

Random numbers, 264, 265
Random sample, 261, 263, 266
*Sample design, 269
Sampling distribution, 262
*Sampling frame, 268
*Self-weighting, 276
Simple random sample, 263
Standard error of the mean, 280
Standard error of the median, 285
*Strata, 270
*Stratification, 270
*Stratified sampling, 270
*Systematic sampling, 269

10.12 References

Among the many published tables of random numbers, one of the most widely used is

RAND CORPORATION, *A Million Random Digits with 100,000 Normal Deviates.* New York: Macmillan Publishing Co., Inc., third printing 1966.

There also exist calculators that are preprogrammed to generate random numbers, and it is fairly easy to program a computer so that a person can generate his or her own random numbers. The following is one of many articles on this subject.

KIMBERLING, C., "Generate Your Own Random Numbers." *Mathematics Teacher*, February 1984.

Interesting material on the early development of tables of random numbers may be found in

BENNETT, D. J., *Randomness*. Cambridge, Mass.: Harvard University Press, 1998.

Derivations of the various standard error formulas and more general formulations (and proof) of the central limit theorem may be found in most textbooks on mathematical statistics. All sorts of information about sampling is given in

COCHRAN, W. G., *Sampling Techniques*, 3rd ed. New York: John Wiley & Sons, Inc., 1977.

SCHAEFFER, R. L., MENDENHALL, W., and OTT, L., *Elementary Survey Sampling*, 4th ed. Boston: PWS-Kent Publishing Co., 1990.

SLONIN, M. J., *Sampling in a Nutshell*. New York: Simon and Schuster, 1973.

WILLIAMS, W. H., *A Sampler on Sampling*. New York: John Wiley & Sons, Inc., 1978.

R.91   Among 20 workers on a picket line, 11 are men and 9 are women. If a television news reporter randomly chooses four of them to be shown on camera, what are the probabilities that this will include

(a) two men and two women;

(b) three men and one woman?

R.92   Find the standard-normal-curve area that lies

(a) to the left of $z = 2.15$;

(b) to the left of $z = -0.62$;

(c) between $z = 1.05$ and $z = 1.85$;

(d) between $z = -0.63$ and $z = 0.63$.

R.93   Use Table I to find α, given that $z_{\alpha} = 1.175$.

R.94   A customs official wants to check 12 of 875 shipments listed on a ship's manifest. Using the 28th, 29th, and 30th columns of the table in Figure 10.2, starting with the 6th row and going down the page, which ones (by number) will the customs official inspect?

R.95   Check in each case whether the condition for the binomial approximation to the hypergeometric distribution is satisfied:

(a) $a = 40, b = 160$, and $n = 8$;

(b) $a = 100, b = 60$, and $n = 10$;

(c) $a = 68, b = 82$, and $n = 12$.

R.96   If the amount of time a tourist spends in a cathedral is a random variable having the normal distribution with $\mu = 23.4$ minutes and $\sigma = 6.8$ minutes, find the probability that a tourist will spend

(a) at most 16.0 minutes in the cathedral;

(b) anywhere from 20.0 to 30.0 minutes in the cathedral.

R.97   If a random sample of size $n = 4$ is to be chosen from a finite population of size $N = 55$, what is the probability of each possible sample?

*R.98   The probability that a baseball player will strike out on any given at bat is 0.36. What is the probability that he will strike out for the first time at his

(a) second at bat;

(b) fifth at bat?

(*Hint*: Use the formula for the geometric distribution.)

R.99   Find the mean of the binomial distribution with $n = 8$ and $p = 0.40$, using

(a) Table V and the formula that defines μ;

(b) the special formula for the mean of a binomial distribution.

R.100   Use the normal distribution to approximate the binomial probability that more than 90 of 100 scorpion stings will cause extreme discomfort if the probability is 0.82 that any one of them will cause extreme discomfort.

R.101 Check in each case whether the conditions for the Poisson approximation to the binomial distribution are satisfied:

(a) $n = 180$ and $p = \frac{1}{9}$;

(b) $n = 480$ and $p = \frac{1}{60}$;

(c) $n = 575$ and $p = \frac{1}{100}$.

R.102 It is known that 6% of all rats carry a certain disease. If we examine a random sample of 120 rats, will this satisfy the condition for using the Poisson approximation to the binomial distribution? If so, use the Poisson distribution to approximate the probability that only 5 of the rats will carry the disease.

R.103 A random variable has a normal distribution with $\sigma = 4.0$. If the probability is 0.9713 that it will take on a value less than 82.6, what is the probability that it will take on a value between 70.0 and 80.0?

*R.104 Convert the 40 samples of Figure 10.7 into 20 samples of size $n = 10$ by combining samples 1 and 2, samples 3 and 4, . . . , and samples 39 and 40. Calculate the mean of each of these 20 samples and determine their mean and their standard deviation. Compare the results with the values we might expect in accordance with the theorem on page 280.

*R.105 A small cruise ship has deluxe outside cabins, standard outside cabins, and inside cabins, and the probabilities are 0.30, 0.60, and 0.10 that a travel agent will receive a reservation for the first, second, or third of these categories. If a travel agent receives nine reservations, what is the probability that four of them will be for a deluxe outside cabin, four will be for a standard outside cabin, and one will be for an inside cabin?

R.106 In a certain community, the response time of an ambulance may be regarded as a random variable having the normal distribution with $\mu = 6.7$ minutes and a standard deviation of $\sigma = 1.5$ minutes. What is the probability that the ambulance will take at most 10.0 minutes to respond to a call?

R.107 A zoo has a large collection of anteaters, including five males and four females. If a veterinarian randomly picks three of them for examination, what are the probabilities that

(a) none of them will be females;

(b) two of them will be females?

R.108 It has been claimed that

(a) if the sample size is increased by 44%, the standard error of the mean is reduced by 20%;

(b) if the standard error of the mean is to be reduced by 20%, the sample size must be increased by 56.25%.

Which of these two statements is correct and which one is false?

*R.109 The size of an animal population is sometimes estimated by the **capture–recapture method.** In this method, n_1 of the animals are captured, marked, and released. Later, n_2 of the animals are captured, x of them are found to be marked, and all this information is used to estimate N, the size of the population. If $n_1 = 3$ rare owls are captured, marked, and released and later $n_2 = 4$ such owls are captured and $x = 1$ of them is found to be marked, for what value of N is the probability of getting $x = 1$ a maximum? (*Hint:* Try $N = 9, 10, 11, 12, 13,$ and 14.)

R.110 Figure R.4 shows the probability density of a continuous random variable that takes on values on the interval from 0 to 3.

(a) Verify that the total area under the curve is equal to 1.

(b) Find the probability that the random variable will take on a value greater than 1.5.

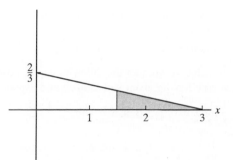

FIGURE R.4
Diagram for Exercise
R.110.

*R.111 Suppose that we want to find the probability that a random variable having the hypergeometric distribution with $n = 14$, $a = 180$, and $b = 120$ will take on the value $x = 5$.

(a) Verify that the binomial distribution with $n = 14$ and $p = \dfrac{180}{180 + 120} = \dfrac{3}{5}$ can be used to approximate this hypergeometric distribution.

(b) Verify that the normal distribution with $\mu = np = 14 \cdot \dfrac{3}{5} = 8.4$ and $\sigma = \sqrt{np(1 - p)} = \sqrt{14(0.6)(0.4)} \approx 1.83$ can be used to approximate the binomial distribution of part (a).

(c) Use the normal distribution with $\mu = 8.4$ and $\sigma = 1.83$ to approximate the hypergeometric probability with $x = 5$, $n = 14$, $a = 180$, and $b = 120$.

R.112 The amount of time that it takes an electrician to repair a ceiling fan can be treated as a random variable having the normal distribution with $\mu = 24.55$ minutes and $\sigma = 3.16$ minutes. Find the probability that it will take an electrician anywhere from 20.00 to 30.00 minutes to repair a ceiling fan.

R.113 How many different samples of size

(a) $n = 3$ can be chosen from among $N = 14$ different magazines at a doctor's office;

(b) $n = 5$ can be chosen for a potential customer from among $N = 24$ houses listed for sale in Scottsdale, Arizona?

 *R.114 During a busy weekend, 60 delivery vans obtained

17.7	23.0	24.9	17.6	26.4	18.0	22.4	25.8	29.5	20.8
23.2	26.7	21.5	23.8	19.1	22.3	27.4	22.4	20.6	23.5
16.5	22.2	21.9	14.4	25.6	23.0	25.4	21.2	16.8	28.4
20.5	21.5	22.6	19.8	20.5	21.7	16.3	18.9	24.0	21.3
22.2	24.8	17.5	18.0	21.4	22.5	20.6	17.7	15.9	22.5
26.7	21.3	24.5	19.3	25.4	20.0	16.5	21.1	23.8	20.5

miles per gallon. Use a computer or a graphing calculator to check whether these data can be looked upon as values of a random variable having a normal distribution.

R.115 The number of blossoms on a rare plant is a random variable having the Poisson distribution with $\lambda = 2.2$. What are the probabilities that such a plant will have

(a) no blossoms;

(b) at most two blossoms?

R.116 What is the finite population correction factor if

(a) $N = 120$ and $n = 30$;

(b) $N = 400$ and $n = 50$?

R.117 Determine in each case whether the following can be probability distributions (defined in each case for the given values of x) and explain your answers:

(a) $f(x) = \dfrac{1}{8}$ for $x = 0, 1, 2, 3, 4, 5, 6$, and 7;

(b) $f(x) = \dfrac{x + 1}{16}$ for $x = 1, 2, 3$, and 4;

(c) $f(x) = \dfrac{(x-1)(x-2)}{20}$ for $x = 2, 3, 4$, and 5.

R.118 Refer to the upper part of the printout of Figure R.5 to find the probabilities that a random variable having the Poisson distribution with $\lambda = 1.5$ will take on

(a) a value less than 3;

(b) the value 3, 4, or 5;

(c) a value greater than 4.

R.119 Repeat Exercise R.118, using the lower part of the printout of Figure R.5, namely, the cumulative probabilities.

R.120 Use the upper part of Figure R.5 to calculate the mean of the Poisson distribution with $\lambda = 1.5$ and, thus, verify the formula $\mu = \lambda$.

R.121 Use the upper part of Figure R.5 and the computing formula on page 226 to calculate the variance of the Poisson distribution with $\lambda = 1.5$ and, thus, verify the formula $\sigma^2 = \lambda$.

R.122 A panel of 300 persons chosen for jury duty included only 30 persons under 25 years of age. For a particular narcotics case, the actual jury of 12 selected from this panel did not contain anyone under the age of 25. The youthful defendant's attorney complained that this jury of 12 is not representative. He argued that the probability of having one of the 12 jurors under 25 years of age should be *many times* the probability of having none of them under 25 years of age.

(a) Find the ratio of these two probabilities, using the hypergeometric distribution.

(b) Find the ratio of these two probabilities, using the binomial distribution as an approximation.

R.123 Random samples of size $n = 2$ are drawn without replacement from the finite population that consists of the numbers 1, 3, 5, and 7. As can easily be verified, $\mu = 4$ for this population and $\sigma = \sqrt{5}$.

(a) List the six possible samples of size $n = 2$ that can be drawn without replacement from the given population, and calculate their means.

```
MTB > PDF c1;
SUBC>    Poisson 1.5.


Poisson with mu = 1.50000

            x          P( X = x)
          0.00           0.2231
          1.00           0.3347
          2.00           0.2510
          3.00           0.1255
          4.00           0.0471
          5.00           0.0141
          6.00           0.0035
          7.00           0.0008
          8.00           0.0001
          9.00           0.0000


MTB > CDF c1;
SUBC>    Poisson 1.5.


Poisson with mu = 1.50000

            x          P( X <= x)
          0.00           0.2231
          1.00           0.5578
          2.00           0.8088
          3.00           0.9344
          4.00           0.9814
          5.00           0.9955
          6.00           0.9991
          7.00           0.9998
          8.00           1.0000
          9.00           1.0000
```

FIGURE R.5
Poisson distribution
with $\lambda = 1.5$.

(b) Assigning each of the samples obtained in part (a) the probability $\frac{1}{6}$, construct the sampling distribution of the mean for random samples of size $n = 2$ drawn without replacement from the given population.

(c) Calculate the mean and the standard deviation of the probability distribution obtained in part (b), and compare the results with the values we would expect in accordance with the formulas on page 280.

*R.124 Among 80 persons interviewed for certain jobs by a government agency, 40 are married, 20 are single, 10 are divorced, and 10 are widowed. In how many ways can a 10 percent stratified sample be chosen from among the persons interviewed if

(a) one-fourth of the sample is to be allocated to each group;

(b) the allocation is proportional?

R.125 The probabilities are 0.22, 0.34, 0.24, 0.13, 0.06, and 0.01 that 0, 1, 2, 3, 4, or 5 of a doctor's patients will come down with the flu while the doctor is out of town during the week after Christmas Day.

(a) Find the mean of this probability distribution.

(b) Use the computing formula to determine the variance of this probability distribution.

R.126 Find the mean of the binomial distribution with $n = 9$ and $p = 0.40$, using

(a) Table V and the formula that defines μ;

(b) the special formula for the mean of a binomial distribution.

R.127 Find the standard deviation of the binomial distribution of Exercise R.126, using

(a) Table V, the result of part (b) of Exercise R.126, and the computing formula for σ^2;

(b) the special formula for the standard deviation of a binomial distribution.

R.128 Use the upper part of the printout of Figure R.6 to determine the probabilities that a random variable having the binomial distribution with $n = 12$ and $p = 0.44$ will take on

FIGURE R.6

Binomial distribution with $n = 12$ and $p = 0.44$.

```
MTB > PDF c1;
SUBC>    Binomial 12 0.44.


Binomial with n = 12 and p = 0.440000

            x          P( X = x)
          0.00          0.0010
          1.00          0.0090
          2.00          0.0388
          3.00          0.1015
          4.00          0.1794
          5.00          0.2256
          6.00          0.2068
          7.00          0.1393
          8.00          0.0684
          9.00          0.0239
         10.00          0.0056
         11.00          0.0008
         12.00          0.0001

MTB > CDF c1;
SUBC>    Binomial 12 0.44.


Binomial with n = 12 and p = 0.440000

            x          P( X <= x)
          0.00          0.0010
          1.00          0.0099
          2.00          0.0487
          3.00          0.1502
          4.00          0.3296
          5.00          0.5552
          6.00          0.7620
          7.00          0.9012
          8.00          0.9696
          9.00          0.9935
         10.00          0.9991
         11.00          0.9999
         12.00          1.0000
```

(a) a value less than 4;

(b) the value 6, 7, or 8;

(c) a value greater than or equal to 9.

R.129 Repeat Exercise R.128, using the lower part of the printout of Figure R.6, namely, the cumulative probabilities.

R.130 Use the normal approximation to the binomial distribution to approximate the probabilities that a random variable having the binomial distribution with $n = 18$ and $p = 0.27$ will take on a value

(a) less than 6;

(b) anywhere from 4 to 8;

(c) greater than 6.

R.131 With reference to Exercise R.130, use the printout of Figure 8.6 to determine the percentage errors of the three approximations. Are these approximations justified by the rule of thumb on page 253?

R.132 Check in each case whether the conditions for the normal approximation to the binomial distribution are satisfied:

(a) $n = 55$ and $p = \frac{1}{5}$;

(b) $n = 105$ and $p = \frac{1}{35}$;

(c) $n = 210$ and $p = \frac{1}{30}$;

(d) $n = 40$ and $p = 0.95$.

R.133 If a random sample of size $n = 3$ is to be chosen from a finite population of size $N = 65$, what is the probability of each possible sample?

R.134 Using the computer printout of Figure 8.5, find the probability that a random variable having the Poisson distribution with $\lambda = 0.6$ will take on a value greater than 2 by

(a) adding the probabilities corresponding to 3, 4, 5, and 6;

(b) working with the cumulative probabilities.

R.135 Among 12 Krugerrands, seven are genuine, three are gold-plated counterfeits, and the other two are pure brass counterfeits. If a not-very-knowledgable buyer randomly picks three of these coins, what is the probability that he will get one of each kind?

11

PROBLEMS OF ESTIMATION

11.1   The Estimation of Means 303
11.2   The Estimation of Means (σ unknown) 308
11.3   The Estimation of Standard Deviations 316
11.4   The Estimation of Proportions 323
11.5   Checklist of Key Terms 329
11.6   References 330

*T*raditionally, problems of statistical inference have been classified as problems of **estimation,** where we try to determine (within reasonable limits) the values of population parameters; **tests of hypotheses,** where we accept or reject assertions about the parameters or the form of populations; and problems of **prediction,** where we forecast future values of random variables. In each case, the inferences are based on sample data, although in some methods not covered in this book, called **Bayesian methods,** inferences are also based on collateral information and/or subjective judgments. In this chapter we shall concentrate on problems of estimation. Tests of hypotheses are treated in subsequent chapters, and problems of prediction are taken up in Chapters 16 and 17.

Problems of estimation are easy to illustrate because they arise just about everywhere—in science, in business, and in everyday life. In science, a psychologist may want to determine the average time that it takes an adult to react to a visual stimulus; in business, a union official may want to know how much variability there is in the amount of time it takes the union's members to get to work; and in everyday life, we may want to find out what percentage of all one-car accidents are due to driver fatigue. Among these three illustrations, the psychologist's problem concerns a population mean, the union official's problem concerns a measure of variation (perhaps, a standard deviation); and our everyday life problem concerns a percentage. Conceptually, all such problems are treated in the same way, but there are differences in the particular methods that are employed. Methods of estimating population means are taken up in Sections 11.1 and 11.2; those relating to measures of variability are treated in Section 11.3; and those relating to the estimation of percentages (also, proportions and probabilities) are discussed in Section 11.4.

11.1 THE ESTIMATION OF MEANS

To illustrate some of the problems we face when we estimate the mean of a population from sample data, let us refer to a study in which industrial designers want to determine the average (mean) time it takes an adult to assemble an "easy to assemble" toy. Using a random sample, they obtain the following data (in minutes) for 36 persons who assembled the toy:

17	13	18	19	17	21	29	22	16	28	21	15
26	23	24	20	8	17	17	21	32	18	25	22
16	10	20	22	19	14	30	22	12	24	28	11

The mean of this sample is $\bar{x} = 19.9$ minutes, and in the absence of any other information, this figure can be used as an estimate of μ, the "true" average time it takes an adult to assemble the given toy.

This kind of estimate is called a **point estimate**, since it consists of a single number, or a single point on the real number scale. Although this is the most common way in which estimates are expressed, it leaves room for many questions. By itself, it does not tell us on how much information the estimate is based, and it does not tell us anything about the possible size of the error. And, of course, we must expect an error. This should be clear from our discussion of the sampling distribution of the mean in Chapter 10, where we saw that the chance fluctuations of the mean (and, hence, its reliability as an estimate of μ) depend on two things—the size of the sample and the size of the population standard deviation σ. Thus, we might supplement the estimate, $\bar{x} = 19.9$ minutes, with the information that it is the mean of a random sample of size $n = 36$, whose standard deviation is $s = 5.73$ minutes. Although this does not tell us the actual value of σ, the sample standard deviation can serve as an estimate of this quantity.

Scientific reports often present sample means in this way, together with the values of n and s, but this does not supply readers of the report with a coherent picture unless they have had some formal training in statistics. To take care of this, we refer to the theory of Sections 10.7 and 10.8, and the definition in Example 9.6, according to which z_α is such that the area to its right under the standard normal curve is equal to α, and hence the area under the standard normal curve between $-z_{\alpha/2}$ and $z_{\alpha/2}$ is equal to $1 - \alpha$. Making use of the fact that for large random samples from infinite populations, the sampling distribution of the

mean is approximately a normal distribution with $\mu_{\bar{x}} = \mu$ and $\sigma_{\bar{x}} = \dfrac{\sigma}{\sqrt{n}}$, we find from Figure 11.1 that the probability is $1 - \alpha$ that the mean of a large random sample from an infinite population will differ from the mean of the population by at most $z_{\alpha/2} \cdot \dfrac{\sigma}{\sqrt{n}}$. In other words,

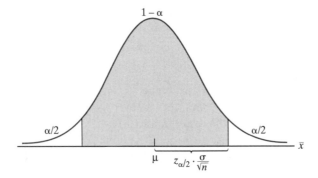

FIGURE 11.1
Sampling distribution
of the mean.

When we use $\bar{x}$ as an estimate of μ, the probability is $1 - \alpha$ that this estimate is "off" either way by at most

*Maximum error
of estimate*

$$E = z_{\alpha/2} \cdot \frac{\sigma}{\sqrt{n}}$$

This result applies when n is large, $n \ge 30$, and the population is infinite (or large enough so that the finite population factor need not be used). The two values that are most commonly, though not necessarily, used for $1 - \alpha$ are 0.95 and 0.99, and the corresponding values of $\alpha/2$ are 0.025 and 0.005. As the reader was asked to show in Exercise 9.16, $z_{0.025} = 1.96$ and $z_{0.005} = 2.575$.

EXAMPLE 11.1   A team of efficiency experts intends to use the mean of a random sample of size $n = 150$ to estimate the average mechanical aptitude of assembly-line workers in a large industry (as measured by a certain standardized test). If, based on experience, the efficiency experts can assume that $\sigma = 6.2$ for such data, what can they assert with probability 0.99 about the maximum error of their estimate?

Solution   Substituting $n = 150$, $\sigma = 6.2$, and $z_{0.005} = 2.575$ into the formula for E, we have

$$E = 2.575 \cdot \frac{6.2}{\sqrt{150}} \approx 1.30$$

Thus, the efficiency experts can assert with probability 0.99 that their error will be at most 1.30.   ∎

Suppose now that the efficiency experts actually collect the necessary data and obtain $\bar{x} = 69.5$. Can they still assert with probability 0.99 that the error of their estimate is at most 1.30? After all, $\bar{x} = 69.5$ differs from μ, the population mean, by at most 1.30 or it does not, and they have no way of knowing whether it is one or the other. Actually, they can make this assertion, but it must be understood that the 0.99 probability applies to the *method* used (getting the sample data and using the formula for E) and not directly to the single problem at hand.

To make this distinction, it has become the custom to use the word "confidence" here instead of "probability."

> **In general, we make probability statements about future values of random variables (say, the potential error of an estimate) and confidence statements once the data have been obtained.**

Accordingly, we would say in our example that the efficiency experts can be 99% confident that the error of their estimate, $\bar{x} = 69.5$, is at most 1.30.

Use of the formula for E involves a complication. We need to know the value of the population standard deviation σ. Since σ is not known in most practical situations, we sometimes replace it with a plausible guess, and, being conservative, this may cause us to overstate the error somewhat. Otherwise, we can replace σ with an estimate, usually the sample standard deviation s. In general, this is considered to be reasonable provided the sample size is sufficiently large, and by sufficiently large we again mean $n \geq 30$.

EXAMPLE 11.2   With reference to the illustration on page 303, what can we assert with 95% confidence about the maximum error if we use $\bar{x} = 19.9$ minutes as an estimate of the average time it takes an adult to assemble the given kind of toy?

Solution   Substituting $n = 36$, $s = 5.73$ for σ, and $z_{0.025} = 1.96$ into the formula for E, we find that we can assert with 95% confidence that the error is at most

$$E = 1.96 \cdot \frac{5.73}{\sqrt{36}} \approx 1.87 \, \text{minutes}$$

Of course, the error is at most 1.87 minutes or it is not, and we do not know which is actually the case, but if we had to bet, 95 to 5 (or 19 to 1) would be fair odds that the error is at most 1.87 minutes.   ∎

The formula for E can also be used to determine the sample size that is needed to attain a desired degree of precision. Suppose that we want to use the mean of a large random sample to estimate the mean of a population, and we want to be able to assert with probability $1 - \alpha$ that the error of this estimate will not exceed some prescribed quantity E. As before, we write $E = z_{\alpha/2} \cdot \dfrac{\sigma}{\sqrt{n}}$,

and upon solving this equation for n we get

Sample size for estimating μ

$$n = \left[\frac{z_{\alpha/2} \cdot \sigma}{E} \right]^2$$

EXAMPLE 11.3   The dean of a college wants to use the mean of a random sample to estimate the average amount of time students take to get from one class to the next, and she wants to be able to assert with probability 0.95 that her error will be at most 0.30 minute. If she knows from studies of a similar kind that it is reasonable to let $\sigma = 1.50$ minutes, how large a sample will she need?

Solution   Substituting $E = 0.30$, $\sigma = 1.50$, and $z_{0.025} = 1.96$ into the formula for n, we get

$$n = \left(\frac{1.96 \cdot 1.50}{0.30} \right)^2 \approx 96.04$$

which we round up to the nearest integer, 97. Thus, a random sample of size $n = 97$ is required for the estimate. (Note that the treatment would have been the same if we had said "she wants to be able to assert with 95% confidence that her error *is* at most 0.30 minute" instead of "she wants to be able to assert with probability 0.95 that her error *will be* at most 0.30 minute." It depends on when the assertion is to be made—after or before she collects the data.) ∎

As can be seen from the formula for n and also from Example 11.3, it has the same shortcoming as the formula for E; that is, we must know (at least approximately) the value of the population standard deviation, σ. For this reason, we sometimes begin with a relatively small sample and then use its standard deviation to see whether more data are required.

Let us now introduce a different way of presenting a sample mean together with an assessment of the error we might be making if we use it to estimate the mean of the population from which the sample came. As on page 303, we shall make use of the fact that, for large random samples from infinite populations, the sampling distribution of the mean is approximately a normal distribution with the mean $\mu_{\bar{x}} = \mu$ and the standard deviation $\sigma_{\bar{x}} = \dfrac{\sigma}{\sqrt{n}}$, so that

$$z = \frac{\bar{x} - \mu}{\sigma/\sqrt{n}}$$

is a value of a random variable having approximately the standard normal distribution. Since the probability is $1 - \alpha$ that a random variable having this distribution will take on a value between $-z_{\alpha/2}$ and $z_{\alpha/2}$, namely, that

$$-z_{\alpha/2} < z < z_{\alpha/2}$$

we can substitute into this inequality the foregoing expression for z and get

$$-z_{\alpha/2} < \frac{\bar{x} - \mu}{\sigma/\sqrt{n}} < z_{\alpha/2}$$

Then, if we multiply each term by $\sigma/\sqrt{n}$, subtract $\bar{x}$ from each term, and finally multiply each term by -1, we get

$$\bar{x} + z_{\alpha/2} \cdot \frac{\sigma}{\sqrt{n}} > \mu > \bar{x} - z_{\alpha/2} \cdot \frac{\sigma}{\sqrt{n}}$$

[Since we multiplied by -1, we had to reverse the inequality signs, as is always the case when we multiply the expressions on both sides of an inequality by a negative number. For instance, where 4 is greater than (to the right of) 3, -4 is less than (to the left of) -3.] The result obtained previously can also be written as

*Confidence
interval for μ*

$$\bar{x} - z_{\alpha/2} \cdot \frac{\sigma}{\sqrt{n}} < \mu < \bar{x} + z_{\alpha/2} \cdot \frac{\sigma}{\sqrt{n}}$$

and we can assert with probability $1 - \alpha$ that it will be satisfied for any given sample. In other words, we can assert with $(1 - \alpha)100\%$ confidence that the interval from $\bar{x} - z_{\alpha/2} \cdot \frac{\sigma}{\sqrt{n}}$ to $\bar{x} + z_{\alpha/2} \cdot \frac{\sigma}{\sqrt{n}}$, determined on the basis of a large random sample, contains the population mean we are trying to estimate. When σ is unknown and n is at least 30, we replace σ by the sample standard deviation s.

An interval like this is called a **confidence interval,** its endpoints are called **confidence limits,** and $1 - \alpha$ or $(1 - \alpha)100\%$ is called the **degree of confidence.** As before, the values used most often for the degree of confidence are 0.95 and 0.99 (or 95% and 99%), and the corresponding values of $z_{\alpha/2}$ are 1.96 and 2.575. In contrast to point estimates, estimates given in the form of a confidence interval are called **interval estimates.** They have the advantage of not requiring further elaboration about their reliability. This is taken care of indirectly by their width and the degree of confidence.

Since n must be large to justify the normal approximation to the sampling distribution of the mean, we refer to a confidence interval calculated by means of the preceding formula as a **large-sample confidence interval for μ.** It is also called a **z interval,** being based on the z statistic that has the standard normal distribution.

EXAMPLE 11.4   With reference to Example 11.1, where we had $n = 150$ and $\sigma = 6.2$, use the added information that the efficiency experts obtained the sample mean $\bar{x} = 69.5$ to calculate a 95% confidence interval for the average mechanical aptitude of assembly line workers in the given industry.

Solution   Substituting $n = 150$, $\sigma = 6.2$, $\bar{x} = 69.5$, and $z_{0.025} = 1.96$ into the confidence interval formula, we get

$$69.5 - 1.96 \cdot \frac{6.2}{\sqrt{150}} < \mu < 69.5 + 1.96 \cdot \frac{6.2}{\sqrt{150}}$$

$$68.5 < \mu < 70.5$$

where the confidence limits are rounded to one decimal. Of course, the statement that the interval from 68.5 to 70.5 contains the true average mechanical aptitude score of assembly line workers in the given industry is either true or false, and we do not know whether it is true or false, but we can be 95% confident that it is true. Why? Because the method we used works 95% of the time. To put it differently, the interval may contain μ or it may not, but if we had to bet, 95 to 5 (or 19 to 1) would be fair odds that it does.   ∎

Had we wanted to construct a 99% confidence interval in Example 11.4, we would have substituted 2.575 instead of 1.96 for $z_{\alpha/2}$, and we would have obtained $68.2 < \mu < 70.8$. The 99% confidence interval is wider than the 95% confidence interval—it goes from 68.2 to 70.8 instead of from 68.5 to 70.5, and this illustrates the important fact that

> **When we increase the degree of confidence, the confidence interval becomes wider and, thus, tells us less about the quantity we are trying to estimate.**

Indeed, we might say that "the surer we want to be, the less we have to be sure of."

11.2   THE ESTIMATION OF MEANS (σ unknown)

In Section 11.1 we assumed that the samples were large enough, $n \geqslant 30$, to approximate the sampling distribution of the mean with a normal distribution and, when necessary, to replace σ with s. To develop corresponding methods that apply in general when σ is unknown, we must assume that the populations we are sampling have roughly the shape of normal distributions. We can then base our methods on the *t* **statistic**

$$t = \frac{\bar{x} - \mu}{s/\sqrt{n}}$$

which is a value of a random variable having the *t* **distribution.** More specifically, this distribution is called the **Student *t* distribution** or **Student's *t* distribution,** as it was first developed by a statistician, W. S. Gosset, who published his work under the pen name "Student." As is shown in Figure 11.2, the shape of this continuous distribution is very similar to that of the standard normal distribution—like the standard normal distribution, it is bell shaped and symmetrical with zero mean. The exact shape of the *t* distribution depends on a parameter called the **number of degrees of freedom,** or simply the **degrees of freedom,** which, for the methods of this section, equals $n - 1$, the sample size less one.

FIGURE 11.2
Standard normal distribution and *t* distribution.

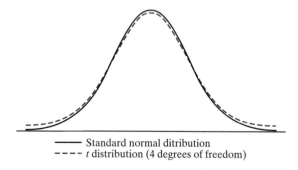

——— Standard normal ditribution
- - - - *t* distribution (4 degrees of freedom)

For the standard normal distribution, we defined $z_{\alpha/2}$ in such a way that the area under the curve to its right equals $\alpha/2$, and, hence, the area under the curve between $-z_{\alpha/2}$ and $z_{\alpha/2}$ equals $1 - \alpha$. As is shown in Figure 11.3, the corresponding values for the t distribution are $-t_{\alpha/2}$ and $t_{\alpha/2}$. Since these values depend on $n - 1$, the number of degrees of freedom, they must be obtained from a special table, such as Table II at the end of this book, or perhaps a computer. Table II contains among others the values of $t_{0.025}$ and $t_{0.005}$ for 1 through 30 and selected larger degrees of freedom. As can be seen, $t_{0.025}$ and $t_{0.005}$ approach the corresponding values for the standard normal distribution when the number of degrees of freedom becomes large.

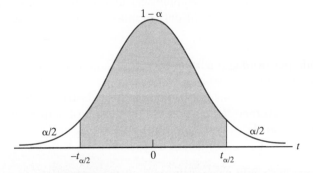

FIGURE 11.3
t distribution.

Proceeding as on page 306, we can assert with probability $1 - \alpha$ that a random variable having the t distribution will take on a value between $-t_{\alpha/2}$ and $t_{\alpha/2}$, namely, that

$$-t_{\alpha/2} < t < t_{\alpha/2}$$

Then, substituting into this inequality the expression for t on page 308, we get

$$-t_{\alpha/2} < \frac{\bar{x} - \mu}{s/\sqrt{n}} < t_{\alpha/2}$$

and the same steps as on pages 306 and 307 yield the following confidence interval for μ:

Confidence interval for μ (σ unknown)

$$\bar{x} - t_{\alpha/2} \cdot \frac{s}{\sqrt{n}} < \mu < \bar{x} + t_{\alpha/2} \cdot \frac{s}{\sqrt{n}}$$

The degree of confidence is $1 - \alpha$, and the only difference between this confidence interval and the z interval with s substituted for σ is that $t_{\alpha/2}$ takes the place of $z_{\alpha/2}$. This confidence interval for μ is usually referred to as a **t interval**, and since most tables of the t distribution give the values of $t_{\alpha/2}$ only for small numbers of degrees of freedom, it has also gone by the name of a **small-sample confidence interval for μ**. Nowadays, computers and other technology provide the values of $t_{\alpha/2}$ for hundreds of degrees of freedom, so this distinction no longer applies.

It must be remembered, though, that for the t interval there is the added assumption that the sample comes from a normal population, or at least from a population having roughly the shape of a normal distribution. This is important, and it will be discussed further in the two examples that follow.

In Example 11.5 we will be given only the values of n, $\bar{x}$, and s, not the original data, so that there is really no way of checking the "normality" of the population sampled. All we can do in that case is hope for the best. In Example 11.6 we will be given the actual data, so that we can form a normal probability plot (see Section 9.3) to judge whether it is reasonable to look upon the data as a sample from a normal population. This requires the use of appropriate technology—computer software or a graphing calculator—but there are alternative procedures that are ordinarily taught only in more advanced courses in statistics.

EXAMPLE 11.5   While performing a certain task under simulated weightlessness, the pulse rate of 12 astronauts increased on the average by 27.33 beats per minute with a standard deviation of 4.28 beats per minute. Construct a 99% confidence interval for the true average increase in the pulse rate of astronauts performing the given task (under the stated condition).

Solution   As we have said previously, without the actual data there is no way of judging the "normality" of the population sampled. Nevertheless, making it clear that the result is subject to the validity of this assumption, we proceed as follows: Substituting $n = 12$, $\bar{x} = 27.33$, $s = 4.28$, and $t_{0.005} = 3.106$ (the entry in Table II for $12 - 1 = 11$ degrees of freedom) into the t interval formula, we get

$$27.33 - 3.106 \cdot \frac{4.28}{\sqrt{12}} < \mu < 27.33 + 3.106 \cdot \frac{4.28}{\sqrt{12}}$$

and, hence,

$$23.49 < \mu < 31.17$$

beats per minute.   ∎

EXAMPLE 11.6   To test the durability of a new paint for white center lines, a highway department painted test strips across heavily traveled roads in eight different locations, and electronic counters showed that they deteriorated after having been crossed by (to the nearest hundred)

142,600	167,800	136,500	108,300
126,400	133,700	162,000	149,400

cars.

(a) Check whether it is reasonable to treat these data as a sample from a normal population.

(b) If so, construct a 95% confidence interval for the average amount of traffic (car crossings) the new white paint can withstand before it deteriorates.

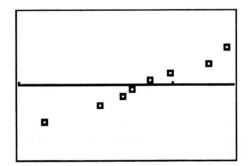

FIGURE 11.4
Normal probability plot for Example 11.6, reproduced from the display screen of a TI-83 graphing calculator.

Solution

(a) Using a graphing calculator, we obtained the normal probability plot shown in Figure 11.4. As can be seen, the pattern of the eight dots may well described as linear, which supports the assumption that the data constitute a sample from a normal population.

(b) The mean and the standard deviation of the $n = 8$ values are $\bar{x} = 140{,}800$ and $s = 19{,}200$ to the nearest 100, and since $t_{0.025} = 2.365$ for $8 - 1 = 7$ degrees of freedom (according to Table II), substitution into the t interval formula yields

$$140{,}800 - 2.365 \cdot \frac{19{,}200}{\sqrt{8}} < \mu < 140{,}800 + 2.365 \cdot \frac{19{,}200}{\sqrt{8}}$$

$$124{,}700 < \mu < 156{,}900$$

This is the desired 95% confidence interval for the average amount of traffic (car crossings) the paint can withstand before it deteriorates. Again, we can't tell for sure whether the interval from 124,700 to 156,900 contains the true average number of car crossings the paint can withstand before it deteriorates, but 95 to 5 would be fair odds that it does. These odds are based on the fact that the method we used—taking a random sample from a normal population and using the formula given previously—works 95% of the time.   ■

Figure 11.5 is a computer printout showing the result we obtained for Example 11.6. The differences between 124,700 and 124,758 and between 156,900 and 156,917 are due to rounding. Indeed, since the original data were given to the nearest hundred, we rounded the mean, the standard deviation, and the confidence limits to the nearest hundred.

FIGURE 11.5
Computer printout for Example 11.6.

```
MTB > TInterval 95.0 cl.


Variable     N      Mean    StDev   SE Mean     95.0 % C.I.
C1           8    140838    19228      6798   ( 124758,  156917)
```

If we had wanted the degree of confidence to be 0.97 instead of 0.95 in Example 11.6, we could not have worked it with the use of Table II, which does not give any values of $t_{0.015}$. Of course, we could have used a computer or a graphing calculator, but let us use here the result that $t_{0.015} = 2.7146$ according to our HEWLETT PACKARD STAT/MATH calculator. The result is

$$140{,}800 - 2.7146 \cdot \frac{19{,}200}{\sqrt{8}} < \mu < 140{,}800 + 2.7146 \cdot \frac{19{,}200}{\sqrt{8}}$$

$$122{,}370 < \mu < 159{,}230$$

which, rounded to the nearest 100, becomes $122{,}400 < \mu < 159{,}200$. As should have been expected, the 97% confidence interval is wider than the corresponding 95% interval.

The method we used earlier to determine the maximum error we risk with $(1 - \alpha)100\%$ confidence when we use a sample mean to estimate the mean of a population is readily adapted to problems in which σ is unknown, provided that the population sampled has roughly the shape of a normal distribution. All we have to do is substitute s for σ and $t_{\alpha/2}$ for $z_{\alpha/2}$ in the formula for the maximum error on page 304.

EXAMPLE 11.7   With reference to Example 11.5, suppose that $\bar{x} = 27.33$ is being used as an estimate of the true average increase in the pulse rate of astronauts performing the given task. What can be said with 99% confidence about the maximum error?

Solution   Substituting $s = 4.28$, $n = 12$, and $t_{0.005} = 3.106$ (the entry in Table II for $12 - 1 = 11$ degrees of freedom) into the modified formula for E, we get

$$E = t_{\alpha/2} \cdot \frac{s}{\sqrt{n}} = 3.106 \cdot \frac{4.28}{\sqrt{12}} \approx 3.84$$

Thus, if we use $\bar{x} = 27.33$ beats per minute as an estimate of the true average increase in the pulse rate of astronauts performing the given task (under the stated conditions), we can assert with 99% confidence that our error is at most 3.84 beats per minute.   ∎

■ E x e r c i s e s

11.1   A study made by a staff officer of an armored division showed that in a random sample of $n = 40$ days the division had on the average 1,126 vehicles in operating condition. Given that $\sigma = 135$ for such data, what can this officer assert with 95% confidence about the maximum error if he uses $\bar{x} = 1{,}126$ as an estimate of the actual daily average number of vehicles that this armored division has in operating condition?

11.2 With reference to Exercise 11.1, construct a 99% confidence interval for the actual daily average number of vehicles that this armored division has in operating condition.

11.3 A study of the annual growth of a certain kind of cactus showed that (under controlled conditions) a random sample of $n = 60$ of these cacti grew on the average 42.8 mm per year. Given that $\sigma = 5.4$ mm for such data, what can one conclude with 99% confidence about the maximum error if $\bar{x} = 42.8$ is used as an estimate of the true average annual growth of this kind of cactus (under the given controls)?

11.4 With reference to Exercise 11.3, construct a 95% confidence interval for the true average annual growth of this kind of cactus.

11.5 In a study of automobile collision insurance costs, a random sample of $n = 35$ repair costs of front-end damage caused by hitting a wall at a specified speed had a mean of $1,438. Given that $\sigma = \$269$ for such data, what can be said with 98% confidence about the maximum error if $\bar{x} = \$1,438$ is used as an estimate of the average cost of such repairs? Also, construct a 90% confidence interval for the average cost of such repairs.

11.6 In a class of $n = 42$ students, where each student was asked to perform a measurement of the specific gravity of aluminum, their results averaged 2.695. Given that $\sigma = 0.045$ for such measurements, with what degree of confidence can they conclude that they are not off by more than 0.010 if they estimate the specific gravity of aluminum as 2.695?

11.7 In an experiment conducted to determine the mean lifetime of certain electronic tubes, a random sample of $n = 36$ of these tubes lasted on the average for 825 hours. Given that $\sigma = 42$ hours for such data, how confident can one be that the error does not exceed 12 hours when using $\bar{x} = 825$ as an estimate of the mean lifetime of this kind of electronic tube?

*11.8 When a sample constitutes an appreciable portion of a finite population, we must use the second standard-error formula on page 280 and, hence, to determine the maximum error we must use the formula

$$E = z_{\alpha/2} \cdot \frac{\sigma}{\sqrt{n}} \cdot \sqrt{\frac{N - n}{N - 1}}$$

Taking a random sample of $n = 200$ from among $N = 800$ delinquent accounts, a CPA conducting an audit of a power company found that the amounts owed had a mean of $48.15 with a standard deviation of $6.19. Using $s = \$6.19$ as an estimate of σ, what can she assert with 95% confidence about the maximum error if she uses $48.15 as an estimate of the average amount owed by all 800 delinquent accounts?

11.9 Before bidding on a contract, a contractor wants to be 95% confident that he is in error by no more than 6 minutes when using the mean of a random sample to estimate the average time it takes for a certain kind of adobe brick to harden. How large a sample will he need if he can assume that $\sigma = 22$ minutes for the time it takes for such brick to harden?

11.10 It is desired to estimate the average number of hours of continuous use until a model 737 airplane will first require repairs. If it can be assumed that $\sigma = 138$ hours for such data, how large a sample is needed to be able to assert with a probability of 0.99 that the sample mean will be off by no more than 40 hours?

11.11 Before purchasing a large shipment of ground pork, a sausage manufacturer wants to be "pretty sure" that his error does not exceed 0.25 ounce when using the mean of a random sample to estimate the actual fat content per pound. If the standard deviation of the fat content is known to be 0.77 ounce per pound, how many one-pound samples would he need if by "pretty sure" he meant 95% confident?

11.12 Rework Exercise 11.11, changing "pretty sure" to mean 99% confident.

11.13 A computer is programmed to yield values of a random variable having a normal distribution whose mean and standard deviation are known only to the programmer. Each of 30 students is asked to use the computer to simulate a random sample of size $n = 5$ and use it to construct a 90% confidence interval for μ. Their results are as follows:

$6.30 < \mu < 8.26,$	$6.50 < \mu < 7.72,$	$6.93 < \mu < 8.01,$
$6.60 < \mu < 8.00,$	$6.51 < \mu < 7.51,$	$6.82 < \mu < 8.66,$
$7.02 < \mu < 8.11,$	$6.94 < \mu < 7.64,$	$6.24 < \mu < 7.26,$
$6.87 < \mu < 8.17,$	$6.77 < \mu < 8.13,$	$6.14 < \mu < 6.82,$
$6.83 < \mu < 7.93,$	$6.66 < \mu < 8.10,$	$6.73 < \mu < 7.49,$
$6.41 < \mu < 7.67,$	$6.76 < \mu < 7.57,$	$6.97 < \mu < 7.47,$
$6.01 < \mu < 7.43,$	$7.15 < \mu < 7.89,$	$6.87 < \mu < 7.81,$
$7.35 < \mu < 7.99,$	$6.60 < \mu < 8.16,$	$6.47 < \mu < 7.81,$
$7.01 < \mu < 8.33,$	$6.97 < \mu < 7.55,$	$6.56 < \mu < 7.48,$
$7.13 < \mu < 8.03,$	$7.39 < \mu < 8.01,$	$5.98 < \mu < 7.68$

(a) How many of these 30 confidence intervals would we expect to contain the mean of the population sampled?

(b) Given that the computer was programmed so that $\mu = 7.30$, how many of the confidence intervals actually contain the mean of the population sampled? Discuss the result.

11.14 With reference to Example 11.4, where we had $n = 150$, $\bar{x} = 69.5$, and $\sigma = 6.2$, suppose that we had asked for a 97% confidence interval.

(a) Find the value of $z_{0.015}$ from Table I.

(b) Use the value of $z_{0.015}$ obtained in part (a) to calculate a 97% z interval for the average mechanical aptitude of assembly-line workers in the given industry.

*11.15 If a sample constitutes an appreciable portion of a finite population, we must make the same correction as in Exercise 11.8 in the formula for z intervals for population means. In a reading achievement test, a random sample of $n = 120$ of $N = 360$ fifth graders in a certain school district averaged 83.4 with a standard deviation of 12.7.

(a) Using $s = 12.7$ as an estimate of σ, construct a 93% z interval for the mean score that all 360 of the fifth graders would have obtained in the reading achievement test.

(b) Use appropriate computer software or a graphing calculator to determine the corresponding t interval.

11.16 Use Table II to find

(a) $t_{0.050}$ for 11 degrees of freedom;

(b) $t_{0.025}$ for 19 degrees of freedom;

(c) $t_{0.010}$ for 14 degrees of freedom;

(d) $t_{0.005}$ for 17 degrees of freedom.

11.17 In a pollution study of the air in a certain downtown area, an EPA technician obtained a mean of 2.34 micrograms of suspended benzene-soluble matter per cubic meter with a standard deviation of 0.48 microgram for a sample of size $n = 9$. Assuming that the population sampled is normal,

(a) construct a 95% confidence interval for the mean of the population sampled;

(b) what can the technician assert with 99% confidence about the maximum error if $\bar{x} = 2.34$ micrograms per cubic meter is used as an estimate of the mean of the population sampled?

11.18 A consumer testing service wants to study the noise level of a new vacuum cleaner. Measuring the noise level of a random sample of $n = 12$ of the machines, it gets the following data (in decibels):

74.0	78.6	76.8	75.5	73.8	75.6
77.3	75.8	73.9	70.2	81.0	73.9

(a) Use a normal probability plot to verify that it is reasonable to treat these data as a sample from a normal population.

(b) Construct a 95% confidence interval for the average noise level of such vacuum cleaners.

11.19 A random sample of $n = 9$ pieces of Manila rope (designed for nautical use) has a mean breaking strength of 41,250 pounds and a standard deviation of 1,527 pounds. Assuming that it is reasonable to treat these data as a sample from a normal population, what can we assert with 95% confidence about the maximum error if $\bar{x} = 41,250$ pounds is used as an estimate of the mean breaking strength of such rope?

11.20 With reference to Exercise 11.19, construct a 98% confidence interval for the mean breaking strength of the given kind of rope.

11.21 Ten bearings manufactured by a certain process have a mean diameter of 0.406 cm with a standard deviation of 0.003 cm. Construct a 99% confidence interval for the mean diameter of bearings manufactured by this process. Assume that the population sampled is normal.

11.22 With reference to Exercise 11.21, what can be claimed with 90% confidence about the maximum error if $\bar{x} = 0.406$ cm is used as an estimate of the mean diameter of bearings manufactured by this process?

11.23 The following are measurements of the thermal efficiency of $n = 18$ diesel engines made by a certain manufacturer:

34.8	30.7	35.0	34.9	33.6	28.7
32.1	28.8	29.0	31.4	31.7	31.8
33.6	30.0	29.7	33.4	28.2	31.6

(a) Use a normal probability plot to verify that it is reasonable to treat these data as a sample from a normal population.

(b) Construct a 99% confidence interval for the average thermal efficiency of such diesel engines.

11.24 To establish the authenticity of an ancient coin, its weight is often of critical importance. Given that four experts independently weighed a Phoenician tetradrachm and got 14.28, 14.34, 14.26, and 14.32 grams, construct a 95% confidence interval for the actual weight of this coin. (Such repeated measurements of one and the same object can usually be looked upon as a sample from a normal population.)

 11.25 Use a computer or a graphing calculator to rework Exercise 11.24.

11.26 Five containers of a commercial solvent, randomly selected from a large production lot, weigh 19.5, 19.3, 20.0, 19.0, and 19.7 pounds. Assuming that these data can be looked upon as a sample from a normal population, construct a 99% t interval for the mean weight of the containers of the solvent in the production lot.

11.27 In setting the type for a book, a compositor made 10, 11, 14, 8, 12, 17, 9, 12, 15, and 12 mistakes in a random sample of ten galleys. Assuming that it is reasonable to approximate the population sampled with a normal distribution, construct a 98% confidence interval for the average number of mistakes that this compositor makes per galley.

 11.28 With reference to Exercise 11.27, change the degree of confidence to 0.93 and use a computer package or a graphing calculator to rework the exercise.

11.29 At seven weather observation posts in the White Mountains, the rainfall during a summer storm was measured as 0.12, 0.14, 0.18, 0.20, 0.15, 0.12, and 0.14 inch. Assuming that it is reasonable to treat these data as a random sample from a normal population, construct a 95% confidence interval for the average rainfall in the White Mountains during that storm.

 11.30 Use a computer package or a graphing calculator to rework Exercise 11.29 with the degree of confidence changed to 0.97.

11.31 A dentist finds in a routine check that 12 prison inmates, a random sample, needed 2, 3, 6, 1, 4, 2, 4, 5, 0, 3, 5, and 1 filling. If she assumes that these data constitue a sample from a population that can be approximated closely with a normal distribution, what can she assert with 99% confidence about the maximum error if she uses the mean of this sample as an estimate of the average number of fillings needed by the inmates of this very large prison?

11.32 The excess length of a 99% t interval for μ, expressed as a percentage of the corresponding 95% t interval, depends only on the size of the sample. What is the excess length when
(a) $n = 10$;
(b) $n = 25$?

11.3 THE ESTIMATION OF STANDARD DEVIATIONS

So far in this chapter we have learned how to judge the maximum error when estimating the mean of a population and how to construct confidence intervals for population means. These are important techniques, but even more important is the fact that the concepts on which they are based—interval estimation and confidence statements about errors—carry over to the estimation of other population parameters. In this section we shall introduce methods of estimating population standard deviations and variances, and in Section 11.4 we shall concern ourselves with the estimation of proportions (also probabilities and percentages); that is, with the estimation of the parameter p of binomial populations.

There are various ways in which we can estimate the standard deviation of a population. In Exercise 10.68 we suggested the possibility of estimating σ in terms of the sample range, in Exercise 11.43 we shall suggest use of the two quartiles, and there even exists a method based on the fact that for a normal population roughly 68% of the values fall within one standard deviation of the mean. Most widely used, however, is the sample standard deviation s, which we used already as an estimate of σ in connection with the z (confidence) interval for μ.

Let us begin here with confidence intervals for σ based on s, which require that the population we are sampling has roughly the shape of a normal distribution. In that case, the statistic

Chi-square statistic

$$\chi^2 = \frac{(n-1)s^2}{\sigma^2}$$

called the **chi-square statistic** (χ is the lowercase Greek letter *chi*), is a value of a random variable having approximately the **chi-square distribution**. The parameter of this important continuous distribution is called the number of degrees of freedom, just like the parameter of the t distribution, and as the chi-square distribution is used here, the number of degrees of freedom is $n - 1$. An example of a chi-square distribution is shown in Figure 11.6. Unlike the normal and t distributions, its domain consists only of the nonnegative real numbers.

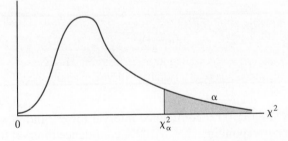

FIGURE 11.6
Chi-square
distribution.

Analogous to z_α and t_α, we now define χ^2_α as the value for which the area under the curve to its right (see Figure 11.6) is equal to α; like t_α, this value depends on the number of degrees of freedom and must be obtained from a table, or perhaps a computer. Thus, $\chi^2_{\alpha/2}$ is such that the area under the curve to its right is $\alpha/2$, while $\chi^2_{1-\alpha/2}$ is such that the area under the curve to its left is $\alpha/2$ (see

Figure 11.7). For instance, $\chi^2_{0.975}$ is the value for which the area under the curve to its left is 0.025. We made this distinction because the chi-square distribution, unlike the normal and t distributions, is not symmetrical. Values of $\chi^2_{0.995}$, $\chi^2_{0.975}$, $\chi^2_{0.025}$, and $\chi^2_{0.005}$ among others are given in Table III at the end of the book for $1, 2, 3, \ldots$, and 30 degrees of freedom.

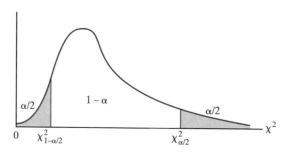

FIGURE 11.7
Chi-square
distribution.

We can now proceed as on pages 306 and 307. Since the probability is $1 - \alpha$ that a random variable having a chi-square distribution will take on a value between $\chi^2_{1-\alpha/2}$ and $\chi^2_{\alpha/2}$, namely, that

$$\chi^2_{1-\alpha/2} < \chi^2 < \chi^2_{\alpha/2}$$

we can substitute into this inequality the expression for χ^2 on page 317 and get

$$\chi^2_{1-\alpha/2} < \frac{(n-1)s^2}{\sigma^2} < \chi^2_{\alpha/2}$$

Then, applying some relatively simple algebra, we can rewrite this inequality as

Confidence interval
for σ^2

$$\frac{(n-1)s^2}{\chi^2_{\alpha/2}} < \sigma^2 < \frac{(n-1)s^2}{\chi^2_{1-\alpha/2}}$$

This is a $(1 - \alpha)100\%$ confidence interval for σ^2, and if we take square roots, we get a corresponding $(1 - \alpha)100\%$ confidence interval for σ. In the past, this kind of confidence interval has been referred to as a **small-sample confidence interval for σ** because most chi-square tables are limited to small numbers of degrees of freedom. As was the case in connection with the t distribution, this no longer applies in view of the general availability of computers and other technology.

It is important to remember, however, that the population sampled must have roughly the shape of a normal distribution. In Example 11.8 we will be given only the values of n and s, so that there is no way of checking on the "normality" of the population sampled. In Example 11.9 we will be given the original data, so that we can form a normal probability plot (see Section 9.3) to judge whether it is reasonable to look upon the data as a sample from a normal population.

EXAMPLE 11.8   With reference to Example 11.5, where we had $n = 12$ and $s = 4.28$ beats per minute, construct a 99% confidence interval for σ, the true standard deviation of the increase in the pulse rate of astronauts performing a given task (under stated conditions).

Solution   As we said on page 310, without the actual data there is no way of judging the "normality" of the population sampled. So we shall have to state again that the result is subject to the validity of the assumption that the data came from a normal population. Then, substituting $n = 12$, $s = 4.28$, and $\chi^2_{0.995} = 2.603$ and $\chi^2_{0.005} = 26.757$ for $12 - 1 = 11$ degrees of freedom, into the confidence interval formula for σ^2, we get

$$\frac{11(4.28)^2}{26.757} < \sigma^2 < \frac{11(4.28)^2}{2.603}$$

$$7.53 < \sigma^2 < 77.41$$

Finally, taking square roots, we get $2.74 < \sigma < 8.80$ beats per minute for the desired 99% confidence interval for σ.   ■

EXAMPLE 11.9   In a study of the effectiveness of a hinge lubricant, a research organization wants to investigate the variability in the number of cycles, openings and closings, before the hinge squeaks. Using $n = 15$ hinges, the number of cycles they got were

4295	4390	4338	4426	4698
4405	4694	4468	4863	4230
4664	4494	4535	4479	4600

(a) Check whether it is reasonable to treat these data as a sample from a normal population.

(b) If so, construct a 95% confidence interval for σ.

Solution   (a) Using MINITAB, we obtained the computer-generated normal probability plot shown in Figure 11.8. As can be seen, the pattern of the 15 dots closely follows that of the straight line, and this constitutes support for the assumption that the data constitute a sample from a normal population.

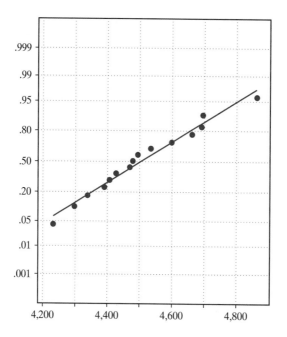

FIGURE 11.8
Computer printout of
normal probability
plot for Example 11.9.

(b) Part of the computer printout of Figure 11.8, which we deleted together with some other nonessential information, showed that the
standard deviation of the given data is $s = 172.3$. Substituting this
value together $n = 15$, and $\chi^2_{0.975} = 5.629$ and $\chi^2_{0.025} = 26.119$ for
$15 - 1 = 14$ degrees of freedom, into the confidence interval formula, we get

$$\frac{14(172.3)^2}{26.119} < \sigma^2 < \frac{14(172.3)^2}{5.629}$$

$$15{,}913 < \sigma^2 < 73{,}836$$

Finally, taking square roots, we get

$$126.1 < \sigma < 271.7 \text{ cycles} \qquad \blacksquare$$

There exists another approach to the construction of confidence intervals
for population standard deviations. For large samples, when $n \geqslant 30$, we can make
use of the theory that the sampling distribution of s can be approximated with
a normal distribution having the mean σ and the standard deviation

$$\sigma_s = \frac{\sigma}{\sqrt{2n}}$$

Then, converting to standard units, we can assert with probability $1 - \alpha$ that

$$-z_{\alpha/2} < \frac{s - \sigma}{\dfrac{\sigma}{\sqrt{2n}}} < z_{\alpha/2}$$

and fairly simple algebra leads to the following **large-sample confidence interval for** σ:

Large-sample confidence interval for σ

$$\frac{s}{1 + \dfrac{z_{\alpha/2}}{\sqrt{2n}}} < \sigma < \frac{s}{1 - \dfrac{z_{\alpha/2}}{\sqrt{2n}}}$$

EXAMPLE 11.10   With reference to Example 4.7, where we showed that $s = 14.35$ minutes for the $n = 110$ waiting times between eruptions of Old Faithful, construct a 95% confidence interval for the standard deviation of the population sampled.

Solution   Substituting $n = 110$, $s = 14.35$, and $z_{0.025} = 1.96$ into the large-sample confidence interval formula for σ, we get

$$\frac{14.35}{1 + \dfrac{1.96}{\sqrt{220}}} < \sigma < \frac{14.35}{1 - \dfrac{1.96}{\sqrt{220}}}$$

and, hence, $12.68 < \sigma < 16.53$ minutes. This means that we are 95% confident (and would consider it fair to give odds of 19 to 1) that the interval from 12.68 minutes to 16.53 minutes contains the true standard deviation of waiting times between eruptions of Old Faithful.

■ Exercises

11.33 The refractive indices of $n = 15$ pieces of glass, randomly selected from a large lot purchased by an optical firm, have a standard deviation of 0.012. Assuming that these measurements can be looked upon as a sample from a normal population, construct a 95% confidence interval for σ, the standard deviation of the population sampled.

11.34 Example 11.6 dealt with the durability of white center lines, and in the solution we obtained $s = 19{,}200$ car crossings for $n = 8$ locations. We also showed that it is reasonable to look upon the data as a sample from a normal population. Construct a 95% confidence interval for σ, which here might measure the consistency of the paint, or perhaps the uniformity of the traffic.

11.35 Exercise 11.18 dealt with the noise level of some new vacuum cleaners, and in the solution the reader was asked to verify that it is reasonable to treat the given data as a sample from a normal population. Calculate s for the $n = 12$ measurements and construct a 99% confidence interval for σ, which here measures the variability of the noise level of the vacuum cleaners.

11.36 Exercise 11.23 dealt with the thermal efficiency of certain diesel engines, and in the solution the reader was asked to verify that it is reasonable to treat the given data as a sample from a normal population. Calculate s for the given $n = 18$ measurements and construct a 98% confidence interval for σ, which here measures the variability of the thermal efficiency of the given kind of engine.

11.37 With reference to Exercise 11.26 and subject to the same assumption, construct a 95% confidence interval for σ^2, the variance of the weights of the containers in the production lot.

11.38 With reference to the exercises noted below and subject to the assumption that their data constitute random samples from normal populations, construct 98% confidence intervals for σ, the respective population standard deviations.
(a) Exercise 11.17, where we had $n = 9$ and $s = 0.48$ microgram;
(b) Exercise 11.21, where we had $n = 10$ and $s = 0.003$ cm.

11.39 With reference to the exercises noted below and subject to the same assumption, construct 99% confidence intervals for σ, the standard deviation of the population sampled.
(a) Exercise 11.24;
(b) Exercise 11.31.

11.40 With reference to the exercises noted below, construct 95% confidence intervals for σ, the standard deviation of the population sampled.
(a) Exercise 11.1, where we had $n = 40$ and $s = 135$ vehicles;
(b) Exercise 11.5, where we had $n = 35$ and $s = \$269$.

11.41 With reference to the exercises noted below, construct 99% confidence intervals for σ, the standard deviation of the population sampled.
(a) Exercise 11.3, where we had $n = 60$ and $s = 5.4$ mm;
(b) Exercise 11.6, where we had $n = 42$ and $s = 0.045$.

11.42 When we deal with very small samples, good estimates of the population standard deviation can often be obtained on the basis of the sample range (the largest sample value minus the smallest). Such quick estimates of σ are given by the sample range divided by the divisor d, which depends on the size of the sample. For samples from populations having roughly the shape of a normal distribution, its values are shown in the following table for $n = 2, 3, \ldots$, and 12:

n	2	3	4	5	6	7	8	9	10	11	12
d	1.13	1.69	2.06	2.33	2.53	2.70	2.85	2.97	3.08	3.17	3.26

For instance, during the monsoon season, there were 8, 11, 9, 5, 6, 12, 7, and 9 thunderstorms in Northern Arizona in eight successive weeks. The range of this sample is $12 - 5 = 7$, and since $d = 2.85$ for $n = 8$, we can estimate the population standard deviation as $\frac{7}{2.85} = 2.46$. This is quite close to the sample standard deviation, which is $s = 2.39$ as can easily be verified.
(a) In Example 11.6 we gave the following figures on the number of car crossings after which certain white center lines showed signs of deterioration: 142,600, 167,800, 136,500, 108,300, 126,400, 133,700, 162,000, and 149,400. Use the range of this sample to estimate σ, the standard deviation of the population sampled, and compare it with $s = 19,228$, the value of the sample standard deviation given in the printout of Figure 11.5.
(b) With reference to Exercise 11.18, use the range to estimate σ for the noise level of the new kind of vacuum cleaner, and compare the result with the sample standard deviation $s = 2.75$.
(c) In Exercise 11.24 we gave 14.28, 14.34, 14.26, and 14.32 grams as measurements of the weight of a Phoenician tetradrachm, and it can easily be verified that $s = 0.0365$ gram for these data. Use the sample range to obtain another estimate of the standard deviation of such weights, and compare the two results.

11.43 Fairly reasonable estimates of the population standard deviation can often be obtained by dividing the **interquartile range,** $Q_3 - Q_1$, by 1.35. For the waiting times between eruptions of Old Faithful we obtained $Q_1 = 69.71$ and $Q_3 = 87.58$ in Example 3.24 and $s = 14.35$ in Example 4.7. Estimate the actual standard deviation of waiting times between eruptions of Old Faithful in terms of these quartiles, and compare the result with the value obtained for s.

11.4   THE ESTIMATION OF PROPORTIONS

In this section we shall deal with **count data,** namely, with data obtained by counting rather than measuring. For instance, we shall concern ourselves with the number of persons who experience a side effect from a flu vaccine, the number of defectives in a shipment of manufactured product, the number of television viewers who like a certain situation comedy, the number of tires that last more than 40,000 miles, and so forth.

In particular, we shall concern ourselves with the estimation of the binomial parameter p, the probability of a success on an individual trial or the proportion of the time an event will occur, or with the estimation of $100p$, the corresponding percentage. Consequently, we will be able to use what we learned about the binomial distribution in Chapter 8; especially, its approximation by the normal distribution.

So far in this chapter, we have followed the convention of using lowercase Greek letters to denote the parameters of populations—μ for population means and σ for population standard deviations. In connection with binomial populations, more rigorous texts use θ (lowercase Greek *theta*) for the probability of a success on an individual trial, but having used p throughout Chapter 8, we shall continue doing so in this chapter and in Chapter 14.

The information that is usually available for the estimation of a population proportion (percentage, or probability) is a **sample proportion,** $\dfrac{x}{n}$, where x is the number of times that an event has occurred in n trials. For example, if a study shows that 54 of 120 cheerleaders, presumably a random sample, suffered what auditory experts call "moderate to severe" damage to their voices, then $\dfrac{x}{n} = \dfrac{54}{120} = 0.45$, and we can use this figure as a point estimate of the true proportion of cheerleaders who are afflicted in this way, or the probability that any one cheerleader will be afflicted in this way. Similarly, a supermarket chain might estimate the proportion of its shoppers who regularly use discount coupons as 0.68 if a random sample of 300 shoppers included 204 who regularly use discount coupons.

To be able to use methods based on the binomial distribution, it will be assumed throughout this section that there is a fixed number of independent trials and that for each trial the probability of a success—the parameter we want to estimate—has the constant value p. Under these conditions, we know from Chapter 8 that the sampling distribution of the number of successes has the mean $\mu = np$ and the standard deviation $\sigma = \sqrt{np(1 - p)}$, and that it can be approximated by a normal distribution so long as np and $n(1 - p)$ are both greater than 5. Usually, this requires that n must be large. For $n = 50$, for example, the normal-curve approximations used in this section may be used so long as $50p > 5$ and $50(1 - p) > 5$, namely, so long as p lies between 0.10 and 0.90. Similarly, for $n = 100$ they may be used so long as p lies between 0.05 and 0.95, and for $n = 200$ they may be used so long as p lies between 0.025 and 0.975. This should give the reader some idea of what we mean here by "n being large."

If we convert to standard units, then for large values of n, the statistic

$$z = \frac{x - np}{\sqrt{np(1 - p)}}$$

is a value of a random variable having approximately the standard normal distribution. If we substitute this expression for z into the inequality

$$-z_{\alpha/2} < z < z_{\alpha/2}$$

(as on page 306), some relatively simple algebraic manipulation will yield

$$\frac{x}{n} - z_{\alpha/2} \cdot \sqrt{\frac{p(1 - p)}{n}} < p < \frac{x}{n} + z_{\alpha/2} \cdot \sqrt{\frac{p(1 - p)}{n}}$$

which looks like a confidence-interval formula for p. Indeed, if we used it repeatedly, the inequality should be satisfied $(1 - \alpha)100\%$ of the time, but observe that the unknown parameter p appears not only in the middle, but also in

$$\sqrt{\frac{p(1 - p)}{n}}$$

to the left of the first inequality sign and to the right of the other. The quantity $\sqrt{\frac{p(1 - p)}{n}}$ is called the **standard error of a proportion,** as it is, in fact, the standard deviation of the sampling distribution of a sample proportion (see Exercise 11.63).

To get around this difficulty and, at the same time, simplify the resulting formula, we substitute

$$\hat{p} = \frac{x}{n} \text{ for } p \text{ in } \sqrt{\frac{p(1 - p)}{n}}$$

where $\hat{p}$ reads "p-hat." (This kind of notation is widely used in statistics. For instance, when we use the mean of a sample to estimate the mean of a population, we might denote it by $\hat{\mu}$, and when we use the standard deviation of a sample to estimate the standard deviation of a population, we might denote it by $\hat{\sigma}$.) Thus, we get the following **$(1 - \alpha)100\%$ large-sample confidence interval for p:**

Large-sample confidence interval for p

$$\hat{p} - z_{\alpha/2} \cdot \sqrt{\frac{\hat{p}(1 - \hat{p})}{n}} < p < \hat{p} + z_{\alpha/2} \cdot \sqrt{\frac{\hat{p}(1 - \hat{p})}{n}}$$

EXAMPLE 11.11   In a random sample, 136 of 400 persons given a flu vaccine experienced some discomfort. Construct a 95% confidence interval for the true proportion of persons who will experience some discomfort from the vaccine.

Solution   Substituting $n = 400$, $\hat{p} = \frac{136}{400} = 0.34$, and $z_{0.025} = 1.96$ into the confidence-interval formula, we get

$$0.34 - 1.96\sqrt{\frac{(0.34)(0.66)}{400}} < p < 0.34 + 1.96\sqrt{\frac{(0.34)(0.66)}{400}}$$

$$0.294 < p < 0.386$$

or, rounding to two decimals, $0.29 < p < 0.39$.   ∎

As we have pointed out before, an interval like this contains the parameter it is intended to estimate or it does not. In any particular instance we do not know which is the case, but the 95% confidence implies that the interval was obtained by a method which works 95% of the time. Note also that for $n = 400$ and p on the interval from 0.29 to 0.39, np and $n(1 - p)$ are both much greater than 5, so there can be no doubt that we are justified in using the normal approximation to the binomial distribution.

When it comes to small samples, we can construct confidence intervals for p by using a special table, but the resulting intervals are usually so wide that they are not of much value. For example, for $x = 4$ and $n = 10$, the 95% confidence interval is $0.12 < p < 0.75$. Clearly, this interval is so wide that it does not tell us very much about the actual value of p.

The large-sample theory presented here can also be used to assess the error we may be making when we use a sample proportion to estimate a population proportion, namely, the binomial parameter p. Proceeding as on page 304, we can assert with probability $1 - \alpha$ that the difference between a sample proportion and p will be at most

$$E = z_{\alpha/2} \cdot \sqrt{\frac{p(1 - p)}{n}}$$

However, since p is unknown, we substitute for it again the sample proportion $\hat{p}$, and we arrive at the result that

If $\hat{p}$ is used as an estimate of p, we can assert with $(1 - \alpha)\,100\%$ confidence that the error is at most

Maximum error of estimate

$$E = z_{\alpha/2} \cdot \sqrt{\frac{\hat{p}(1 - \hat{p})}{n}}$$

Like the confidence-interval formula on page 324, this formula requires that n must be large enough to use the normal approximation to the binomial distribution.

EXAMPLE 11.12   In a random sample of 250 persons interviewed while exiting from polling places all over a state, 145 said that they voted for the reelection of the incumbent governor. With 99% confidence, what can we say about the maximum error if we use $\hat{p} = \frac{145}{250} = 0.58$ as an estimate of the actual proportion of the vote that the incumbent governor will get?

Solution   Substituting $n = 250$, $\hat{p} = 0.58$, and $z_{0.005} = 2.575$ into the formula for E, we get

$$E = 2.575\sqrt{\frac{(0.58)(0.42)}{250}} \approx 0.080$$

Thus, if we use $\hat{p} = 0.58$ as an estimate of the actual proportion of the vote the incumbent governor will get, we can assert with 99% confidence that our error is at most 0.080.   ∎

With reference to this example, note also that for $n = 250$ the normal approximation to the binomial distribution is justified for any value of p between 0.02 and 0.98.

As in Section 11.1, we can use the formula for the maximum error to determine how large a sample is needed to attain a desired degree of precision. If we want to use a sample proportion to estimate a population proportion p, and we want to be able to assert with probability $1 - \alpha$ that our error will not exceed some prescribed quantity E, we write as before

$$E = z_{\alpha/2} \cdot \sqrt{\frac{p(1 - p)}{n}}$$

Upon solving this equation for n, we get

Sample size

$$n = p(1 - p)\left[\frac{z_{\alpha/2}}{E}\right]^2$$

This formula cannot be used as is, because it involves the quantity p we are trying to estimate. However, since $p(1 - p)$ *increases* from 0 to $\frac{1}{4}$ when p increases from 0 to $\frac{1}{2}$ or decreases from 1 to $\frac{1}{2}$, we can proceed as follows:

If we have some information about the values that p might assume, we substitute for it in the formula for n whichever of these values is closest to $\frac{1}{2}$; if we have no information about the values that p might assume, we substitute $\frac{1}{4}$ for $p(1 - p)$ in the formula for n.

In either case, since the value we obtain for n may well be larger than necessary, we can say that the probability is *at least* $1 - \alpha$ that our error will not exceed E.

EXAMPLE 11.13   Suppose that a state highway department wants to estimate what proportion of all trucks operating between two cities carry too heavy a load, and it wants to be able to assert with a probability of at least 0.95 that its error will not exceed 0.04. How large a sample will it need if

(a) it knows that the true proportion lies somewhere on the interval from 0.10 to 0.25;

(b) it has no idea what the true value might be?

Solution

(a) Substituting $z_{0.025} = 1.96$, $E = 0.04$, and $p = 0.25$ into the formula for n, we get

$$n = (0.25)(0.75)\left(\frac{1.96}{0.04}\right)^2 \approx 450.19$$

and we round this up to the next integer, 451.

(b) Substituting $z_{0.025} = 1.96$, $E = 0.04$, and $p(1 - p) = \frac{1}{4}$ into the formula for n, we get

$$n = \frac{1}{4}\left(\frac{1.96}{0.04}\right)^2 = 600.25$$

and we round this up to the next integer, 601.   ∎

Example 11.13 illustrates how some knowledge about p can substantially reduce the sample size needed to attain a desired degree of precision. Note also that in a problem like this we round up, if necessary, to the nearest integer.

Exercises

11.44 In accordance with the rule that np and $n(1 - p)$ must both be greater than 5, for what values of p can we use the normal approximation to the binomial distribution when
(a) $n = 400$;
(b) $n = 500$?

11.45 In accordance with the rule that np and $n(1 - p)$ must both be greater than 5, for what values of n can we use the normal approximation to the binomial distribution when
(a) $p = 0.04$;
(b) $p = 0.92$?

11.46 In a random sample of 200 eligible voters interviewed in a large city, 126 objected to the use of public funds for the construction of a new professional football stadium. Construct a 95% confidence interval for the corresponding proportion for the population sampled.

11.47 With reference to Exercise 11.46, what can we say with 99% confidence about the maximum error if $\dfrac{x}{n} = \dfrac{126}{200} = 0.63$ is used as an estimate of the proportion of all eligible voters in that city who are against the use of public funds for the construction of a new professional football stadium?

11.48 Among the 300 fish caught in Woods Canyon Lake, 42 were inedible as a result of the chemical pollution of the environment. Construct a 99% confidence interval for the corresponding true proportion.

11.49 With reference to Exercise 11.48, what can we say with 95% confidence about the maximum error if $\frac{x}{n} = \frac{42}{300} = 0.14$ is used as an estimate of the corresponding true proportion?

11.50 In a random sample of 120 cheerleaders, 54 had suffered moderate to severe damage to their voices. With 90% confidence, what can we assert about the maximum error if the sample proportion $\frac{54}{120} = 0.45$ is used as an estimate of the true proportion of cheerleaders who are afflicted in this way?

11.51 A random sample of 300 shoppers at a large supermarket includes 234 who regularly use discount coupons. Construct a 98% confidence interval for the probability that any one shopper at that supermarket, randomly chosen for an interview, will confirm that he or she regularly uses discount coupons.

11.52 In a random sample of 1,200 adults interviewed nationwide, only 324 felt that the salaries of certain government officials should be raised. Construct a 95% confidence interval for the actual percentage of adults who share that opinion.

11.53 In a random sample of 400 television viewers interviewed nationwide, 152 had seen a certain controversial program. With 99% confidence, what can we assert about the maximum error if we use $\frac{152}{400} \cdot 100 = 38\%$ as an estimate of the corresponding true percentage?

11.54 In a random sample of 360 high school seniors in a Western state, 252 said that they expect to continue their education at an in-state college. Construct a 95% confidence interval for the corresponding true percentage.

11.55 With reference to Exercise 11.54, what can we assert with 99% confidence about the maximum error if we use $\frac{252}{360} \cdot 100 = 70\%$ as an estimate of the true percentage of high school seniors in that state who expect to continue their education in an in-state college?

11.56 In a random sample of 140 supposed UFO sightings, 119 could easily be explained in terms of natural phenomena. Construct a 99% confidence interval for the probability that a supposed UFO sighting will easily be explained in terms of natural phenomena.

11.57 In a random sample of 80 persons convicted in U.S. District Courts on narcotics charges, 36 received probation. With 98% confidence, what can we say about the maximum error if we use $\frac{36}{80} = 0.45$ as an estimate of the probability that a person convicted in a U.S. District Court on narcotics charges will receive probation?

*11.58 When a sample constitutes more than 5% of a finite population, and the sample itself is large, we use the same finite population correction factor as in Section 10.7, and hence the following confidence limits for p:

$$\hat{p} \pm z_{\alpha/2} \cdot \sqrt{\frac{\hat{p}(1 - \hat{p})(N - n)}{n(N - 1)}}$$

Here N is, as before, the size of the population sampled.

   (a) Among the $N = 360$ families in an apartment complex, a random sample of $n = 100$ families is interviewed, and it is found that 34 of them have children of college age. Use the preceding formula to construct a 95% confidence interval for the proportion of all the families living in the apartment complex who have children of college age.

   (b) With reference to Exercise 11.51, suppose that $N = 1,800$ persons shop at the supermarket on a regular basis. Use this added information to recalculate the confidence interval asked for in that exercise.

11.59 A political pollster is engaged by a politician to estimate the proportion of registered voters in her district who plan to vote for her in the next election. Find the sample size needed if she wants, with at least 95% confidence, the poll to be accurate to within
   (a) 8 percentage points;
   (b) 2 percentage points.

11.60 Suppose that we want to estimate what proportion of all drivers exceed the maximum speed limit on a stretch of I-17 near the Rock Springs exit. How large a sample will we need so that the error of our estimate is to be at most 0.05 with at least
   (a) 90% confidence;
   (b) 95% confidence;
   (c) 99% confidence?

11.61 Rework Exercise 11.60, given that the proportion in question is at least 0.60.

11.62 A national manufacturer wants to determine what percentage of purchases of razor blades for use by men is actually made by women. How large a sample of men shaving with razor blades will the manufacturer need to be at least 98% confident that the sample percentage will not be off by more than 2.5 percentage points and
   (a) nothing is known about the true percentage;
   (b) there is good reason to believe that the true percentage is at most 30%?

11.63 Since the proportion of successes is simply the number of successes divided by n, the mean and the standard deviation of the sampling distribution of the proportion of successes may be obtained by dividing by n the mean and the standard deviation of the sampling distribution of the number of successes. Use this argument to verify the standard error formula given on page 324.

11.5 Checklist of Key Terms *(with page references to their definitions)*

Bayesian methods, 302
Chi-square distribution, 317
Chi-square statistic, 317
Confidence, 305
Confidence interval, 307
Confidence limits, 307
Count data, 323
Degree of confidence, 307
Degrees of freedom, 308, 317
Estimation, 302
Interquartile range, 323
Interval estimate, 307
Large-sample confidence interval, 307, 321, 324

Number of degrees of freedom, 308, 317
Point estimate, 303
Sample proportion, 323
Small-sample confidence interval, 309, 318
Standard error of a proportion, 324
Standard error of a standard deviation, 320
Student *t* distribution, 308
t distribution, 308
t interval, 309
t statistic, 308
z interval, 307

11.6 References

An informal introduction to interval estimation is given under the heading of "How to be precise though vague" in

MORONEY, M. J., *Facts from Figures.* London: Penguin Books, Ltd., 1956

and also in

GONICK, L., and SMITH, WOOLCOTT, *A Cartoon Guide to Statistics.* New York: Harper Collins Publishers, 1993.

Detailed discussions of the chi-square and t distributions may be found in most textbooks on mathematical statistics, and more detailed tables of these distributions are given in

PEARSON, E. S., and HARTLEY, H. O., *Biometrika Tables for Statisticians*, Vol. I. New York: John Wiley & Sons, Inc., 1968.

Tables of confidence intervals for proportions, including those for small samples, were first published in Vol. 26 (1934) of Biometrika. Nowadays, they may be found, for instance, in

MAXWELL, E. A., *Introduction to Statistical Thinking.* Upper Saddle River, N. J.: Prentice Hall, 1983.

12

TESTS OF HYPOTHESES: MEANS

12.1   Tests of Hypotheses 332

12.2   Significance Tests 337

12.3   Tests Concerning Means 345

12.4   Tests Concerning Means (σ unknown) 349

12.5   Differences Between Means 354

12.6   Differences Between Means (σ's unknown) 357

12.7   Differences Between Means (Paired data) 359

12.8   Checklist of Key Terms 365

12.9   References 365

*I*n the introduction to Chapter 11 we gave three examples of statistical inference, which were all problems of estimation—one from science, one from business, and one from everyday life. However, they would have been tests of hypotheses if the psychologist had wanted to decide whether the average time it takes an adult to react to a visual stimulus is really 0.38 second, if the union official had wanted to check whether the standard deviation of the commuting time of his union members is really 10.3 minutes, and if we had wanted to see whether it is true that 38% of all one-car accidents are due to driver fatigue.

The first of these examples concerns a population mean, the second concerns a population standard deviation, and the third concerns a population percentage. Conceptually, all three of these problems are treated in the same way, but, as in connection with the original problems of estimation, there are differences in the particular methods that are employed. After a general introduction to tests of hypotheses in Sections 12.1 and 12.2, the remainder of this chapter will be devoted to tests concerning the mean of one population, or the means of two populations. Tests concerning population standard deviations will be treated in Chapter 13, and tests concerning percentages (proportions, or probabilities) will be dealt with in Chapter 14. Subsequent chapters will deal with other, specialized, tests of hypotheses.

12.1 TESTS OF HYPOTHESES

In the preceding introduction we referred to the three decision problems as tests of hypotheses without actually giving a formal definition of what we mean here by a hypothesis. In general,

A statistical hypothesis is an assertion or conjecture about a parameter, or parameters, of a population (or populations); it may also concern the type, or nature, of a population (or populations).

With regard to the second part of this definition, we shall see in Section 14.6 how we can test whether it is reasonable to treat a population sampled as being a binomial population, a Poisson population, or perhaps a normal population. In this chapter we shall be concerned only with hypotheses about population parameters; in particular, the mean of one population or the means of two populations.

To develop procedures for testing statistical hypotheses, we must always know exactly what to expect when a hypothesis is true, and it is for this reason that we often hypothesize the opposite of what we hope to prove. Suppose, for instance, that we suspect that a dice game is not honest. If we formulate the hypothesis that the dice are crooked, everything would depend on how crooked they are, but if we assume that they are perfectly balanced, we could calculate all the necessary probabilities and take it from there. Also, if we want to show that one method of teaching computer programming is more effective than another, we would hypothesize that the two methods are equally effective; if we want to show that one diet is healthier than another, we hypothesize that they are equally healthy; and if we want to show that a new copper-bearing steel has a higher yield strength than ordinary steel, we hypothesize that the two yield strengths are the same. Since we hypothesize that there is no difference in the effectiveness of the two teaching methods, medically no difference between the two diets, and no difference in the yield strength of the two kinds of steel, we call hypotheses like these **null hypotheses** and denote them by H_0. In effect, the term "null hypothesis" is used for any hypothesis set up primarily to see whether it can be rejected.

The idea of setting up a null hypothesis is common even in nonstatistical thinking. It is precisely what we do in criminal proceedings, where an accused is presumed to be innocent until his guilt has been established beyond a reasonable doubt. The presumption of innocence is a null hypothesis.

The hypothesis that we use as an alternative to the null hypothesis, namely, the hypothesis that we accept when the null hypothesis is rejected, is appropriately called an **alternative hypothesis** and is denoted by H_A. It must always be formulated together with the null hypothesis, for otherwise we would not know when to reject H_0. For instance, if the psychologist of the example on page 331 tests the null hypothesis $\mu = 0.38$ second against the alternative hypothesis $\mu > 0.38$ second, he would reject the null hypothesis only if he gets a sample mean that is much greater than 0.38 second. On the other hand, if he uses the alternative hypothesis $\mu \neq 0.38$ second, he would reject the null hypothesis if he gets a sample mean that is much greater than, or much less than, 0.38 second.

As in the preceding illustration, alternative hypotheses usually specify that the population mean (or whatever other parameter may be of concern) is less than, greater than, or not equal to the value assumed under the null hypothesis. For any given problem, the choice of one of these alternatives depends on what we hope to be able to show, or perhaps on where we want to put the burden of proof.

EXAMPLE 12.1 The average drying time of a manufacturer's paint is 20 minutes. Investigating the effectiveness of a modification in the chemical composition of his paint, the manufacturer wants to test the null hypothesis $\mu = 20$ minutes against a suitable alternative, where μ is the average drying time of the modified paint.

(a) What alternative hypothesis should the manufacturer use if he wants to make the modification only if it actually decreases the drying time of the paint?

(b) What alternative hypothesis should the manufacturer use if the new process is actually cheaper and he wants to make the modification unless it actually increases the drying time of the paint?

Solution (a) He should use the alternative hypothesis $\mu < 20$ and make the modification only if the null hypothesis can be rejected.

(b) He should use the alternative hypothesis $\mu > 20$ and make the modification unless the null hypothesis is rejected.   ∎

In general, if the test of a hypothesis concerns the parameter μ, its value assumed under the null hypothesis is denoted by μ_0, and the null hypothesis, itself, is $\mu = \mu_0$.

To illustrate in detail the problems we face when testing a statistical hypothesis, let us refer again to the reaction-time example on page 331, and let us suppose that the psychologist wants to test the null hypothesis

$$H_0: \quad \mu = 0.38 \text{ second}$$

against the alternative hypothesis

$$H_A: \quad \mu \neq 0.38 \text{ second}$$

where μ is the mean reaction time of an adult to the visual stimulus. To perform this test, the psychologist decides to take a random sample of $n = 40$ adults with the intention of accepting the null hypothesis if the mean of the sample falls anywhere between 0.36 second and 0.40 second; otherwise, he will reject it.

This provides a clear-cut criterion for accepting or rejecting the null hypothesis, but unfortunately it is not infallible. Since the decision is based on a sample, there is the possibility that the sample mean may be less than 0.36 second or greater than 0.40 second even though the true mean is 0.38 second. There is also the possibility that the sample mean may fall between 0.36 second and 0.40

second, even though the true mean is, say, 0.39 second. Thus, before adopting the criterion (and, for that matter, any decision criterion) it would seem wise to investigate the chances that it may lead to a wrong decision.

Assuming that it is known from similar studies that $\sigma = 0.08$ second for this kind of data, let us first investigate the possibility of falsely rejecting the null hypothesis. Thus, assume for the sake of argument that the true average reaction time is 0.38 second; then find the probability that the sample mean will be less than or equal to 0.36 or greater than or equal to 0.40. The probability that this will happen purely due to chance is given by the sum of the areas of the two tinted regions of Figure 12.1, and it can readily be determined by approximating the sampling distribution of the mean with a normal distribution. Assuming that the population sampled may be looked upon as being infinite, which seems reasonable in this case, we have

$$\sigma_{\bar{x}} = \frac{\sigma}{\sqrt{n}} = \frac{0.08}{\sqrt{40}} \approx 0.0126$$

and it follows that the dividing lines of the criterion, in standard units, are

$$z = \frac{0.36 - 0.38}{0.0126} = -1.59 \quad \text{and} \quad z = \frac{0.40 - 0.38}{0.0126} = 1.59$$

It follows from Table I that the area in each tail of the sampling distribution of Figure 12.1 is $0.5000 - 0.4441 = 0.0559$, and hence the probability of getting a value in either tail of the sampling distribution is $2(0.0559) = 0.1118$, or 0.11 rounded to two decimals.

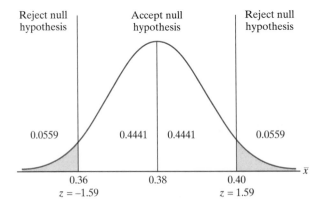

FIGURE 12.1
Test criterion and sampling distribution of $\bar{x}$ with $\mu = 0.38$ second.

Let us now consider the other possibility, where the test fails to detect that the null hypothesis is false; namely, that $\mu \neq 0.38$ second. Thus, assume for the sake of argument that the true average reaction time is 0.41 second. Now, getting a sample mean on the interval from 0.36 second to 0.40 second would lead to the erroneous acceptance of the null hypothesis that $\mu = 0.38$ second. The probability that this will happen purely due to chance is given by the area of the tinted region of Figure 12.2. The mean of the sampling distribution is now 0.41 second, but its standard deviation is, as before,

$$\sigma_{\bar{x}} = \frac{0.08}{\sqrt{40}} \approx 0.0126$$

and the dividing lines of the criterion, in standard units, are

$$z = \frac{0.36 - 0.41}{0.0126} \approx -3.97 \quad \text{and} \quad z = \frac{0.40 - 0.41}{0.0126} \approx -0.79$$

Since the area under the curve to the left of -3.97 is negligible, it follows from Table I that the area of the tinted region of Figure 12.2 is $0.5000 - 0.2852 = 0.2148$ or 0.21 rounded to two decimals. This is the probability of erroneously accepting the null hypothesis when actually $\mu = 0.41$. It will be up to the psychologist to decide whether the 0.11 probability of erroneously rejecting the null hypothesis $\mu = 0.38$ and the 0.21 probability of erroneously accepting it when actually $\mu = 0.41$ are acceptable risks.

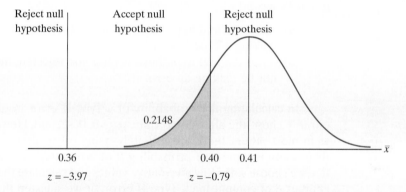

FIGURE 12.2
Test criterion and sampling distribution of $\bar{x}$ with $\mu = 0.41$ second.

Reject null hypothesis   Accept null hypothesis   Reject null hypothesis

0.2148

0.36   0.40   0.41
$z = -3.97$ $z = -0.79$

The situation described here is typical of tests of hypotheses, and it may be summarized as in the following table:

	Accept H_0	Reject H_0
H_0 is true	Correct decision	Type I error
H_0 is false	Type II error	Correct decision

If the null hypothesis H_0 is true and accepted or false and rejected, the decision is in either case correct; if it is true and rejected or false and accepted, the decision is in either case in error. The first of these errors is called a **Type I error** and the probability of committing it is designated by the Greek letter α (*alpha*); the second is called a **Type II error** and the probability of committing it is designated by the Greek letter β (beta). Thus, in our example we showed that for the given test criterion $\alpha = 0.11$ and $\beta = 0.21$ when $\mu = 0.41$.

The scheme just outlined is reminiscent of what we did in Section 7.2. Analogous to the decision that the director of the research division of the pharmaceutical company had to make in Example 7.9, now the psychologist must decide whether to accept or reject the null hypothesis $\mu = 0.38$. It is difficult to carry this analogy much further, though, because in actual practice we can seldom associate cash values with the various possibilities, as we did in Example 7.9.

EXAMPLE 12.2   Suppose that the psychologist has actually taken the sample and obtained $\bar{x} = 0.408$. What will he decide and will it be in error if

   (a) $\mu = 0.38$ second;
   (b) $\mu = 0.42$ second?

Solution   Since $\bar{x} = 0.408$ exceeds 0.40, the psychologist will reject the null hypothesis $\mu = 0.38$ second.

   (a) Since the null hypothesis is true and rejected, the psychologist will be making a Type I error.
   (b) Since the null hypothesis is false and rejected, the psychologist will not be making an error.   ∎

In calculating the probability of a Type II error in our illustration, we arbitrarily chose the alternative value $\mu = 0.41$ second. However, in this problem, as in most others, there are infinitely many other alternatives, and for each of them there is a positive probability β of erroneously accepting H_0. So, in practice, we choose some key alternative values and calculate the corresponding probabilities β of committing a Type II error, or we sidestep the issue by proceeding in a way that will be explained in Section 12.2.

If we do calculate β for various alternative values of μ and plot these probabilities as in Figure 12.3, we obtain a curve that is called the **operating characteristic curve,** or simply the **OC-curve,** of the test criterion. Since the probability of a Type II error is the probability of accepting H_0 when it is false, we "completed the picture" in Figure 12.3 by labeling the vertical scale "Probability of accepting H_0 and plotting at $\mu = 0.38$ second the probability of accepting H_0 when it is true, namely, $1 - \alpha = 1 - 0.11 = 0.89$.

Examination of the curve of Figure 12.3 shows that the probability of accepting H_0 is greatest when H_0 is true, and that it is still fairly high for small departures from $\mu = 0.38$. However, for larger and larger departures from $\mu = 0.38$ in either direction, the probabilities of failing to detect them and accepting H_0 become smaller and smaller. In Exercise 12.10 the reader will be asked to verify some of the probabilities plotted in Figure 12.3.

If we had plotted the probabilities of rejecting H_0 instead of those of accepting H_0, we would have obtained the graph of the **power function** of the test criterion instead of its operating characteristic curve. The concept of an OC-curve

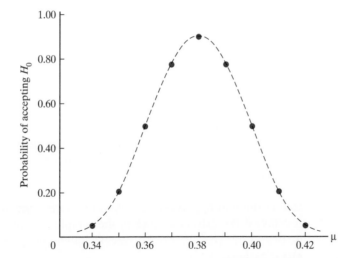

FIGURE 12.3
Operating
characteristic curve.

is used more widely in applications, especially in industrial applications, while the concept of a power function is used more widely in matters that are of theoretical interest. A detailed study of operating characteristic curves and power functions is beyond the scope of this text, and the purpose of our example is mainly to show how statistical methods can be used to measure and control the risks one is exposed to when testing hypotheses. Of course, the methods discussed here are not limited to the particular problem concerning the average reaction time to a visual stimulus—H_0 could have been the hypothesis that the average age at which women get a divorce is 28.5, the hypothesis that an antibiotic is 87% effective, the hypothesis that a computer-assisted method of instruction will, on the average, raise a student's score on a standard achievement test by 7.4 points, and so forth.

12.2 SIGNIFICANCE TESTS

In the problem dealing with the reaction time of adults to a visual stimulus, we had less trouble with Type I errors than with Type II errors, because we formulated the null hypothesis as a **simple hypothesis** about the parameter μ; that is, we formulated it so that μ took on a single value, the value $\mu = 0.38$ second, and the corresponding value of a Type I error could be calculated.[†] Had we formulated instead a **composite hypothesis** about the parameter μ, say, $\mu \neq 0.38$ second, $\mu < 0.38$ second, or $\mu > 0.38$ second, where in each case μ can take on more than one possible value, we could not have calculated the probability of a Type I error without specifying how much μ differs from, is less than, or is greater than 0.38 second.

[†]Note that we are applying the term "simple hypothesis" to hypotheses about specific parameters. Some statisticians use the term "simple hypothesis" only when the hypothesis completely specifies the population.

In the same illustration, the alternative hypothesis was the composite hypothesis $\mu \neq 0.38$ second, and it took quite some work to calculate the probabilities of Type II errors (for various alternative values of μ) shown in the OC-curve of Figure 12.3. Since this is typical of most practical situations (that is, alternative hypotheses are usually composite), let us demonstrate how Type II errors can often be sidestepped altogether.

Studies have shown that in a certain city licensed drivers average 0.9 traffic ticket per year, but a social scientist suspects that drivers over 65 years of age average more than 0.9 traffic ticket per year. So she checks the records of a random sample of licensed drivers over 65 in the given city, and bases her decision on the following criterion:

Reject the null hypothesis $\mu = 0.9$ (and accept the alternative hypothesis $\mu > 0.9$) if the drivers over 65 in the sample average, say, at least 1.2 traffic tickets per year; otherwise, reserve judgment (perhaps, pending further investigation).

If one reserves judgment as in this criterion, there is no possibility of committing a Type II error—no matter what happens, the null hypothesis is never accepted. This would seem all right in the preceding example, where the social scientist wants to see primarily whether her suspicion is justified; namely, whether the null hypothesis can be rejected. If it cannot be rejected, this does not mean that she must necessarily accept it. Indeed, her suspicion may not be completely resolved.

The procedure we have outlined here is called a **significance test,** or a **test of significance.** If the difference between what we expect under the null hypothesis and what we observe in a sample is too large to be reasonably attributed to chance, we reject the null hypothesis. If the difference between what we expect and what we observe is so small that it may well be attributed to chance, we say that the result is **not statistically significant,** or simply that it is **not significant.** We then accept the null hypothesis or reserve judgment, depending on whether a definite decision one way or the other must be reached.

Since "significant" is often used interchangeably with "meaningful" or "important" in everyday language, it must be understood that we are using it here as a technical term. Specifically, the word "significant" is used in situations in which a null hypothesis is rejected. If a result is statistically significant, this does not mean that it is necessarily of any great importance, or that it is of any practical value. Suppose, for instance, that the psychologist of our reaction-time example has actually taken the sample, as in Example 12.2, and obtained $\bar{x} = 0.408$. According to the criterion on page 333 this result is statistically significant, meaning that the difference between $\bar{x} = 0.408$ and $\mu = 0.38$ is too large to be attributed to chance. It is possible, though, that no one may care about this result; not even a lawyer involved in a litigation where reaction times may be of critical relevance in determining a client's liability. His reaction might be that the whole thing is simply not worth bothering about.

Returning to the original criterion used in the reaction-time example, the one on page 333, we could convert it into that of a significance test by writing

Reject the null hypothesis $\mu = 0.38$ second (and accept the alternative $\mu \neq 0.38$ second) if the mean of the 40 sample values is less than or equal to 0.36 second or greater than or equal to 0.40 second; reserve judgment if the sample mean falls anywhere between 0.36 second and 0.40 second.

So far as the rejection of the null hypothesis is concerned, the criterion has remained unchanged and the probability of a Type I error is still 0.11. However, so far as its acceptance is concerned, the psychologist is now playing it safe by reserving judgment.

Reserving judgment in a significance test is similar to what happens in court proceedings where the prosecution does not have sufficient evidence to get a conviction, but where it would be going too far to say that the defendent definitely did not commit the crime. In general, whether one can afford the luxury of reserving judgment in any given situation depends entirely on the nature of the situation. If a decision must be reached one way or the other, there is no way of avoiding the risk of committing a Type II error.

Since most of the remainder of this book is devoted to significance tests—indeed, most statistical problems that are not problems of estimation or prediction deal with tests of this kind—it will help to perform such tests by proceeding systematically as outlined in the following five steps. The first of these may look simple and straightforward, yet it presents the greatest difficulties to most beginners.

1. We formulate a simple null hypothesis and an appropriate alternative hypothesis that is to be accepted when the null hypothesis is rejected.

In the reaction-time example the null hypothesis was $\mu = 0.38$ second and the alternative hypothesis was $\mu \neq 0.38$ second. We choose this alternative as an illustration; in actual practice, it would reflect the psychologist's intent to reject the null hypothesis if 0.38 second is either too high or too low. We refer to this kind of alternative as a **two-sided alternative.** In the traffic-ticket example the null hypothesis was $\mu = 0.9$ ticket and the alternative hypothesis was $\mu > 0.9$ (to confirm the social scientist's suspicion that licensed drivers over 65 average more than 0.9 traffic ticket per year). This is called a **one-sided alternative.** We can also write a one-sided alternative with the inequality going the other way. For instance, if we hope to show that the average time required to do a certain job is less than 15 minutes, we would test the null hypothesis $\mu = 15$ minutes against the alternative hypothesis $\mu < 15$ minutes.

This is not the first time that we concerned ourselves with the formulation of hypotheses. Prior to Example 12.1 we mentioned some of the things that must be taken into account when choosing H_A, but throughout this chapter, so far, the null hypothesis has always been specified.

Basically, there are two things we must watch in connection with H_0. First, whenever possible we formulate null hypotheses as simple hypotheses about the parameters with which we are concerned; second, we formulate null hypotheses in such a way that their rejection proves whatever point we hope to make. As we have pointed out before, we choose null hypotheses as simple hypotheses so that we can calculate, or specify, the probabilities of Type I errors. We saw how this works in the reaction-time example. The reason for choosing null hypotheses so that their rejection proves whatever point we hope to make is that, in general, it is much easier to prove that something is false than to prove that it is true. Suppose, for instance, that somebody claims that all 6,000 male students attending a certain university weigh at least 145 pounds. To show that this claim is true, we literally have to weigh each of the 6,000 students; however, to show that it is false, we have only to find one student who weighs less than 145 pounds, and that should not be too difficult.

EXAMPLE 12.3   A bakery machine fills boxes with crackers, averaging 454 grams (roughly one pound) of crackers per box.

   (a) If the management of the bakery is concerned about the possibility that the actual average is different from 454 grams, what null hypothesis and what alternative hypothesis should it use to put this to a test?
   (b) If the management of the bakery is concerned about the possibility that the actual average is less than 454 grams, what null hypothesis and what alternative hypothesis should it use to put this to a test?

Solution   **(a)** The words "different from" suggest that the hypothesis $\mu \neq 454$ grams is needed together with the only other possibility, namely, the hypothesis $\mu = 454$ grams. Since the second of these hypotheses is a simple hypothesis, and its rejection (and the acceptance of the other hypothesis) confirms the management's concern, we follow the two rules on page 339 by writing

$$H_0: \quad \mu = 454 \text{ grams}$$

$$H_A: \quad \mu \neq 454 \text{ grams}$$

   (b) The words "less than" suggest that we need the hypothesis $\mu < 454$ grams, but for the other hypothesis there are many possibilities, including $\mu \geq 454$ grams, $\mu = 454$ grams, and, say, $\mu = 456$ grams. Two of these (and many others) are simple hypotheses, but since it would be to the bakery's disadvantage to put too many crackers into the boxes, a sensible choice would be

$$H_0: \quad \mu = 454 \text{ grams}$$

$$H_A: \quad \mu < 454 \text{ grams}$$

Note that the null hypothesis is a simple hypothesis and that its rejection (and the acceptance of the alternative) confirms the management's suspicion.   ∎

It is important to add that H_0 and H_A must be formulated before any data are actually collected, or at least without looking at the data. In particular, the choice of a one-sided alternative or a two-sided alternative should not be suggested by the data. However, it often happens that we are presented with data before we had the opportunity to contemplate the hypotheses, and in such situations we must try to assess the motives (or objectives) without using the data. If there is any doubt whether a situation calls for a one-sided or a two-sided alternative, the scrupulous action calls for a two-sided alternative.

Like the first step given on page 339, the second step looks simple and straightforward, but it is not without complications.

2. We specify the probability of a Type I error.

When H_0 is a simple hypothesis this can always be done, and we usually set the probability of a Type I error, also called the **level of significance,** at $\alpha = 0.05$ or $\alpha = 0.01$. Testing a simple hypothesis at the 0.05 (or 0.01) level of significance simply means that we are fixing the probability of rejecting H_0 when it is true at 0.05 (or 0.01).

The decision to use 0.05, 0.01, or some other value depends mostly on the consequences of committing a Type I error. Although it may seem desirable to make the probability of a Type I error small, we cannot make it too small, since this would tend to make the probabilities of serious Type II errors too large, and make it difficult, perhaps too difficult, to get significant results. To some extent, the choice of 0.05 or 0.01, and not, say, 0.08 and 0.03, is dictated by the availability of statistical tables. However, with the general availability of computers and various kinds of statistical calculators, this restriction no longer applies.

There are situations where we cannot, or do not want to, specify the probability of a Type I error. This could happen when we do not have enough information about the consequences of Type I errors, or when one person processes the data while another person makes the decisions. What can be done in that case is discussed on page 347.

After the null hypothesis, the alternative hypothesis, and the probability of a Type I error have been specified, the next step is

3. Based on the sampling distribution of an appropriate statistic, we construct a criterion for testing the null hypothesis against the chosen alternative hypothesis at the specified level of significance.

Note that in the response-time example we interchanged steps 2 and 3. First we specified the criterion and then we calculated the probability of a Type I error, but that is not what we do in actual practice. Finally,

4. We calculate the value of the statistic on which the decision is to be based.

and

5. We decide whether to reject the null hypothesis, whether to accept it, or whether to reserve judgment.

In the response-time example we rejected the null hypothesis $\mu = 0.38$ second for values of $\bar{x}$ less than or equal to 0.36 and also for values of $\bar{x}$ greater than or equal to 0.40. Such a criterion is referred to as a **two-sided criterion,** which goes here with the two-sided alternative hypothesis $\mu \neq 0.38$ second. In the traffic-ticket example we rejected the null hypothesis $\mu = 0.9$ ticket for values of $\bar{x}$ greater than or equal to 1.2, and we refer to this criterion as a **one-sided criterion.** It went with the one-sided alternative hypothesis $\mu > 0.9$ ticket.

In general, a test is called a **two-sided test** or a **two-tailed test** if the criterion on which it is based is two sided; namely, if the null hypothesis is rejected for values of the **test statistic** falling into either tail of its sampling distribution. Correspondingly, a test is called a **one-sided test** or a **one-tailed test** if the criterion on which it is based is one sided; namely, if the null hypothesis is rejected for values of the test statistic falling into one specified tail of its sampling distribution. By "test statistic" we mean the statistic (for instance, the sample mean) on which the test is based. Although there are exceptions, two-tailed tests are usually used in connection with two-sided alternative hypotheses, and one-tailed tests are usually used in connection with one-sided alternative hypotheses.

As part of the third step we must also specify whether the alternative to rejecting the null hypothesis is to accept it or to reserve judgment. This, as we have said, depends on whether we must make a decision one way or the other, or whether the circumstances permit that we delay a decision pending further study. In exercises and examples, the phrase "whether or not" will sometimes be used to indicate that a decision must be reached one way or the other.

In connection with the fifth step, let us point out that we often accept null hypotheses with the tacit hope that we are not exposed to overly high risks of committing serious Type II errors. Of course, if necessary we can calculate enough probabilities of Type II errors to get an overall picture from the operating characteristic curve of the test criterion.

Before we consider various special tests for means in the remainder of this chapter, let us point out that the concepts we have introduced here are not limited to tests concerning population means; they apply equally to tests concerning other parameters, or tests concerning the nature, or form, of populations.

■ Exercises

12.1   An ambulance service is considering replacing its ambulances with new equipment. If μ_0 is the average weekly maintenance cost of one of the old ambulances and μ is the average weekly maintenance cost it can expect for one of the new ones, it wants to test the null hypothesis $\mu = \mu_0$.
  (a) What alternative hypothesis should it use if it wants to buy the new ambulances only if it can be shown that this will reduce the average weekly maintenance cost?
  (b) What alternative hypothesis should it use if it is anxious to buy the new ambulances (which have some other nice features) unless it can be shown that they will actually increase the average weekly maintenance cost?

12.2   A large restaurant has a waiter whom the manager suspects of making on the average more mistakes than all of its other waiters. If μ_0 is the average daily number of mistakes made by all the other waiters and μ is the daily average number of mistakes made by the waiter who is under suspicion, the manager of the restaurant wants to test the null hypothesis $\mu = \mu_0$.

(a) If the manager of the restaurant has decided to let the waiter go only if the suspicion is confirmed, what alternative hypothesis should she use?

(b) If the manager of the restaurant has decided to let the waiter go unless he actually averages fewer mistakes than all the other waiters, what alternative hypothesis should she use?

12.3 Rework Example 12.2, supposing that the mean of the psychologist's sample is $\bar{x} = 0.365$ second.

12.4 A botanist wants to test the null hypothesis that the mean diameter of the flowers of a certain plant is 8.5 cm. He decides to take a random sample and to accept the null hypothesis if the mean of the sample falls between 8.2 cm and 8.8 cm. If the mean of the sample is less than or equal to 8.2 cm or greater than or equal to 8.8 cm, he will reject the null hypothesis and otherwise he will accept it. What decision will he make and will it be in error if
(a) $\mu = 8.5$ cm and he gets $\bar{x} = 9.1$ cm;
(b) $\mu = 8.5$ cm and he gets $\bar{x} = 8.3$ cm;
(c) $\mu = 8.7$ cm and he gets $\bar{x} = 9.1$ cm;
(d) $\mu = 8.7$ cm and he gets $\bar{x} = 8.3$ cm?

12.5 Suppose that a psychological testing service is asked to check whether an executive is emotionally fit to assume the presidency of a large corporation. What type of error would it commit if it erroneously rejects the null hypothesis that the executive is fit for the job? What type of error would it commit if it erroneously accepts the null hypothesis that the executive is fit for the job?

12.6 Suppose we want to test the null hypothesis that an antipollution device for cars is effective. Explain under what conditions we would commit a Type I error and under what conditions we would commit a Type II error.

12.7 Whether an error is a Type I error or a Type II error depends on how we formulate the null hypothesis. To illustrate this, rephrase the null hypothesis of Exercise 12.6 so that the Type I error becomes a Type II error, and vice versa.

12.8 For a given population with $\sigma = 21.0$ cm we want to test the null hypothesis $\mu = 260$ cm against the alternative hypothesis $\mu \neq 260$ cm on the basis of a random sample of size $n = 36$. If the null hypothesis is rejected when $\bar{x}$ is less than or equal to 253 cm or greater or equal to 267 cm and otherwise it is accepted, find
(a) the probability of a Type I error;
(b) the probability of a Type II error when $\mu = 263.5$ cm.

12.9 For a given population with $\sigma = \$12$, we want to test the null hypothesis $\mu = \$75$ on the basis of a random sample of size $n = 100$. If the null hypothesis is rejected when $\bar{x}$ is greater than or equal to $\$76.50$ and otherwise it is accepted, find
(a) the probability of a Type I error;
(b) the probability of a Type II error when $\mu = \$75.3$;
(c) the probability of a Type II error when $\mu = \$77.22$.

12.10 With reference to the operating characteristic curve of Figure 12.3, verify that the probabilities of Type II errors are
(a) 0.78 when $\mu = 0.37$ second or $\mu = 0.39$ second;
(b) 0.50 when $\mu = 0.36$ second or $\mu = 0.40$ second;
(c) 0.06 when $\mu = 0.34$ second or $\mu = 0.42$ second.

12.11 Suppose that in the response-time example the criterion is changed so that the null hypothesis $\mu = 0.38$ second is rejected if the sample mean is less than or equal to 0.355 or greater than or equal to 0.405; otherwise, the null hypothesis is accepted. The sample size is still $n = 40$ and the population standard deviation is still $\sigma = 0.08$.

(a) How does this affect the probability of a Type I error?

(b) How does this affect the probability of a Type II error when $\mu = 0.41$?

12.12 Suppose that in the response-time example the sample size is increased to $n = 60$, while everything else remains unchanged.

(a) How does this affect the probability of a Type I error?

(b) How does this affect the probability of a Type II error when $\mu = 0.41$ second?

12.13 In a study designed to compare the IQ of persons of Hispanic and Asian ancestry, an educator gets a substantial difference between the respective sample means, and she concludes that the difference between the corresponding population means is statistically significant. Comment.

12.14 The mean age of Mr. and Mrs. Miller's three children is 15.9 years while the mean age of Mr. and Mrs. Brown's children is 12.8 years. Does it make any sense to ask whether the difference between these two means is significant?

12.15 In a certain experiment, a null hypothesis is rejected at the 0.05 level of significance. Does this mean that the probability is at most 0.05 that the null hypothesis is true?

12.16 In a study of extrasensory perception, 280 persons were asked to predict patterns on cards drawn at random from a deck. If two of them scored better than could be expected at the 0.01 level of significance, comment on the conclusion that these two persons must have extraordinary powers.

12.17 During the production of spring-loaded postal scales, samples are obtained at regular intervals of time to check at the 0.05 level of significance whether the production process is under control. Is there cause for alarm if in 80 such samples the null hypothesis that the production process is under control is rejected

(a) three times;

(b) seven times?

12.18 It has been claimed that on the average 2.6 workers are absent from an assembly line. If an efficiency expert is asked to put this to a test, what null hypothesis and what alternative hypothesis should he use?

12.19 With reference to Exercise 12.18, would the efficiency expert use a one-tailed test or a two-tailed test if he is going to base his decision on the mean of a random sample?

12.20 The manufacturer of a blood pressure medication claims that on the average the medication will lower a person's blood pressure by more than 20 mm. If a medical team suspects this claim, what null hypothesis and what alternative hypothesis should it use to put this to a test?

12.21 With reference to Exercise 12.20, would the medical team use a one-tailed test or a two-tailed test if it intends to base its decision on the mean of a random sample?

12.22 A production-line inspector is concerned that the average "fill" going into boxes of cornflakes might not be the desired 24 ounces. Performing an appropriate statistical test based on a sample, she concludes that μ is significantly different from 24 ounces. Comment.

12.23 With reference to Exercise 12.22, suppose that the inspector's sample has the mean $\bar{x} = 24.02$ ounces, and since it enabled her to reject the null hypothesis $\mu = 24$ ounces, she orders expensive adjustments in the machinery. Comment.

12.24 Suppose that an unscrupulous manufacturer wants "scientific proof" that a totally useless chemical additive will improve the mileage yield of gasoline.

(a) If a research group performs one experiment to investigate the additive, using the 0.05 level of significance, what is the probability that they will come up with "significant results" (which the manufacturer can use to promote the additive even though it is totally ineffective)?

(b) If two independent research groups investigate the additive, both using the 0.05 level of significance, what is the probability that at least one of them will come up with "significant results," even though the additive is totally ineffective?

(c) If 32 independent research groups investigate the additive, with all of them using the 0.05 level of significance, what is the probability that at least one of them will come up with "significant results," even though the additive is totally ineffective?

12.25 Suppose that a manufacturer of pharmaceuticals would like to find a new ointment to reduce swellings. It tries 20 different medications and tests for each one whether it reduces swellings at the 0.10 level of significance.

(a) What is the probability that at least one of them will "prove effective," even though all of them are totally useless?

(b) What is the probability that more than one will "prove effective," even though all of them are totally useless?

12.3 TESTS CONCERNING MEANS

Having used tests concerning means to illustrate the basic principles of hypothesis testing, let us now demonstrate how we proceed in practice. Actually, we shall depart somewhat from the procedure used in Sections 12.1 and 12.2. In the response-time example as well as in the traffic-ticket example we stated the test criterion in terms of $\bar{x}$—in the first case we rejected the null hypothesis for $\bar{x} \leq 0.36$ or $\bar{x} \geq 0.40$, and in the second case we rejected it for $\bar{x} \geq 1.2$. Now we shall base it on the statistic

Statistic for test
concerning mean

$$z = \frac{\bar{x} - \mu_0}{\sigma/\sqrt{n}}$$

where μ_0 is the value of the mean assumed under the null hypothesis. The reason for working with standard units, or z-values, is that it enables us to formulate criteria that are applicable to a great variety of problems, not just one.

The test of this section is essentially a **large-sample test;** that is, we require that the samples are large enough, $n \geq 30$, so that the sampling distribution of the mean can be approximated closely with a normal distribution and z is a value of a random variable having approximately the standard normal distribution. (In the special case where we sample a normal population, z is a value of a random variable having the standard normal distribution regardless of the size of n.) Alternatively, we refer to this test as a **z test** or as a **one-sample z test** to distinguish it from the test we shall discuss in Section 12.5. Sometimes the z test is referred to as a test concerning a mean with σ known to emphasize this essential feature.

Thus using z-values (standard units), we can base tests of the null hypothesis $\mu = \mu_0$ on the test criteria shown in Figure 12.4. Depending on the alternative hypotheses, the dividing lines of the test criteria, also called the **critical values,** are $-z_\alpha$ or z_α for the one-sided alternatives and $-z_{\alpha/2}$ and $z_{\alpha/2}$ for the two-sided

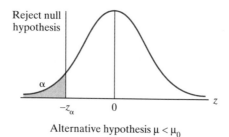

Alternative hypothesis $\mu < \mu_0$

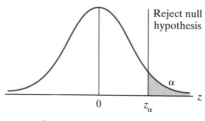

Alternative hypothesis $\mu > \mu_0$

FIGURE 12.4
Test criteria for z test
concerning a
population mean.

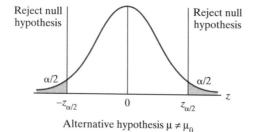

Alternative hypothesis $\mu \neq \mu_0$

alternative. As before, z_α and $z_{\alpha/2}$ are such that the area to their right under the standard normal distribution is α and $\alpha/2$. Symbolically, we can formulate these test criteria as follows:

Alternative hypothesis	Reject the null hypothesis if	Accept the null hypothesis or reserve judgment if
$\mu < \mu_0$	$z \leq -z_\alpha$	$z > -z_\alpha$
$\mu > \mu_0$	$z \geq z_\alpha$	$z < z_\alpha$
$\mu \neq \mu_0$	$z \leq -z_{\alpha/2}$ or $z \geq z_{\alpha/2}$	$-z_{\alpha/2} < z < z_{\alpha/2}$

If the level of significance is 0.05, the dividing lines are -1.645 or 1.645 for the one-sided alternatives, and -1.96 and 1.96 for the two-sided alternative; if the level of significance is 0.01, the dividing lines are -2.33 or 2.33 for the one-sided alternatives, and -2.575 and 2.575 for the two-sided alternative. All these values come directly from Table I.

EXAMPLE 12.4   An oceanographer wants to test, on the basis of the mean of a random sample of size $n = 35$ and at the 0.05 level of significance, whether the average depth of the ocean in a certain area is 72.4 fathoms, as has been recorded. What will she decide if she gets $\bar{x} = 73.2$ fathoms, and she can assume from information gathered in similar studies that $\sigma = 2.1$ fathoms?

Solution

1. H_0: $\mu = 72.4$ fathoms
   H_A: $\mu \neq 72.4$ fathoms
2. $\alpha = 0.05$
3. Reject the null hypothesis if $z \leq -1.96$ or $z \geq 1.96$, where

$$z = \frac{\bar{x} - \mu_0}{\sigma/\sqrt{n}}$$

   and otherwise accept it (or reserve judgment).
4. Substituting $\mu_0 = 72.4$, $\sigma = 2.1$, $n = 35$, and $\bar{x} = 73.2$ into the formula for z, we get

$$z = \frac{73.2 - 72.4}{2.1/\sqrt{35}} \approx 2.25$$

5. Since $z = 2.25$ exceeds 1.96, the null hypothesis must be rejected; to put it another way, the difference between $\bar{x} = 73.2$ and $\mu = 72.4$ is significant.   ∎

If the oceanographer had used the 0.01 level of significance in this example, she would not have been able to reject the null hypothesis because $z = 2.25$ falls between -2.575 and 2.575. This illustrates the importance of specifying the level of significance before any calculations have actually been made. This will spare us the temptation of later choosing a level of significance that happens to suit our purpose.

In problems like this, we often accompany the calculated value of the test statistic with the corresponding **p-value;** namely, with the probability of getting a difference between $\bar{x}$ and μ_0 that is numerically greater than or equal to the one which is actually observed. For instance, in Example 12.4 the p-value is given by the total area under the standard normal curve to the left of $z = -2.25$, and to the right of $z = 2.25$, and Table I tells us that it equals $2(0.5000 - 0.4878) = 0.0244$. This practice is not new by any means, but it has been advocated more widely in recent years in view of the general availability of computers. For many distributions, computers can provide p-values that are not directly available from tables.

Quoting p-values is the method referred to on page 341 for problems where we cannot, or do not want to, specify the level of significance. This applies, for example, to problems in which we study a set of data without having to reach a decision, or when we process a set of data to enable someone else to make a decision. p-values are provided by just about all statistical software and also by graphing calculators, making it unnecessary to compare results with tabular values and making it possible to use levels of significance for which critical values are not tabulated. Of course, if it is necessary to make a decision, we still have the responsibility to specify the level of significance before we collect (or look at) the data.

In general, p-values may be defined as follows:

Corresponding to an observed value of a test statistic, the p-value is the lowest level of significance for which the null hypothesis could have been rejected.

In Example 12.4 the p-value was 0.0244 and we could have rejected the null hypothesis at the 0.0244 level of significance. Of course, we could have rejected it for any level of significance greater than that, as we did for $\alpha = 0.05$.

If we want to base tests of significance on p-values instead of critical values obtained from tables, steps 1 and 2 remain the same, but steps 3, 4, and 5 must be modified as follows:

3′. We specify the test statistic.

4′. We calculate the value of the specified test statistic and the corresponding p-value from the sample data.

5′. We compare the p-value obtained in step 4′ with the level of significance specified in step 2. If the p-value is less than or equal to the level of significance, the null hypothesis must be rejected; otherwise, we accept the null hypothesis or reserve judgment.

EXAMPLE 12.5   Rework Example 12.4, using the p-value rather than the critical value approach.

Solution   Steps 1 and 2 remain the same as in Example 12.4, but steps 3, 4, and 5 are replaced by the following:

3′. The test statistic is

$$z = \frac{\bar{x} - \mu_0}{\sigma/\sqrt{n}}$$

4′. Substituting $\mu_0 = 72.4$, $\sigma = 2.1$, $n = 35$, and $\bar{x} = 73.2$ into the formula for z, we get

$$z = \frac{73.2 - 72.4}{2.1/\sqrt{35}} \approx 2.25$$

and from Table I we find that the p-value, the area under the curve to the left of -2.25 and to the right of 2.25, is $2(0.5000 - 0.4878) = 0.0244$.

5′. Since 0.0244 is less than $\alpha = 0.05$, the null hypothesis must be rejected. ∎

As we have indicated previously the p-value approach can be used to advantage when we study data without having to reach a decision. To illustrate, consider the plight of a social scientist exploring the relationship between family economics and school performance. He could be testing hundreds of hypotheses involving dozens of variables. The work is very complicated, and there are no immediate policy consequences. In this situation, the social scientist can tabulate the tests of hypotheses according to their p-values. Those tests leading to the lowest p-values are the most provocative, and they will certainly be the subject of future discussion. The social scientist need not actually accept or reject the hypotheses, and the use of p-values furnishes a convenient alternative.

12.4   TESTS CONCERNING MEANS (σ unknown)

When we do not know the value of the population standard deviation, we must assume as in Section 11.2 that the population we are sampling has roughly the shape of a normal distribution. We can then base our decision on the statistic

Statistic for test concerning mean (σ unknown)

$$t = \frac{\bar{x} - \mu_0}{s/\sqrt{n}}$$

which is a value of a random variable having the t distribution (see page 308) with $n - 1$ degrees of freedom. (If the assumption about the population cannot be met and n is large, we can use instead the z test of Section 12.3 with s substituted for σ. If the assumption about the population cannot be met and n is small, $n < 30$, we may have to use one of the alternative tests described in Chapter 18.)

Tests based on the t statistic are referred to as **t tests,** and to distinguish between the one for tests concerning one mean and the one given in Section 12.6, we refer to the former as a **one-sample t test.** (Since most tables of critical values for the one-sample t test are limited to small numbers of degrees of freedom, small values of $n - 1$, the one-sample t test has also been referred to as a **small-sample test concerning means.** Of course, with easy access to computers and other technology, this distinction no longer applies.)

The criteria for the one-sample t test are very much like those shown in Figure 12.4 and in the table on page 346. Now, however, the curves represent t distributions instead of normal distributions, and z, z_α, and $z_{\alpha/2}$ are replaced by t, t_α, and $t_{\alpha/2}$. As defined on page 309, t_α and $t_{\alpha/2}$ are values for which the area to their right under the t distribution curve are α and $\alpha/2$. For relatively small numbers of degrees of freedom and α equal to $0.10, 0.05,$ and 0.01, the critical values may be obtained from Table II; for larger numbers of degrees of freedom and other values of α this requires appropriate computer software, a graphing calculator, or a special statistical calculator.

EXAMPLE 12.6   The yield of alfalfa from a random sample of six test plots is 1.4, 1.8, 1.1, 1.9, 2.2, and 1.2 tons per acre.

(a) Check whether these data can be looked upon as a sample from a normal population.

(b) If so, test at the 0.05 level of significance whether this supports the contention that the average yield for this kind of alfalfa is 1.5 tons per acre.

Solution

(a) This is difficult to decide for such a small sample, but the normal probability plot of Figure 12.5 shows no appreciable departure from linearity.

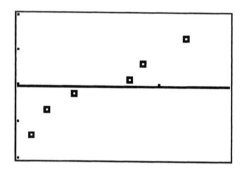

FIGURE 12.5
Normal probability plot for Example 12.6 reproduced from display screen of TI-83 graphing calculator.

(b) 1. H_0: $\mu = 1.5$
    H_A: $\mu \neq 1.5$
2. $\alpha = 0.05$
3. Reject the null hypothesis if $t \leq -2.571$ or $t \geq 2.571$, where

$$t = \frac{\bar{x} - \mu_0}{s/\sqrt{n}}$$

and 2.571 is the value of $t_{0.025}$ for $6 - 1 = 5$ degrees of freedom; otherwise, state that the data support the contention.

4. First, calculating the mean and the standard deviation of the given data, we get $\bar{x} = 1.6$ and $s = 0.434$. Then, substituting these values together with $n = 6$ and $\mu_0 = 1.5$ into the formula for t, we obtain

$$t = \frac{1.6 - 1.5}{0.434/\sqrt{6}} \approx 0.56$$

5. Since $t = 0.56$ falls on the interval from -2.571 to 2.571, the null hypothesis cannot be rejected; in other words, the data support the contention that the average yield of the given kind of alfalfa is 1.5 tons per acre.   ∎

Figure 12.6 shows a solution of Example 12.6 with the use of a graphing calculator. It confirms the values we obtained for $\bar{x}$, s, and t, and it shows that the p-value is 0.5965 rounded to four decimals. Since 0.5965 exceeds 0.05, we conclude, as before, that the null hypothesis cannot be rejected.

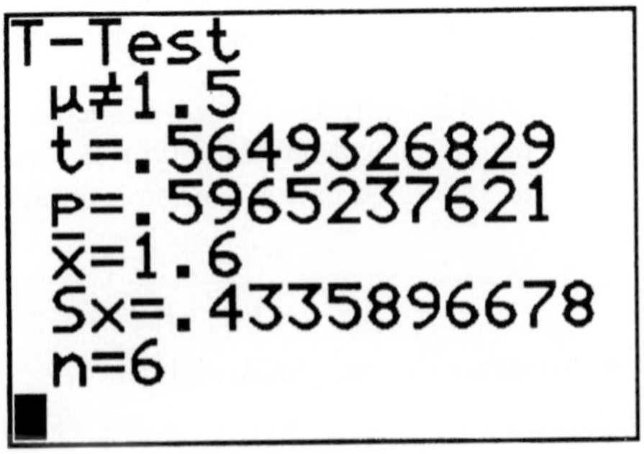

FIGURE 12.6
Solution of Example 12.6 reproduced from the display screen of a TI-83 graphing calculator.

```
T-Test
µ≠1.5
t=.5649326829
p=.5965237621
x̄=1.6
Sx=.4335896678
n=6
■
```

■ Exercises

12.26 A law student wants to check on her professor's claim that convicted embezzlers spend on the average 12.3 months in jail. So she decides to test the null hypothesis $\mu = 12.3$ against the alternative hypothesis $\mu \neq 12.3$ at the 0.05 level of significance, using a random sample of $n = 35$ such cases from court files. What will she conclude if she gets $\bar{x} = 11.5$ months and uses the five-step significance test described on pages 339 and 341, knowing that $\sigma = 3.8$ months.

12.27 Rework Exercise 12.26, using the p-value approach to reach the decision.

12.28 In a study of new sources of food, it is reported that a pound of a certain kind of fish yields on the average 3.52 ounces of FPC (fish-protein concentrate) used to enrich various food products, with the standard deviation $\sigma = 0.07$ ounce. To check whether $\mu = 3.52$ ounces is correct, a dietician decides to use the alternative hypothesis $\mu \neq 3.52$ ounces, a random sample of size $n = 32$, and the 0.05 level of significance. What will he conclude if he gets a sample mean of 3.55 ounces of FPC (per pound of fish)?

12.29 Rework Exercise 12.28 with the level of significance changed to 0.01.

12.30 With reference to Exercise 12.28, find the p-value corresponding to the observed sample mean, and use it to verify the conclusions arrived at in Exercises 12.28 and 12.29.

12.31 According to the norms established for a reading comprehension test, eighth graders should average 83.2 with the standard deviation $\sigma = 8.6$. A district superintendent feels that the eighth graders in his district are above average in reading comprehension, but he lacks proof. So he decides to test the null hypothesis $\mu = 83.2$ against a suitable alternative at the 0.01 level of significance, using the five-step format described on pages 339 and 341 and a random sample of 45 eighth graders from his district. What can he conclude if $\bar{x} = 86.7$?

12.32 Rework Exercise 12.31, using the p-value approach to arrive at the decision.

12.33 A horticulturist knows from experience that the honeybees visiting her orchard weigh 0.87 gram on the average with $\sigma = 0.15$ gram. Feeling that this year's honeybees look bigger, she decides to weigh a random sample of $n = 50$ of the bees all together and she gets an average weight of 0.91 gram per bee. Using the 0.01 level of significance, what can she conclude about her impressions that this year's bees are larger?

12.34 With reference to Exercise 12.33, could the horticulturist have rejected the null hypothesis at the 0.05 level of significance? Also, what is the lowest level of significance at which the horticulturist could have rejected the null hypothesis?

12.35 If we wish to test the null hypothesis $\mu = \mu_0$ in such a way that the probability of a Type I error is α and the probability of a Type II error is β for the specified alternative value $\mu = \mu_A$, we must take a random sample of size n, where

$$n = \frac{\sigma^2(z_\alpha + z_\beta)^2}{(\mu_A - \mu_0)^2}$$

if the alternative hypothesis is one sided, and

$$n = \frac{\sigma^2(z_{\alpha/2} + z_\beta)^2}{(\mu_A - \mu_0)^2}$$

if the alternative hypothesis is two sided.

Suppose that we want to test the null hypothesis $\mu = 540$ mm against the alternative hypothesis $\mu < 540$ mm for a population whose standard deviation is $\sigma = 88$ mm. How large a sample will we need if the probability of a Type I error is to be 0.05 and the probability of a Type II error is to be 0.01 when $\mu = 520$ mm? Determine also for what values of $\bar{x}$ the null hypothesis will be rejected.

12.36 Suppose we want to test the null hypothesis $\mu = 260$ pounds against the alternative hypothesis $\mu > 260$ pounds for all the athletes in a professional football league. If it can be assumed that $\sigma = 18$, how large a sample will we need if the probability of a Type I error is to be 0.05 and the probability of a Type II error is to be 0.05 when $\mu = 270$ pounds?

12.37 Suppose that we want to test the null hypothesis $\mu = \$650$ against the alternative hypothesis $\mu \neq \$650$ for a population whose standard deviation is $\sigma = \$26$. How large a sample will we need if the probability of a Type I error is to be 0.05 and the probability of a Type II error is to be 0.20 for $\mu = \$660$? Determine also for what values of $\bar{x}$ the null hypothesis will be rejected.

12.38 A random sample of $n = 12$ graduates of a secretarial school typed on the average $\bar{x} = 78.2$ words per minutes with a standard deviation of $s = 7.9$ words per minute. Assuming that such data can be looked upon as a random sample from a normal population, use the one-sample t test to test the null hypothesis $\mu = 80$ words per minute against the alternative hypothesis $\mu < 80$ words per minute for graduates of this secretarial school. Use the 0.05 level of significance.

12.39 A soft-drink vending machine, tested $n = 9$ times, yielded a mean cup fill of 6.2 ounces with a standard deviation of 0.15 ounce. Assuming that these data can be looked upon as a random sample from a normal population, test the null hypothesis $\mu = 6.0$ ounces against the alternative hypothesis $\mu > 6.0$ ounces at the 0.01 level of significance.

 12.40 Use appropriate computer software or a graphing calculator to find the p-value corresponding to the value of the test statistic obtained in Exercise 12.39.

12.41 A new tranquilizer given to $n = 16$ patients reduced their pulse rate on the average by 4.36 beats per minute with a standard deviation of 0.36 beat per minute. Assuming that such data can be looked upon as a random sample from a normal population, use the 0.10 level of significance to test the pharmaceutical company's claim that on the average its new tranquilizer reduces a patient's pulse rate by 4.50 beats per minute.

12.42 Use appropriate computer software or a graphing calculator to find the p-value corresponding to the value of the test statistic obtained in Exercise 12.41.

12.43 A large group of senior citizens enrolled for adult evening classes at a university. To get a quick check on whether there has been an increase from last year's average age of 65.4 years, the director of the program takes a random sample of 15 of the enrollees, getting 68, 62, 70, 64, 61, 58, 65, 86, 88, 62, 60, 71, 60, 84, and 61 years. Using these data, he wants to perform a one-sample t test to test the null hypothesis $\mu = 65.4$ against the alternative hypothesis $\mu > 65.4$.
   (a) Use appropriate computer software or a graphing calculator to see whether he can look upon the data as a sample from a normal population.
   (b) If so, perform the one-sample t test at the 0.05 level of significance.

12.44 Five measurements of the tar content of a certain kind of cigarette yielded 14.5, 14.2, 14.4, 14.8 and 14.1 mg/cg (milligrams per cigarette).
   (a) To check whether these data can be looked upon as a sample from a normal population, as required for the one-sample t test, use appropriate computer software or a graphing calculator to obtain a normal probability plot.
   (b) If there is no evident departure from normality, use the 0.05 level of significance to test the null hypothesis $\mu = 14.1$ mg/cg against the alternative hypothesis $\mu > 14.1$ mg/cg.

12.45 A random sample from a company's very extensive files shows that orders for a certain piece of machinery were filled in

11	9	8	11	9	10	9	10	10	12
10	13	10	10	11	10	10	10	12	10

days. Since a normal probability plot showed a distinct linear pattern, the one-sample t test can be used to test the null hypothesis $\mu = 9.5$ against the alternative hypothesis $\mu > 9.5$. Making use of the fact that $\bar{x} = 10.25$ and $s = 1.16$ for these data, perform the test at the 0.01 level of significance.

12.46 With reference to Exercise 12.45, use appropriate computer software or a graphing calculator to find the p-value corresponding to the value of the test statistic.

12.47 Forty mature dogs of a certain breed weighed

66.2	59.2	70.8	58.0	64.3	50.7	62.5	58.4
48.7	52.4	51.0	35.7	62.6	52.3	41.2	61.1
52.9	58.8	64.1	48.9	74.3	50.3	55.7	55.5
51.8	55.8	48.9	51.8	63.1	44.6	47.0	49.0
62.5	45.0	78.6	54.2	72.2	52.4	60.5	46.8

ounces. Since a normal probability plot showed a distinct linear pattern, the one-sample t test can be used to test the kennel club's claim that $\mu = 58.0$ ounces against the alternative hypothesis $\mu < 58.0$ ounces. Making use of the fact that $\bar{x} = 56.0$ ounces and $s = 9.12$ ounces for these data, perform this one-sample t test at the 0.05 level of significance. Use $t_{0.05} = 1.685$ for 39 degrees of freedom.

12.48 With reference to Exercise 12.47, use appropriate computer software or a graphing calculator to find the p-value corresponding to the value obtained for the test statistic.

12.5 DIFFERENCES BETWEEN MEANS

There are many problems in which we must decide whether an observed difference between two sample means can be attributed to chance, or whether it is indicative of the fact that the two samples came from populations with unequal means. For instance, we may want to know whether there really is a difference in the mean gasoline consumption of two kinds of cars, when sample data show that one kind averaged 24.6 miles per gallon while, under the same conditions, the other kind averaged 25.7 miles per gallon. Similarly, we may want to decide on the basis of sample data whether men can perform a certain task faster than women, whether one kind of ceramic insulator is more brittle than another, whether the average diet in one country is more nutritious than that in another country, and so on.

The method we shall use to test whether an observed difference between two sample means can be attributed to chance, or whether it is statistically significant, is based on the following theory: If $\bar{x}_1$ and $\bar{x}_2$ are the means of two independent random samples, then the mean and the standard deviation of the sampling distribution of the statistic $\bar{x}_1 - \bar{x}_2$ are

$$\mu_{\bar{x}_1 - \bar{x}_2} = \mu_1 - \mu_2 \quad \text{and} \quad \sigma_{\bar{x}_1 - \bar{x}_2} = \sqrt{\frac{\sigma_1^2}{n_1} + \frac{\sigma_2^2}{n_2}}$$

where μ_1, μ_2, σ_1 and σ_2 are the means and the standard deviations of the two populations sampled. It is customary to refer to the standard deviation of this sampling distribution as the **standard error of the difference between two means.**

By "independent" samples we mean that the selection of one sample is in no way affected by the selection of the other. Thus, the theory does not apply to "before and after" kinds of comparisons, nor does it apply, say, if we want to compare the daily caloric consumption of husbands and wives. A special method for comparing the means of dependent samples is explained in Section 12.7.

Then, if we limit ourselves to large samples, $n_1 \geq 30$ and $n_2 \geq 30$, we can base tests of the null hypothesis $\mu_1 - \mu_2 = \delta$, (delta, the Greek letter for lowercase d) on the statistic

Statistic for test concerning difference between two means

$$z = \frac{\bar{x}_1 - \bar{x}_2 - \delta}{\sqrt{\dfrac{\sigma_1^2}{n_1} + \dfrac{\sigma_2^2}{n_2}}}$$

which is a value of a random variable having approximately the standard normal distribution. Note that we obtained this formula for z by converting to standard units, namely, by subtracting from $\bar{x}_1 - \bar{x}_2$ the mean of its sampling distribution, which under the null hypothesis is $\mu_1 - \mu_2 = \delta$, and then dividing by the standard deviation of its sampling distribution.

Depending on whether the alternative hypothesis is $\mu_1 - \mu_2 < \delta$, $\mu_1 - \mu_2 > \delta$, or $\mu_1 - \mu_2 \neq \delta$, the criteria we use for the corresponding tests are shown in Figure 12.7.

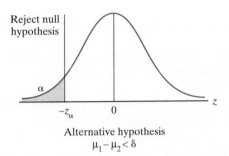

Alternative hypothesis
$\mu_1 - \mu_2 < \delta$

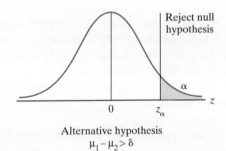

Alternative hypothesis
$\mu_1 - \mu_2 > \delta$

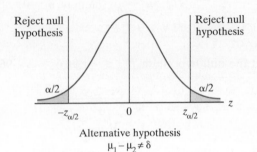

FIGURE 12.7
Test criteria for two-sample z test.

Alternative hypothesis
$\mu_1 - \mu_2 \neq \delta$

Note that these criteria are like the criteria of Figure 12.4 with $\mu_1 - \mu_2$ substituted for μ and δ substituted for μ_0. Analogous to the table on page 346, the criteria for tests of the null hypothesis $\mu_1 - \mu_2 = \delta$ are as follows:

Alternative hypothesis	Reject the null hypothesis if	Accept the null hypothesis or reserve judgment if
$\mu_1 - \mu_2 < \delta$	$z \leq -z_\alpha$	$z > -z_\alpha$
$\mu_1 - \mu_2 > \delta$	$z \geq z_\alpha$	$z < z_\alpha$
$\mu_1 - \mu_2 \neq \delta$	$z \leq -z_{\alpha/2}$ or $z \geq z_{\alpha/2}$	$-z_{\alpha/2} < z < z_{\alpha/2}$

Although δ can be any constant, it is worth noting that in the great majority of problems its value is zero, and we test the null hypothesis of "no difference," namely, the null hypothesis $\mu_1 - \mu_2 = 0$ (or simply $\mu_1 = \mu_2$).

The test we have described here, the **two-sample z test,** is essentially a large-sample test. It is exact only when both of the populations we are sampling are normal populations. Sometimes it is referred to as a **test concerning difference between means with σ_1 and σ_2 known** to emphasize this essential feature.

EXAMPLE 12.7   In a study to test whether or not there is a difference between the average heights of adult females in two different countries, random samples of size $n_1 = 120$ and $n_2 = 150$ yielded $\bar{x}_1 = 62.7$ inches and $\bar{x}_2 = 61.8$ inches. Extensive studies of a similar kind have shown that it is reasonable to let $\sigma_1 = 2.50$ inches and $\sigma_2 = 2.62$ inches. Test at the 0.05 level of significance whether the difference between these two sample means is significant.

Solution

1. In view of the "whether or not" in the formulation of the problem, we use

$$H_0: \quad \mu_1 = \mu_2 \text{ (namely, } \delta = 0)$$
$$H_A: \quad \mu_1 \neq \mu_2$$

2. $\alpha = 0.05$
3. Reject the null hypothesis if $z \leq -1.96$ or $z \geq 1.96$, where

$$z = \frac{\bar{x}_1 - \bar{x}_2 - \delta}{\sqrt{\dfrac{\sigma_1^2}{n_1} + \dfrac{\sigma_2^2}{n_2}}}$$

with $\delta = 0$; otherwise, accept the null hypothesis or reserve judgment.
4. Substituting $n_1 = 120$, $n_2 = 150$, $\bar{x}_1 = 62.7$, $\bar{x}_2 = 61.8$, $\sigma_1 = 2.50$, $\sigma_2 = 2.62$. and $\delta = 0$ into the formula for z, we get

$$z = \frac{62.7 - 61.8}{\sqrt{\dfrac{(2.50)^2}{120} + \dfrac{(2.62)^2}{150}}} \approx 2.88$$

5. Since $z = 2.88$ exceeds 1.96, the null hypothesis must be rejected; in other words, the difference between $\bar{x}_1 = 62.7$ and $\bar{x}_2 = 61.8$ is statistically significant. (Whether it is also of any practical significance, say, to a manufacturer of women's clothes, is another matter.) ∎

If the person who did this analysis had not been asked to make a decision, he or she would simply have reported that the p-value corresponding to the value of the test statistic is $2(0.5000 - 0.4980) = 0.0040$.

Let us add that there is a certain awkwardness about comparing means when the population standard deviations are unequal. Consider, for example, two normal populations with the means $\mu_1 = 50$ and $\mu_2 = 52$ and the standard deviations $\sigma_1 = 5$ and $\sigma_2 = 15$. Although the second population has a larger mean, it is much more likely to produce a value below 40, as can easily be verified. An investigator faced with a situation like this ought to decide whether the comparison of μ_1 and μ_2 really addresses whatever is of any relevance.

12.6   DIFFERENCES BETWEEN MEANS (σ's unknown)

Similar to what we did in Section 12.4, when σ_1 and σ_2 are unknown, we can base a significance test of the difference between two means on an appropriate t statistic. For this test, we must assume again that the populations we are sampling have roughly the shape of normal distributions. Moreover, we must assume that they have equal standard deviations; namely, that $\sigma_1 = \sigma_2$. Then, we base tests of the null hypothesis $\mu_1 - \mu_2 = \delta$, and $\mu_1 = \mu_2$ in particular, on the statistic

Statistic for test concerning difference between means (σ's unknown)

$$t = \frac{\bar{x}_1 - \bar{x}_2 - \delta}{s_p\sqrt{\dfrac{1}{n_1} + \dfrac{1}{n_2}}} \quad \text{where} \quad s_p = \sqrt{\frac{(n_1 - 1)s_1^2 + (n_2 - 1)s_2^2}{n_1 + n_2 - 2}}$$

which is a value of a random variable having the t distribution with $(n_1 - 1) + (n_2 - 1) = n_1 + n_2 - 2$ degrees of freedom. (If the assumptions about the populations cannot be met and the n's are large, we can use instead the z test of Section 12.5 with s_1 and s_2 substituted for σ_1 and σ_2. If the assumptions about the populations cannot be met and the n's are small, we may have to use one of the alternative tests described in Chapter 18.)

Tests based on this new t statistic are referred to as **two-sample t tests.** (Since most tables of critical values for the two-sample t test are limited to small numbers of degrees of freedom, small values of $n_1 + n_2 - 2$, the two-sample t test has also been referred to as a **small-sample test concerning the difference between means.** As before in connection with the one-sample t test, with easy access to computers and other technology, this distinction no longer applies.)

The criteria for the two-sample t test are very much like those shown in Figure 12.7 and in the table on page 356. As in connection with the one-sample t test, the curves now represent t distributions instead of normal distributions, and z, z_α, and $z_{\alpha/2}$ are replaced by t, t_α, and $t_{\alpha/2}$.

As can be seen from its definition, the calculation of the statistic for the two-sample t test consists of two steps. First we calculate the value of s_p, called the **pooled standard deviation,** which is an estimate of $\sigma_1 = \sigma_2$, the by-assumption-equal population standard deviations. Then we substitute it together with the $\bar{x}$'s and n's into the formula for t. The example that follows illustrates this procedure.

EXAMPLE 12.8   The following random samples are measurements of the heat-producing capacity (in millions of calories per ton) of specimens of coal from two mines:

Mine 1:	8,380	8,180	8,500	7,840	7,990
Mine 2:	7,660	7,510	7,910	8,070	7,790

Use the 0.05 level of significance to test whether the difference between the means of these two samples is significant.

Solution   Normal probability plots show that there is no reason to suspect the assumption that the data constitute samples from normal populations. Also, a test that will be described later shows in Example 13.3 that there is no reason to suspect the assumption that $\sigma_1 = \sigma_2$.

1. H_0: $\mu_1 = \mu_2$
   H_A: $\mu_1 \neq \mu_2$
2. $\alpha = 0.05$
3. Reject the null hypothesis if $t \leq -2.306$ or $t \geq 2.306$, where t is given by the formula on page 357 with $\delta = 0$, and 2.306 is the value of $t_{0.025}$ for $5 + 5 - 2 = 8$ degrees of freedom; otherwise, state that the difference between the means of the two samples is not significant.
4. The means and the standard deviations of the two samples are $\bar{x}_1 = 8,178$, $\bar{x}_2 = 7,788$, $s_1 = 271.1$, and $s_2 = 216.8$. Substituting the values of s_1 and s_2 together with $n_1 = n_2 = 5$ into the formula for s_p, we get

$$s_p = \sqrt{\frac{4(271.1)^2 + 4(216.8)^2}{8}} \approx 245.5$$

and, hence,

$$t = \frac{8,178 - 7,788}{245.5\sqrt{\dfrac{1}{5} + \dfrac{1}{5}}} \approx 2.51$$

5. Since $t = 2.51$ exceeds 2.306, the null hypothesis must be rejected; in other words, we conclude that the difference between the two sample means is significant.   ∎

```
MTB > TwoSample c1 c2;
SUBC>    Alternative 0;
SUBC>    Pooled.

Twosample T for C1 vs C2
       N       Mean       StDev
C1   5       8178         271
C2   5       7788         217
T-Test mu C1 = mu C2 (vs not =): T= 2.51   P=0.036   DF=  8
Both use Pooled StDev =   245
```

FIGURE 12.8
Computer printout for
Example 12.8.

Figure 12.8 is a MINITAB printout for Example 12.8, modified by deleting some details that are of no relevance to our example. It confirms our calculations, including the value we obtained for t, and it shows that the p-value corresponding to $t = 2.51$ (and the two-sided alternative hypothesis $\mu_1 \neq \mu_2$) is 0.036. Since this p-value is less than $\alpha = 0.05$, it reconfirms that the null hypothesis must be rejected.

12.7   DIFFERENCES BETWEEN MEANS (Paired data)

The methods of Sections 12.5 and 12.6 can be used only when the two samples are independent. Therefore, they cannot be used when we deal with "before and after" kinds of comparisons, the ages of husbands and wives, bank robber arrests and convictions in various jurisdictions, interest rates charged and paid by financial institutions, first-half and second-half pass completions by quarterbacks, cars stocked and cars sold by used-car dealers, and numerous other kinds of situations in which data are naturally paired. To handle this kind of data, we work with the (signed) differences between the pairs and test whether they can be looked upon as a random sample from a population with the mean $\mu = \delta$, usually $\mu = 0$. The tests we use for this purpose are the one-sample z test of Section 12.3 and the one-sample t test of Section 12.4, whichever is appropriate.

EXAMPLE 12.9   Following are the average weekly losses of man-hours due to accidents in ten industrial plants before and after the installation of an elaborate safety program:

45 and 36	73 and 60	46 and 44	124 and 119	33 and 35,
57 and 51	83 and 77	34 and 29	26 and 24	17 and 11

Use the 0.05 level of significance to test whether the safety program is effective.

Solution   The differences between the respective pairs are 9, 13, 2, 5, −2, 6, 6, 5, 2, and 6, and a normal probability plot (not displayed here) shows a distinct linear pattern. Thus, we can use the one-sample t test and proceed as follows:

1. H_0: $\mu = 0$
   H_A: $\mu > 0$ (The alternative is that on the average there were more accidents "before" than "after.")

2. $\alpha = 0.05$

3. Reject the null hypothesis if $t \geq 1.833$, where

$$t = \frac{\bar{x} - \mu_0}{s/\sqrt{n}}$$

and 1.833 is the value of $t_{0.05}$ for $10 - 1 = 9$ degrees of freedom; otherwise, accept the null hypothesis or reserve judgment (as the situation may demand).

4. First, calculating the mean and the standard deviation of the ten differences, we get $\bar{x} = 5.2$ and $s = 4.08$. Then, substituting these values together with $n = 10$ and $\mu_0 = 0$ into the formula for t, we have

$$t = \frac{5.2 - 0}{4.08/\sqrt{10}} \approx 4.03$$

5. Since $t = 4.03$ exceeds 1.833, the null hypothesis must be rejected; in other words, we have shown that the industrial safety program is effective.   ∎

When the one-sample t test is used in a problem like this, it is referred to as the **paired-sample t test.**

Exercises

12.49 Random samples showed that 40 executives in the insurance industry claimed on the average 9.4 business lunches as deductible expenses, while 50 bank executives claimed on the average 7.9 business lunches as deductible expenses. If, on the basis of collateral information, it can be assumed that $\sigma_1 = \sigma_2 = 3.0$ for such data, test at the 0.05 level of significance whether the difference between these two sample means is significant.

12.50 Rework Exercise 12.49, using the sample standard deviations, $s_1 = 3.3$ and $s_2 = 2.9$, instead of the assumed values of σ_1 and σ_2.

12.51 An investigation of two kinds of photocopying equipment showed that a random sample of 60 failures of one kind of equipment took on the average 84.2 minutes to repair, while a random sample of 60 failures of another kind of equipment took on the average 91.6 minutes to repair. If, on the basis of collateral information, it can be assumed that $\sigma_1 = \sigma_2 = 19.0$ minutes for such data, test at the 0.02 level of significance whether the difference between these two sample means is significant.

12.52 Rework Exercise 12.51, using the sample standard deviations, $s_1 = 19.4$ minutes and $s_2 = 18.8$ minutes, instead of the assumed value of σ_1 and σ_2.

12.53 With reference to Exercise 12.52, find the p-value corresponding to the value obtained for the z statistic.

12.54 To test the claim that the resistance of electric wire can be reduced by at most 0.050 ohm by alloying, a random sample of size $n_1 = 35$ of the standard wire yielded $\bar{x}_1 = 0.135$ ohm and a random sample of $n_2 = 35$ of the alloyed wire yielded $\bar{x}_2 = 0.082$ ohm. If, on the basis of collateral information, it can be assumed that $\sigma_1 = 0.006$ ohm and $\sigma_2 = 0.006$ ohm, what can we conclude about the claim if the probability of a Type I error is to be 0.01?

12.55 Rework Exercise 12.54, using the sample standard deviations, $s_1 = 0.004$ and $s_2 = 0.005$ instead of the assumed value of σ_1 and σ_2.

12.56 With reference to Exercise 12.54, find the p-value corresponding to the value obtained for the z statistic.

*12.57 If we substitute

$$ z = \frac{\bar{x}_1 - \bar{x}_2 - \delta}{\sqrt{\dfrac{\sigma_1^2}{n_1} + \dfrac{\sigma_2^2}{n_2}}} $$

into $-z_{\alpha/2} < z < z_{\alpha/2}$, and manipulate the inequality algebraically so that the middle term is δ, we obtain a $(1 - \alpha)100\%$ confidence interval for $\delta = \mu_1 - \mu_2$ that applies under the same conditions as the two-sample z test of Section 12.5. Use this method and the figures given in Exercise 12.51 to construct a 95% confidence interval for the difference between the true average repair times of the two kinds of photocopying equipment.

*12.58 Use the method of Exercise 12.57 and the figures given in Exercise 12.54 to construct a 98% confidence interval for the true average reduction due to alloying of the resistance of the electric wire.

12.59 Random samples of 12 measurements each of the hydrogen content (in percent number of atoms) of gases collected from the eruptions of two volcanos yielded $\bar{x}_1 = 41.2$, $\bar{x}_2 = 45.8$, $s_1 = 5.2$, and $s_2 = 6.7$. Assuming that the conditions underlying the two-sample t test can be met, decide at the 0.05 level of significance whether to accept or reject the null hypothesis that there is no difference in the mean hydrogen content of gases from the two eruptions.

 12.60 With reference to Exercise 12.59, determine the p-value corresponding the value obtained for the t statistic. Use it to determine whether the null hypothesis could have been rejected at the 0.10 level of significance.

12.61 In the comparison of two kinds of paint, a consumer testing service found that four 1-gallon cans of Brand A covered on the average 514 square feet with a standard deviation of 32 square feet, while four 1-gallon cans of Brand B covered on the average 487 square feet with a standard deviation of 27 square feet. Assuming that the conditions required for the two-sample t test can be met, test at the 0.02 level of significance whether the difference between these two sample means is significant.

 12.62 With reference to Exercise 12.61, what is the smallest level of significance at which the null hypothesis could have been rejected?

12.63 Six guinea pigs injected with 0.5 mg of a medication took on the average 15.4 seconds to fall asleep with a standard deviation of 2.2 seconds, while six other guinea pigs injected with 1.5 mg of the same medication took on the average 10.6 seconds to fall asleep with a standard deviation of 2.6 seconds. Assuming that the two samples are independent random samples and that the requirements for the two-sample t test can be met, test at the 0.05 level of significance whether or not this increase in dosage will, in general, reduce the average time it takes a guinea pig to fall asleep by 2.0 seconds.

12.64 Following are two computer-generated random samples from normal populations:

Sample 1.	46.0	34.1	33.2	41.3	34.9	36.2	37.3	45.5
	46.9	43.0	33.7	39.2	27.7	32.7	37.6	39.7
Sample 2.	37.1	40.7	41.0	51.2	45.2	46.1	31.0	39.2
	39.5	46.8	41.7	36.4	41.3	38.2	46.3	45.7

Use the two-sample t test to test the null hypothesis $\mu_1 = \mu_2$ against the alternative hypothesis $\mu_1 \neq \mu_2$ at the 0.05 level of significance.

 12.65 With reference to Exercise 12.64, use appropriate computer software or a graphing calculator to find the p-value corresponding to the value of the test statistic obtained in that exercise. Also, use this p-value to confirm the decision reached in Exercise 12.64.

12.66 Following are two computer-generated random samples of size $n = 12$ from normal populations:

Sample 1.	31.7	30.3	29.3	30.1	29.2	31.5
	28.9	31.1	27.6	29.3	32.0	30.9
Sample 2.	29.4	29.7	26.7	28.0	30.8	29.8
	28.3	28.6	28.2	29.4	31.2	29.8

Use the two-sample t test to test the null hypothesis $\mu_1 = \mu_2$ against the alternative hypothesis $\mu_1 \neq \mu_2$ at the 0.10 level of significance. State assumptions.

 12.67 With reference to Exercise 12.66, use appropriate computer software or a graphing calculator to find the p-value corresponding to the value of the test statistic obtained in that exercise. Also, use this p-value to confirm the result of Exercise 12.66.

12.68 Following are two computer-generated random sample of size $n_1 = 6$ and $n_2 = 8$ from normal populations:

Sample 1.	51.6	45.5	49.3	53.8	52.6	49.5		
Sample 2.	42.5	38.5	39.6	32.8	41.0	39.0	36.7	41.9

Use the two-sample t test to test the null hypothesis $\mu_1 - \mu_2 = 5$ against the alternative hypothesis $\mu_1 - \mu_2 > 5$ at the 0.025 level of significance. State assumptions.

 12.69 With reference to Exercise 12.68, use appropriate computer software or a graphing calculator to find the p-value corresponding to the value of the test statistics obtained in that exercise. Also, use this p-value to confirm the decision reached in Exercise 12.68.

12.70 To compare two kinds of bumper guards, ten of each kind were mounted on a certain make midsize car. Then each car was run into a concrete wall at 5 miles per hour, and the following are the costs of the repairs (in dollars):

Bumper guard A:	545	495	506	447	530
	510	487	539	559	531
Bumper guard B:	536	475	513	558	546
	514	517	473	562	529

Assuming that the assumptions underlying the two-sample t test can be met, use the 0.05 level of significance to test whether the difference between the corresponding sample means is significant.

12.71 Following are measurements of the wing span of two varieties of sparrows in millimeters:

Variety 1:	162	159	154	176	165	164	145	157	128
Variety 2:	147	180	153	135	157	153	141	138	161

Assuming that the conditions underlying the two-sample t test can be met, test at the 0.05 level of significance whether the difference between the means of these two random samples is significant.

12.72 Following are the gasoline mileages (per gallon) obtained with eight tankfuls each of two kinds of gasoline:

Gasoline A: 26.3   24.6   26.7   26.7   27.1   26.3   26.1   25.0

Gasoline B: 27.2   26.7   26.6   27.3   29.9   28.7   28.1   26.4

Assuming that the conditions underlying the two-sample t test can be met, test the null hypothesis $\mu_1 = \mu_2$ against the alternative hypothesis $\mu_1 < \mu_2$ at the 0.05 level of significance.

12.73 With reference to Exercise 12.72, use appropriate computer software or a graphing calculator to find the p-value corresponding to the value obtained for the test statistic in that exercise. Also, use this p-value to confirm the result obtained in Exercise 12.72.

12.74 A dietician wants to compare the fat content per serving of two kinds of ice cream. Following are measurements of the grams per serving in random samples of size $n_1 = n_2 = 7$:

Ice cream X: 22.5   21.9   20.4   20.7   19.9   21.1   22.2

Ice cream Y: 14.2   11.3   14.0   11.6   12.8   11.6   13.8

Assuming that the conditions underlying the two-sample t test can be met, test the null hypothesis $\mu_1 - \mu_2 = 10$ against the alternative hypothesis $\mu_1 - \mu_2 < 10$ at the 0.05 level of significance.

12.75 With reference to Exercise 12.74, use appropriate computer software or a graphing calculator to find the p-value corresponding to the value of the t statistic obtained in that exercise. Use this p-value to determine whether the null hypothesis of Exercise 12.74 could have been rejected at the 0.01 level of significance.

12.76 If we substitute the expression for t on page 357 into $-t_{\alpha/2} < t < t_{\alpha/2}$, and manipulate the inequality algebraically so that the middle term is δ, we obtain a $(1 - \alpha)100\%$ confidence interval for $\delta = \mu_1 - \mu_2$ that applies under the same conditions as the two-sample t test of Section 12.6. Use this method and the data of Example 12.8 to construct a 95% confidence interval for the difference between the average heat-producing capacities of coal from the two mines. (Actually, this work was deleted from the computer printout of Figure 12.8.)

12.77 Use the method of Exercise 12.76 and the data of Exercise 12.74 to construct a 99% confidence interval for the difference between the true average fat content of servings of the two kinds of ice cream.

12.78 The following data were obtained in an experiment designed to check whether there is a systematic difference in the weights (in grams) obtained with two scales:

Rock specimen	Scale I	Scale II
1	12.13	12.17
2	17.56	17.61
3	9.33	9.35
4	11.40	11.42
5	28.62	28.61
6	10.25	10.27
7	23.37	23.42
8	16.27	16.26
9	12.40	12.45
10	24.78	24.75

Assuming that the differences between the respective weights can be looked upon as a random sample from a normal population, test at $\alpha = 0.01$ whether to reject or accept the null hypothesis $\delta = 0$.

12.79 With reference to Exercise 12.78, use appropriate computer software or a graphing calculator to find the p-value corresponding to the value obtained for the t statistic in that exercise. Using this p-value, decide whether the null hypothesis of Exercise 12.78 could have been rejected at the 0.05 level of significance.

12.80 In a study of the effectiveness of physical exercise in weight reduction, a random sample of 36 persons engaged in a prescribed program of physical exercise for one month. The results (in pounds) were as follows:

Before	After	Before	After	Before	After
209	196	178	171	169	170
212	207	180	177	192	190
158	159	180	180	211	203
193	183	245	229	188	190
201	194	222	219	190	195
199	197	170	164	153	152
183	179	165	162	201	199
179	173	243	231	144	140
179	180	202	197	169	175
187	190	213	205	174	170
196	197	201	201	236	227
201	196	164	166	188	185

Using the one-sample z test with s substituted for σ, test at the 0.01 level of significance whether the program of physical exercise is effective in reducing weight.

12.81 Following are the ratings of two supervisors of the performance of a random sample of 24 employees on a scale from 1 to 25:

Supervisor A	Supervisor B	Supervisor A	Supervisor B
25	23	24	24
23	22	24	22
21	23	25	25
22	20	20	15
15	17	16	16
22	22	19	18
17	20	21	19
18	15	17	17
25	22	23	22
21	23	19	19
20	23	19	15
17	16	20	18

Assuming that the differences between the ratings can be looked upon as a random sample from a normal population, test at the 0.05 level of significance whether to accept or reject the null hypothesis $\delta = 0$.

 12.82 With reference to Exercise 12.81, use appropriate computer software or a graphing calculator to find the p-value corresponding to the value obtained for the t statistic in that exercise. Verify that it confirms the result obtained in Exercise 12.81.

12.8 Checklist of Key Terms *(With page references to their definitions)*

Alternative hypothesis, 332
Composite hypothesis, 337
Critical value, 345
Level of significance, 341
Null hypothesis, 332
OC-curve, 336
One-sample t test, 349
One-sample z test, 345
One-sided alternative, 339
One-sided criterion 342
One-sided test, 342
One-tailed test, 342
Operating characteristic curve, 336
Paired-sample t test, 360
Pooled standard deviation, 358
Power function, 336
p-value, 347, 348

Significance test, 338
Simple hypothesis, 337
Standard error of difference between means, 354
Statistical hypothesis, 332
Statistically significant, 338
Test of significance, 338
Test statistic, 342
Two-sample t test, 357
Two-sample z test, 356
Two-sided alternative, 339
Two-sided criterion 341
Two-sided test, 342
Two-tailed test, 342
Type I error, 335
Type II error, 336

12.9 References

Some easy reading on tests of hypotheses may be found in

BROOK, R. J., ARNOLD, G. C., HASSARD, T. H., and PRINGLE, R. M., eds., *The Fascination of Statistics*. New York: Marcel Dekker, Inc., 1986.

GONICK, L., and SMITH, W., *The Cartoon Guide to Statistics*. New York: HarperCollins Publishers, Inc., 1993.

A detailed treatment of significance tests, the choice of the level of significance, p-values, and so forth may be found in Chapters 26 and 29 of

FREEDMAN, D., PISANI, R., and PURVES, R., *Statistics*. New York: Norton & Company, Inc., 1978.

13

TESTS OF HYPOTHESES: STANDARD DEVIATIONS

13.1   Tests Concerning Standard Deviations 367

13.2   Tests Concerning Two Standard Deviations 370

13.3   Checklist of Key Terms 375

13.4   References 376

*I*n Chapter 12 we learned how to perform tests of hypotheses about means—the mean of one population and the means of two populations. These tests are useful statistical techniques, but even more important are the concepts on which they are based: statistical hypotheses, null hypotheses, and alternative hypotheses, Type I and Type II errors, tests of significance, level of significance, *p*-values, and above all the concept of statistical significance. Also of importance is the awareness of the assumptions on which such tests are based.

As we shall see here and in subsequent chapters, all these concepts carry over to tests about other population parameters, tests about the nature (or form) of populations, and even tests about the randomness of samples. In this chapter we shall concentrate on tests about population standard deviations. These tests are not only important in their own right, but they must sometimes be used before tests about other parameters can be performed. This was the case, for example, in connection with the two-sample *t* test, which requires that the populations sampled have equal standard deviations.

In this chapter, Section 13.1 will be devoted to tests about the standard deviation of one population, and Section 13.2 deals with tests about the standard deviations of two populations.

13.1 TESTS CONCERNING STANDARD DEVIATIONS

The tests we shall consider in this section concern the problem of whether a population standard deviation equals a specified constant σ_0. This kind of test may be required whenever we study the uniformity of a product, process, or operation: for instance, if we must judge whether a certain kind of glass is sufficiently homogeneous for making delicate optical equipment, whether the knowledge of a group of students is sufficiently uniform so that they can be taught in one class, whether a lack of uniformity in some workers' performance may call for stricter supervision, and so forth.

The test of the null hypothesis $\sigma = \sigma_0$, that a population standard deviation equals a specified constant, is based on the same assumption, the same statistic, and the same sampling theory as the confidence interval for σ^2 on page 318. Again assuming that we are dealing with a random sample from a normal population (or at least a population having roughly the shape of a normal distribution), we use the **chi-square statistic**

Statistic for test concerning standard deviation

$$\chi^2 = \frac{(n-1)s^2}{\sigma_0^2}$$

which is like the one on page 317 with σ replaced by σ_0. As before, the sampling distribution of this statistic is the chi-square distribution with $n-1$ degrees of freedom.

The test criteria are shown in Figure 13.1; depending on the alternative hypothesis, the critical values are $\chi^2_{1-\alpha}$ and χ^2_α for the one-sided alternatives, and they are $\chi^2_{1-\alpha/2}$ and $\chi^2_{\alpha/2}$ for the two-sided alternative. Symbolically, we can formulate these criteria for testing the null hypothesis $\sigma = \sigma_0$ as follows:

Alternative hypothesis	Reject the null hypothesis if	Accept the null hypothesis or reserve judgment if
$\sigma < \sigma_0$	$\chi^2 \leq \chi^2_{1-\alpha}$	$\chi^2 > \chi^2_{1-\alpha}$
$\sigma > \sigma_0$	$\chi^2 \geq \chi^2_\alpha$	$\chi^2 < \chi^2_\alpha$
$\sigma \neq \sigma_0$	$\chi^2 \leq \chi^2_{1-\alpha/2}$ or $\chi^2 \geq \chi^2_{\alpha/2}$	$\chi^2_{1-\alpha/2} < \chi^2 < \chi^2_{\alpha/2}$

The values of $\chi^2_{0.995}$, $\chi^2_{0.99}$, $\chi^2_{0.975}$, $\chi^2_{0.95}$, $\chi^2_{0.05}$, $\chi^2_{0.025}$, $\chi^2_{0.01}$, and $\chi^2_{0.005}$ are given in Table III at the end of the book for $1, 2, 3, \ldots$, and 30 degrees of freedom.

Alternative hypothesis $\sigma < \sigma_0$

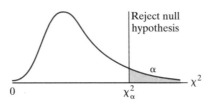

Alternative hypothesis $\sigma > \sigma_0$

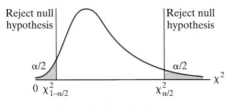

Alternative hypothesis $\sigma \neq \sigma_0$

FIGURE 13.1
Criteria for tests concerning standard deviations.

EXAMPLE 13.1   To judge certain safety features of a car, an engineer must know whether the reaction time of drivers to a given emergency situation has a standard deviation of 0.010 second, or whether it is greater than 0.010 second. What can she conclude at the 0.05 level of significance if she gets the following random sample of $n = 15$ reaction times?

0.32	0.30	0.31	0.28	0.30
0.31	0.28	0.31	0.29	0.28
0.30	0.29	0.27	0.29	0.29

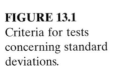 **Solution**   The normal probability plot of Figure 13.2 shows a distinct linear pattern, and hence justifies the assumption that the population sampled has the shape of a normal distribution.

1. H_0: $\sigma = 0.010$
   H_A: $\sigma > 0.010$

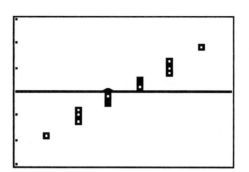

FIGURE 13.2
Normal probability plot for Example 13.1 reproduced from display screen of TI-83 graphing calculator.

2. $\alpha = 0.05$

3. Reject the null hypothesis if $x^2 \geq 23.685$, where

$$x^2 = \frac{(n-1)s^2}{\sigma_0^2}$$

and 23.685 is the value of $x_{0.05}^2$ for $15 - 1 = 14$ degrees of freedom; otherwise, accept it.

4. Calculating the standard deviation of the sample data, we get $s = 0.014$, and substituting this value together with $n = 15$ and $\sigma_0 = 0.010$ into the formula for x^2, we get

$$x^2 = \frac{14(0.014)^2}{(0.010)^2} \approx 27.44$$

5. Since $x^2 = 27.44$ exceeds 23.685, the null hypothesis must be rejected; in other words, the engineer can conclude that the standard deviation of the reaction time of drivers to the given emergency situation is greater than 0.010 second.   ∎

Since most tables of critical values for chi-square tests are limited to small numbers of degrees of freedom, the test we have described here has often been referred to as a **small-sample test concerning standard deviations.** As we have said before, with easy access to computers and other technology, this distinction no longer applies.

When n is large, $n \geq 30$, tests of the null hypothesis $\sigma = \sigma_0$ can be based on the same theory as the large-sample confidence interval for σ given in Section 11.3. That is, we use the statistic

Statistic for large-sample test concerning standard deviation

$$z = \frac{s - \sigma_0}{\sigma_0/\sqrt{2n}}$$

which is a value of a random variable having the standard normal distribution. Thus, the criteria for this large-sample test of the null hypothesis $\sigma = \sigma_0$ are like those shown in Figure 12.4 and in the table on page 346; the only difference is that μ and μ_0 are replaced by σ and σ_0.

EXAMPLE 13.2   The specifications for the mass production of certain springs require, among other things, that the standard deviation of their compressed lengths should not exceed 0.040 cm. If a random sample of size $n = 35$ from a certain production lot has $s = 0.053$, does this constitute evidence at the 0.01 level of significance for the null hypothesis $\sigma = 0.040$ or for the alternative hypothesis $\sigma > 0.040$?

Solution
1. H_0: $\sigma = 0.040$
   H_A: $\sigma > 0.040$
2. $\alpha = 0.01$
3. The null hypothesis must be rejected if $z \geq 2.33$, where

$$z = \frac{s - \sigma_0}{\sigma_0/\sqrt{2n}}$$

and otherwise it must be accepted.
4. Substituting $n = 35$, $s = 0.053$, and $\sigma_0 = 0.040$ into the formula for z, we get

$$z = \frac{0.053 - 0.040}{0.040/\sqrt{70}} \approx 2.72$$

5. Since $z = 2.72$ exceeds 2.33, the null hypothesis must be rejected; in other words, the data show that the production lot does not meet specifications.   ■

The p-value corresponding to $z = 2.72$ is $0.5000 - 0.4967 = 0.0033$, and since this is less than 0.01, it would also have led to the rejection of the null hypothesis.

13.2   TESTS CONCERNING TWO STANDARD DEVIATIONS

In this section we shall discuss tests concerning the equality of two standard deviations. Among other applications, it is often used in connection with the two-sample t test, where it has to be assumed that the two populations sampled have equal standard deviations. For instance, in Example 12.8, which dealt with the heat-producing capacity of coal from two mines, we had $s_1 = 271.1$ (millions of calories per ton) and $s_2 = 216.8$. Despite what may seem to be a large difference, we assumed that the corresponding population standard deviations were equal. Now we shall put this to a rigorous test.

Given independent random sample of size n_1 and n_2 from populations having roughly the shape of normal distributions and the standard deviations σ_1 and σ_2, we base tests of the null hypothesis $\sigma_1 = \sigma_2$ on the **F statistic**:

Statistics for test concerning the equality of two standard deviations

$$F = \frac{s_1^2}{s_2^2} \quad \text{or} \quad F = \frac{s_2^2}{s_1^2} \quad \text{depending on } H_A$$

where s_1 and s_2 are the corresponding sample standard deviations. Based on the assumption that the populations sampled have roughly the shape of normal distributions and the null hypothesis $\sigma_1 = \sigma_2$, it can be shown that such ratios, appropriately called **variance ratios,** are values of random variables having the **F distribution.** This continuous distribution depends on two parameters, called the **numerator and denominator degrees of freedom.** They are $n_1 - 1$ and $n_2 - 1$, or $n_2 - 1$ and $n_1 - 1$, depending on which of the two sample variances go into the numerator and the denominator of the F statistic.

If we based all the tests on the statistic

$$F = \frac{s_1^2}{s_2^2}$$

we could reject the null hypothesis $\sigma_1 = \sigma_2$ for $F \leq F_{1-\alpha}$ when the alternative hypothesis is $\sigma_1 < \sigma_2$ and for $F \geq F_\alpha$ when the alternative hypothesis is $\sigma_1 > \sigma_2$. In this notation, $F_{1-\alpha}$ and F_α are defined in the same way in which we defined the critical values $\chi^2_{1-\alpha}$ and χ^2_α for the chi-square distribution. Unfortunately, things are not as simple as that. Since there exists a fairly straightforward mathematical relationship between $F_{1-\alpha}$ and F_α (see Exercise 13.18), most F tables give only values corresponding to right-hand tails with α less than 0.50; for instance, Table IV at the end of the book contains only values of $F_{0.05}$ and $F_{0.01}$. For this reason, we use

$$F = \frac{s_2^2}{s_1^2} \quad \text{or} \quad F = \frac{s_1^2}{s_2^2}$$

depending on whether the alternative hypothesis is $\sigma_1 < \sigma_2$ or $\sigma_1 > \sigma_2$, and in either case we reject the null hypothesis for $F \geq F_\alpha$ (see Figure 13.3). When the

FIGURE 13.3
Criteria for tests
concerning the
equality of two
standard deviations.

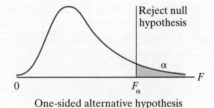

One-sided alternative hypothesis

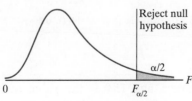

Two-sided alternative hypothesis

alternative hypothesis is $\sigma_1 \neq \sigma_2$, we use the greater of the two variance ratios,

$$F = \frac{s_1^2}{s_2^2} \quad \text{or} \quad F = \frac{s_2^2}{s_1^2}$$

and reject the null hypothesis for $F \geq F_{\alpha/2}$ (see Figure 13.3). In all these tests the degrees of freedom are $n_1 - 1$ and $n_2 - 1$, or $n_2 - 1$ and $n_1 - 1$, depending on which sample variance goes into the numerator and which one goes into the denominator. Symbolically, these criteria for testing the null hypothesis $\sigma_1 = \sigma_2$ are summarized in the following table:

Alternative hypothesis	Test statistic	Reject the null hypothesis if	Accept the null hypothesis or reserve judgment if
$\sigma_1 < \sigma_2$	$F = \dfrac{s_2^2}{s_1^2}$	$F \geq F_\alpha$	$F < F_\alpha$
$\sigma_1 > \sigma_2$	$F = \dfrac{s_1^2}{s_2^2}$	$F \geq F_\alpha$	$F < F_\alpha$
$\sigma_1 \neq \sigma_2$	*The larger of the two ratios*	$F \geq F_{\alpha/2}$	$F < F_{\alpha/2}$

The degrees of freedom are as indicated above.

EXAMPLE 13.3   In Example 12.8 we had $s_1 = 271.1$ and $s_2 = 216.8$ for two independent random samples of size $n_1 = 5$ and $n_2 = 5$, and in the solution we justified the assumption that the data constitute samples from normal populations. Use the 0.02 level of significance to test whether there is any evidence that the standard deviations of the two populations sampled are not equal.

Solution   Having already justified the assumption about normality, we proceed as follows:

1. H_0: $\sigma_1 = \sigma_2$
   H_A: $\sigma_1 \neq \sigma_2$
2. $\alpha = 0.02$
3. Reject the null hypothesis if $F \geq 16.0$, where

$$F = \frac{s_1^2}{s_2^2} \quad \text{or} \quad F = \frac{s_2^2}{s_1^2}$$

whichever is larger, and 16.0 is the value of $F_{0.01}$ for $5 - 1 = 4$ and $5 - 1 = 4$ degrees of freedom; otherwise, accept the null hypothesis.

4. Since $s_1 = 271.1$ and $s_2 = 216.8$, we substitute these values into the first of the two variance ratios and get

$$F = \frac{(271.1)^2}{(216.8)^2} \approx 1.56$$

5. Since $F = 1.56$ does not exceed 16.0, the null hypothesis cannot be rejected; there is no reason for not using the two-sample t test in Example 12.8. ∎

In a problem like this where the alternative hypothesis is $\sigma_1 \neq \sigma_2$, Table IV limits us to the 0.02 and 0.10 levels of significance. Had we wanted to use another level of significance, we would have had to refer to a more extensive table, a computer, or some other technology. A HEWLETT PACKARD STAT/MATH calculator yields the p-value 0.3386, so the result would have been the same for just about any reasonable level of significance.

EXAMPLE 13.4   It is desired to determine whether there is less variability in the gold plating done by company 1 than in the gold plating done by company 2. If independent random samples yielded $s_1 = 0.033$ mil (based on $n_1 = 12$) and $s_2 = 0.061$ mil (based on $n_2 = 10$), test the null hypothesis $\sigma_1 = \sigma_2$ against the alternative hypothesis $\sigma_1 < \sigma_2$ at the 0.05 level of significance.

Solution   Assuming that the populations sampled have roughly the shape of normal distributions, we proceed as follows.

1. H_0: $\sigma_1 = \sigma_2$
    H_A: $\sigma_1 < \sigma_2$
2. $\alpha = 0.05$
3. Reject the null hypothesis if $F \geq 2.90$, where

$$F = \frac{s_2^2}{s_1^2}$$

and 2.90 is the value of $F_{0.05}$ for $10 - 1 = 9$ and $12 - 1 = 11$ degrees of freedom; otherwise, accept the null hypothesis or reserve judgment.
4. Substituting $s_1 = 0.033$ and $s_2 = 0.061$ into the formula for F, we get

$$F = \frac{(0.061)^2}{(0.033)^2} \approx 3.42$$

5. Since $F = 3.42$ exceeds 2.90, the null hypothesis must be rejected; in other words, we conclude that there is less variability in the gold plating done by company 1. ∎

Since the procedure described in this section is very sensitive to departures from the underlying assumptions, it must be used with considerable caution. To put it differently, we say that the test is not **robust.**

Exercises

13.1 In a laboratory experiment, $s = 0.0086$ for $n = 10$ determinations of the specific heat of iron. Assuming that the population sampled has roughly the shape of a normal distribution, use the 0.05 level of significance to test the null hypothesis $\sigma = 0.0100$ for such determinations against the alternative hypothesis $\sigma < 0.0100$.

13.2 In a random sample of the amounts of time that $n = 18$ women took to complete the written test for their driver's licenses, the standard deviation was $s = 3.8$ minutes. Assuming that these data can be looked upon as a sample from a population having roughly the shape of a normal distribution, test the null hypothesis $\sigma = 2.7$ minutes against the alternative hypothesis $\sigma \neq 2.7$ minutes at the 0.01 level of significance.

13.3 Use appropriate technology to find the p-value corresponding to the test statistic obtained in Exercise 13.2, and use it to rework that exercise at the 0.03 level of significance.

13.4 Records show that the standard deviation of measurements made on sheet metal stampings by experienced inspectors is $\sigma = 0.41$ square inch. If a new inspector measures $n = 12$ stampings getting a standard deviation of 0.56 square inch, test at the 0.05 level of significance whether he is making satisfactory measurements or whether the variability of his measurements is excessive.

13.5 Use appropriate technology to find the p-value corresponding to the test statistic obtained in Exercise 13.4, and use it to rework that exercise at the 0.015 level of significance.

13.6 With reference to Exercise 12.39, where we had $n = 9$ and $s = 0.15$ ounce, assume that the data on which the exercise is based can be looked upon as a random sample from a normal population, and test the null hypothesis $\sigma = 0.10$ ounce against the alternative hypothesis $\sigma > 0.10$ ounce at the 0.01 level of significance.

13.7 In Exercise 12.44 we gave five measurements of the tar content of a certain kind of cigarette, for which it can be shown that $s = 0.27$ mg/cg. Assuming that the reader was able to verify in part (a) of that exercise that the data can be looked upon as a random sample from a normal population, test the null hypothesis $\sigma = 0.18$ against the alternative hypothesis $\sigma > 0.18$ at the 0.05 level of significance.

13.8 In Exercise 12.28 a dietician took a random sample of size $n = 32$ to investigate the yield of fish-protein concentrate in a pound of a certain kind of fish. Given that the standard deviation of her sample was $s = 0.095$, test the null hypothesis $\sigma = 0.07$ against the alternative hypothesis $\sigma \neq 0.07$ at the 0.01 level of significance.

13.9 In Exercise 12.31 we cited a reading comprehension test, where the norms included the standard deviation $\sigma = 8.6$ for the scores of eighth graders. If the $n = 45$ eighth graders referred to in that exercise had the standard deviation $s = 6.9$ and the district superintendent wants to test the null hypothesis $\sigma = 8.6$ against the alternative hypothesis $\sigma \neq 8.6$, what can he conclude at the 0.05 level of significance?

13.10 With reference to Exercise 12.33, where we had $n = 50$ and $\sigma = 0.15$ gram, suppose that the weights of the fifty honeybees had the standard deviation $s = 0.12$. What can the horticulturist conclude at the 0.05 level of significance if she tests the null hypothesis $\sigma = 0.15$ against the alternative hypothesis $\sigma < 0.15$?

13.11 With reference to Exercise 13.10, find the p-value corresponding to the value obtained for the z statistic, and use it to decide whether the null hypothesis could have been rejected at the 0.01 level of significance.

13.12 It has been reported that the constant annual growth of certain miniature fruit trees in their fifth to tenth years has the standard deviation $\sigma = 0.75$ inch. If a nursery owner got $s = 0.92$ for $n = 40$ such trees, test the null hypothesis $\sigma = 0.75$ against the alternative hypothesis $\sigma \neq 0.75$ at the 0.05 level of significance.

13.13 Two different lighting techniques are compared by measuring the intensity of light at selected locations by both methods. If $n_1 = 12$ measurements of the first technique have the standard deviation $s_1 = 2.6$ foot-candles, $n_2 = 16$ measurements of the second technique have the standard deviation $s_2 = 4.4$ foot-candles, and it can be assumed that both samples may be regarded as independent random samples from normal populations, test at the 0.05 level of significance whether the two lighting techniques are equally variable or whether the first technique is less variable than the second.

13.14 The amounts of time required by Dr. L. to do routine insurance checkups on $n_1 = 25$ patients have the standard deviation $s_1 = 4.2$ minutes, while the amounts of time required by Dr. M. to do the same procedure on $n_2 = 21$ patients have the standard deviation $s_2 = 3.0$ minutes. Assuming that these data constitute independent random samples from normal populations, test at the 0.05 level of significance whether the amounts of time required by these two doctors for this procedure are equally variable or whether they are more variable for Dr. L.

13.15 With reference to Exercise 12.70, test at the 0.02 level of significance whether there is any reason to doubt the assumption that $\sigma_1 = \sigma_2$ as required by the two-sample t test.

13.16 With reference to Exercise 12.71, test at the 0.10 level of significance whether there is any reason to doubt the assumption that $\sigma_1 = \sigma_2$ as required by the two-sample t test.

13.17 With reference to Exercise 12.72, test at the 0.02 level of significance whether there is any reason to doubt the assumption that $\sigma_1 = \sigma_2$ as required by the two-sample t test.

*13.18 If $F_{1-\alpha}$ is a critical value of F for n_1 and n_2 degrees of freedom and F_α is a critical value of F for n_2 and n_1 degrees of freedom, then

$$F_{1-\alpha} = \frac{1}{F_\alpha}$$

Note that the degrees of freedom are interchanged. If in Example 13.4 we had wanted to reject the null hypothesis for

$$F = \frac{s_1^2}{s_2^2} \leq F_{0.95}$$

what would have been the critical value of $F_{0.95}$?

13.3 Checklist of Key Terms *(with page references to their definitions)*

Chi-square statistic, 367
Denominator degrees of freedom, 371
F distribution, 371
F statistic, 370

Numerator degrees of freedom, 371
Robust, 373
Variance ratio, 371

13.4 References

Theoretical discussions of the chi-square and F distributions may be found in most textbooks on mathematical statistics; for instance, in

MILLER, I., and MILLER, M., *John E. Freund's Mathematical Statistics,* 6th ed. Upper Saddle River, N. J.: Prentice Hall, 1999.

For more detailed tables of the chi-square and F distributions, see, for example,

PEARSON, E. S., and HARTLEY, H. O., *Biometrika Tables for Statisticians,* Vol. I. New York: John Wiley & Sons, Inc., 1968.

14

TESTS OF HYPOTHESES BASED ON COUNT DATA

14.1   Tests Concerning Proportions 378

14.2   Tests Concerning Proportions (Large samples) 379

14.3   Differences Between Proportions 382

14.4   The Analysis of an $r \times c$ Table 386

14.5   Goodness of Fit 401

14.6   Checklist of Key Terms 408

14.7   References 408

*F*or the most part, Chapters 11 through 13 dealt with inferences based on measurements. We used measurements to estimate the means of populations and their standard deviations, and we also used them to test hypotheses about these parameters. The only exception was Section 11.4, where we used sample proportions to estimate population proportions, percentages, and probabilities. Such data were referred to as count data, since they are obtained by performing counts rather than measurements.

In this chapter we shall use count data in tests of hypotheses. Sections 14.1 and 14.2 deal with tests concerning proportions, which serve also as tests concerning percentages (proportions multiplied by 100) and as tests concerning probabilities (proportions in the long run). These tests are based on the observed number of successes in n trials, or the observed proportion of successes in n trials, and it will be assumed throughout that these trials are independent and that the probability of a success is the same for each trial. In other words, it will be assumed that we are testing hypotheses about the parameter p of binomial populations. In Sections 14.3 and 14.4 we shall study tests about two or more population proportions, and in Section 14.5 we shall generalize the discussion to the multinomial case, where there are more than two possible outcomes for each trial. This kind of problem might arise, for example, when the fuel economy of cars is rated low, average, or high, or when counts are grouped into a two-way classification pertaining, say, to professional advancement and education. Finally, Section 14.6 deals with the comparison of observed distributions and distributions that might be expected according to theory or assumptions.

14.1 TESTS CONCERNING PROPORTIONS

The tests we shall discuss here and in Section 14.2 make it possible, for example, to decide on the basis of sample data whether it is true that the proportion of tenth graders who can name the two senators of their state is only 0.28, whether it is true that 12% of the information supplied by the IRS to taxpayers is in error, or whether the probability is really 0.25 that a flight from Seattle to San Francisco will be late.

Whenever possible, such tests are based directly on tables of binomial probabilities, or on information about binomial probabilities obtained with the use of computers or other technology. Furthermore, these tests are simplest when using the p-value approach with steps 3′, 4′, and 5′ (see page 348), since this will greatly reduce the required amount of work.

EXAMPLE 14.1 It has been claimed that more than 70% of the students attending a large state university are opposed to a plan to increase student fees in order to build new parking facilities. If 15 of 18 students selected at random at that university are opposed to the plan, test the claim at the 0.05 level of significance.

Solution

1. H_0: $p = 0.70$
   H_A: $p > 0.70$
2. $\alpha = 0.05$
3′. The test statistic is the observed number of students in the sample who oppose the plan.
4′. The test statistic is $x = 15$, and Table V shows that the p-value, the probability of 15 or more "successes" for $n = 18$ and $p = 0.70$, is $0.105 + 0.046 + 0.013 + 0.002 = 0.166$.
5′. Since 0.166 is greater than 0.05, the null hypothesis cannot be rejected; in other words, the data do not support the claim that more than 70% of the students at the given university are opposed to the plan. ■

EXAMPLE 14.2 It has been claimed that 38% of all shoppers can identify a highly advertised trade mark. If, in a random sample, 25 of 45 shoppers were able to identify the trade mark, test at the 0.05 level of significance whether to accept or reject the null hypothesis $p = 0.38$.

```
MTB > CDF c1;
SUBC>    Binomial 45 .38.

Binomial with n = 45 and p = 0.380000

         x          P( X <= x)
      24.00           0.9875
      25.00           0.9944
```

FIGURE 14.1
Computer printout for
Example 14.2.

Solution   Since Table V does not give the binomial probabilities for $p = 0.38$ or $n > 20$, we could use the National Bureau of Standards table referred to on page 232 or appropriate technology.

> **1.** H_0: $p = 0.38$
> H_A: $p \neq 0.38$
> **2.** $\alpha = 0.05$
> **3'.** The test statistic is the observed number of shoppers in the sample who can identify the trade mark.
> **4'.** The test statistic is $x = 25$ and for this two-tailed test the p-value is twice the smaller of the probabilities for $x \leq 25$ and $x \geq 25$. Since the probability of $x \leq 25$ is 0.9944 according to Figure 14.1 and the probability of $x \geq 25$ is 1 minus the probability of $x \leq 24$, namely, $1 - 0.9875 = 0.0125$, the p-value is $2(0.0125) = 0.0250$.
> **5'.** Since 0.0250 is less than 0.05, the null hypothesis must be rejected. The correct percentage of shoppers who can identify the trade mark is not 38%; in fact, it is greater than 38%.   ∎

14.2   TESTS CONCERNING PROPORTIONS (Large samples)

When n is large enough to justify the normal-curve approximation to the binomial distribution, $np > 5$ and $n(1 - p) > 5$, tests of the null hypothesis $p = p_0$ can be based on the statistic

*Statistic for
large-sample test
concerning proportion*

$$z = \frac{x - np_0}{\sqrt{np_0(1 - p_0)}}$$

which is a value of a random variable having approximately the standard normal distribution. Since x is a discrete random variable, many statisticians prefer to make the continuity correction that represents x by the interval from $x - \frac{1}{2}$ to $x + \frac{1}{2}$, and hence use the alternative statistic

*Statistic for large-
sample test
concerning proportion
(with continuity
correction)*

$$z = \frac{x \pm \frac{1}{2} - np_0}{\sqrt{np_0(1 - p_0)}}$$

where the + sign is used when $x < np_0$ and the − sign is used when $x > np_0$. Note that the continuity correction does not even have to be considered when, without it, the null hypothesis cannot be rejected. Otherwise, it will have to be considered mainly when, without it, the value we obtain for z is very close to the critical value (or one of the critical values) of the test criterion.

The criteria for this large-sample test are again like those of Figure 12.4 with p and p_0 substituted for μ and μ_0. Analogous to the table on page 346, the criteria for tests of the null hypothesis $p = p_0$ are as follows:

Alternative hypothesis	Reject the null hypothesis if	Accept the null hypothesis or reserve judgment if
$p < p_0$	$z \leq -z_\alpha$	$z > -z_\alpha$
$p > p_0$	$z \geq z_\alpha$	$z < z_\alpha$
$p \neq p_0$	$z \leq -z_{\alpha/2}$ or $z \geq z_{\alpha/2}$	$-z_{\alpha/2} < z < z_{\alpha/2}$

EXAMPLE 14.3   To test a nutritionist's claim that at least 75% of the preschool children in a certain large county have protein-deficient diets, a sample survey revealed that 206 of 300 preschool children in that county had protein-deficient diets. Test the null hypothesis $p = 0.75$ against the alternative hypothesis $p < 0.75$ at the 0.01 level of significance.

Solution   Since $np = 300(0.75) = 225$ and $n(1 - p) = 300(0.25) = 75$ are both greater than 5, we can use the large-sample test based on the normal approximation to the binomial distribution.

1. H_0: $p = 0.75$
   H_A: $p < 0.75$
2. $\alpha = 0.01$
3. Reject the null hypothesis if $z \leq -2.33$, where

$$z = \frac{x - np_0}{\sqrt{np_0(1 - p_0)}}$$

otherwise, accept the null hypothesis or reserve judgment.
4. Substituting $x = 206$, $n = 300$, and $p_0 = 0.75$ into the formula for z, we get

$$z = \frac{206 - 300(0.75)}{\sqrt{300(0.75)(0.25)}} \approx -2.53$$

5. Since -2.53 is less than -2.33, the null hypothesis must be rejected. In other words, we conclude that less than 75% of the preschool children in that county have protein deficient diets. (Had we used the continuity correction, we would have obtained $z = -2.47$, and the conclusion would have been the same.) ∎

■ E x e r c i s e s

14.1 A medical research worker claims that only 10% of all persons exposed to a certain amount of radiation will feel any ill effects. If, in a random sample, 5 of 18 persons exposed to that much radiation feel any ill effects, test the null hypothesis $p = 0.10$ against the alternative hypothesis $p > 0.10$ at the 0.05 level of significance.

14.2 The manufacturer of a spot remover claims that her product removes 90% of all spots. If, in a random sample, the spot remover removes only 13 of 16 spots, is this sufficient evidence to refute her claim, namely, to reject the null hypothesis $p = 0.90$ against the alternative hypothesis $p < 0.90$ at the 0.05 level of significance?

14.3 A social scientist claims that among persons living in urban areas 50% are opposed to capital punishment (while the others are in favor of it or undecided). Test the null hypothesis $p = 0.50$ against the alternative hypothesis $p \neq 0.50$ at the 0.10 level of significance if in a random sample of $n = 20$ persons living in urban areas, 14 are against capital punishment.

14.4 In a study of aviophobia, a psychologist claims that 27% of all women are afraid of flying. If 18 women are interviewed, how few or how many of them must be afraid of flying so that the null hypothesis $p = 0.27$ can be rejected against the alternative hypothesis $p \neq 0.27$ at the 0.05 level of significance? Use the printout of the binomial distribution with $n = 18$ and $p = 0.27$ given in Figure 8.6.

14.5 A state highway official claims that at least 22% of all cars in the state do not meet requirements concerning brakes, headlights, direction signals, etc. Use the probabilities given in Figure 9.26 to test this claim at the 0.01 level of significance if a random sample of 30 cars included only 4 that did not meet all these requirements.

14.6 Suppose we want to test the "honesty" of a coin on the basis of the number of heads we will get in 20 flips of the coin. How few or how many heads will we have to get so that we can reject the null hypothesis $p = 0.50$ against the alternative hypothesis $p \neq 0.50$ at the
(a) 0.05 level of significance;
(b) 0.01 level of significance?

14.7 With reference to Example 14.1, where we tested the null hypothesis $p = 0.70$ against the alternative hypothesis $p > 0.70$ at the 0.05 level of significance, how many of the 18 students would have to be opposed to the increase in the student fees so that the null hypothesis could be rejected?

14.8 An airline claims that at most 5% of all of its lost luggage is never found. Use the binomial probabilities in Figure 8.4 to test the null hypothesis $p = 0.05$ against the alternative hypothesis $p > 0.05$ at the 0.05 level of significance if, in a random sample of $n = 150$ pieces of lost luggage, 13 were never found. Could we have used the large-sample test of Section 14.2 in this exercise?

14.9 In a random sample of 600 cars making right turns at a certain intersection, 157 pulled into the wrong lane. Test the claim that 30% of all drivers make this mistake, using the
(a) 0.05 level of significance;
(b) 0.01 level of significance.

14.10 In a sample survey of retired persons, 214 out of 400 stated that they prefer living in an apartment to living in a one-family home. Test the null hypothesis that the true proportion of retired persons who feel this way is 0.52 against the alternative hypothesis that this figure is incorrect one way or the other, using the 0.05 level of significance.

14.11 A committee investigating the problem claims that at least 36% of all accidents in elementary schools are due at least in part to improper supervision. If a random sample of 300 such accidents included 94 that were due at least in part to improper supervision, does this support the committee's claim? To answer this question, test the null hypothesis $p = 0.36$ against the alternative hypothesis $p < 0.36$ at the 0.05 level of significance, using
(a) the formula for z without the continuity correction;
(b) the formula for z with the continuity correction.

14.12 A food chemist claims that 10% of the bricks of a certain kind of cheddar cheese show signs of mold after three months under standard refrigeration. To check on this claim, 40 bricks of this cheese were kept for three months under standard refrigeration, and it was found that 9 of them showed signs of mold.
(a) Use a computer printout of the binomial distribution with $n = 40$ and $p = 0.10$ to test the null hypothesis $p = 0.10$ against the alternative hypothesis $p \neq 0.10$ at the 0.01 level of significance.
(b) Suppose that someone else checking on the food chemist's claim forgot about the condition for using the normal approximation to the binomial distribution and used the large-sample test (both with and without the continuity correction) instead of the test based on binomial probabilities. Would he have reached the same decision?

14.13 In the construction of tables of random numbers there are various ways of testing for possible departures from randomness. One of these consists of checking whether there are as many even digits (0, 2, 4, 6, or 8) as there are odd digits (1, 3, 5, 7, or 9). Thus, count the number of even digits among the 350 digits in the first 10 rows of the sample page of random numbers reproduced in Figure 10.2, and test at the 0.05 level of significance whether there is any significant sign of a lack of randomness.

14.14 For each of 500 simulated random samples, a statistics class determined a 95% confidence interval for the mean, and it found that only 464 of them contained the mean of the population sampled; the other 36 did not. At the 0.01 level of significance, is there any real evidence to doubt that the method employed yields 95% confidence intervals?

14.3 DIFFERENCES BETWEEN PROPORTIONS

After we presented tests concerning means in Chapter 12 and tests concerning standard deviations in Chapter 13, we learned how to perform tests concerning the means of two populations and tests concerning the standard deviations of two populations. Continuing this pattern, we shall now present a test concerning two population proportions.

There are many problems where we must decide whether an observed difference between two sample proportions can be attributed to chance, or whether it is indicative of the fact that the corresponding population proportions are not equal. For instance, we may want to decide on the basis of sample data whether there is a difference between the actual proportions of persons with and without flu shots who catch the disease, or we may want to test on the basis of samples whether two manufacturers of electronic equipment ship equal proportions of defectives.

The method we shall use here to test whether an observed difference between two sample proportions is significant, or whether it can be attributed to chance, is based on the following theory: If x_1 and x_2 are the numbers of successes obtained in n_1 trials of one kind and n_2 of another, the trials are all independent, and the corresponding probabilities of a success are, respectively, p_1 and p_2, then the sampling distribution of the difference

$$\frac{x_1}{n_1} - \frac{x_2}{n_2}$$

has the mean $p_1 - p_2$ and the standard deviation

$$\sqrt{\frac{p_1(1 - p_1)}{n_1} + \frac{p_2(1 - p_2)}{n_2}}$$

It is customary to refer to this standard deviation as the **standard error of the difference between two proportions.**

When we test the null hypothesis $p_1 = p_2 (= p)$ against an appropriate alternative hypothesis, the sampling distribution of the difference between two sample proportions has the mean $p_1 - p_2 = 0$, and its standard deviation can be written as

$$\sqrt{p(1 - p)\left(\frac{1}{n_1} + \frac{1}{n_2}\right)}$$

where p is usually estimated by **pooling** the data and substituting for p the combined sample proportion

$$\hat{p} = \frac{x_1 + x_2}{n_1 + n_2}$$

which, as before, reads "p-hat." Then, converting to standard units, we obtain the statistic

Statistic for test concerning difference between two proportions

$$z = \frac{\dfrac{x_1}{n_1} - \dfrac{x_2}{n_2}}{\sqrt{\hat{p}(1 - \hat{p})\left(\dfrac{1}{n_1} + \dfrac{1}{n_2}\right)}} \quad \text{with} \quad \hat{p} = \frac{x_1 + x_2}{n_1 + n_2}$$

which, for large samples, is a value of a random variable having approximately the standard normal distribution. To make this formula appear more compact, we can substitute in the numerator $\hat{p}_1$ for x_1/n_1 and $\hat{p}_2$ for x_2/n_2.

The test criteria are again like those shown in Figure 12.4 with p_1 and p_2 substituted for μ and μ_0. Analogous to the table on page 346, the criteria for tests of the null hypothesis $p_1 = p_2$ are as follows:

Alternative hypothesis	Reject the null hypothesis if	Accept the null hypothesis or reserve judgment if
$p_1 < p_2$	$z \leq -z_\alpha$	$z > -z_\alpha$
$p_1 > p_2$	$z \geq z_\alpha$	$z < z_\alpha$
$p_1 \neq p_2$	$z \leq -z_{\alpha/2}$ or $z \geq z_{\alpha/2}$	$-z_{\alpha/2} < z < z_{\alpha/2}$

EXAMPLE 14.4   A study showed that 56 of 80 persons who saw a spaghetti sauce advertised on television during a situation comedy and 38 of 80 other persons who saw it advertised during a football game remembered the brand name two hours later. At the 0.01 level of significance, what can we conclude about the claim that it is more cost effective to advertise this product during a situation comedy rather than during a football game? Assume that the cost of running the ad on the two programs is the same.

Solution

1. H_0: $p_1 = p_2$
   H_A: $p_1 > p_2$
2. $\alpha = 0.01$
3. Reject the null hypothesis if $z \geq 2.33$, where z is given by the formula on page 383; otherwise, accept it or reserve judgment.
4. Substituting $x_1 = 56$, $x_2 = 38$, $n_1 = 80$, $n_2 = 80$, and

$$\hat{p} = \frac{56 + 38}{80 + 80} = 0.5875$$

   into the formula for z, we get

$$z = \frac{\dfrac{56}{80} - \dfrac{38}{80}}{\sqrt{(0.5875)(0.4125)\left(\dfrac{1}{80} + \dfrac{1}{80}\right)}} \approx 2.89$$

5. Since $z = 2.89$ exceeds 2.33, the null hypothesis must be rejected; in other words, we conclude that advertising the spaghetti sauce during a situation comedy is more cost effective than advertising it during a football game. ∎

■ Exercises

14.15 One method of seeding clouds was successful in 57 of 150 attempts, while another method was successful in 33 of 100 attempts. At the 0.05 level of significance, can we conclude that the first method is better than the second?

14.16 One mail solicitation for a charity brought 412 responses to 5,000 letters and another, more expensive, mail solicitation brought 312 responses to 3,000 letters. Use the 0.01 level of significance to test the null hypothesis that the two solicitations are equally effective against the alternative that the more expensive one is more effective.

14.17 In a random sample of visitors to the Heard Museum in Phoenix, Arizona, 22 of 100 families from New England and 33 of 120 families from California purchased some Indian jewelry in the gift shop. Use the 0.05 level of significance to test the null hypothesis $p_1 = p_2$, that there is no difference between the corresponding population proportions, against the alternative hypothesis $p_1 \neq p_2$.

14.18 To test the effectiveness of a new seasickness remedy, a random sample of 60 passengers on a cruise ship were given a pill containing the medication and a random sample of 60 other passengers were given a placebo containing only sugar on the evening before anticipated rough seas. If 21 of the passengers in the first group and 30 of the passengers in the second group got seasick during the night, what can we conclude at the 0.01 level of significance about the effectiveness of the new remedy?

14.19 The service department of a Chrysler dealership offers fresh donuts to customers in its waiting room. When it supplied 12 dozen donuts from Bakery A, it found that 96 were eaten completely while the others were partially eaten and discarded. When it supplied 12 dozen donuts from Bakery B it found that 105 were eaten completely while the others were partially eaten and discarded. At the 0.05 level of significance test whether the difference between the corresponding sample proportions is significant.

14.20 A random sample of 100 high school students were asked whether they would turn to their parents for help with a homework assignment in mathematics, and another random sample of 100 high school students were asked the same question with regard to a homework assignment in English. If 62 students in the first sample and 44 students in the second sample would turn to their parents for help, test at the 0.05 level of significance whether the difference between the two sample proportions, $\frac{62}{100}$ and $\frac{44}{100}$, may be attributed to chance.

14.21 In a random sample of 200 marriage license applications recorded in 1987, 62 of the women were at least one year older than the men, and in a random sample of 300 marriage license applications recorded in 1997, 99 of the women were at least one year older than the men. At the 0.01 level of significance, is the upward trend statistically significant?

14.22 In a random sample of 250 persons who skipped breakfast, 102 reported that they experienced midmorning fatigue, and in a random sample of 250 persons who ate breakfast, 73 reported that they experienced midmorning fatigue. At the 0.05 level of significance, does this prove that midmorning fatigue is more prevalent among persons who skip breakfast?

14.23 To test the null hypothesis that the difference between two population proportions equals some constant δ (lowercase Greek *delta*), we can use the statistic

Test concerning difference between two proportions

$$z = \frac{\hat{p}_1 - \hat{p}_2 - \delta}{\sqrt{\dfrac{\hat{p}_1(1 - \hat{p}_1)}{n_1} + \dfrac{\hat{p}_2(1 - \hat{p}_2)}{n_2}}}$$

which, for large samples, is a value of a random variable having approximately the standard normal distribution. As before, $\hat{p}_1$ and $\hat{p}_2$ denote the sample proportions x_1/n_1 and x_2/n_2. The test criteria are again like those of Figure 12.7 with p_1 and p_2 substituted for μ_1 and μ_2.

In a true–false test, a test item is considered to be good if it discriminates between well-prepared and poorly prepared students. If 205 of 250 well-prepared students and 137 of 250 poorly prepared students answer a certain test item correctly, test at the 0.05 level of significance whether to accept the null hypothesis $p_1 - p_2 = 0.20$ or the alternative hypothesis $p_1 - p_2 > 0.20$.

14.24 A medical research worker wants to study how male and female rats react to the injection of a certain toxic substance. If 72 of 200 male rats and 48 of 200 female rats reacted strongly to the injection, use the method of Exercise 14.23 to test at the 0.01 level of significance whether the corresponding true percentage for male rats exceeds that for female rats by 10%.

14.4 THE ANALYSIS OF AN $r \times c$ TABLE

The method we shall describe in this section applies to several kinds of problems that differ conceptually but are analyzed in the same way. First let us consider a problem that is an immediate generalization of the kind of problem we studied in Section 14.3. Suppose that independent random samples of single, married, and widowed or divorced persons were asked whether "friends and social life" or "job or primary activity" contributes most to their general well-being, and that the results were as follows:

	Single	Married	Widowed or divorced
Friends and social life	47	59	56
Job or primary activity	33	61	44
Total	80	120	100

Here we have samples of size $n_1 = 80$, $n_2 = 120$, and $n_3 = 100$ from three binomial populations, and we want to determine whether the differences among the proportions of persons choosing "friends and social life" are statistically significant. The hypothesis we shall have to test is the null hypothesis $p_1 = p_2 = p_3$, where p_1, p_2, and p_3 are the corresponding true proportions for the three binomial populations. The alternative hypothesis will have to be that p_1, p_2, and p_3 are not all equal.

The binomial distribution applies only when each trial has two possible outcomes. When there are more than two possible outcomes, we use the multinomial distribution (see Section 8.6) instead of the binomial distribution, provided the trials are all independent, the number of trials is fixed, and for each possible outcome the probability does not change from trial to trial. To illustrate such a multinomial situation, suppose that in the preceding example there had been the third alternative "health and physical condition," and the result had been as shown in the following table:

	Single	Married	Widowed or divorced	
Friends and social life	41	49	42	132
Job or primary activity	27	50	33	110
Health and physical condition	12	21	25	58
	80	120	100	300

As before, there are three separate samples, the column totals are the fixed sample sizes, but each trial (each person interviewed) allows for three different outcomes. Note that the row totals, $41 + 49 + 42 = 132$, $27 + 50 + 33 = 110$, and $12 + 21 + 25 = 58$ depend on the responses of the persons interviewed, and hence on chance. In general, a table like this, with r rows and c columns, is called an **$r \times c$** ("r by c") **table.** In particular, the preceding one is referred to as a 3×3 table.

The null hypothesis we shall want to test in this multinomial situation is that for each of the three choices ("friends and social life," "job or primary activity," and "health and physical condition") the probabilities are the same for each of the three groups of persons interviewed. Symbolically, if p_{ij} is the probability of obtaining a response belonging to the ith row and the jth column of the table, the null hypothesis is

$$H_0: \quad p_{11} = p_{12} = p_{13}, p_{21} = p_{22} = p_{23}, \text{ and}$$
$$p_{31} = p_{32} = p_{33}$$

where the p's must add up to 1 for each column. More compactly, we can write this null hypothesis as

$$H_0: \quad p_{i1} = p_{i2} = p_{i3} \quad \text{for } i = 1, 2, \text{ and } 3$$

The alternative hypothesis is that the p's are not all equal for at least one row, namely,

H_A: p_{i1}, p_{i2}, and p_{i3} are not all equal for at least one value of i

Before we show how all these problems are analyzed, let us mention one more situation where the method of this section applies. What distinguishes it from the preceding examples is that the column totals as well as the row totals are left to chance. To illustrate, suppose that we want to investigate whether there is a relationship between the test scores of persons who have gone through a certain job-training program and their subsequent performance on the job. Suppose, furthermore, that a random sample of 400 cases taken from very extensive files yielded the following result:

		Performance			
		Poor	*Fair*	*Good*	
	Below average	67	64	25	156
Test score	*Average*	42	76	56	174
	Above average	10	23	37	70
		119	163	118	400

Here there is only one sample, the **grand total** of 400 is its fixed size, and each trial (each case chosen from the files) permits nine different outcomes. It is mainly in connection with problems like this that $r \times c$ tables are referred to as **contingency tables.**

The purpose of the investigation that led to the table immediately above was to see whether there is a relationship between the test scores of persons who have gone through the job-training program and their subsequent performance on the job. In general, the hypotheses we test in the analysis of a contingency table are

H_0: The two variables under consideration are independent.

H_A: The two variables are not independent.

Despite the differences we have described, the analysis of an $r \times c$ table is the same for all three of our examples, and we shall illustrate it here in detail by analyzing the second example, the one that dealt with the different factors contributing to one's well-being. If the null hypothesis is true, we can combine the three samples and estimate the probability that any one person will choose "friends and social life" as the factor that contributes most to his or her well-being as

$$\frac{41 + 49 + 42}{300} = \frac{132}{300}$$

Hence, among the 80 single persons and the 120 married persons we can expect, respectively, $\dfrac{132}{300} \cdot 80 = \dfrac{132 \cdot 80}{300} = 35.2$ and $\dfrac{132}{300} \cdot 120 = \dfrac{132 \cdot 120}{300} = 52.8$ to choose "friends and social life" as the factor contributing most to their well-being. Note that in both cases we obtained the expected frequency by multiplying the row total by the column total and then dividing by the grand total for the entire table. Indeed, the argument that led to this result can be used to show that in general

> **The expected frequency for any cell of an $r \times c$ table can be obtained by multiplying the total of the row to which it belongs by the total of the column to which it belongs and then dividing by the grand total for the entire table.**

With this rule we get expected frequencies of $\dfrac{110 \cdot 80}{300} \approx 29.3$ and $\dfrac{110 \cdot 120}{300} = 44.0$ for the first and second cells of the second row.

It is not necessary to calculate all the expected frequencies in this way, as it can be shown (see Exercises 14.50 and 14.51) that the sum of the expected frequencies for any row or column equals the sum of the corresponding observed frequencies. Thus, we can get some of the expected frequencies by subtraction from row or column totals. For instance, for our example we get

$$132 - 35.2 - 52.8 = 44.0$$

for the third cell of the first row

$$110 - 29.3 - 44.0 = 36.7$$

for the third cell of the second row, and

$$80 - 35.2 - 29.3 = 15.5$$
$$120 - 52.8 - 44.0 = 23.2$$
$$100 - 44.0 - 36.7 = 19.3$$

for the three cells of the third row. These results are summarized in the following table, where the expected frequencies are shown in parentheses below the corresponding observed frequencies:

	Single	Married	Widowed or divorced
Friends and social life	41 (35.2)	49 (52.8)	42 (44.0)
Job or primary activity	27 (29.3)	50 (44.0)	33 (36.7)
Health and physical condition	12 (15.5)	21 (23.2)	25 (19.3)

To test the null hypothesis under which the **expected cell frequencies** were calculated, we compare them with the **observed cell frequencies.** It stands to reason that the null hypothesis should be rejected if the discrepancies between the observed and expected frequencies are large, and that it should be accepted (or at least that we reserve judgment) if the discrepancies between the observed and expected frequencies are small.

Denoting the observed frequencies by the letter o and the expected frequencies by the letter e, we base this comparison on the following **chi-square statistic:**

Statistic for analysis of $r \times c$ table

$$\chi^2 = \sum \frac{(o - e)^2}{e}$$

If the null hypothesis is true, this statistic is a value of a random variable having approximately the chi-square distribution (see page 317) with $(r - 1)(c - 1)$ degrees of freedom. When $r = 3$ and $c = 3$ as in our example, the number of degrees of freedom is $(3 - 1)(3 - 1) = 4$. Note that after we had calculated four of the expected frequencies with the rule on page 389, all the remaining ones were automatically determined by subtraction from row or column totals.

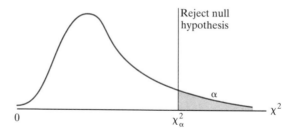

FIGURE 14.2
Criterion for χ^2 test.

Since we shall want to reject the null hypothesis when the discrepancies between the o's and e's are large, we use the one-tailed test criterion of Figure 14.2; symbolically, we reject the null hypothesis at the level of significance α if $\chi^2 \geq \chi^2_\alpha$ for $(r - 1)(c - 1)$ degrees of freedom. Remember, though, that this test is only an approximate large-sample test, and it is recommended that it not be used when one (or more) of the expected frequencies is less than 5. (When this is the case, we can sometimes salvage the situation by combining some of the cells, rows, or columns so that none of the expected cell frequencies will be less than 5. In that case there is a corresponding loss in the number of degrees of freedom.)

EXAMPLE 14.5   With reference to the problem dealing with the factors contributing most to one's well-being, test at the 0.01 level of significance whether for each of the three alternatives the probabilities are the same for persons who are single, married, or widowed or divorced.

Solution   1. H_0: $p_{i1} = p_{i2} = p_{i3}$ for $i = 1, 2,$ and 3.
H_A: $p_{i1}, p_{i2},$ and p_{i3} are not all equal for at least one value of i.

2. $\alpha = 0.01$

3. Reject the null hypothesis if $\chi^2 \geq 13.277$, where

$$\chi^2 = \sum \frac{(o - e)^2}{e}$$

and 13.277 is the value of $\chi^2_{0.01}$ for $(3 - 1)(3 - 1) = 4$ degrees of freedom; otherwise, accept the null hypothesis or reserve judgment.

4. Substituting the observed and expected frequencies summarized in the table on page 389 into the formula for χ^2, we get

$$\chi^2 = \frac{(41 - 35.2)^2}{35.2} + \frac{(49 - 52.8)^2}{52.8} + \frac{(42 - 44.0)^2}{44.0}$$
$$+ \frac{(27 - 29.3)^2}{29.3} + \frac{(50 - 44.0)^2}{44.0} + \frac{(33 - 36.7)^2}{36.7}$$
$$+ \frac{(12 - 15.5)^2}{15.5} + \frac{(21 - 23.2)^2}{23.2} + \frac{(25 - 19.3)^2}{19.3}$$
$$\approx 5.37$$

5. Since $\chi^2 = 5.37$ is less than 13.277, the null hypothesis cannot be rejected; that is, we conclude that for each of the three alternatives the probabilities are the same for persons who are single, married, or widowed or divorced. ∎

A MINITAB printout of the preceding chi-square analysis is shown in Figure 14.3. The difference between the values of χ^2 obtained previously and in Figure 14.3 is due to rounding. The printout also shows that the p-value

FIGURE 14.3
Computer printout for the analysis of Example 14.5.

```
MTB > ChiSquare c1 c2 c3.


Expected counts are printed below observed counts

              C1         C2         C3      Total
    1         41         49         42        132
           35.20      52.80      44.00

    2         27         50         33        110
           29.33      44.00      36.67

    3         12         21         25         58
           15.47      23.20      19.33

Total         80        120        100        300

ChiSq =   0.956 +   0.273 +   0.091 +
          0.186 +   0.818 +   0.367 +
          0.777 +   0.209 +   1.661 = 5.337
df  = 4,  p = 0.255
```

corresponding to the value of the chi-square statistic is 0.255. Since this exceeds 0.05, the specified level of significance, we conclude as before that the null hypothesis cannot be rejected.

Some statisticians prefer the alternative formula

$$\chi^2 = \sum \frac{o^2}{e} - n$$

where n is the grand total of the frequencies for the entire table. This alternative formula does simplify the calculations, but the original formula shows more clearly how χ^2 is actually affected by the discrepancies between the o's and the e's.

EXAMPLE 14.6   Use this alternative formula to recalculate χ^2 for Example 14.5.

Solution

$$\chi^2 = \frac{41^2}{35.2} + \frac{49^2}{52.8} + \frac{42^2}{44.0} + \frac{27^2}{29.3} + \frac{50^2}{44.0}$$

$$+ \frac{33^2}{36.7} + \frac{12^2}{15.5} + \frac{21^2}{23.2} + \frac{25^2}{19.3} - 300$$

$$\approx 5.37$$

This agrees with the result obtained before.   ■

Before we test for independence in our third example, the one dealing with test scores and on the job performance, let us demonstrate first that the rule on page 389 for calculating expected cell frequencies applies also to this kind of situation. Under the null hypothesis of independence, the probability of randomly choosing the file of a person whose test score is below average and whose on the job performance is poor is given by the product of the probability of choosing the file of a person whose test score is below average and the probability of choosing the file of a person whose on the job performance is poor. Using the total of the first row, the total of the first column, and the grand total for the entire table to estimate these probabilities, we get

$$\frac{67 + 64 + 25}{400} = \frac{156}{400}$$

for the probability of choosing the file of a person whose test score is below average. Similarly, we get

$$\frac{67 + 42 + 10}{400} = \frac{119}{400}$$

for the probability of choosing the file of a person whose on the job performance is poor. Hence we estimate the probability of choosing the file of a person whose test score is below average and whose on the job performance is poor as $\dfrac{156}{400} \cdot \dfrac{119}{400}$, and in a sample of size 400 we would expect to find

$$400 \cdot \frac{156}{400} \cdot \frac{119}{400} = \frac{156 \cdot 119}{400} \approx 46.4$$

persons who fit this distribution. Observe that in the final step of these calcula-tions, $\dfrac{156 \cdot 119}{400}$ is precisely the product of the total of the first row and the total of the first column divided by the grand total for the entire table. This illustrates that the rule on page 393 for calculating expected cell frequencies applies also to the example where the row totals as well as the column totals depend on chance.

EXAMPLE 14.7 With reference to the problem dealing with test scores in a job-training program and on-the-job performance, test at the 0.01 level of significance whether these two kinds of assessments are independent.

Solution

1. H_0: Test scores and on-the-job performance are independent.
   H_A: Test scores and on-the-job performance are not independent.
2. $\alpha = 0.01$
3. Reject the null hypothesis if $\chi^2 \geq 13.277$, where

$$\chi^2 = \sum \frac{(o - e)^2}{e}$$

and 13.277 is the value of $\chi^2_{0.01}$ for $(3 - 1)(3 - 1) = 4$ degrees of free-dom; otherwise, accept the null hypothesis or reserve judgment.

4. Multiplying row totals by column totals and then dividing by the grand total for the entire table, we obtain the expected cell frequencies shown in the following table in parentheses underneath the corresponding observed cell frequencies:

		Performance		
		Poor	*Fair*	*Good*
	Below average	67 (46.4)	64 (63.6)	25 (46.0)
Qualification test score	*Average*	42 (51.8)	76 (70.9)	56 (51.3)
	Above average	10 (20.8)	23 (28.5)	37 (20.7)

Then, substituting the observed frequencies and the expected frequencies into the original formula for χ^2, we get

$$\chi^2 = \frac{(67 - 46.4)^2}{46.4} + \frac{(64 - 63.6)^2}{63.6} + \frac{(25 - 46.0)^2}{46.0}$$
$$+ \frac{(42 - 51.8)^2}{51.8} + \frac{(76 - 70.9)^2}{70.9} + \frac{(56 - 51.3)^2}{51.3}$$
$$+ \frac{(10 - 20.8)^2}{20.8} + \frac{(23 - 28.5)^2}{28.5} + \frac{(37 - 20.7)^2}{20.7}$$

$$\approx 40.89$$

5. Since $\chi^2 = 40.89$ exceeds 13.277, the null hypothesis must be rejected; that is, we conclude that there is a relationship between the test scores and on-the-job performance.   ∎

Right after Example 14.5 we repeated the work with a computer printout of the chi-square analysis. This time we shall repeat the work with the use of a graphing calculator. At the top of Figure 14.4 we find the observed data entered into **MATRIX [A],** and at the bottom we find the results of the chi-square analysis, $\chi^2 = 41.01$ rounded to two decimals and the corresponding p-value of 0.000000027. (This value of chi-square differs from the one obtained previously due to rounding, and the p-value cannot be taken too literally since our statistic has only approximately the chi-square distribution with 4 degrees of freedom.) Although the middle part of Figure 14.4 is not really needed, it does show the expected cell frequencies in **MATRIX [B].**

In the analysis of $r \times c$ tables, the special case where $r = 2$ and the column totals are fixed sample sizes has many important applications. Here, we are testing, in fact, for significant differences among c sample proportions, and we can simplify the notation by letting $p_1, p_2, \ldots,$ and p_c denote the corresponding population proportions. Also, for $c = 2$ we have here an alternative method for testing the significance of the difference between two proportions (as in Section 14.3), but it applies only when the alternative hypothesis is $p_1 \neq p_2$. In that case, the relationship between the z statistic of Section 14.3 and the χ^2 statistic of this section is $z^2 = \chi^2$, as the reader will be asked to verify in Exercise 14.49 for the data of Example 14.4.

A noteworthy, though undesirable, feature of the chi-square analysis of an $r \times c$ table is that the χ^2 statistic is not affected by interchanges of rows and/or columns. This makes it wasteful of information whenever the row categories and/or column categories reflect a definite order; that is, when we deal with ordered categorical data. This was the case, for instance, in Example 14.7, where the test scores ranged from below average to average to above average, and the on-the-job performance ratings ranged from poor to fair to good. To avoid this shortcoming of the chi-square analysis of $r \times c$ tables, statisticians have developed alternative procedures. In these procedures, numbers replace the ordered categories. Usually, but not necessarily, these numbers are consecutive integers, preferably integers that will make the arithmetic as simple as possible. (For instance, for three ordered categories we might use the integers -1, 0, and 1.) We shall not go into any details about this, but an illustration may be found in

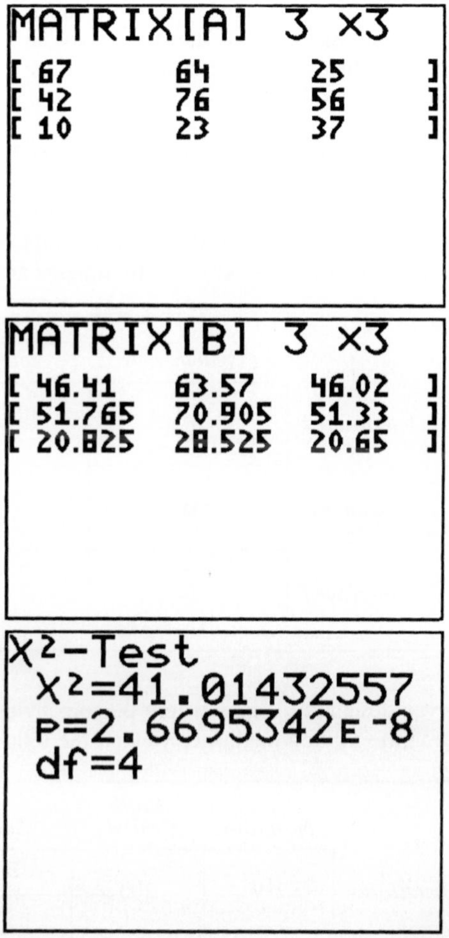

```
MATRIX[A]  3 ×3
[ 67      64      25     ]
[ 42      76      56     ]
[ 10      23      37     ]
```

```
MATRIX[B]  3 ×3
[ 46.41    63.57    46.02 ]
[ 51.765   70.905   51.33 ]
[ 20.825   28.525   20.65 ]
```

```
X²-Test
  X²=41.01432557
  P=2.6695342E-8
  df=4
```

FIGURE 14.4
Analysis of the $r \times c$ table of Example 14.7 reproduced from the display screen of a TI-83 graphing calculator.

Example 17.2. Also, two books dealing with the analysis of ordinal categorical data are listed among the references at the end of this chapter.

■ E x e r c i s e s

14.25 Use the alternative formula on page 392 to recalculate the value of the chi-square statistic obtained in Example 14.7.

14.26 Suppose that we ask 100 Chevrolet dealers, 100 Ford dealers, and 100 Chrysler dealers whether they expect sales to go down, remain the same, or go up, or whether they cannot tell. What hypotheses shall we want to test if we intend to perform a chi-square analysis of the resulting 4×3 table?

14.27 Suppose that we take a random sample of 400 persons living in federal housing projects and classify each one according to whether he or she has part-time employment, full-time employment, or no employment, and also according to whether he or she has 0, 1, 2, 3, or 4 or more children. What hypotheses shall we want to test if we are going to perform a chi-square analysis of the resulting 3×5 table?

14.28 Suppose that we take random samples of voters in Memphis, Nashville, Knoxville, and Chattanooga and ask them whether they favor an increase in the state tax on cigarettes. What hypotheses shall we want to test if we are going to perform a chi-square analysis of the resulting 2 × 4 table?

14.29 Of a group of 200 persons suffering anxiety disorders, 100 received psychotherapy and 100 received psychological counseling. A panel of psychiatrists determined after six months whether their condition had deteriorated, remained unchanged, or improved. The results are shown in the following table. Use the 0.05 level of significance to test whether the two kinds of treatments are equally effective.

	Psychotherapy	*Psychological counseling*
Deteriorated	8	11
Unchanged	58	62
Improved	34	27

14.30 A sample survey designed to show where persons living in different parts of the country buy nonprescription medicines yielded the following results:

	Northeast	*North Central*	*South*	*West*
Drugstores	219	200	181	180
Grocery Stores	39	52	89	60
Others	42	48	30	60

Test the null hypothesis that, so far as nonprescription medicines are concerned, the buying habits of persons living in the given parts of the country are the same. Use the 0.01 level of significance.

14.31 Repeat Exercise 14.30, using appropriate computer software or a graphing calculator. Use the *p*-value to verify the conclusion obtained in Exercise 14.30.

14.32 A research organization, interested in testing whether the proportions of sons taking up the occupations of their fathers are equal for a selected group of occupations, took random samples of size 200, 150, 180, and 100, respectively, in which the fathers are doctors, bankers, teachers, and lawyers, and obtained the following results:

	Doctors	Bankers	Teachers	Lawyers
Same Occupation	37	22	26	23
Different occupation	163	128	154	77

Use the 0.05 level of significance to test whether the differences among the four sample proportions, $\frac{37}{200} = 0.185$, $\frac{22}{150} \approx 0.147$, $\frac{26}{180} \approx 0.144$, and $\frac{23}{100} = 0.23$, are significant.

14.33 The dean of a large university wants to determine whether there is a connection between academic rank and a faculty member's opinion concerning a proposed curriculum change. Interviewing a sample of 80 instructors, 140 assistant professors, 100 associate professors, and 80 professors, he obtains the results shown in the following table:

	Instructor	Assistant Professor	Associate Professor	Professor
Against	8	19	15	12
Indifferent	40	41	24	16
For	32	80	61	52

Use the 0.01 level of significance to test the null hypothesis that there are really no differences in opinion concerning the curriculum change among the four groups.

14.34 Repeat Exercise 14.33, using appropriate computer software or a graphing calculator. Use the p-value to verify the conclusion obtained in Exercise 14.33.

14.35 Decide on the basis of the information given in the following table, the result of a sample survey conducted at a large state university, whether there is a relationship between students' interest and ability in studying a foreign language. Use the 0.05 level of significance.

		Ability		
		low	average	high
Interest	low	28	17	15
	average	20	40	20
	high	12	28	40

14.36 In a study to determine whether there is a relationship between bank employees' standard of dress and their professional advancement, a random sample of size $n = 500$ yielded the results shown in the following table:

| | **Speed of advancement** | | |
	Slow	*Average*	*Fast*
Very well dressed	38	135	129
Well dressed	32	68	43
Poorly dressed	13	25	17

Use the 0.025 level of significance to test whether there really is a relationship between bank employees' standard of dress and their professional advancement.

 14.37 Repeat either Exercise 14.35 or Exercise 14.36, using appropriate computer software or a graphing calculator and changing the level of significance to 0.01.

14.38 Tests of the fidelity and the selectivity of 190 radios, a random sample, produced the results shown in the following table:

| | | **Fidelity** | | |
		Low	*Average*	*High*
	Low	7	12	31
Selectivity *Average*		35	59	18
	High	15	13	0

Use the 0.01 level of significance to test the null hypothesis that fidelity is independent of selectivity.

 14.39 Repeat Exercise 14.38, using appropriate computer software or a graphing calculator. Use the *p*-value to verify the conclusion reached in that exercise.

14.40 A large manufacturer hires many handicapped workers. To see whether their handicaps affect their performance, the personnel manager obtained the following sample data, where the column totals are fixed sample sizes:

	Deaf	Blind	Other handicap	No handicap
Above average	11	3	14	36
Performance Average	24	11	39	134
Below average	5	6	7	30

Explain why a "standard" chi-square analysis with $(3 - 1)(4 - 1) = 6$ degrees of freedom cannot be performed.

14.41 With reference to Exercise 14.40, combine the first three columns and then test at the 0.05 level of significance whether handicaps affect the workers' performance.

14.42 If the analysis of a contingency table shows that there is a relationship between the two variables under consideration, the strength of the relationship may be measured by the **contingency coefficient**

Contingency coefficient

$$C = \sqrt{\frac{\chi^2}{\chi^2 + n}}$$

where χ^2 is the value of the chi-square statistic obtained for the table and n is the grand total of all the frequencies. This coefficient takes on values between 0 (corresponding to independence) and a maximum value less than 1 depending on the size of the table; for instance, it can be shown that for a 3×3 table the maximum value of C is $\sqrt{2/3} \approx 0.82$.

Calculate C for Example 14.7, which dealt with the test scores and the on-the-job performance of persons who have gone through the job-training program, where we had $n = 400$ and $\chi^2 = 40.89$.

14.43 Find the value of C (see Exercise 14.42) for the contingency table of Exercise 14.35.

14.44 Find the value of C (see Exercise 14.42) for the contingency table of Exercise 14.38.

14.45 Tests were made to compare the ability of three materials to withstand extreme temperature changes, with the following results:

	Material A	Material B	Material C
Number crumbled	43	28	23
Number not crumbled	77	52	77

Test at the 0.01 level of significance whether the chances of crumbling are the same for all three materials.

14.46 The following table shows the results of a study in which random samples of the members of five large unions were asked whether they are for or against a certain political candidate:

	Union 1	Union 2	Union 3	Union 4	Union 5
For the candidate	74	81	69	75	91
Against the candidate	26	19	31	25	9

At the 0.01 level of significance, can we conclude that the differences among the five sample proportions are significant?

14.47 Rework Exercise 14.17 by analyzing the data like a 2 × 2 table, and verify that the value obtained here for x^2 equals the square of the value obtained originally for z.

14.48 Rework Exercise 14.19 by analyzing the data like a 2 × 2 table, and verify that the value obtained here for x^2 equals the square of the value obtained originally for z.

14.49 Rework Example 14.4 by analyzing the data like a 2 × 2 table, and verify that the value obtained here for x^2 equals the square of the value obtained originally for z.

14.50 Verify that if the expected frequencies for an $r \times c$ table are calculated with the rule on page 389, the sum of the expected frequencies for any row equals the sum of the corresponding observed frequencies.

14.51 Verify that if the expected frequencies for an $r \times c$ table are calculated with the rule on page 389, the sum of the expected frequencies for any column equals the sum of the corresponding observed frequencies.

14.5   GOODNESS OF FIT

In this section we shall consider another application of the chi-square criterion, in which we compare an observed frequency distribution with a distribution we might expect according to theory or assumptions. We refer to such a comparison as a test of **goodness of fit.**

To illustrate, suppose that the management of an airport wants to check an air-traffic controllers' claim that the number of radio messages received per minute is a random variable having the Poisson distribution with the mean $\lambda = 1.5$. If correct, this might require hiring additional personnel. Appropriate recording devices yielded the following data on the number of radio messages received in a random sample of 200 one-minute intervals:

0	0	1	1	5	3	1	1	2	0	1	0	3	1	0	2	2	2	2	0	1	2	2	0	1
0	2	1	0	2	3	1	1	3	1	0	0	0	1	0	1	2	0	1	3	1	0	0	0	3
0	0	0	1	2	2	0	0	3	0	3	1	1	1	5	2	2	0	2	4	1	1	1	4	2
3	0	0	0	1	0	1	1	1	0	1	0	2	0	2	2	0	1	1	0	2	2	2	1	4
0	0	2	2	0	1	2	0	2	0	0	2	1	1	1	2	1	2	4	0	0	2	0	0	2
0	2	0	0	2	1	3	0	2	0	1	1	3	0	1	0	2	1	0	1	2	2	3	1	1
4	0	1	2	0	0	1	0	2	2	1	0	3	1	1	0	1	0	0	3	2	1	3	1	0
0	2	3	0	0	3	3	0	2	1	0	3	0	0	2	1	1	1	0	1	0	0	2	3	3

Summarized, these data yield

Number of radio messages	Observed frequency
0	70
1	57
2	45
3	21
4	5
5	2

If the air-traffic controllers' claim is true, we get the corresponding expected frequencies by multiplying the probabilities in Figure 14.5 by 200 (after combining "5 or more" into one category.) This yields

Number of radio messages	Expected frequency
0	44.6
1	66.9
2	50.2
3	25.1
4	9.4
5 or more	3.7

```
MTB > PDF c1;
SUBC>    Poisson 1.5.


Poisson with mu = 1.50000

             x           P( X = x)
          0.00           0.2231
          1.00           0.3347
          2.00           0.2510
          3.00           0.1255
          4.00           0.0471
          5.00           0.0141
          6.00           0.0035
          7.00           0.0008
          8.00           0.0001

MTB >
```

FIGURE 14.5
Computer printout for
Poisson probabilities
with $\lambda = 1.5$.

To test whether the discrepancies between the observed frequencies and the expected frequencies can be attributed to chance, we use the same chi-square statistic as in Section 14.4:

*Statistic for test of
goodness of fit*

$$\chi^2 = \sum \frac{(o - e)^2}{e}$$

calculating $\dfrac{(o - e)^2}{e}$ separately for each class of the distribution. Then, if the value we get for χ^2 is greater than or equal to χ^2_α, we reject the null hypothesis on which the expected frequencies are based at the level of significance α. The number of degrees of freedom is $k - m - 1$, where k is the number of terms

$$\frac{(o - e)^2}{e}$$

added in the formula for χ^2, and m is the number of parameters of the probability distribution (in this case the Poisson distribution) that have to be estimated from the sample data.

EXAMPLE 14.8   Based on the two sets of frequencies given previously, test the air-traffic controllers' claim at the 0.01 level of significance.

Solution

1. H_0: The population sampled has the Poisson distribution with $\lambda = 1.5$.
   H_A: The population sampled does not have the Poisson distribution with $\lambda = 1.5$.
2. $\alpha = 0.01$

3. Since the expected frequency corresponding to "5 or more" is less than 5, the classes corresponding to 4 and "5 or more" must be combined and $k = 5$; also, since none of the parameters of the Poisson distributed had to be estimated from the data, $m = 0$. Therefore, reject the null hypothesis if $\chi^2 \geq 13.277$, where

$$\chi^2 = \sum \frac{(o - e)^2}{e}$$

and 13.277 is the value of $\chi^2_{0.01}$ for $k - m - 1 = 5 - 0 - 1 = 4$ degrees of freedom; otherwise, accept the null hypothesis or reserve judgment.

4. Substituting the observed frequencies and the expected frequencies into the formula for χ^2, we get

$$\chi^2 = \frac{(70 - 44.6)^2}{44.6} + \frac{(57 - 66.9)^2}{66.9} + \frac{(45 - 50.2)^2}{50.2}$$
$$+ \frac{(21 - 25.1)^2}{25.1} + \frac{(7 - 13.1)^2}{13.1}$$
$$\approx 20.0$$

5. Since 20.0 exceeds 13.277, the null hypothesis must be rejected; we conclude that either the population does not have a Poisson distribution or it has a Poisson distribution with λ different from 1.5. ∎

To check whether a Poisson distribution with λ different from 1.5 might provide a better fit, let us calculate the mean of the observed distribution, getting

$$\frac{0 \cdot 70 + 1 \cdot 57 + 2 \cdot 45 + 3 \cdot 21 + 4 \cdot 5 + 5 \cdot 2}{200} = \frac{240}{200} = 1.2$$

Thus, let us see what happens when we use $\lambda = 1.2$ instead of $\lambda = 1.5$.

FIGURE 14.6

Computer printout of Poisson probabilities with $\lambda = 1.2$.

```
MTB > PDF c1;
SUBC>    Poisson 1.2.


Poisson with mu = 1.20000

              x         P( X = x)
           0.00          0.3012
           1.00          0.3614
           2.00          0.2169
           3.00          0.0867
           4.00          0.0260
           5.00          0.0062
           6.00          0.0012
           7.00          0.0002
           8.00          0.0000

MTB >
```

For $\lambda = 1.2$, we get the expected frequencies by multiplying the probabilities in Figure 14.6 by 200 (after combining "5 or more" into one category). This yields

Number of radio messages	Expected frequency
0	60.2
1	72.3
2	43.4
3	17.3
4	5.2
5 or more	1.5

EXAMPLE 14.9 Based on the observed frequencies on page 401 and the expected frequencies given immediately above, test at the 0.01 level of significance whether the data constitute a sample from a Poisson population.

Solution
1. H_0: The population sampled has a Poisson distribution.
   H_A: The population sampled does not have a Poisson distribution.
2. $\alpha = 0.01$
3. Since the expected frequency corresponding to "5 or more" is again less than 5, the classes corresponding to 4 and "5 or more" must be combined and $k = 5$; also, since the parameter λ was estimated from the data, $m = 1$. Therefore, reject the null hypothesis if $\chi^2 \geq 11.345$, where

$$\chi^2 = \sum \frac{(o - e)^2}{e}$$

   and 11.345 is the value of $\chi^2_{0.01}$ for $k - m - 1 = 5 - 1 - 1 = 3$ degrees of freedom; otherwise, accept the null hypothesis or reserve judgment.
4. Substituting the observed frequencies and the expected frequencies into the formula for χ^2, we get

$$\chi^2 = \frac{(70 - 60.2)^2}{60.2} + \frac{(57 - 72.3)^2}{72.3} + \frac{(45 - 43.4)^2}{43.4}$$
$$+ \frac{(21 - 17.3)^2}{17.3} + \frac{(7 - 6.7)^2}{6.7}$$
$$\approx 5.7$$

5. Since 5.7 is less than 11.345, the null hypothesis cannot be rejected. Using a HEWLETT PACKARD statistical calculator, we found that the p-value corresponding to $\chi^2 = 5.7$ and 3 degrees of freedom is 0.1272, so the fit is not too bad, but we would be inclined to reserve judgment about the nature of the population.   ∎

The method illustrated in this section is used quite generally to test how well distributions, expected on the basis of theory or assumptions, fit, or describe, observed data. In the exercises that follow we shall test also whether observed distributions have (at least approximately) the shape of binomial and normal distributions.

Exercises

14.52 To see whether a die is balanced, it was rolled 360 times, and the following results were obtained:

Number	Frequency
1	65
2	57
3	49
4	70
5	68
6	51

At the 0.05 level of significance, do these results support the hypothesis that the die is balanced?

14.53 Ten years' data show that in a given city there were no bank robberies in 57 months, one bank robbery in 36 months, two bank robberies in 15 months, and three or more bank robberies in 12 months. At the 0.05 level of significance, does this substantiate the claim that the probabilities of 0, 1, 2, or 3 or more bank robberies in any one month are 0.40, 0.30, 0.20, and 0.10?

14.54 Following is the distribution of the number of females in 160 litters, each consisting of four mice:

Number of females	Number of litters
0	12
1	37
2	55
3	47
4	9

Test at the 0.01 level of significance whether these data may be looked upon as random samples from a binomial population with $n = 4$ and $p = 0.50$.

14.55 A quality control engineer takes daily samples of three tractors coming off an assembly line and on 150 working days he obtained the data summarized in the following table:

Number of tractors requiring adjustments	Number of days
0	75
1	61
2	13
3	1

Test at the 0.01 level of significance whether these data may be looked upon as random samples from a binomial population. (*Hint*: Calculate the mean of this distribution and use the formula $\mu = np$ to estimate p.)

14.56 Each day a surgeon schedules at most four operations, and his activities on 300 days are summarized in the following table:

Number of surgeries	Number of days
0	2
1	10
2	33
3	136
4	119

Test at the 0.05 level of significance whether these data may be looked upon as random samples from a binomial population with $n = 4$ and $p = 0.70$.

14.57 With reference to Exercise 4.56, test at the 0.05 level of significance whether the data may be looked upon as random samples from a binomial population. (*Hint*: Calculate the mean of the given distribution and use the formula $\mu = np$ to estimate p.)

14.58 Following is the distribution of the numbers of bears spotted on 100 sightseeing tours in Denali National Park:

Number of bears	Number of tours
0	70
1	23
2	7
3	0

Given that for the Poisson distribution with $\lambda = 0.5$ the probabilities of 0, 1, 2, and 3 "successes" are, respectively, 0.61, 0.30, 0.08, and 0.01, test at the 0.05 level of significance whether the data on bear sightings may be looked upon as a random sample from a Poisson population with $\lambda = 0.5$.

14.59 Following is the distribution of the number of power failures reported in a Western city on 300 days:

Number of power failures	Number of days
0	9
1	43
2	64
3	62
4	42
5	36
6	22
7	14
8	6
9	2

Use appropriate computer software or a graphing calculator to get the necessary Poisson probabilities to test at the 0.05 level of significance whether the daily number of power failures in this city is a random variable having the Poisson distribution with $\lambda = 3.2$.

14.60 The following is the distribution of the readings obtained with a Geiger counter of the number of particles emitted by a radioactive substance in 100 successive 40-second intervals:

Number of particles	Frequency
5–9	1
10–14	10
15–19	37
20–24	36
25–29	13
30–34	2
35–39	1

(a) Verify that the mean and the standard deviation of this distribution are $\bar{x} = 20$ and $s = 5$.
(b) Find the probabilities that a random variable having a normal distribution with $\mu = 20$ and $\sigma = 5$ will take on a value less than 9.5, between 9.5 and 14.5, between 14.5 and 19.5, between 19.5 and 24.5, between 24.5 and 29.5, between 29.5 and 34.5, and greater than 34.5.
(c) Find the expected normal curve frequencies for the various classes by multiplying the probabilities obtained in part (b) by the total frequency, and then test at the 0.05 level of significance whether the data may be looked upon as a random sample from a normal population.

14.61 Following is the distribution of the number of minutes it took 80 persons to complete a certain tax form:

Time required to complete form (minutes)	Frequency
10–14	8
15–19	28
20–24	27
25–29	12
30–34	4
35–39	1

As can easily be verified, the mean of this distribution is $\bar{x} = 20.7$ and its standard deviation is $s = 5.4$. To test the null hypothesis that these data constitute a random sample from a normal population, proceed with the following steps:

(a) Find the probabilities that a random variable having a normal distribution with $\mu = 20.7$ and $\sigma = 5.4$ will take on a value less than 14.5, between 14.5 and 19.5, between 19.5 and 24.5, between 24.5 and 29.5, between 29.5 and 34.5, and greater than 34.5.

(b) Changing the first and last classes of the distribution to "14 or less" and "35 or more," find the expected normal curve frequencies corresponding to the six classes of the distribution by multiplying the probabilities obtained in part (a) by the total frequency of 80.

(c) Test at the 0.05 level of significance whether the given data may be looked upon as a random sample from a normal population.

14.6 Checklist of Key Terms (with page references to their definitions)

Cell, 389
Chi-square statistic, 390
Contingency coefficient, 399
Contingency table, 388
Expected cell frequencies, 389, 390
Goodness of fit, 401

Grand total, 388
Observed cell frequencies, 389, 390
Pooling, 383
$r \times c$ table, 387
Standard error of difference between
  two proportions, 383

14.7 References

The theory that underlies the various tests of this chapter is discussed in most textbooks on mathematical statistics; for instance, in

MILLER, I., and MILLER, M., *John E. Freund's Mathematical Statistics,* 6th ed. Upper Saddle River, N. J.: Prentice Hall, 1999.

Details about contingency tables may be found in

EVERITT, B. S., *The Analysis of Contingency Tables*. New York: John Wiley & Sons, Inc., 1977.

Recent research on the analysis of r × c tables with ordered categories can be found in

AGRESTI, A., *Analysis of Ordinal Categorical Data*. New York: John Wiley & Sons, Inc., 1984.

GOODMAN, L. A., *The Analysis of Cross-Classified Data Having Ordered Categories.* Cambridge, Mass.: Harvard University Press, 1984.

R.136 In 18 rounds on various golf courses, two professional golfers scored 71 and 73, 74 and 70, 69 and 69, 72 and 71, 75 and 73, 70 and 71, 74 and 72, 68 and 69, 70 and 69, 71 and 71, 77 and 74, 73 and 74, 70 and 70, 69 and 71, 74 and 71, 75 and 73, 72 and 72, and 76 and 72. Assuming that the differences in their scores can be looked upon as a random sample from a normal population, test at the 0.05 level of significance whether the two golfers are equally good.

R.137 Six packages of sunflower seeds randomly selected from a large shipment weighed, respectively, 15.9, 15.5, 16.2, 15.8, 16.0, and 15.7 ounces. Use a normal probability plot generated by means of a computer or a graphing calculator to justify the assumption that these data constitute a sample from a normal population.

R.138 With reference to Exercise R.137, what can we assert with 99% confidence about the maximum error if we use the mean of the sample, $\bar{x} = 15.85$ ounces, to estimate the mean of the population sampled?

R.139 A research worker wants to determine what percentage of the farm workers in a certain area where cotton is grown are illegal aliens. How large a sample will she need if she feels that the actual percentage is at most 15% and she wants to be able to assert with 95% confidence that her estimate is off by at most 2.5%?

R.140 What null hypothesis and what alternative hypothesis should we use if we want to test the claim that on the average children attending elementary schools in an urban school district live more than a mile from the school that they attend?

R.141 Measurements of the amount of chloroform (micrograms per liter) in $n = 36$ specimens of the drinking water of a city yielded $\bar{x} = 34.8$. If it can be assumed that $\sigma = 4.9$ for such data, construct a 98% confidence interval for the average amount of chloroform in the drinking water of that city.

R.142 Test runs with eight models of an experimental engine showed that they operated for 25, 18, 31, 19, 32, 27, 24, and 28 minutes with a gallon of a certain kind of fuel. Estimate the standard deviation of the population sampled using

(a) the sample standard deviation;

(b) the sample range and the method described in Exercise 11.42.

R.143 In an experiment, an interviewer of job applicants is asked to write down her initial impression (favorable or unfavorable) after two minutes and her final impression at the end of the interview. Use the following data and the 0.01 level of significance to test the interviewer's claim that her initial and final impressions are the same 85% of the time:

		Initial impression	
		Favorable	*Unfavorable*
Final impression	*Favorable*	184	32
	Unfavorable	56	128

R.144 The following are the pull strengths (in pounds) required to break the bond of two kinds of glue:

Glue 1:   25.3   20.2   21.1   27.0   16.9   30.1
            17.8   22.9   27.2   20.0

Glue 2:   24.9   22.5   21.8   23.6   19.8   21.6
            20.4   22.1

As a preliminary to the two-sample *t* test, use the 0.02 level of significance to test whether it is reasonable to assume that the corresponding population standard deviations are equal.

R.145 Based on the results of $n = 14$ trials, we want to test the null hypothesis $p = 0.30$ against the alternative hypothesis $p > 0.30$. If we reject the null hypothesis when the number of successes is eight or more and otherwise we accept it, find the probability of a

(a) Type I error;

(b) Type II error when $p = 0.40$;

(c) Type II error when $p = 0.50$;

(d) Type II error when $p = 0.60$.

 R.146 In five track meets, a high jumper cleared 84, $81\frac{1}{2}$, 82, $80\frac{1}{2}$, and 83 inches. Use a normal probability plot generated by means of a computer or a graphing calculator to justify the assumption that these data constitute a sample from a normal population.

R.147 With reference to Exercise R.146, show that the null hypothesis $\mu = 78$ can be rejected against the alternative hypothesis $\mu > 78$ at the 0.01 level of significance.

R.148 With reference to Exercises R.146 and R.147, show that the null hypothesis $\mu = 78$ can no longer be rejected against the alternative hypothesis $\mu > 78$ at the 0.01 level of significance if the fifth figure is recorded incorrectly as 93 instead of 83. Explain the apparent paradox that even though the difference between $\bar{x}$ and μ has increased, it is no longer significant.

 R.149 With reference to the preceding exercises, use a normal probability plot generated by means of a computer or a graphing calculator to show that the one-sample *t* test should not have been used in Exercise R.148.

R.150 In a study of parents' feeling about a required course in sex education taught to their children, a random sample of 360 parents are classified according to whether they have one, two, or three or more children in the school system and also whether they feel that the course is poor, adequate, or good. The results are shown in the following table:

	Number of children		
	1	*2*	*3 or more*
Poor	48	40	12
Adequate	55	53	29
Good	57	46	20

Test at the 0.05 level of significance whether there is a relationship between parents' reaction to the course and the number of children they have in the school system.

R.151 A political pollster wants to determine the proportion of the population that favors a regulatory change in the use of marijuana by cancer patients. How large a sample will she need if she wants to be able to assert with a probability of at least 0.90 that the sample proportion will differ from the population proportion by at most 0.04?

R.152 A bank is considering replacing its ATMs with a new model. If μ_0 is the average length of time that its old machines function between repairs, against what alternative hypothesis should it test the null hypothesis $\mu = \mu_0$, where μ is the corresponding average length of time for the new machines, if

(a) it does not want to replace the old machines unless the new machines prove to be superior;

(b) it wants to replace the old machines unless the new machines actually turn out to be inferior?

R.153 In a study conducted at a large airport, 81 persons in a random sample of 300 persons who had just gotten off a plane and 32 persons in a random sample of 200 persons who were about to board a plane admitted that they were afraid of flying. Use the z statistic to test at the 0.01 level of significance whether the difference between the corresponding sample proportions is significant.

R.154 With reference to Exercise R.153, use the χ^2 statistic to rework this exercise, and verify that the value obtained for χ^2 equals the square of the value obtained for z in Exercise R.153.

R.155 To find out whether the inhabitants of two South Pacific islands may be regarded as having the same racial ancestry, an anthropologist determined the cephalic indices of six adult males from each island, getting $\bar{x}_1 = 77.4$ and $\bar{x}_2 = 72.2$ and the corresponding standard deviations $s_1 = 3.3$ and $s_2 = 2.1$. Assuming that the data constitute independent random samples from normal populations, test the null hypothesis $\sigma_1 = \sigma_2$ against the alternative hypothesis $\sigma_1 \neq \sigma_2$ at the 0.10 level of significance (as a preliminary to the two-sample t test.)

R.156 With reference to Exercise R.155 and its result, use the two-sample t test to test at the 0.05 level of significance whether the difference between the two sample means is significant.

R.157 In 12 test runs over a marked course, a newly designed motorboat averaged 33.6 seconds with a standard deviation of 2.3 seconds. Assuming that it is reasonable to treat the data as a random sample from a normal population, test the null hypothesis $\mu = 35.0$ seconds against the alternative hypothesis $\mu < 35.0$ seconds at the 0.025 level of significance.

R.158 The following table shows how many times the departure of a daily flight from Vancouver to San Francisco was delayed in 50 weeks:

Delays per week	Number of weeks
0	12
1	16
2	13
3	8
4	1

Use the 0.05 level of significance to test the null hypothesis that the departure of the flight is delayed 10% of the time; namely, the null hypothesis that the data constitute a random sample from a binomial population with $n = 7$ and $p = 0.10$.

R.159 With reference to Exercise R.158, use the 0.05 level of significance to test the null hypothesis that the data constitute a random sample from a binomial population with $n = 7$. (*Hint:* Estimate p by calculating the mean of the given distribution and then using the formula $\mu = np$.)

R.160 In a random sample of 10 rounds of golf played on her home course, a golf professional averaged 70.8 with a standard deviation of 1.28. Assuming that her scores can be looked upon as a random sample from a normal population, use the 0.01 level of significance to test the null hypothesis $\sigma = 1.0$ against the alternative hypothesis that her game is actually less consistent.

R.161 In a random sample of 120 persons shopping at Fashion Square, 84 made at least one purchase. Construct a 95% confidence interval for the probability that any one person shopping at Fashion Square will make at least one purchase.

R.162 In an election for County Treasurer, the Independent candidate received 10,361 votes (about 48%) and the Republican candidate received 11,225 votes (about 52%). Is it reasonable to ask whether the difference between these two percentages is significant?

R.163 In accordance with the rule that np and $n(1 - p)$ must both be greater than 5, for what values of p can we use the normal approximation to the binomial distribution when $n = 400$?

R.164 In a study of the length of young-of-the-year fresh-water drumfish in Lake Erie, it was found that the lengths of 60 of them had the standard deviation 10.4 mm. Construct a 95% confidence interval for the corresponding population parameter σ.

R.165 In order to evaluate the clinical effects of a certain steroid in treating chronically underweight persons, a random sample of 60 such persons were given 25-mg dosages over a period of 12 weeks, while another random sample of 60 such persons were given 50-mg dosages over the same period of time. The results showed that those in the first group gained on the average 8.5 pounds while those in the second group gained on the average 11.3 pounds. If previous tests have shown that $\sigma_1 = \sigma_2 = 1.4$ pounds, test at the 0.05 level of significance whether the difference between the two sample means is significant.

R.166 A random sample taken as part of a study in nutrition showed that 400 young adults in an Australian city averaged a protein intake of 1.274 g per kilogram of body weight. Given that $\sigma = 0.22$ g per kilogram of body weight, what can one assert with 95% confidence about the maximum error if $\bar{x} = 1.274$ g is used as an estimate of the mean protein intake per kilogram of body weight for the population sampled?

 R.167 A microbiologist found 13, 17, 7, 11, 15, and 9 microorganisms in six cultures. Use appropriate computer software or a graphing calculator to obtain a normal probability plot and, thus, verify that these data may be looked upon as a sample from a normal population.

R.168 With reference to Exercise R.167, construct a 99% confidence interval for the mean of the population sampled.

R.169 A team of physicians is asked to check whether a highly priced athlete is physically fit to play in the NFL. What type of error would be committed if the hypothesis that the athlete is physically fit to play in the NFL is erroneously accepted? What type of error would be committed if the hypothesis that the athlete is physically fit to play in the NFL is erroneously rejected?

R.170 To compare the action of two phosphorescent coatings of airplane instrument dials, a technician coats five dials with Coating A, five others with Coating B, illuminates them with ultraviolet light, and records the number of minutes that each dial glows after the light source has been turned off. His results were as follows (in minutes):

Coating A:	54	55	49	60	52
Coating B:	60	65	61	54	55

Assuming that these data may be looked upon as independent random samples from two normal populations with equal standard deviations, test the null hypothesis $\mu_A = \mu_B$ against the alternative hypothesis $\mu_A < \mu_B$ at the 0.05 level of significance.

R.171 A testing laboratory wants to estimate the average lifetime of a multitoothed cutting tool. If a random sample of size $n = 6$ showed tool lives of 2,470, 2,520, 2,425, 2,505, 2,440, and 2,400 pieces, and it can be assumed that these data constitute a random sample from a normal population, what can they assert with 99% confidence about the maximum error if the mean of this sample is used as an estimate of the mean of the population sampled?

R.172 In a study of the reading habits of financial advisors, it is desired to estimate the average number of financial reports they read per week. Assuming that it is reasonable to use $\sigma = 3.4$, how large a sample would be required if one wants to be able to assert with probability 0.99 that the sample mean will not be off by more than 1.2?

R.173 Ten measurements of the time required to kill a centipede with a given insecticide have the standard deviation $s = 3.5$ seconds. Assuming that these measurements constitute a random sample from a normal population, construct a 95% confidence interval for σ, the actual standard deviation of the time required by the insecticide to kill a centipede.

R.174 In a random sample of the members of a large regional teachers' association, it was found that of 400 high school English teachers only 212 majored in English in college. Construct a 95% confidence interval for the true proportion of high school English teachers belonging to this association who majored in English in college.

R.175 Following is the distribution of the daily emission of sulfur oxides by an industrial plant:

Tons of sulfur oxides	Frequency
5.0– 8.9	3
9.0–12.9	10
13.0–16.9	14
17.0–20.9	25
21.0–24.9	17
25.0–28.9	9
29.0–32.9	2

As can easily be verified, the mean of this distribution is $\bar{x} = 18.85$ and its standard deviation is $s = 5.55$. To test the null hypothesis that these data constitute a random sample from a normal population, proceed with the following steps:

(a) Find the probabilities that a random variable having the normal distribution with $\mu = 18.85$ and $\sigma = 5.5$ will take on a value less than 8.95, between 8.95 and 12.95, between 12.95 and 16.95, between 16.95 and 20.95, between 20.95 and 24.95, between 24.95 and 28.95, and greater than 28.95.

(b) Changing the first and last classes of the distribution to "8.95 or less" and "28.95 or more," find the expected normal curve frequencies corresponding to the seven classes of the distribution by multiplying the probabilities obtained in part (a) by the total frequency of 80.

(c) Test at the 0.05 level of significance whether the given data may be looked upon as a random sample from a normal population.

R.176 In a multiple-choice test administered to high school sophomores after a visit to a natural history museum, 23 of 80 boys and 19 of 80 girls, both random samples, identified a geode as an Italian pastry. Test at the 0.05 level of significance whether the difference between the corresponding sample proportions is significant.

R.177 In a random sample of $n = 25$ servings of a breakfast cereal, the sugar content averaged 10.42 grams with a standard deviation of 1.76 grams. Assuming that these data constitute a sample from a normal population, construct a 95% confidence interval for σ, the standard deviation of the population sampled.

R.178 The following table shows how samples of the residents of three federally financed housing projects replied to the question whether they would continue to live there if they had the choice:

	Project 1	Project 2	Project 3
Yes	63	84	69
No	37	16	31

Test at the 0.01 level of significance whether the differences among the three sample proportions (of "yes" answers) may be attributed to chance.

R.179 An undercover government agent wants to determine what percentage of vendors at flea markets keep records for income tax purposes. How large a random sample will he need if he wants to be able to assert with 95% confidence that the error of his estimate, the sample percentage, is at most 6%.

(a) he has no idea about the true value;

(b) he is certain that the true percentage is at least 60%.

R.180 Among 210 persons with alcohol problems admitted to the psychiatric emergency room of a hospital, 36 were admitted on a Monday, 19 on a Tuesday, 18 on a Wednesday, 24 on a Thursday, 33 on a Friday, 40 on a Saturday, and 40 on a Sunday. Use the 0.05 level of significance to test the null hypothesis that this psychiatric emergency room can expect equally many persons with alcohol problems on each day of the week.

R.181 In a study of the relationship between family size and intelligence, 40 "only children" had an average IQ of 101.5 and 50 "first borns" in families with two children had an average IQ of 105.9. If it can be assumed that $\sigma_1 = \sigma_2 = 5.9$ for such data, test at the 0.01 level of significance whether the difference between the two sample means is significant.

R.182 A geneticist found that in independent random samples of 100 men and 100 women there were 31 men and 24 women with an inherited blood disorder. Can she conclude at the 0.01 level of significance that the corresponding true proportion for men is significantly greater than that for women?

(a) Comment on the formulation of this question.

(b) Restate the question as it should have been asked, and answer it by performing the appropriate test.

15 ANALYSIS OF VARIANCE

15.1   Differences Among *k* Means: An Example 418

15.2   The Design of Experiments: Randomization 422

15.3   One-Way Analysis of Variance 424

15.4   Multiple Comparisons 430

15.5   The Design of Experiments: Blocking 436

15.6   Two-Way Analysis of Variance 438

15.7   Two-Way Analysis of Variation Without Interaction 438

15.8   The Design of Experiments: Replication 443

15.9   Two-Way Analysis of Variation with Interaction 443

15.10   The Design of Experiments: Further Considerations 450

15.11   Checklist of Key Terms 457

15.12   References 458

*I*n this chapter we shall generalize the material in Sections 12.5 and 12.6 by considering problems in which we must decide whether observed differences among more than two sample means can be attributed to chance, or whether they are indicative of real differences among the means of the populations sampled. For instance, we may want to decide on the basis of sample data whether there really is a difference in the effectiveness of three methods of teaching a foreign language. We may also want to compare the average yield per acre of eight varieties of wheat, we may want to judge whether there really is a difference in the average mileage obtained with five kinds of gasoline, or we may want to determine whether there really is a difference in the durability of four exterior house paints. The method we use for the analysis of problems like this is called an **analysis of variance,** or an **ANOVA** for short.

Beyond this, an analysis of variance can be used to sort out several questions at the same time. For instance, with regard to the first of our four examples in the preceding paragraph, we might also ask whether the observed differences among the sample means are really due to the differences in teaching the foreign language and not due to the quality of the teaching or the merits of the textbooks being used, or perhaps the intelligence of the students being taught. Similarly, with regard to the different varieties of wheat, we might ask whether the differences we observe in their yield are really due to their quality and not due to the use of different fertilizers, differences in the composition of the soil, or perhaps differences in the amount of irrigation that is applied to the soil. Considerations like this lead to the important subject of **experimental design;** namely, to the problem of planning experiments in such a way that meaningful questions can be asked and put to a test.

Following an introductory example in Section 15.1 and the discussion of **randomization** in Section 15.2, we shall present the **one-way analysis of variance** in Section 15.3, followed by a discussion of **multiple comparisons** in Section 15.4. The latter are designed to sort out interpretations when an analysis of variance leads to significant results. Subsequently, the notion of **blocking** in Section 15.5 leads to the analysis of **two-way experiments** in Section 15.6. Various related topics are introduced in the remainder of the chapter.

15.1 DIFFERENCES AMONG k MEANS: AN EXAMPLE

To illustrate the kind of situation in which we might perform an analysis of variance, consider the following part of a study of the calcium contamination of river water. The data are the amounts of calcium (average parts per million) measured at three different locations along the Mississippi River:

Location M:	42	37	41	39	43	41
Location N:	37	40	39	38	41	39
Location O:	32	28	34	32	30	33

As can easily be verified, the means of these three samples are 40.5, 39.0, and 31.5, and what we would like to know is whether the differences among them are significant or whether they can be attributed to chance.

In a problem like this, we denote the means of the k populations sampled by $\mu_1, \mu_2, \ldots,$ and μ_k, and test the null hypothesis $\mu_1 = \mu_2 = \cdots = \mu_k$ against the alternative hypothesis that these μ's are not all equal.[†] This null hypothesis would be supported when the differences among the sample means are small, and the alternative hypothesis would be supported when at least some of the differences among the sample means are large. Thus, we need a measure of the discrepancies among the $\bar{x}$'s, and with it a rule that tells us when the discrepancies are so large that the null hypothesis can be rejected.

[†]For work later in this chapter, it will be desirable to write the k means as $\mu_1 = \mu + \alpha_1, \mu_2 = \mu + \alpha_2, \ldots,$ and $\mu_k = \mu + \alpha_k$, where

$$\mu = \frac{\mu_1 + \mu_2 + \cdots + \mu_k}{k}$$

is called the **grand mean** and the α's, whose sum is zero (see Exercise 15.27), are called the **treatment effects.** In this notation, we test the null hypothesis $\alpha_1 = \alpha_2 = \cdots = \alpha_k = 0$ against the alternative hypothesis that the α's are not all equal to zero.

To begin with, let us make two assumptions that are critical to the method by which we shall analyze our problem:

The data constitute random samples from normal populations. These normal populations all have the same standard deviation.

In our illustration we have samples from $k = 3$ populations, one for each location, and we shall assume that these populations are normal populations with the same standard deviation σ. The three populations may or may not have equal means; indeed, this is precisely what we hope to discover with an analysis of variance.

Of course, the value of σ is unknown, but in an analysis of variance we shall estimate σ^2, the population variance, in two different ways, and then base the decision whether or not to reject the null hypothesis on the ratio of these two estimates. The first of these two estimates will be based on the variation *among* the $\bar{x}$'s, and it will tend to be greater than what we might expect when the null hypothesis is *false*. The second estimate will be based on the variation *within* the samples, and hence it will not be affected by the null hypothesis being true or false.

Let us begin with the first of the two estimates of σ^2 by calculating the variance of the $\bar{x}$'s. Since the mean of the three $\bar{x}$'s is

$$\frac{40.5 + 39.0 + 31.5}{3} = 37.0$$

substitution into the formula that defines the sample variance (see page 83) yields

$$s_{\bar{x}}^2 = \frac{(40.5 - 37.0)^2 + (39.0 - 37.0)^2 + (31.5 - 37.0)^2}{3 - 1}$$

$$= 23.25$$

where the subscript $\bar{x}$ serves to indicate that $s_{\bar{x}}^2$ is the variance of the sample means.

If the null hypothesis is true, we can look upon the three samples as samples from one and the same population and, hence, upon $s_{\bar{x}}^2$ as an estimate of $\sigma_{\bar{x}}^2$, the square of the standard error of the mean. Now, since

$$\sigma_{\bar{x}} = \frac{\sigma}{\sqrt{n}}$$

for random samples of size n from an infinite population, we can look upon $s_{\bar{x}}^2$ as an estimate of

$$\sigma_{\bar{x}}^2 = \left(\frac{\sigma}{\sqrt{n}}\right)^2 = \frac{\sigma^2}{n}$$

and, therefore, upon $n \cdot s_{\bar{x}}^2$ as an estimate of σ^2. For our illustration, we thus have $n \cdot s_{\bar{x}}^2 = 6(23.25) = 139.5$ as an estimate of σ^2, the common variance of the three populations sampled.

This is the estimate of σ^2 based on the variation among the sample means, and if σ^2 were known, we could compare $n \cdot s_{\bar{x}}^2$ with σ^2 and reject the null hypothesis if $n \cdot s_{\bar{x}}^2$ is much greater than σ^2. However, in actual practice σ^2 is unknown, and we have no choice but to obtain another estimate of σ^2 that is not affected by the null hypothesis being true or false. As we said before, such a second estimate would be based on the variation within the samples, as measured by s_1^2, s_2^2, and s_3^2. The values of these three sample variances are $s_1^2 = 4.7$, $s_2^2 = 2.0$, and $s_3^2 = 4.7$ for our example, but rather than choose one of them, we *pool* (average) them, getting

$$\frac{s_1^2 + s_2^2 + s_3^2}{3} = \frac{4.7 + 2 + 4.7}{3} = 3.8$$

as our second estimate of σ^2.

We now have two estimates of σ^2, $n \cdot s_{\bar{x}}^2 = 139.5$ and $\dfrac{s_1^2 + s_2^2 + s_3^2}{3} = 3.8$, and it should be observed that the first estimate, based on the variation among the sample means, is much greater than the second, based on the variation within the samples. This suggests that the three population means are probably not all equal; namely, that the null hypothesis ought to be rejected. To put this comparison on a rigorous basis, we use the **F statistic**

Statistic for test concerning differences among means

$$F = \frac{\text{estimate of } \sigma^2 \text{ based on the variation among the } \bar{x}\text{'s}}{\text{estimate of } \sigma^2 \text{ based on the variation within the samples}}$$

If the null hypothesis is true and the assumptions we made are valid, the sampling distribution of this statistic is the **F distribution,** which we met earlier in Chapter 13, where we also used it to compare two variances and referred to the F statistic as a **variance ratio.** Since the null hypothesis will be rejected only when F is large (that is, when the variation among the $\bar{x}$'s is too great to be attributed to chance), we base our decision on the criterion of Figure 15.1. For $\alpha = 0.05$ or 0.01, the values of F_α may be looked up in Table IV at the end of the book, and if we compare the means of k random samples of size n, the **numerator and denominator degrees of freedom** are, respectively, $k - 1$ and $k(n - 1)$.

FIGURE 15.1
Test criterion based on F distribution

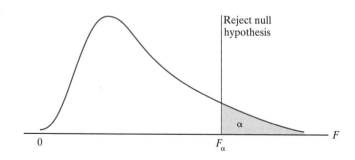

Reject null hypothesis

α

0 F_α F

Returning to our numerical illustration, we find that $F = \dfrac{139.5}{3.8} \approx 36.7$ rounded to one decimal, and since this exceeds 6.36, the value of $F_{0.01}$ for $k - 1 = 3 - 1 = 2$ and $k(n - 1) = 3(6 - 1) = 15$ degrees of freedom, the null hypothesis must be rejected at the 0.01 level of significance. In other words, the differences among the three sample means are too large to be attributed to chance. (As we shall see later, the p-value for $F = 36.7$ and 2 and 15 degrees of freedom is actually less than 0.000002.)

The technique we have described in this section is the simplest form of an analysis of variance. Although we could go ahead and perform F tests for differences among k means without further discussion, it will be instructive to look at the problem from a somewhat different analysis-of-variance point of view, and we shall do so in Section 15.3.

■ Exercises

15.1   Samples of peanut butter produced by three different companies were tested for aflatoxin content (parts per billion), with the following results:

Company 1:   0.5   6.3   1.1   2.7   4.3   4.3
Company 2:   2.5   1.8   3.6   5.2   1.2   0.7
Company 3:   3.3   1.5   0.4   4.2   2.2   1.0

(a) Calculate $n \cdot s_{\bar{x}}^2$ for these data, the mean of the variances of the three samples, and the value of F.
(b) Assuming that the data constitute random samples from three normal populations with the same standard deviation, test at the 0.05 level of significance whether the differences among the three sample means can be attributed to chance.

15.2   An agronomist planted three test plots each with four varieties of wheat and obtained the following yields (in pounds per plot):

Variety A: 65   64   60
Variety B: 55   56   63
Variety C: 56   59   59
Variety D: 62   59   62

(a) Calculate $n \cdot s_{\bar{x}}^2$ for these data, the mean of the variances of the four samples, and the value of F.
(b) Assuming that these data constitute random samples from four normal populations with the same standard deviation, test at the 0.01 level of significance whether the differences among the four sample means can be attributed to chance.

15.3   What are the numerator and denominator degrees of freedom of the F distribution when we compare the means of
(a) $k = 6$ random samples of size $n = 10$;
(b) $k = 5$ random samples of size $n = 8$?

15.4   Following are the caloric values of the fat content of meals served at three elementary schools:

School 1:   150   139   148   138   119   140   145   125
School 2:   132   128   126   140   145   136   148   125
School 3:   147   142   139   155   136   130   134   145

(a) Given that $\bar{x}_1 = 138$, $\bar{x}_2 = 135$, $\bar{x}_3 = 141$, $s_1^2 = 118.29$, $s_2^2 = 76.29$, and $s_3^2 = 64.00$, calculate $n \cdot s_{\bar{x}}^2$ for these data, the mean of the variances of the three samples, and the value of F.

(b) Assuming that these data constitute random samples from three normal populations with the same standard deviation, test at the 0.05 level of significance whether the differences among the three sample means can be attributed to chance.

15.5 Following are the fourth-grade reading comprehension scores on a standardized test obtained by random samples of students from three large schools:

School 1:	81	83	77	72	86	92	83	78	80	75
School 2:	73	112	66	104	95	81	62	76	129	90
School 3:	84	89	81	76	79	83	85	74	80	78

Explain why the method described in Section 15.1 should probably not be used to test for significant differences among the three sample means.

15.2 THE DESIGN OF EXPERIMENTS: RANDOMIZATION

Suppose that we are asked to compare the cleansing action of three detergents on the basis of the following whiteness readings made on 15 swatches of white cloth, which were first soiled with India ink and then washed in an agitator-type machine with the respective detergents:

Detergent X:	77	81	71	76	80
Detergent Y:	72	58	74	66	70
Detergent Z:	76	85	82	80	77

The means of the three samples are 77, 68, and 80, and an analysis of variance showed that the means of the three populations sampled are not all equal.

It would seem natural to conclude that the three detergents are not equally effective, but a moment's reflection will show that this conclusion is not so "natural" at all. For all we know, the swatches cleaned with detergent Y may have been more soiled than the others, the washing times may have been longer for detergent Z, there may have been differences in water hardness or water temperature, and even the instruments used to make the whiteness readings may have gone out of adjustment after the readings for detergents X and Z were made.

It is entirely possible, of course, that the differences among the three sample means, 77, 68, and 80, are due largely to differences in the quality of the three detergents, but we have just listed several other factors that could be responsible. It is important to remember that *a significance test may show that differences among sample means are too large to be attributed to chance, but it cannot say why the differences occurred.*

In general, if we want to show that one factor (among various others) can be considered the cause of an observed phenomenon, we must somehow make sure that none of the other factors can reasonably be held responsible. There are various ways in which this can be done; for instance, we can conduct a rigorously **controlled experiment** in which all variables except the one of concern are held fixed. To do this in the example dealing with the three detergents, we might soil the swatches with exactly equal amounts of India ink, always use exactly the same washing time, use water of exactly the same temperature and hardness, and inspect the measuring instruments after each use. Under these rigid conditions, significant differences among the sample means cannot be due to differently soiled swatches, or differences in washing time, water temperature, water hardness, or measuring instruments. On the positive side, the differences show that the detergents are not all equally effective if they are used in this narrowly restricted way. Of course, we cannot say whether the same differences would exist if the washing time is longer or shorter, if the water has a different temperature or hardness, and so on.

In most cases, "overcontrolled" experiments like the one just described do not really provide us with the kind of information we want. Also, such experiments are rarely possible in actual practice; for example, it would have been difficult in our illustration to be sure that the instruments really were measuring identically on repeated washings or that some other factor, not thought of or properly controlled, was not responsible for the observed differences in whiteness. So, we look for alternatives. At the other extreme we can conduct an experiment in which none of the extraneous factors is controlled, but in which we protect ourselves against their effects by **randomization.** That is, we design, or plan, the experiment in such a way that the variations caused by these extraneous factors can all be combined under the general heading of "chance."

In our example dealing with the three detergents, we could accomplish this by numbering the swatches (which need not be equally soiled) from 1 to 15, specifying the random order in which they are to be washed and measured, and randomly selecting the five swatches that are to be washed with each of the three detergents. When all the variations due to uncontrolled extraneous factors can thus be included under the heading of chance variation, we refer to the design of the experiment as a **completely randomized design.**

As should be apparent, randomization protects against the effects of the extraneous factors only in a probabilistic sort of way. For instance, in our example it is possible, though very unlikely, that detergent X will be randomly assigned to the five swatches that happen to be the least soiled, or that the water happens to be coldest when we wash the five swatches with detergent Y. It is partly for this reason that we often try to control some of the factors and randomize the others, and thus use designs that are somewhere between the two extremes that we have described.

Randomization protects against the effects of factors that cannot be completely controlled, but it does not relieve a person designing an experiment from the responsibility of designing it carefully simply because randomization will be used. In our example, a serious effort should be made to prepare the swatches as equally soiled as possible.

Finally, we must point out that randomization should be used even when all extraneous factors are carefully controlled. In our example, even if careful steps have been taken to control the amount of India ink with which the swatches are soiled, the wash temperature, the water hardness, and so on, assigning the swatches to the detergents should still be randomized.

15.3 ONE-WAY ANALYSIS OF VARIANCE

An **analysis of variance** expresses a measure of the total variation in a set of data as a sum of terms, each attributed to a specific source, or cause, of variation. Here, we shall describe this with reference to the calcium contamination example and, as we shall see, *the approach is different, but otherwise we accomplish exactly what we accomplished in Section 15.1*. In that example there were two such sources of variation: (1) the differences in location along the Mississippi River, and (2) chance, which in problems of this kind is called the **experimental error.** When there is only one source of variation other than chance, we refer to the analysis as a **one-way analysis of variance.** Other versions of the analysis of variance will be treated later in this chapter.

As a measure of the total variation of kn observations consisting of k samples of size n, we shall use the **total sum of squares**[†]

$$SST = \sum_{i=1}^{k} \sum_{j=1}^{n} (x_{ij} - \bar{x}..)^2$$

where x_{ij} is the jth observation of the ith sample ($i = 1, 2, \ldots, k$ and $j = 1, 2, \ldots, n$), and $\bar{x}..$ is the **grand mean,** the mean of all the kn measurements or observations. Note that if we divide the total sum of squares SST by $kn - 1$, we get the variance of the combined data.

If we let $\bar{x}_{i.}$ denote the mean of the ith sample (for $i = 1, 2, \ldots, k$), we can now write the following identity that forms the basis for a one-way analysis of variance:[‡]

Identity for one-way analysis of variance

$$SST = n \cdot \sum_{i=1}^{k} (\bar{x}_{i.} - \bar{x}..)^2 + \sum_{i=1}^{k} \sum_{j=1}^{n} (x_{ij} - \bar{x}_{i.})^2$$

It is customary to refer to the first term on the right, the quantity that measures the variation among the sample means, as the **treatment sum of squares** $SS(Tr)$, and to the second term, which measures the variation within the individual samples, as the **error sum of squares** SSE. The choice of the word "treatment" is explained by the origin of many analysis-of-variance techniques in agricultural experiments where different fertilizers, for example, are regarded as different

---

[†]The use of double subscripts and double summations is treated briefly in Section 3.8.
[‡]This identity may be derived by writing the total sum of squares as

$$SST = \sum_{i=1}^{k} \sum_{j=1}^{n} (x_{ij} - \bar{x}..)^2 = \sum_{i=1}^{k} \sum_{j=1}^{n} [(\bar{x}_{i.} - \bar{x}..) + (x_{ij} - \bar{x}_{i.})]^2$$

and then expanding the squares $[(\bar{x}_{i.} - \bar{x}..) + (x_{ij} - \bar{x}_{i.})]^2$ by means of the binomial theorem and simplifying algebraically.

treatments applied to the soil. So, we shall refer to the three locations along the Mississippi River as three treatments, and in other problems we may refer to four different nationalities as four different treatments, five different advertising campaigns as five different treatments, and so on. The word "error" in "error sum of squares" pertains to the experimental error, or chance.

In this notation, the identity for a one-way analysis of variance reads

$$SST = SS(Tr) + SSE$$

and since its proof requires quite a bit of algebraic manipulation, let us merely verify it numerically for the example of Section 15.1. Substituting the original data, the three sample means, and the grand mean (see pages 418 and 419) into the formulas for the three sums of squares, we get

$$
\begin{aligned}
SST = {}& (42 - 37)^2 + (37 - 37)^2 + (41 - 37)^2 + (39 - 37)^2 \\
& + (43 - 37)^2 + (41 - 37)^2 + (37 - 37)^2 + (40 - 37)^2 \\
& + (39 - 37)^2 + (38 - 37)^2 + (41 - 37)^2 + (39 - 37)^2 \\
& + (32 - 37)^2 + (28 - 37)^2 + (34 - 37)^2 + (32 - 37)^2 \\
& + (30 - 37)^2 + (33 - 37)^2 \\
= {}& 336
\end{aligned}
$$

$$
\begin{aligned}
SS(Tr) = {}& 6[(40.5 - 37)^2 + (39.0 - 37)^2 + (31.5 - 37)^2] \\
= {}& 279
\end{aligned}
$$

$$
\begin{aligned}
SSE = {}& (42 - 40.5)^2 + (37 - 40.5)^2 + (41 - 40.5)^2 + (39 - 40.5)^2 \\
& + (43 - 40.5)^2 + (41 - 40.5)^2 + (37 - 39.0)^2 + (40 - 39.0)^2 \\
& + (39 - 39.0)^2 + (38 - 39.0)^2 + (41 - 39.0)^2 + (39 - 39.0)^2 \\
& + (32 - 31.5)^2 + (28 - 31.5)^2 + (34 - 31.5)^2 + (32 - 31.5)^2 \\
& + (30 - 31.5)^2 + (33 - 31.5)^2 \\
= {}& 57
\end{aligned}
$$

and it can be seen that

$$SS(Tr) + SSE = 279 + 57 = 336 = SST$$

Although this may not be apparent immediately, what we have done here is very similar to what we did in Section 15.1. Indeed, $SS(Tr)$ divided by $k - 1$ equals the quantity that we denoted by $n \cdot s_{\bar{x}}^2$ and put into the numerator of the F statistic on page 420. Called the **treatment mean square,** it measures the variation among the sample means and it is denoted by $MS(Tr)$. Thus,

$$MS(Tr) = \frac{SS(Tr)}{k - 1}$$

and for the calcium contamination example we get $MS(Tr) = \frac{279}{2} = 139.5$. This equals the value we got for $n \cdot s_{\bar{x}}^2$ on page 419.

Similarly, SSE divided by $k(n-1)$ equals the mean of the k sample variances, $\frac{1}{3}(s_1^2 + s_2^2 + s_3^2)$ in our example, which we put into the denominator of the F statistic on page 420. Called the **error mean square,** it measures the variation within the samples and it is denoted by MSE. Thus,

$$MSE = \frac{SSE}{k(n-1)}$$

and for the calcium contamination example we get $MSE = \dfrac{57}{3(6-1)} = 3.8$. This equals the value we got for $\frac{1}{3}(s_1^2 + s_2^2 + s_3^2)$ on page 419.

Since F was defined on page 420 as the ratio of these two measures of the variation among the sample means and within the samples, we can now write

Statistic for test concerning differences among means

$$F = \frac{MS(Tr)}{MSE}$$

In practice, we display the work required for the determination of F in the following kind of table, called an **analysis of variance table:**

Source of variation	Degrees of freedom	Sum of squares	Mean square	F
Treatments	$k-1$	$SS(Tr)$	$MS(Tr)$	$\dfrac{MS(Tr)}{MSE}$
Error	$k(n-1)$	SSE	MSE	
Total	$kn-1$	SST		

The degrees of freedom for treatments and error are the numerator and denominator degrees of freedom referred to on page 420. Note that they equal the quantities we divide into the sums of squares to obtain the corresponding mean squares.

After we have calculated F, we proceed as in Section 15.1. Assuming again that the data constitute samples from normal populations with the same standard deviation, we reject the null hypothesis

$$\mu_1 = \mu_2 = \cdots = \mu_k$$

against the alternative hypothesis that these μ's are not all equal, or the null hypothesis

$$\alpha_1 = \alpha_2 = \cdots = \alpha_k = 0$$

against the alternative hypothesis that these treatment effects are not all equal to zero, if the value of F is greater than or equal to F_α for $k - 1$ and $k(n - 1)$ degrees of freedom.

EXAMPLE 15.1   Use the sums of squares calculated on page 425 to construct an analysis of variance table for the calcium contamination example, and test at the 0.01 level of significance whether the differences among the means obtained for the three locations along the Mississippi River are significant.

Solution   Since $k = 3$, $n = 6$, $SST = 336$, $SS(Tr) = 279$, and $SSE = 57$, we get $k - 1 = 2$, $k(n - 1) = 15$, $MS(Tr) = \frac{279}{2} = 139.5$, $MSE = \frac{57}{15} = 3.8$, and $F = \frac{139.5}{3.8} = 36.71$ rounded to two decimals. All these results are summarized in the following table:

Source of variation	Degrees of freedom	Sum of squares	Mean square	F
Treatments	2	279	139.5	36.71
Error	15	57	3.8	
Total	17	336		

Note that the total degrees of freedom, $kn - 1$, is the sum of the degrees of freedom for treatments and error.

Finally, since $F = 36.71$ exceeds 6.36, the value of $F_{0.01}$ for 2 and 15 degrees of freedom, we conclude as in Section 15.1 that the null hypothesis must be rejected.   ∎

The numbers that we used in the example dealing with the calcium contamination of Mississippi River water were intentionally rounded so that the calculations would be easy. In actual practice, the calculation of the sums of squares can be quite tedious unless we use the following computing formulas in which $T_{i.}$ denotes the sum of the values for the ith treatment (that is, the sum of the values in the ith sample), and $T_{..}$ denotes the **grand total** of all the data:

Computing formulas for sums of squares (equal sample sizes)

$$SST = \sum_{i=1}^{k} \sum_{j=1}^{n} x_{ij}^2 - \frac{1}{kn} \cdot T_{..}^2$$

$$SS(Tr) = \frac{1}{n} \cdot \sum_{i=1}^{k} T_{i.}^2 - \frac{1}{kn} \cdot T_{..}^2$$

and by subtraction

$$SSE = SST - SS(Tr)$$

EXAMPLE 15.2   Use these computing formulas to verify the sums of squares obtained on page 425.

Solution   First calculating the various totals, we get

$$T_{1.} = 42 + 37 + 41 + 39 + 43 + 41 = 243$$
$$T_{2.} = 37 + 40 + 39 + 38 + 41 + 39 = 234$$
$$T_{3.} = 32 + 28 + 34 + 32 + 30 + 33 = 189$$
$$T_{..} = 243 + 234 + 189 = 666$$

and

$$\sum \sum x^2 = 42^2 + 37^2 + 41^2 + \cdots + 32^2 + 30^2 + 33^2$$
$$= 24{,}978$$

Then, substituting these totals together with $k = 3$ and $n = 6$ into the formulas for the sums of squares, we get

$$SST = 24{,}978 - \frac{1}{18}(666)^2$$
$$= 24{,}978 - 24{,}642$$
$$= 336$$

$$SS(Tr) = \frac{1}{6}(243^2 + 234^2 + 189^2) - 24{,}642$$
$$= 24{,}921 - 24{,}642$$
$$= 279$$

and

$$SSE = 336 - 279 = 57$$

As can be seen, these results are identical with the ones obtained before.   ■

A MINITAB printout of the analysis of variance of this section is shown in Figure 15.2. Besides the degrees of freedom, the sums of squares, the mean squares, and the value of F and the corresponding p-value, it provides information of no relevance here that has been deleted. The p-value is given as 0.000 rounded to three decimals; according to our HEWLETT PACKARD STAT/MATH calculator it is 0.0000017 rounded to seven decimals.

FIGURE 15.2
Analysis of variance of the calcium contamination data.

```
MTB > AOVOneway C1-C3.


Analysis of Variance
Source      DF         SS        MS        F        P
Factor       2     279.00    139.50    36.71    0.000
Error       15      57.00      3.80
Total       17     336.00
```

The method discussed here applies only when the sample sizes are all equal, but minor modifications make it applicable also when the sample sizes are not all equal. If the ith sample is of size n_i, the computing formulas for the sums of squares become

Computing formulas for sums of squares (unequal sample sizes)

$$SST = \sum_{i=1}^{k} \sum_{j=1}^{n_i} x_{ij}^2 - \frac{1}{N} \cdot T_{..}^2$$

$$SS(Tr) = \sum_{i=1}^{k} \frac{T_{i.}^2}{n_i} - \frac{1}{N} \cdot T_{..}^2$$

$$SSE = SST - SS(Tr)$$

where $N = n_1 + n_2 + \cdots + n_k$. The only other change is that the total number of degrees of freedom is $N - 1$, and the degrees of freedom for treatments and error are $k - 1$ and $N - k$.

EXAMPLE 15.3   A laboratory technician wants to compare the breaking strength of three kinds of thread and originally he had planned to repeat each determination six times. Not having enough time, however, he has to base his analysis on the following results (in ounces):

Thread 1:   18.0   16.4   15.7   19.6   16.5   18.2
Thread 2:   21.1   17.8   18.6   20.8   17.9   19.0
Thread 3:   16.5   17.8   16.1

Assuming that these data constitute random samples from three normal populations with the same standard deviation, perform an analysis of variance to test at the 0.05 level of significance whether the differences among the sample means are significant.

Solution

1. H_0: $\mu_1 = \mu_2 = \mu_3$
   H_A: The μ's are not all equal.
2. $\alpha = 0.05$
3. Reject the null hypothesis if $F \geq 3.89$, where F is to be determined by an analysis of variance and 3.89 is the value of $F_{0.05}$ for $k - 1 = 3 - 1 = 2$ and $N - k = 15 - 3 = 12$ degrees of freedom; otherwise, accept the null hypothesis or reserve judgment.
4. Substituting $n_1 = 6, n_2 = 6, n_3 = 3, N = 15, T_{1.} = 104.4, T_{2.} = 115.2,$ $T_{3.} = 50.4, T_{..} = 270.0,$ and $\Sigma\Sigma\, x^2 = 4{,}897.46$ into the computing formulas for the sums of squares, we get

$$SST = 4{,}897.46 - \frac{1}{15}(270.0)^2 = 37.46$$

$$SS(Tr) = \frac{104.4^2}{6} + \frac{115.2^2}{6} + \frac{50.4^2}{3} - \frac{1}{15}(270.0)^2$$

$$= 15.12$$

and

$$SSE = 37.46 - 15.12 = 22.34$$

Since the degrees of freedom are $k - 1 = 3 - 1 = 2$, $N - k = 15 - 3 = 12$, and $N - 1 = 14$, we then get

$$MS(Tr) = \frac{15.12}{2} = 7.56 \quad MSE = \frac{22.34}{12} = 1.86 \text{ and } F = \frac{7.56}{1.86} = 4.06$$

and all these results are summarized in the following analysis-of-variance table:

Source of variation	Degrees of freedom	Sum of squares	Mean square	F
Treatments	2	15.12	7.56	4.06
Error	12	22.34	1.86	
Total	14	37.46		

5. Since $F = 4.06$ exceeds 3.89, the null hypothesis must be rejected; in other words, we conclude that there is a difference in the strength of the three kinds of thread.   ∎

If the level of significance had not been specified in this example, we could have noted that $F = 4.06$ falls between 3.89 and 6.93, the values of $F_{0.05}$ and $F_{0.01}$ for 2 and 12 degrees of freedom, and we could simply have stated for the p-value that $0.01 < p < 0.05$. Or, using the same calculator as in Example 15.2, we would have found that the p-value is 0.0450 rounded to four decimals.

15.4   MULTIPLE COMPARISONS

An analysis of variance provides a method for determining whether the differences among k sample means are significant. It does not, however, tell us which means are significantly different from which others. Consider, for example, the following data on the amounts of time (in minutes) it took a certain person to drive to work on five days, selected at random, along each of four different routes:

Route 1:	25	26	25	25	28
Route 2:	27	27	28	26	26
Route 3:	28	29	33	30	30
Route 4:	28	29	27	30	27

First we shall have to see whether the differences among the four sample means 25.8, 26.8, 30.0, and 28.2, are significant. To this end we performed an analysis of variance with the use of a graphing calculator, with the results shown in Figure 15.3. Since the display screen of the TI-83 graphing calculator is fairly small, only part of the results are shown immediately, as in the upper part of Figure 15.3. The rest of the results are obtained by scrolling down, and they are shown in the lower part of Figure 15.3. Thus, we find that $F = 8.74$ rounded to two decimals and that the corresponding p-value for the F distribution with 3 and 16 degrees of freedom is 0.001 rounded to three decimals. This means that the differences among the four sample means are significant, most certainly at the 0.01 level of significance.

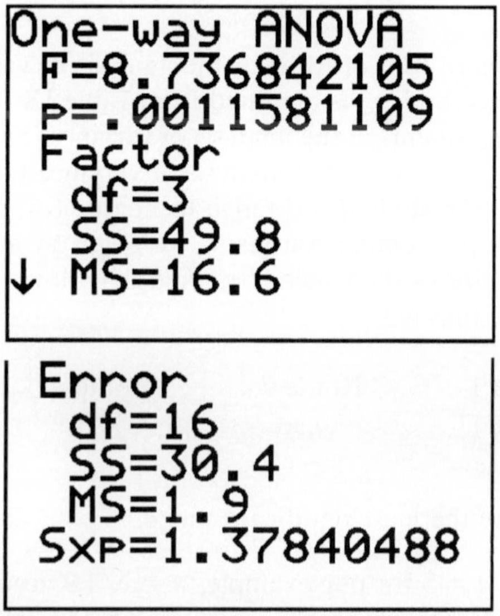

FIGURE 15.3
Analysis of variance of the travel-time data reproduced from the display screen of a TI-83 graphing calculator.

```
One-way ANOVA
 F=8.736842105
 P=.0011581109
 Factor
  df=3
  SS=49.8
↓ MS=16.6
```

```
Error
 df=16
 SS=30.4
 MS=1.9
Sxp=1.37840488
```

While the route with the lowest average driving time, Route 1, certainly appears to be faster than Route 3, we are not so sure whether we can declare Route 1 to be faster also than Route 2. It is true that 25.8 is less than 26.8, but at this time we have no idea whether the difference between these two means is significant. Of course, we could perform an ordinary two-sample t test to compare these two routes, but there are altogether $\binom{4}{2} = 6$ possible pairs, and if we perform that many tests there is a good chance that we may commit at least one Type I error. [If the t tests are performed at the 0.05 level of significance, the probability is $1 - (0.95)^6$ or about 0.26 of committing at least one Type I error.]

To control the Type I error probabilities when conducting comparisons like these, an area of study called **multiple comparisons** has been developed in fairly recent years. This is a complicated topic that is widely misunderstood, and there are a few issues that have not been clarified even by the experts. Here we present one such method, based on what is called the **studentized range.**[†]

As we shall explain it here, this test applies only when the samples are all of the same size. The books referred to on page 458 also explain how to handle the case where the sample sizes are not all equal, and they discuss various alternative multiple comparisons tests, named after the statisticians who contributed most to their development.

The studentized range procedure is designed to control the overall probability of making at least one Type I error when comparing the various pairs of means. It is based on the argument that the difference between the means of two treatments (say, treatments i and j) is significant if

$$|\bar{x}_i - \bar{x}_j| \geq \frac{q_\alpha}{\sqrt{n}} \cdot s$$

where s is the square root of MSE in the analysis of variance, α is the overall level of significance, and q_α is obtained from Table IX for the given values of k (the number of treatments in the analysis of variance) and df (the number of degrees of freedom for error in the analysis of variance table).

When using the studentized range technique (or, for that matter, any one of the many multiple comparison tests), we begin by arranging the treatments according to the size of their means, ranked from low to high. For our driving-time example, we thus get

Route 1	Route 2	Route 4	Route 3
25.8	26.8	28.2	30.0

Then we calculate the least significant range, $\dfrac{q_\alpha}{\sqrt{n}} \cdot s$, for the studentized range technique. Since $n = 5$ for our example, $s = \sqrt{1.9}$ according to Figure 15.3, or 1.38 rounded to two decimals, and $q_{0.05} = 4.05$ for $k = 4$ and df $= 16$ in Table IX, we get

$$\frac{q_\alpha}{\sqrt{n}} \cdot s = \frac{4.05}{\sqrt{5}} \cdot 1.38 = 2.50$$

rounded to two decimals.

Calculating the absolute values of the differences between the means of all possible pairs of routes, we get 1.0 for Routes 1 and 2, 2.4 for Routes 1 and 4, 4.2 for Routes 1 and 3, 1.4 for Routes 2 and 4, 3.2 for Routes 2 and 3, and 1.8 for Routes 4 and 3. As can be seen, only those for Routes 1 and 3 and those for Routes 2 and 3 exceed 2.50 and, hence, are significant. To summarize all this

[†]**Studentizing** is the process of dividing a statistic by a statistically independent estimate of scale. The expression is derived from the *nom de plume* of W. S. Gosset, who first introduced the process in 1907 by discussing the distribution of the mean divided by the sample standard deviation.

information we draw a line under all sets of treatments for which the differences between the means are not significant, and for our example we thus get

Route 1	Route 2	Route 4	Route 3
25.8	26.8	28.2	30.0

Being interested in minimizing the driving time, this tells us that Routes 1, 2, and 4 *as a group* are preferable to Route 3, and that Routes 3 and 4 *as a group* are less desirable than the other two. To go further than that, we may have to consider some other factors; perhaps, the beauty of the scenery along the way.

■ E x e r c i s e s

15.6 An experiment is performed to determine which of three golf ball brands, A, B, or C, will attain the greatest distance when driven from a tee. Criticize the experiment if
(a) one golf pro hits all the brand A balls, another hits all the brand B balls, and a third hits all the brand C balls;
(b) all brand A balls are hit first, brand B balls next, and brand C balls last.

15.7 A botanist wants to compare three kinds of tulip bulbs having, respectively, red, white, and yellow flowers. She has four bulbs of each kind and plants them in a flower bed in the following pattern, where R, W, and Y denote the three colors:

R	R	R	R
W	W	W	W
Y	Y	Y	Y

When the plants reach maturity, she measures their height and performs an analysis of variance. Criticize this experiment and indicate how it might be improved.

15.8 To compare three weight-reducing diets, five of 15 persons are assigned randomly to each of the diets. After they have been on these diets for two weeks, a one-way analysis of variance is performed on their weight losses to test the null hypothesis that the three diets are equally effective. It has been claimed that this procedure cannot yield a valid conclusion because the five persons who originally weighed the most might be assigned to the same diet. Verify that the probability of this happening by chance is about 0.001.

15.9 With reference to the preceding exercise, suppose that five of the 15 persons are assigned randomly to each of the three diets, but it is discovered subsequently that the five persons who originally weighed the most are all assigned to the same diet. Should the one-way analysis of variance still be performed?

15.10 Rework part (b) of Exercise 15.1 by performing an analysis of variance, using the computing formulas to obtain the necessary sums of squares. Compare the values of F obtained here and in part (a) of Exercise 15.1.

15.11 Use appropriate computer software or a graphing calculator to rework Exercise 15.1.

15.12 Rework part (b) of Exercise 15.4 by performing an analysis of variance, using the computing formulas to obtain the necessary sums of squares. Compare the values of F obtained here and in part (a) of Exercise 15.4.

 15.13 Use appropriate computer software or a graphing calculator to rework Exercise 15.4.

15.14 The following are the numbers of mistakes made on five occasions by four word-processor operators, setting a technical report:

Operator 1:	10	13	9	11	12
Operator 2:	11	13	8	16	12
Operator 3:	10	15	13	11	15
Operator 4:	15	7	11	12	9

Assuming that the necessary assumptions can be met, perform an analysis of variance and decide at the 0.05 level of significance whether the differences among the four sample means can be attributed to chance.

15.15 The following data show the yields of soybeans (in bushels per acre) planted two inches apart on essentially similar plots with the rows 20, 24, 28, and 32 inches apart:

20 in.	24 in.	28 in.	32 in.
23.1	21.7	21.9	19.8
22.8	23.0	21.3	20.4
23.2	22.4	21.6	19.3
23.4	21.1	20.2	18.5
23.6	21.9	21.6	19.1
21.7	23.4	23.8	21.9

Assuming that these data constitute random samples from four normal populations with the same standard deviation, perform an analysis of variance to test at the 0.01 level of significance whether the differences among the four sample means can be attributed to chance.

 15.16 Use appropriate computer software or a graphing calculator to rework Exercise 15.15.

15.17 A large marketing firm uses many photocopy machines, several of each of four different models. During the last six months, the office manager has tabulated for each machine the average number of minutes per week that it is out of service due to repairs, resulting in the following data:

Model G:	56	61	68	42	82	70	
Model H:	74	77	92	63	54		
Model K:	25	36	29	56	44	48	38
Model M:	78	105	89	112	61		

Assuming that the necessary assumptions can be met, perform an analysis of variance to decide whether the differences among the means of the four samples can be attributed to chance. Use $\alpha = 0.01$. (*Hint*: Make use of the results that the totals for the four samples are 379, 360, 276, and 445, that the grand total is 1,460, and that $\Sigma\Sigma x^2 = 104,500$.)

15.18 When used with three different lubricants, a specific group of machine parts show the following weight losses (in milligrams) due to friction:

Lubricant X: 10   13   12   10   14   8   12   13
Lubricant Y: 9   8   12   9   8 11   7   6   8   11   9
Lubricant Z: 6   7   7   5   9   8   4   10

Assuming that these data constitute random samples from three normal populations with the same standard deviation, perform an analysis of variance to decide whether the differences among the three sample means can be attributed to chance. Use the 0.01 level of significance.

15.19 To study its performance, a newly designed motorboat was timed over a marked course under various wind and water conditions. Assuming that the necessary conditions can be met, use the following data (in minutes) to test at the 0.05 level of significance whether the difference among the three sample means is significant:

Calm conditions: 26   19   16   22
Moderate conditions: 25   27   25   20   18   23
Choppy conditions: 23   25   28   31   26

 15.20 Use appropriate computer software or a graphing calculator to rework
(a) Exercise 15.18;
(b) Exercise 15.19.

 15.21 The following values are the percentages of the previous year's fruit yield for apple trees managed under eight different spraying schedules.

Schedule						
A	130	98	128	106	139	121
B	142	133	122	131	132	141
C	114	141	95	123	118	140
D	77	99	84	76	70	75
E	109	86	113	101	103	112
F	148	143	111	142	131	100
G	149	129	134	108	119	126
H	92	129	111	103	107	125

Assuming that the necessary assumptions can be met, use appropriate computer software to conduct an analysis of variance with $\alpha = 0.05$.

15.22 In Example 15.1 we did an analysis of variance for the data given originally on page 418, where the means for the three locations along the Mississippi River were 40.5, 39.0, and 31.5. Use the studentized range method to perform a multiple comparisons test at the 0.01 level of significance, and discuss the results assuming that low calcium contamination is desirable.

15.23 As a continuation of Exercise 15.15, use the studentized range method to perform a multiple comparisons test at the 0.01 level of significance and interpret the results.

15.24 As a continuation of Exercise 15.21, use the studentized range method to perform a multiple comparisons test at the 0.05 level of significance and interpret the results.

15.25 An analysis of variance and a subsequent multiple comparisons test of the performance of four real estate persons yielded the following results:

Mr. Brown   Ms. Jones   Mr. Black   Mrs. Smith
_____   _____

where Mrs. Smith had the highest average sales. Interpret the results.

15.26 An analysis of variance and a subsequent multiple comparisons test of the fat content of five frozen dinners yielded the following results:

A   C   B   F   D   E
_____

where A has the most fat and E has the least. Interpret these results, given that these foods are on a list of recommendations for a low-fat diet.

15.27 With reference to the footnote to page 418, verify that the sum of the treatment effects, the α's, is equal to zero.

15.28 Verify symbolically that for a one-way analysis of variance

(a) $\dfrac{SS(Tr)}{k-1} = n \cdot s_{\bar{x}}^2$;

(b) $\dfrac{SSE}{k(n-1)} = \dfrac{1}{k} \cdot \sum_{i=1}^{k} s_i^2$, where s_i^2 is the variance of the ith sample.

15.5   THE DESIGN OF EXPERIMENTS: BLOCKING

To introduce another concept that is of importance in the design of experiments, suppose that a reading comprehension test is given to random samples of eighth graders from each of four schools, and that the results are

School A: 87   70   92
School B: 43   75   56
School C: 70   66   50
School D: 67   85   79

The means of these four samples are 83, 58, 62, and 77, and since the differences among them are very large, it would seem reasonable to conclude that there are some real differences in the average reading comprehension of eighth graders in the four schools. This does not follow, however, from a one-way analysis of variance. We get

Source of variation	Degrees of freedom	Sum of squares	Mean square	F
Treatments	3	1,278	426	2.90
Error	8	1,176	147	
Total	11	2,454		

and since $F = 2.90$ is less than 4.07, the value of $F_{0.05}$ for 3 and 8 degrees of freedom, the null hypothesis (that the population means are all equal) cannot be rejected at the 0.05 level of significance.

The reason for this is that there are not only considerable differences among the four means, but also very large differences among the values within the samples. In the first sample they range from 70 to 92, in the second sample from 43 to 75, in the third sample from 50 to 70, and in the fourth sample from 67 to 85. Giving this some thought, it would seem reasonable to conclude that these differences within the samples may well be due to differences in ability, an extraneous factor (we might call it a "nuisance" factor) that was randomized by taking a random sample of eighth graders from each school. Thus, variations due to differences in ability were included in the experimental error; this "inflated" the error sum of squares that went into the denominator of the F statistic, and the results were not significant.

To avoid this kind of situation, we could hold the extraneous factor fixed, but this will seldom give us the information we want. In our example, we could limit the study to eighth graders with a scholastic grade-point average (GPA) of 90 or above, but then the results would apply only to eighth graders with a GPA of 90 or above. Another possibility is to vary the known source of variability (the extraneous factor) deliberately over as wide a range as necessary, and to do it in such a way that the variability it causes can be measured and, hence, eliminated from the experimental error. This means that we should plan the experiment in such a way that we can perform a **two-way analysis of variance,** in which the total variability of the data is partitioned into three components attributed, respectively, to treatments (in our example, the four schools), the extraneous factor, and experimental error.

As we shall see later, this can be accomplished in our example by randomly selecting from each school one eighth grader with a low GPA, one eighth grader with a typical GPA, and one eighth grader with a high GPA, where "low," "typical," and "high" are presumably defined in a rigorous way. Suppose, then, that we proceed in this way and get the results shown in the following table:

	Low GPA	Typical GPA	High GPA
School A	71	92	89
School B	44	51	85
School C	50	64	72
School D	67	81	86

What we have done here is called **blocking,** and the three levels of GPA are called **blocks.** In general, blocks are the levels at which we hold an extraneous factor fixed, so that we can measure its contribution to the total variability of the data by means of a two-way analysis of variance. In the scheme we chose for our example, we are dealing with **complete blocks**—they are complete in the sense that each treatment appears the same number of times in each block. There is one eighth grader from each school in each block.

Suppose, furthermore, that the order in which the students are tested may have some effect on the results. If the order is randomized within each block (that is, for each level of GPA), we refer to the design of the experiment as a **randomized block design.**

15.6 TWO-WAY ANALYSIS OF VARIANCE

The analysis of experiments where blocking is used to reduce the error sum of squares requires a **two-way analysis of variance.** In this kind of analysis the variables under consideration are referred to as "treatments" and "blocks" even though it applies also to **two-factor experiments,** where both variables are of material concern.

Before we go into any details, let us point out that there are essentially two ways of analyzing such two-variable experiments, and they depend on whether the two variables are independent, or whether there is an **interaction.** Suppose, for instance, that a tire manufacturer is experimenting with different kinds of treads, and he finds that one kind is especially good for use on dirt roads while another kind is especially good for use on hard pavement. If this is the case, we say that there is an interaction between road conditions and tread design. On the other hand, if each of the treads is affected equally by the different road conditions, we would say that there is no interaction and that the two variables (road conditions and tread design) are independent. The latter case will be taken up first in Section 15.7, while a method that is suitable also for testing for interactions will be described in Section 15.9.

15.7 TWO-WAY ANALYSIS OF VARIANCE WITHOUT INTERACTION

To formulate the hypotheses to be tested in the two-variable case, let us write μ_{ij} for the population mean that corresponds to the ith treatment and the jth block. In our example, μ_{ij} is the average reading comprehension score in the ith school for eighth graders with grade point average level j. We express this as

$$\mu_{ij} = \mu + \alpha_i + \beta_j$$

As in the footnote to page 418, μ is the grand mean (the average of all the population means μ_{ij}), and the α_i are the treatment effects (whose sum is zero). Correspondingly, we refer to the β_j as the **block effects** (whose sum is also zero), and write the two null hypotheses we want to test as

$$\alpha_1 = \alpha_2 = \cdots = \alpha_k = 0 \quad \text{and} \quad \beta_1 = \beta_2 = \cdots = \beta_n = 0$$

The alternative to the first null hypothesis (which in our illustration amounts to the hypothesis that the average reading comprehension of eighth graders is the same in all four schools) is that the treatment effects α_i are not all zero; the alternative to the second null hypothesis (which in our illustration amounts to the hypothesis that the average reading comprehension of eighth graders is the same for all three levels of GPA) is that the block effects β_j are not all zero.

To test the second of the null hypotheses, we need a quantity, similar to the treatment sum of squares, that measures the variation among the block means

(58, 72, and 83 for the data on page 437). So, if we let $T_{.j}$ denote the total of all the values in the jth block, substitute it for $T_{i.}$ in the computing formula for $SS(Tr)$ on page 429, sum on j instead of i, and interchange n and k, we obtain, analogous to $SS(Tr)$ the **block sum of squares**

Computing formula for block sum of squares

$$SSB = \frac{1}{k} \cdot \sum_{j=1}^{n} T_{.j}^2 - \frac{1}{kn} \cdot T_{..}^2$$

In a two-way analysis of variance (with no interaction) we compute SST and $SS(Tr)$ according to the formulas on page 429, SSB according to the formula immediately above, and then we get SSE by subtraction. Since

$$SST = SS(Tr) + SSB + SSE$$

we have

Error sum of squares (two-way analysis of variance)

$$SSE = SST - [SS(Tr) + SSB]$$

Observe that the error sum of squares for a two-way analysis of variance does not equal the error sum of squares for a one-way analysis of variance performed on the same data, even though we denote both with the symbol SSE. In fact, we are now partitioning the error sum of squares for the one-way analysis of variance into two terms: the block sum of squares, SSB, and the remainder that is the new error sum of squares, SSE.

We can now construct the following analysis-of-variance table for a two-way analysis of variance (with no interaction):

Source of variation	Degrees of freedom	Sum of squares	Mean square	F
Treatments	$k - 1$	$SS(Tr)$	$MS(Tr) = \dfrac{SS(Tr)}{k-1}$	$\dfrac{MS(Tr)}{MSE}$
Blocks	$n - 1$	SSB	$MSB = \dfrac{SSB}{n-1}$	$\dfrac{MSB}{MSE}$
Error	$(k-1)(n-1)$	SSE	$MSE = \dfrac{SSE}{(k-1)(n-1)}$	
Total	$kn - 1$	SST		

The mean squares are again the sums of squares divided by their respective degrees of freedom, and the two F values are the mean squares for treatments and blocks divided by the mean square for error. Also, the degrees of freedom for blocks are $n - 1$ (like those for treatments with n substituted for k), and the degrees of freedom for error are found by subtracting the degrees of freedom for treatments and blocks from $kn - 1$, the total number of degrees of freedom:

$$(kn - 1) - (k - 1) - (n - 1) = kn - k - n + 1$$
$$= (k - 1)(n - 1)$$

Thus, in the significance test for treatments the numerator and denominator degrees of freedom for F are $k - 1$ and $(k - 1)(n - 1)$, and in the significance test for blocks the numerator and denominator degrees of freedom for F are $n - 1$ and $(k - 1)(n - 1)$.

EXAMPLE 15.4   In the example that we used to illustrate the need for blocking, we gave the following data to compare the reading comprehension scores of eighth graders in four different schools using low, typical, and high grade-point averages as blocks:

	Low GPA	Typical GPA	High GPA
School A	71	92	89
School B	44	51	85
School C	50	64	72
School D	67	81	86

Assuming that the data consist of independent random samples from normal populations all having the same standard deviation, test at the 0.05 level of significance whether the differences among the means obtained for the four schools (treatments) are significant, and also whether the differences among the means obtained for the three levels of GPA (blocks) are significant.

 olution

1. H_0's $\alpha_1 = \alpha_2 = \alpha_3 = \alpha_4 = 0$
   $\beta_1 = \beta_2 = \beta_3 = 0$
   H_A's   The treatment effects are not all equal to zero; the block effects are not all equal to zero.
2. $\alpha = 0.05$ for both tests.
3. For treatments, reject the null hypothesis if $F \geqslant 4.76$, where F is to be determined by a two-way analysis of variance and 4.76 is the value of $F_{0.05}$ for $k - 1 = 4 - 1 = 3$ and $(k - 1)(n - 1) = (4 - 1)(3 - 1) = 6$ degrees of freedom. For blocks, reject the null hypothesis if $F \geqslant 5.14$, where F is to be determined by a two-way analysis of variance and 5.14 is the value of $F_{0.05}$ for $n - 1 = 3 - 1 = 2$ and $(k - 1)(n - 1) = (4 - 1)(3 - 1) = 6$ degrees of freedom. If either null hypothesis cannot be rejected, accept it or reserve judgment.

4. Substituting $k = 4, n = 3, T_{1.} = 252, T_{2.} = 180, \quad T_{3.} = 186, T_{4.} = 234,$ $T_{.1} = 232, T_{.2} = 288, T_{.3} = 332, T_{..} = 852,$ and $\Sigma\Sigma\, x^2 = 63{,}414$ into the computing formulas for the sums of squares, we get

$$SST = 63{,}414 - \tfrac{1}{12}(852)^2$$

$$= 63{,}414 - 60{,}492$$

$$= 2{,}922$$

$$SS(Tr) = \tfrac{1}{3}[252^2 + 180^2 + 186^2 + 234^2] - 60{,}492$$

$$= 1{,}260$$

$$SSB = \tfrac{1}{4}[232^2 + 288^2 + 332^2] - 60{,}492$$

$$= 1{,}256$$

and

$$SSE = 2{,}922 - [1{,}260 + 1{,}256]$$

$$= 406$$

Since the degrees of freedom are $k - 1 = 4 - 1 = 3,\ n - 1 = 3 - 1 = 2,$ $(k - 1)(n - 1) = (4 - 1)(3 - 1) = 6,$ and $kn - 1 = 4 \cdot 3 - 1 = 11,$ we then get $MS(Tr) = \dfrac{1{,}260}{3} = 420,\ MSB = \dfrac{1{,}256}{2} = 628,\ MSE = \dfrac{406}{6} = 67.67,\ F = \dfrac{420}{67.67} = 6.21$ for treatments, and $F = \dfrac{628}{67.67} = 9.28$ for blocks. All these results are summarized in the following analysis-of-variance table:

Source of variation	Degrees of freedom	Sum of squares	Mean square	F
Treatments	3	1,260	420	6.21
Blocks	2	1,256	628	9.28
Error	6	406	67.67	
Total	11	2,922		

5. For treatments, since $F = 6.21$ exceeds 4.76, the null hypothesis must be rejected; for blocks, since $F = 9.28$ exceeds 5.14, the null hypothesis must be rejected. In other words, we conclude that the average reading comprehension of eighth graders is not the same for the four schools, and also that it is not the same for the three levels of GPA. ∎

Observe that by blocking we were able to show that the differences among the means obtained for the four schools are significant, whereas without blocking, in the experiment described on pages 436 and 437, the differences among the means were not significant.

Needless to say, perhaps, there exists extensive software for performing a two-way analysis of variance. The MINITAB printout shown in Figure 15.4 does not provide the values of F and the corresponding p-values, but to calculate the values of F we need only the mean squares. The p-values were not needed since the levels of significance were specified, and the values of $F_{0.05}$ were available from Table IV.

FIGURE 15.4
Computer printout for Example 15.4.

```
MTB > Twoway c3 c1 c2.


Analysis of Variance for C3
Source          DF        SS        MS
C1               3    1260.0     420.0
C2               2    1256.0     628.0
Error            6     406.0      67.7
Total           11    2922.0
```

As we pointed out earlier, a two-way analysis of variance can also be used in the analysis of a two-factor experiment, where both variables (factors) are of material concern. It could be used, for example, in the analysis of the following data collected in an experiment designed to test whether the range of a missile flight (in miles) if affected by differences among launchers and also by differences among fuels.

	Fuel 1	Fuel 2	Fuel 3	Fuel 4
Launcher X	45.9	57.6	52.2	41.7
Launcher Y	46.0	51.0	50.1	38.8
Launcher Z	45.7	56.9	55.3	48.1

Note that we used a different format for this table to distinguish between two factor experiments (where uncontrolled extraneous factors are usually randomized over the entire experiment) and experiments where we deal with treatments and blocks.

Also, when a two-way analysis of variance is used in this way, we usually call the two variables **factor A** and **factor B** (instead of treatments and blocks and write SSA and MSA instead of $SS(Tr)$ and $MS(Tr)$; we still use SSB and MSB but now the B stands for factor B instead of blocks.

15.8 THE DESIGN OF EXPERIMENTS: REPLICATION

In Section 15.5 we showed how we can increase the amount of information to be gained from an experiment by blocking, that is, by eliminating the effect of an extraneous factor. Another way to increase the amount of information to be gained from an experiment is to increase the volume of the data. For instance, in the example on page 436, we might increase the size of the samples and give the reading comprehension test to twenty eighth graders from each school instead of three. For more complicated designs, the same thing can be accomplished by executing the entire experiment more than once, and this is called **replication.** With reference to the example on page 436, we might conduct the experiment (select and test twelve eighth graders) in one week, and then replicate (repeat) the entire experiment in the next week.

Conceptually, replication does not present any difficulties, but computationally it does, and we mentioned it here only because it is required for the work in Section 15.9. Furthermore, if an experiment requiring a two-way analysis of variance is replicated, it may then require a three-way analysis of variance, since replication, itself, could be a source of variation in the data. This would be the case, for instance, in our example dealing with the reading comprehension scores, if it got very hot and humid during the second week, thus making it difficult for the students to concentrate.

15.9 TWO-WAY ANALYSIS OF VARIATION WITH INTERACTION

When we first mentioned the concept of interaction, we cited an experiment where a tire manufacturer discovers that one kind of tread is especially good for use on dirt roads while another kind of tread is especially good for use on hard pavement. A similar situation arises when a farmer finds that one variety of corn does best with one kind of fertilizer while another variety does best with a different kind of fertilizer; or when it is observed that one person makes the fewest mistakes with one word processor while another person makes the fewest mistakes with a different word processor.

To consider a numerical example, let us refer to the two-factor experiment on page 442, the one that dealt with the effect of three different launchers and four different fuels on the range of certain missiles. If we analyzed these data by the method of Section 15.7, we would partition SST, a measure of the total variation among the data, into three components that are attributed, respectively, to the different launchers, the different fuels, and error (or chance). If there are interactions, which is quite possible, the variations they cause would be concealed because they are included as part of *SSE*, the error sum of squares. To isolate a sum of squares that can be attributed to interaction, we need some other way of measuring chance variation, and we shall do this by repeating the entire experiment. Suppose, then, that this yields the data shown in the following table:

	Fuel 1	Fuel 2	Fuel 3	Fuel 4
Launcher X	46.1	55.9	52.6	44.3
Launcher Y	46.3	52.1	51.4	39.6
Launcher Z	45.8	57.9	56.2	47.6

which we call Replicate 2 to distinguish it from the data on page 442, which we now call Replicate 1. Combining the two replications in one table, we get

	Fuel 1	Fuel 2	Fuel 3	Fuel 4
Launcher X	45.9, 46.1	57.6, 55.9	52.2, 52.6	41.7, 44.3
Launcher Y	46.0, 46.3	51.0, 52.1	50.1, 51.4	38.8, 39.6
Launcher Z	45.7, 45.8	56.9, 57.9	55.3, 56.2	48.1, 47.6

where the first value in each cell comes from Replicate 1 and the second value comes from Replicate 2.

Now we can represent chance variation by the variation *within* the 12 cells of the table, and in general our new error sum of squares becomes

$$SSE = \sum_{i=1}^{k} \sum_{j=1}^{n} \sum_{h=1}^{r} (x_{ijh} - \bar{x}_{ij.})^2$$

where x_{ijh} is the value corresponding to the *i*th treatment, the *j*th block, and the *h*th replicate, and $\bar{x}_{ij.}$ is the mean of the values in the cell corresponding to the *i*th treatment and the *j*th block.

Replacing the two values in each cell of the table immediately above by their mean, we get

	Fuel 1	Fuel 2	Fuel 3	Fuel 4
Launcher X	46	56.75	52.4	43
Launcher Y	46.15	51.55	50.75	39.2
Launcher Z	45.75	57.40	55.75	47.85

and this is what our data would look like after chance variation has been removed. In other words, the only variation that is left is due to treatments, blocks, and interaction, and if we performed a two-way analysis of variance as in Section 15.7, we would get corresponding treatment, block, and **interaction sums of squares,** with the latter being what used to be the error sum of squares. Actually, all this is done only conceptually. If we actually performed a two-way analysis of variance with the means replacing the two values in each cell, we would find that each of the sums of squares is divided by a factor of 2. Correspondingly, if there had been r replications, each sum of squares would have been divided by a factor of r.

As in Sections 15.3 and 15.7, there exist computing formulas for the various sums of squares in a two-way analysis of variance with interaction. However, since the necessary calculations are unwieldy to say the least, this work is just about always done with the use of computers. This is precisely what we shall do here, getting the required degrees of freedom, sums of squares and mean squares from the MINITAB printout shown in Figure 15.5.

FIGURE 15.5
Computer printout for the two-way analysis of variance with interaction.

```
MTB > Twoway c3 c1 c2.


Analysis of Variance for C3
Source         DF        SS        MS
C1              2    91.503    45.752
C2              3   570.825   190.275
Interaction     6    50.937     8.489
Error          12     7.775     0.648
Total          23   721.040
```

EXAMPLE 15.5   Analyze the replicated missile-range data on page 444, using a two-way analysis of variance with interaction and the 0.05 level of significance.

Solution   Copying the degrees of freedom, the sums of squares, and the mean squares from Figure 15.5, we find that the F values for treatments, blocks, and interaction are, respectively, $F_{Tr} = \dfrac{45.752}{0.648} = 70.6$, $F_B = \dfrac{190.275}{0.648} = 293.6$, and $F_I = \dfrac{8.489}{0.648} = 13.1$, all rounded to one decimal. All this information is displayed in the following analysis-of-variance table:

Source of variation	Degrees of freedom	Sum of squares	Mean square	F
Treatments	2	91.503	45.752	70.6
Blocks	3	570.825	190.275	293.6
Interaction	6	50.937	8.489	13.1
Error	12	7.775	0.648	
Total	23	721.040		

Since 70.6 exceeds 3.89, the value of $F_{0.05}$ for 2 and 12 degrees of freedom, 293.6 exceeds 3.49, the value of $F_{0.05}$ for 3 and 12 degrees of freedom, and 13.1 exceeds 3.00, the value of $F_{0.05}$ for 6 and 12 degrees of freedom, differences due to all three of these sources of variation are significant. The corresponding p-values (not given in the printout of Figure 15.5) are, respectively, 0.00000023, 0.000000000017, and 0.0001. They were obtained by means of a HEWLETT PACKARD STAT/MATH calculator.   ■

In an analysis like this, we could have included replication as another source of variation. In fact, we tried this, and as it turned out, the corresponding value obtained for F was not significant. It might have had an effect, say, if the two replications had been performed under distinctly different weather conditions.

■ **Exercises**

15.29 To compare the amounts of time that three television stations allot to commercials, a research worker measured the time devoted to commercials in random samples of 15 shows on each station. To her dismay, she discovered that there is so much variation within the samples—for one station the figures vary from 8 to 35 minutes—that it is virtually impossible to get significant results. Is there a way in which she might be able to overcome this obstacle?

15.30 To compare five word processors, A, B, C, D and E, four persons, 1, 2, 3, and 4, were timed in preparing a certain report on each of the machines. The results (in minutes) are shown in the following table:

	1	2	3	4
A	49.1	48.2	52.3	57.0
B	47.5	40.9	44.6	49.5
C	76.2	46.8	50.1	55.3
D	50.7	43.4	47.0	52.6
E	55.8	48.3	82.6	57.8

Explain why these data should not be analyzed by the method of Section 15.7.

15.31 The following are the cholesterol contents (in milligrams per package) that four laboratories obtained for 6-ounce packages of three very similar diet foods:

	Laboratory 1	Laboratory 2	Laboratory 3	Laboratory 4
Diet food A	3.7	2.8	3.1	3.4
Diet food B	3.1	2.6	2.7	3.0
Diet food C	3.5	3.4	3.0	3.3

Perform a two-way analysis of variance, using the 0.01 level of significance for both tests.

15.32 Four different, although supposedly equivalent, forms of a standardized achievement test in science were given to each of five students, and the following are the scores that they obtained:

	Student C	Student D	Student E	Student F	Student G
Form 1	77	62	52	66	68
Form 2	85	63	49	65	76
Form 3	81	65	46	64	79
Form 4	88	72	55	60	66

Perform a two-way analysis of variance, using the 0.01 level of significance for both tests.

15.33 A laboratory technician measured the breaking strength of each of five kinds of linen threads by using four different measuring instruments, I_1, I_2, I_3, and I_4, and obtained the following results (in ounces):

	I_1	I_2	I_3	I_4
Thread 1	20.9	20.4	19.9	21.9
Thread 2	25.0	26.2	27.0	24.8
Thread 3	25.5	23.1	21.5	24.4
Thread 4	24.8	21.2	23.5	25.7
Thread 5	19.6	21.2	22.1	21.1

Perform a two-way analysis of variance, using the 0.05 level of significance for both tests.

15.34 The following are the numbers of defectives produced by four workmen operating, in turn, three different machines:

		Workman		
	B_1	B_2	B_3	B_4
A_1	35	38	41	32
Machine A_2	31	40	38	31
A_3	36	35	43	25

Perform a two-way analysis of variance, using the 0.05 level of significance for both tests.

15.35 Three operators operated each of four different bonding machines (used to bond fine wire in the manufacture of integrated circuits). The operators were assigned to the bonding machines at random. The experiment was then repeated with a new randomization of operators to bonding machines. After all bonds were made, each was tested for bond strength by measuring the number of grams of force required to break the bond. The results of the experiment are as follows:

	Replicate 1				Replicate 2			
Bonder	A	B	C	D	A	B	C	D
Operator 1	11.8	9.6	12.6	10.2	10.6	11.9	9.8	9.9
Operator 2	10.4	12.4	11.0	10.5	12.0	10.3	10.0	11.6
Operator 3	9.6	10.2	11.4	3.1	11.8	9.9	9.1	5.8

Perform an analysis of variance to test the null hypotheses concerning operators, bonders, and the operator–bonder interaction at the 0.01 level of significance.

15.36 In an experiment designed to evaluate three detergents, a laboratory ran three loads of washing at each combination of detergents and water temperatures and obtained the following whiteness readings:

	Detergent A	Detergent B	Detergent C
Cold Water	45, 39, 46	43, 46, 41	55, 48, 53
Warm Water	37, 32, 43	40, 37, 46	56, 51, 53
Hot Water	42, 42, 46	44, 45, 38	46, 49, 42

Use the 0.01 level of significance to test for differences among the detergents, differences due to water temperature, and differences due to interactions.

15.37 A consumer-products-testing service wants to compare the quality of 24 cakes baked in its kitchen with each of four different mixes prepared according to three different recipes (varying the amounts of fresh ingredients added), once by Chef X and once by Chef Y. They ask a taste-tester to rate the cakes on a scale from 1 to 100, yielding the following results, where in each case the first figure pertains to the cake baked by Chef X and the second figure pertains to the cake baked by Chef Y:

	Mix A	Mix B	Mix C	Mix D
Recipe 1	66, 62	70, 68	74, 68	73, 67
Recipe 2	68, 61	71, 73	74, 70	66, 61
Recipe 3	75, 68	69, 71	67, 63	70, 66

Use the 0.05 level of significance to test for differences due to the different recipes, differences due to the different mixes, and differences due to a recipe–mix interaction.

15.38 Repeat the analysis of the data of Exercise 15.37, allowing also for differences due to the different chefs.

15.10 THE DESIGN OF EXPERIMENTS: FURTHER CONSIDERATIONS

In Section 15.5 we saw how blocking can be used to eliminate the variability due to one extraneous factor from the experimental error, and, in principle, several extraneous sources of variation can be handled in the same way. The only real problem is that this may inflate the size of an experiment beyond practical bounds. Suppose, for instance, that in the example dealing with the reading comprehension of eighth graders we would also like to eliminate whatever variability there may be due to differences in age (12, 13, or 14) and in sex. Allowing for all possible combinations of GPA, age, and sex, we will have to use $3 \cdot 3 \cdot 2 = 18$ different blocks, and if there is to be one eighth grader from each school in each block, we will have to select and test $18 \cdot 4 = 72$ eighth graders in all. If we also wanted to eliminate whatever variability there may be due to ethnic background, for which we might consider five categories, this would raise the required number of eighth graders to $72 \cdot 5 = 360$.

In this section we will show how problems like this can sometimes be resolved, at least in part, by planning experiments as **Latin squares.** At the same time, we also hope to impress upon the reader that it is through proper design that experiments can be made to yield a wealth of information. To give an example, suppose that a market research organization wants to compare four ways of packaging a breakfast food, but it is concerned about possible regional differences in the popularity of the breakfast food, and also about the effects of promoting the breakfast food in different ways. So, it decides to test market the different kinds of packaging in the northeastern, southeastern, northwestern, and southwestern parts of the United States and to promote them with discounts, lotteries, coupons, and two-for-one sales. Thus, there are $4 \cdot 4 = 16$ blocks (combinations of regions and methods of promotion) and it would take $16 \cdot 4 = 64$ market areas (cities) to promote each kind of packaging once within each block. Moreover, the test markets must be separated from each other so that the promotion methods do not interfere with each other, and the United States simply does not have 64 sufficiently widely separated test markets. It is of interest to note, however, that with proper planning 16 market areas (cities) will suffice. To illustrate, let us consider the following arrangement, called a **Latin square,** in which the letters A, B, C, and D represent the four kinds of packaging:

	Discounts	Lotteries	Coupons	2-for-1 sales
Northeast	A	B	C	D
Southeast	B	C	D	A
Northwest	C	D	A	B
Southwest	D	A	B	C

In general, a Latin square is a square array of the letters $A, B, C, D, \ldots$, of the English (Latin) alphabet, which is such that each letter occurs once and only once in each row and in each column.

The preceding Latin square, looked upon as an experimental design, requires that discounts be used with packaging A in a city in the Northeast, with packaging B in a city in the Southeast, with packaging C in a city in the Northwest, and with packaging D in a city in the Southwest; that lotteries be used with packaging B in a city in the Northeast, with packaging C in a city in the Southeast, with packaging D in a city in the Northwest, and with packaging A in a city in the Southwest; and so on. Note that each kind of promotion is used once in each region and once with each kind of packaging; each kind of packaging is used once in each region and once with each kind of promotion; and each region is used once with each kind of packaging and once with each kind of promotion. As we shall see, this will enable us to perform an analysis of variance leading to significance tests for all three variables.

The analysis of an $r \times r$ Latin square is very similar to a two-way analysis of variance. The total sum of squares and the sums of squares for rows and columns are calculated in the same way in which we previously calculated SST, $SS(Tr)$, and SSB, but we must find an extra sum of squares that measures the variability due to the variable represented by the letters $A, B, C, D, \ldots$, namely, a new treatment sum of squares. The formula for this sum of squares is

Treatment sum of squares for Latin square

$$SS(Tr) = \frac{1}{r} \cdot (T_A^2 + T_B^2 + T_C^2 + \cdots) - \frac{1}{r^2} \cdot T_{..}^2.$$

where T_A is the total of the observations corresponding to treatment A, T_B is the total of the observations corresponding to treatment B, and so forth. Finally, the error sum of squares is again obtained by subtraction:

Error sum of squares for Latin square

$$SSE = SST - [SSR + SSC + SS(Tr)]$$

where SSR and SSC are the sums of squares for rows and columns.

We can now construct an analysis-of-variance table for the analysis of an $r \times r$ Latin square. The mean squares are again the sums of squares divided by their respective degrees of freedom, and the three F-values are the mean squares for rows, columns, and treatments divided by the mean square for error. The degrees of freedom for rows, columns, and treatments are all $r - 1$, and, by subtraction, the degree of freedom for error is

$$(r^2 - 1) - (r - 1) - (r - 1) - (r - 1) = r^2 - 3r + 2 = (r - 1)(r - 2)$$

Thus, for each of the three significance tests the numerator and denominator degrees of freedom for F are $r - 1$ and $(r - 1)(r - 2)$.

Source of variation	Degrees of freedom	Sum of squares	Mean square	F
Rows	$r-1$	SSR	$MSR = \dfrac{SSR}{r-1}$	$\dfrac{MSR}{MSE}$
Columns	$r-1$	SSC	$MSC = \dfrac{SSC}{r-1}$	$\dfrac{MSC}{MSE}$
Treatments	$r-1$	SS(Tr)	$MS(Tr) = \dfrac{SS(Tr)}{r-1}$	$\dfrac{MS(Tr)}{MSE}$
Error	$(r-1)(r-2)$	SSE	$MSE = \dfrac{SSE}{(r-1)(r-2)}$	
Total	r^2-1	SST		

EXAMPLE 15.6   Suppose that in the breakfast-food study referred to in this section, the market research organization gets the data shown in the following table, where the figures are a week's sales in 10 thousands:

	Discounts	Lotteries	Coupons	2-for-1 sales
Northeast	A 48	B 38	C 42	D 53
Southeast	B 39	C 43	D 50	A 54
Northwest	C 42	D 50	A 47	B 44
Southwest	D 46	A 48	B 46	C 52

Assuming that the necessary assumptions can be met, analyze this Latin square with each of the tests performed at the 0.05 level of significance.

Solution   1. H_0's: The row, column, and treatment effects (defined as in the footnote to page 418 and on page 438) are all equal to zero.
H_A's: The respective effects are not all equal to zero.

2. $\alpha = 0.05$ for each test.

3. For rows, columns, or treatments, reject the null hypothesis if $F \geq 4.76$, where the F's are obtained by means of an analysis of variance, and 4.76 is the value of $F_{0.05}$ for $r - 1 = 4 - 1 = 3$ and $(r - 1)(r - 2) = (4 - 1)(4 - 2) = 6$ degrees of freedom.

4. Substituting $r = 4, T_{1.} = 181, T_{2.} = 186, T_{3.} = 183, T_{4.} = 192, T_{.1} = 175, T_{.2} = 179, T_{.3} = 185, T_{.4} = 203, T_A = 197, T_B = 167, T_C = 179, T_D = 199, T_{..} = 742$, and $\Sigma\Sigma\, x^2 = 34{,}756$ into the computing formulas for the sums of squares, we get

$$SST = 34{,}756 - \tfrac{1}{16}(742)^2 = 34{,}756 - 34{,}410.25 = 345.75$$

$$SSR = \tfrac{1}{4}[181^2 + 186^2 + 183^2 + 192^2] - 34{,}410.25 = 17.25$$

$$SSC = \tfrac{1}{4}[175^2 + 179^2 + 185^2 + 203^2] - 34{,}410.25 = 114.75$$

$$SS(Tr) = \tfrac{1}{4}[197^2 + 167^2 + 179^2 + 199^2] - 34{,}410.25 = 174.75$$

$$SSE = 345.75 - [17.25 + 114.75 + 174.75] = 39.00$$

The remainder of the work is shown in the following analysis-of-variance table:

Source of variation	Degrees of freedom	Sum of squares	Mean square	F
Rows (regions)	3	17.25	$\dfrac{17.25}{3} = 5.75$	$\dfrac{5.75}{6.5} \approx 0.88$
Columns (promotion methods)	3	114.75	$\dfrac{114.75}{3} = 38.25$	$\dfrac{38.25}{6.5} \approx 5.88$
Treatments (packaging)	3	174.75	$\dfrac{174.75}{3} = 58.25$	$\dfrac{58.25}{6.5} \approx 8.96$
Error	6	39.00	$\dfrac{39.00}{6} = 6.5$	
Total	15	345.75		

5. For rows, since $F = 0.88$ is less than 4.76, the null hypothesis cannot be rejected; for columns, since $F = 5.88$ exceeds 4.76, the null hypothesis must be rejected; for treatments, since $F = 8.96$ exceeds 4.76, the null hypothesis must be rejected. In other words, we conclude that differences in promotion and packaging, but not the different regions, affect the breakfast food's sales. ∎

There are many other experimental designs besides the ones we have discussed, and they serve a great variety of special purposes. Widely used, for example, are the **incomplete block designs,** which apply when it is impossible to have each treatment in each block.

The need for such a design arises, for example, when we want to compare 13 kinds of tires but, of course, cannot put them all on a test car at the same time. Numbering the tires from 1 to 13, we might use the following experimental design:

Test run	Tires				Test run	Tires			
1	1	2	4	10	*8*	8	9	11	4
2	2	3	5	11	*9*	9	10	12	5
3	3	4	6	12	*10*	10	11	13	6
4	4	5	7	13	*11*	11	12	1	7
5	5	6	8	1	*12*	12	13	2	8
6	6	7	9	2	*13*	13	1	3	9
7	7	8	10	3					

Here there are 13 test runs, or blocks, and since each kind of tire appears together with each other kind of tire once within the same block, the design is referred to as a **balanced imcomplete block design.** The fact that each kind of tire appears together with each other kind of tire once within the same block is important; it facilitates the statistical analysis because it assures that we have the same amount of information for comparing each pair of tires. In general, the analysis of incomplete block designs is fairly complicated, and we shall not go into it here, as it has been our purpose only to demonstrate what can be accomplished by the careful design of an experiment.

Exercises

15.39 An agronomist wants to compare the yield of 15 varieties of corn, and at the same time study the effects of four different fertilizers and three methods of irrigation. How many test plots must he plant if each variety of corn is to be grown in one test plot with each possible combination of fertilizers and methods of irrigation?

15.40 Suppose that we want to compare the number of defective pieces produced by five machine operators working on four different machine parts (1, 2, 3, and 4) in two different shifts (I and II).

(a) If the machine operators are to be regarded as different treatments, list the blocks (combinations of machine parts and shifts) that would be required if each part is to be produced in each shift.

(b) How many observations would be required if each machine operator is to work twice on each machine part during each shift?

15.41 A pharmaceutical manufacturer wants to market a new cold remedy that is actually a combination of four medications, and wants to experiment first with two dosages for each medication. If A_L and A_H denote the low and high dosage of medication A, B_L and B_H the low and high dosage of medication B, C_L and C_H the low and high dosage of medication C, and D_L and D_H the low and high dosage of medication D, list the 16 preparations that must be tested if each dosage of each medication is to be used once in combination with each dosage of each of the other medications.

15.42 Making use of the fact that each of the letters must occur once and only once in each row and each column, complete the following Latin squares:

(a)
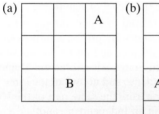

(b)
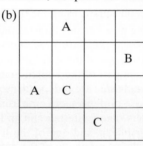

(c)
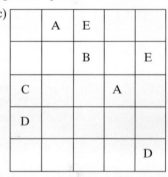

 *15.43 To compare four different brands of golf balls, A, B, C, and D, each kind was driven by each of four golf pros, P_1, P_2, P_3, and P_4, using once each four different drivers, D_1, D_2, D_3, and D_4. The distances from the tee to the points where the balls came to rest (in yards) were as shown in the following table:

	D_1		D_2		D_3		D_4	
P_1	D	231	B	215	A	261	C	199
P_2	C	234	A	300	B	280	D	266
P_3	A	301	C	208	D	247	B	255
P_4	B	253	D	258	C	210	A	290

Assuming that the necessary assumptions can be met, use a computer to analyze this Latin square, using the 0.05 level of significance for each test.

15.44 The sample data in the following 3 × 3 Latin square are the scores in an American history test obtained by nine college students of various ethnic backgrounds and of various professional interests, who were taught by instructors A, B, and C:

Ethnic background

	Mexican	German	Polish
Law	A 75	B 86	C 69
Medicine	B 95	C 79	A 86
Engineering	C 70	A 83	B 93

Assuming that the necessary assumptions can be met, use a computer to analyze this Latin square, using the 0.05 level of significance for each test.

15.45 Among nine persons interviewed in a poll, three are Easterners, three are Southerners, and three are Westerners. By profession, three of them are teachers, three are lawyers, and three are doctors, and no two of the same profession come from the same part of the United States. Also, three are Democrats, three are Republicans, and three are Independents, and no two of the same political affiliation are of the same profession or come from the same part of the United States. If one of the teachers is an Easterner and an Independent, another teacher is a Southerner and a Republican, and one of the lawyers is a Southerner and a Democrat, what is the political affiliation of the doctor who is a Westerner? (*Hint:* Construct a 3 × 3 Latin square. This exercise is a simplified version of a famous problem posed by R. A. Fisher in his classical work *The Design of Experiments.*)

15.46 To test their ability to make decisions under pressure, the nine senior executives of a company are to be interviewed by each of four psychologists. As it takes a psychologist a full day to interview three of the executives, the schedule for the interviews is arranged as follows, where the nine executives are denoted by A, B, C, D, E, F, G, H, and I:

Day	Psychologist	Executives		
March 2	I	B	C	?
March 3	I	E	F	G
March 4	I	H	I	A
March 5	II	C	?	H
March 6	II	B	F	A
March 9	II	D	E	?
March 10	III	D	G	A
March 11	III	C	F	?
March 12	III	B	E	H
March 13	IV	B	?	I
March 16	IV	C	?	A
March 17	IV	D	F	H

Replace the six question marks with the appropriate letters, given that each of the nine executives is to be interviewed together with each of the other executives once and only once on the same day. Note that this will make the arrangement a balanced incomplete block design, which may be important because each executive is tested together with each other executive once under identical conditions.

15.47 A newspaper regularly prints the columns of seven writers but has room for only three in each edition. Complete the following schedule, in which the writers are numbered 1–7, so that each writer's column appears three times per week, and a column of each writer appears together with a column of each other writer once per week.

Day	Writers		
Monday	1	2	3
Tuesday	4		
Wednesday	1	4	5
Thursday	2		
Friday	1	6	7
Saturday	5		
Sunday	2	4	6

15.11 Checklist of Key Terms *(with page references to their definitions)*

Analysis of variance, 417
Analysis of variance table, 426
ANOVA, 417
Balanced incomplete plot design, 454
Block effects, 438
Block sum of squares, 439
Blocking, 437
Blocks, 437
Complete blocks, 437
Completely randomized design, 423
Controlled experiment, 422
Denominator degrees of freedom, 420
Error mean square, 426
Error sum of squares, 424
Experimental design, 418
Experimental error, 424
Factors, 443
Grand mean, 418, 424
Grand total, 427
Incomplete block design, 454
Interaction, 438
  sum of squares, 445

Latin square, 450
Multiple comparisons, 418, 431
Numerator degrees of freedom, 420
One-way analysis of variance, 218, 424
Randomization, 418, 423
Randomized block design, 438
Replication, 441
Studentized range, 432
Total sum of squares, 424
Treatment effect, 418
Treatment mean square, 425
Treatment sum of squares, 424
Treatments, 425
Two-factor experiment, 438
Two-way analysis of variance, 437, 438
  with interaction, 443
  without interaction, 438
Two-way experiment, 418
Variance ratio, 420

15.12 References

The following are some of the many books that have been written on the subject of analysis of variance:

GUENTHER, W. C., *Analysis of Variance.* Upper Saddle River, N. J.: Prentice Hall, Inc., 1964.

SNEDECOR, G. W., and COCHRAN, W. G., *Statistical Methods,* 6th ed. Ames, Iowa: Iowa State University Press, 1973.

Problems relating to the design of experiments are also treated in the preceding books and in

ANDERSON, V. L., and McLEAN, R. A., *Design of Experiments: A Realistic Approach.* New York: Marcel Dekker. Inc., 1974.

BOX, G. E. P., HUNTER, W. G., and HUNTER, J. S., *Statistics for Experimenters.* New York: John Wiley & Sons, Inc., 1978.

COCHRAN, W. G., and COX, G. M., *Experimental Design,* 2nd ed. New York: John Wiley & Sons, Inc., 1957.

FINNEY, D. J., *An Introduction to the Theory of Experimental Design.* Chicago: The University of Chicago Press, 1960.

FLEISS, J., *The Design and Analysis of Clinical Experiments,* New York: John Wiley & Sons, Inc., 1986.

HICKS, C. R., *Fundamental Concepts in the Design of Experiments,* 2nd ed. New York: Holt, Rinehart and Winston, 1973.

ROMANO, A., *Applied Statistics for Science and Industry.* Boston: Allyn and Bacon, Inc., 1977.

A table of Latin squares for r = 3, 4, 5, . . . , and 12 may be found in the aforementioned book by W. G. Cochran and G. M. Cox.

Informally, some questions of experimental design are discussed in Chapters 18 and 19 of

BROOK, R. J., ARNOLD, G. C., HASSARD, T. H., and PRINGLE, R. M., eds., *The Fascination of Statistics.* New York: Marcel Dekker, Inc., 1986.

The topic of multiple comparisons is treated in detail in

FEDERER, W. T., *Experimental Design, Theory and Application.* New York: Macmillan Publishing Co., Inc., 1955.

HOCHBERG, Y., and TAMHANE, A., *Multiple Comparison Procedures.* New York: John Wiley & Sons, Inc., 1987.

16

REGRESSION

16.1   Curve Fitting   460

16.2   The Method of Least Squares   462

16.3   Regression Analysis   473

*16.4   Multiple Regression   482

*16.5   Nonlinear Regression   486

16.6   Checklist of Key Terms   496

16.7   References   496

*I*n many statistical investigations, the main goal is to establish relationships that make it possible to predict one or more variables in terms of others. For instance, studies are made to predict the future sales of a product in terms of its price, a person's weight loss in terms of the number of weeks he or she has been on an 800-calories-per-day diet, family expenditures on medical care in terms of family income, the per capita consumption of certain food items in terms of their nutritional value and the amount of money spent advertising them on television, and so forth.

Of course, it would be ideal if we could predict one quantity exactly in terms of another, but this is seldom possible. In most instances we must be satisfied with predicting averages or expected values. For instance, we cannot predict exactly how much money a specific college graduate will earn ten years after graduation, but given suitable data we can predict the average earnings of all college graduates ten years after graduation. Similarly, we can predict the average yield of a variety of wheat in terms of the total rainfall in July, and we can predict the expected grade-point average of a student starting law school in terms of his or

459

her IQ. This problem of predicting the average value of one variable in terms of the known value of another variable (or the known values of other variables) is called the problem of **regression.** This term dates back to Francis Galton (1822–1911), who used it first in connection with a study of the relationship between the heights of fathers and sons.

In Sections 16.1 and 16.2 we present a general introduction to curve fitting and the method that is most widely used, the **method of least squares.** Then, in Section 16.3, we discuss questions concerning inferences based on straight lines fit to paired data. Problems in which predictions are based on several variables and problems in which the relationship between two variables is not linear are treated in the two optional sections, Sections 16.4 and 16.5.

16.1   CURVE FITTING

Whenever possible, we try to express, or approximate, relationships between known quantities and quantities that are to be predicted in terms of mathematical equations. This has been very successful in the natural sciences, where it is known, for instance, that at a constant temperature the relationship between the volume, y, and the pressure, x, of a gas is given by the formula

$$y = \frac{k}{x}$$

where k is a numerical constant. Also, it has been shown that the relationship between the size of a culture of bacteria, y, and the length of time, x, it has been exposed to certain environmental conditions is given by the formula

$$y = a \cdot b^x$$

where a and b are numerical constants. More recently, equations like these have also been used to describe relationships in the behavioral sciences, the social sciences, and other fields. For instance, the first of the preceding equations is often used in economics to describe the relationship between price and demand, and the second has been used to describe the growth of one's vocabulary or the accumulation of wealth.

Whenever we use observed data to arrive at a mathematical equation that describes the relationship between two variables—a procedure known as **curve fitting**—we must face three kinds of problems:

We must decide what kind of curve, and hence what kind of "predicting" equation to use.

We must find the particular equation that is "best" in some sense.

We must investigate certain questions regarding the merits of the particular equation, and of predictions made from it.

The second of these problems is discussed in some detail in Section 16.2, and the third in Section 16.3.

The first kind of problem is usually decided by direct inspection of the data. We plot the data on ordinary (arithmetic) graph paper, sometimes on special graph paper with special scales, and we decide by visual inspection upon the kind of curve (a straight line, a parabola, . . .) that best describes the overall pattern of the data. There are methods by which this can be done more objectively, but they are fairly advanced and they will not be discussed in this book.

So far as our work here is concerned, we shall concentrate mainly on **linear equations** in two unknowns. They are of the form

$$y = a + bx$$

where a is the y-intercept (the value of y for $x = 0$) and b is the slope of the line (namely, the change in y which accompanies an increase of one unit in x).[†] Linear equations are useful and important not only because many relationships are actually of this form, but also because they often provide close approximations to relationships that would otherwise be difficult to describe in mathematical terms.

The term "linear equation" arises from the fact that the graph of $y = a + bx$ is a straight line. That is, all pairs of values of x and y that satisfy an equation of the form $y = a + bx$ constitute points that fall on a straight line. In practice, the values of a and b are usually estimated from observed data, and once they have been determined, we can substitute values of x into the equation and calculate the corresponding predicted values of y.

To illustrate, suppose that we are given data on a midwestern county's production of wheat, y (in bushels per acre), and its annual rainfall, x (in inches measured from September through August), and that by the method of Section 16.2 we obtain the predicting equation

$$y = 0.23 + 4.42x$$

(see Exercise 16.8). The corresponding graph is shown in Figure 16.1, and it should be observed that for any pair of values of x and y that are such that $y = 0.23 + 4.42x$, we get a point (x, y) that falls on the line. Substituting $x = 6$, for instance, we find that when there is an annual rainfall of 6 inches, we can expect a yield of

$$y = 0.23 + 4.42 \cdot 6 = 26.75$$

bushels per acre; similarly, substituting $x = 12$, we find that when there is an annual rainfall of 12 inches, we can expect a yield of

$$y = 0.23 + 4.42 \cdot 12 = 53.27$$

bushels per acre. The points (6, 26.75) and (12, 53.27) lie on the straight line of Figure 16.1, and this is true for any other points obtained in the same way.

[†]In other branches of mathematics, linear equations in two unknowns are often written as $y = mx + b$, but $y = a + bx$ has the advantage that it lends itself more easily to generalizations—for instance, as in $y = a + bx + cx^2$ or as in $y = a + b_1 x_1 + b_2 x_2$.

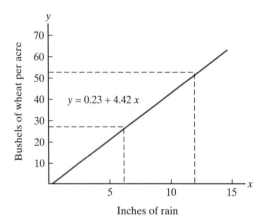

FIGURE 16.1
Graph of linear
equation.

16.2   THE METHOD OF LEAST SQUARES

Once we have decided to fit a straight line to a given set of data, we face the second kind of problem, namely, that of finding the equation of the particular line
which in some sense provides the best possible fit. To illustrate what is involved,
let us consider the following sample data obtained in a study of the relationship
between the length of time that a person has been exposed to a high level of
noise and the sound frequency range to which his or her ears will respond. Here
x is the length of time (rounded to the nearest week) that a person has been living near a major airport directly in the flight path of departing jets, and y is his
or her hearing range (in thousands of cycles per second):

Number of weeks x	Hearing range y
47	15.1
56	14.1
116	13.2
178	12.7
19	14.6
75	13.8
160	11.9
31	14.8
12	15.3
164	12.6
43	14.7
74	14.0

In Figure 16.2, these twelve **data points,** (x, y), are plotted in what is called
a **scattergram.** We did this with the use of a computer, even though it would have

been easy enough to do by hand. As can be seen, the points do not all fall on a straight line, but the overall pattern of the relationship is reasonably well described as being linear. At least, there is no noticeable departure from linearity, so we feel justified in deciding that a straight line is a suitable description of the underlying relationship.

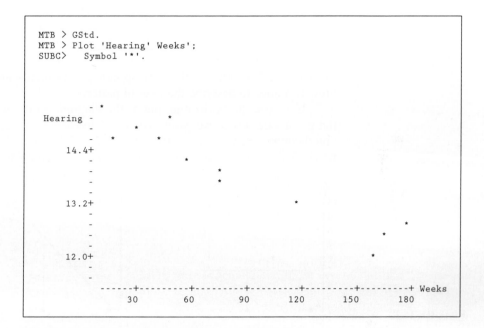

FIGURE 16.2
Computer printout of the hearing range data.

We now face the problem of finding the equation of the line that in some sense provides the best fit to the data and that, it is hoped, will later yield the best possible predictions of y from x. Logically speaking, there is no limit to the number of straight lines that can be drawn on a piece of graph paper. Some of these lines would fit the data so poorly that we could not consider them seriously, but many others would seem to provide more or less good fits, and the problem is to find the one line that fits the data best in some well-defined sense. If all the points actually fall on a straight line there is no problem, but this is an extreme case rarely encountered in practice. In general, we have to be satisfied with a line having certain desirable properties, short of perfection.

The criterion that, today, is used almost exclusively for defining a "best" fit dates back to the early part of the nineteenth century and the work of the French mathematician Adrien Legendre; it is known as the **method of least squares.** As it will be used here, this method requires that the line that we fit to our data be such that the sum of the squares of the vertical distances from the points to the line is a minimum.

To explain why this is done, let us refer to the following data, which might represent the numbers of correct answers, x and y, that four students got on two parts of a multiple-choice test:

x	y
4	6
9	10
1	2
6	2

In Figure 16.3 we plotted the corresponding data points, and through them we drew two lines to describe the overall pattern.

If we use the horizontal line in the diagram on the left to "predict" y for the given values of x, we would get $y = 5$ in each case, and the errors of these "predictions" are $6 - 5 = 1$, $10 - 5 = 5$, $2 - 5 = -3$, and $2 - 5 = -3$. In Figure 16.3, they are the vertical deviations from the points to the line.

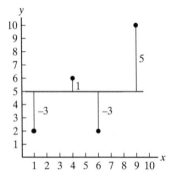 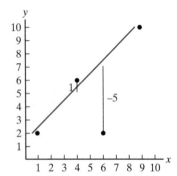

FIGURE 16.3
Two lines fit to the four data points.

The sum of these errors is $1 + 5 + (-3) + (-3) = 0$, but this is no indication of their size and we find ourselves in a position similar to that on page 82 that led to the definition of the standard deviation. Squaring the errors as we squared the deviations from the mean on page 82, we find that the sum of the squares of the errors is $1^2 + 5^2 + (-3)^2 + (-3)^2 = 44$.

Now let us consider the line in the diagram on the right, which was drawn so that it passes through the points $(1, 2)$ and $(9, 10)$; as can easily be verified, its equation is $y = 1 + x$. Judging by eye, this line seems to provide a much better fit than the horizontal line in the diagram on the left, and if we use it to "predict" y for the given values of x, we would get $1 + 4 = 5$, $1 + 9 = 10$, $1 + 1 = 2$, and $1 + 6 = 7$. The errors of these "predictions," which in the figure on the right are also the vertical distances from the points to the line, are $6 - 5 = 1$, $10 - 10 = 0$, $2 - 2 = 0$, and $2 - 7 = -5$.

The sum of these errors is $1 + 0 + 0 + (-5) = -4$, which is numerically greater than the sum we obtained for the errors made with the other line of Figure 16.3, but this is of no consequence. The sum of the squares of the errors is now $1^2 + 0^2 + 0^2 + (-5)^2 = 26$, and this is much less than the 44 that we obtained before. In this sense, the line on the right provides a much better fit to the data than the horizontal line on the left.

We can even go one step further and ask for the equation of the line for which the sum of the squares of the errors (the sum of the squares of the vertical deviations from the points to the line) is a minimum. In Exercise 16.11 the reader will be asked to verify that the equation of such a line is $y = \frac{15}{17} + \frac{14}{17}x$ for our example. We refer to it as a **least-squares line.**

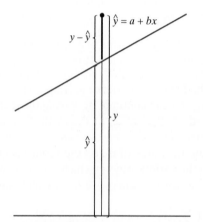

FIGURE 16.4
The difference between y and $\hat{y}$.

To show how the equation of such a line is actually obtained for a given set of **data points,** let us consider n pairs of numbers $(x_1, y_1), (x_2, y_2), \dots,$ and (x_n, y_n) that might represent the thrust and the speed of n rockets, the height and weight of n persons, the reading rate and the reading comprehension of n students, the cost of salvaging n ship wrecks and the values of the recovered treasure, or the age and the repair costs of n automobiles. If we write the equation of the line as $\hat{y} = a + bx$, where the symbol $\hat{y}$ (y-hat) is used to distinguish between observed values of y and the corresponding values $\hat{y}$ on the line, the least-squares criterion requires that we minimize the sum of the squares of the differences between the y's and the $\hat{y}$'s (see Figure 16.4). This means that we must find the numerical values of the constants a and b appearing in the equation $\hat{y} = a + bx$ for which

$$\sum (y - \hat{y})^2 = \sum [y - (a + bx)]^2$$

is as small as possible. As it takes calculus or fairly tedious algebra to find the expressions for a and b that minimize $\sum (y - \hat{y})^2$, let us merely state the result that they are given by the solutions for a and b of the following system of two linear equations:

$$\sum y = na + b\left(\sum x\right)$$
$$\sum xy = a\left(\sum x\right) + b\left(\sum x^2\right)$$

In these equations, called the **normal equations,** n is the number of pairs of observations, $\sum x$ and $\sum y$ are the sums of the observed x's and y's, $\sum x^2$ is the sum of the squares of the x's, and $\sum xy$ is the sum of the products obtained by multiplying each x by the corresponding y.

EXAMPLE 16.1   Given that $n = 12$, $\Sigma\, x = 975$, $\Sigma\, x^2 = 117{,}397$, $\Sigma\, y = 166.8$, $\Sigma\, y^2 = 2{,}331.54$, and $\Sigma\, xy = 12{,}884.4$ for the hearing-range data on page 462, set up the normal equations for getting a least-squares line.

Solution   Substituting $n = 12$ and four of the five sums of squares into the expressions for the normal equations, we get

$$166.8 = 12a + 975b$$
$$12{,}884.4 = 975a + 117{,}397b$$

(Note that we did not need $\Sigma\, y^2$ for this example, and we gave it here together with the other summations for use in a future example.)   ■

If the reader has had some experience solving systems of linear equations in elementary algebra, he or she can continue by solving these two equations for a and b using either the **method of elimination** or the method based on the use of **determinants.** Alternatively, one can solve the two normal equations symbolically for a and b, and then substitute the value of n and the required summations into the resulting formulas. Among the various ways in which we can write these formulas, most convenient, perhaps, is the format where we use as building blocks the quantities

$$S_{xx} = \Sigma\, x^2 - \frac{1}{n}\left(\Sigma\, x\right)^2 \quad \text{and} \quad S_{xy} = \Sigma\, xy - \frac{1}{n}\left(\Sigma\, x\right)\left(\Sigma\, y\right)$$

and then write the computing formulas for a and b as

Solutions of normal equations

$$b = \frac{S_{xy}}{S_{xx}}$$

$$a = \frac{\Sigma\, y - b\left(\Sigma\, x\right)}{n}$$

where we first calculate b and then substitute its value into the formula for a. (Note that this is the same S_{xx} that we used on page 84 in the computing formula for the sample standard deviation.)

EXAMPLE 16.2   Use these formulas to calculate a and b for the hearing-range example.

Solution   First substituting $n = 12$ and the required summation given in Example 16.1 into the formulas for S_{xx} and S_{xy}, we get

$$S_{xx} = 117{,}397 - \frac{1}{12}\,(975)^2 = 38{,}178.25$$

and

$$S_{xy} = 12{,}884.4 - \frac{1}{12}\,(975)\,(166.8) = -668.1$$

Then, $b = \dfrac{-668.1}{38{,}178.24} \approx -0.0175$ and $a = \dfrac{166.8 - (-0.0175)(975)}{12} \approx 15.3$, both

rounded to three significant figures, and the equation of the least-squares line can be written as

$$\hat{y} = 15.3 - 0.0175x$$ ∎

What we have done here may well be described as a mere exercise in arithmetic, for we seldom, if ever, go through all these details in determining a least-squares line. Nowadays, the required summations can be obtained even with the most primitive garden-variety kind of handheld calculator, and the values of a and b can be obtained with any kind of statistical technology. Indeed, the trickiest part of the whole operation is that of entering the data and, if necessary, making corrections, unless we are using a computer or a graphing calculator, where the data can be displayed and edited.

Observe also that when b is negative, as it is in Example 16.2, the least-squares line has a *downward slope* going from left to right. In other words, the relationship between x and y is such that y decreases when x increases, as can be seen also from Figure 16.2. On the other hand, when b is positive, this means that the least-squares line has an *upward slope* going from left to right, namely, that y increases when x increases. Finally, when b equals zero, the least-squares line is horizontal and knowledge of x is of no help in estimating, or predicting, a value of y.

EXAMPLE 16.3   Use a graphing calculator to rework Example 16.2, without actually utilizing the summations given in Example 16.1.

FIGURE 16.5
Data for Example 16.3.

L1	L2	L3	3
47	15.1		
56	14.1		
116	13.2		
178	12.7		
19	14.6		
75	13.8		
160	11.9		

L3(1)=

31	14.8	
12	15.3	
164	12.6	
43	14.7	
74	14	

L2(13) =

Solution   Figure 16.5 shows the original data entered in a graphing calculator. The display screen is too small to show all the data, but the remainder of the data was obtained by scrolling. Then the command **STAT CALC 8** yields the results shown in Figure 16.6. Rounding to three significant figures, as before, we get $a = 15.3$ and $b = -0.0175$, and the equation of the least-squares line is, of course, the same

$$\hat{y} = 15.3 - 0.0175x$$ ■

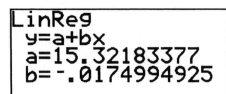

```
LinReg
 y=a+bx
 a=15.32183377
 b=-.0174994925
```

FIGURE 16.6
Solution of Example 16.3.

Had we used a computer for this example, MINITAB would have yielded the printout shown in Figure 16.7. The equation of the least-squares line, called here the **regression equation** (which will be explained later), is again $y = 15.3 - 0.0175$, and the coefficients a and b are given in the column headed "Coef" as 15.3218 and -0.017499. Some of the additional details in this printout will be used later on.

FIGURE 16.7
MINITAB printout for Example 16.3.

```
MTB > GPro.
MTB > Regress 'y' 1 'x';
SUBC>    Constant.



The regression equation is
y = 15.3 - 0.0175 x

Predictor         Coef        Stdev      t-ratio          p
Constant       15.3218       0.1845        83.04      0.000
x             -0.017499     0.001865        -9.38      0.000

s = 0.3645       R-sq = 89.8%       R-sq(adj) = 88.8%

Analysis of Variance

SOURCE         DF           SS           MS          F          p
Regression      1       11.691       11.691      88.00      0.000
Error          10        1.329        0.133
Total          11       13.020
```

EXAMPLE 16.4   Use the least-squares equation obtained in Example 16.2 or 16.3 to estimate the hearing range of a person who has been exposed to the airport noise (as described on page 462) for

(a) one year;
(b) two years.

Solution

(a) Substituting $x = 52$ into $\hat{y} = 15.3 - 0.0175x$, we get
$\hat{y} = 15.3 - 0.0175(52) = 14.4$ thousand cycles per second rounded to three significant figures.

(b) Substituting $x = 104$ into this equation, we get $\hat{y} = 15.3 - 0.0175(104) = 13.5$ thousand cycles per second rounded to three significant figures. ∎

When we make an estimate like this, or a prediction, we cannot really expect that we will always hit the answer right on the nose. With reference to our example, it would be very unreasonable to expect that every person who has been exposed to airport noise for a given length of time will have exactly the same hearing range. To make meaningful estimates or predictions based on least-squares lines, we must look upon the values of $\hat{y}$ obtained by substituting given values of x as averages, or expected values. Interpreted in this way, we refer to least-squares lines as **regression lines,** or better as **estimated regression lines,** since the values of a and b are estimates based on sample data and, hence, can be expected to vary from sample to sample. Questions relating to the goodness of such estimates will be discussed in Section 16.3.

In the discussion of this section we have considered only the problem of fitting a straight line to paired data. More generally, the method of least squares can also be used to fit other kinds of curves and to derive predicting equations in more than two unknowns. The problem of fitting curves other than straight lines by the method of least squares will be discussed briefly in Section 16.5 and some examples of predicting equations in more than two unknowns will be given in Section 16.4. Both of these sections are marked optional.

■ E x e r c i s e s

16.1 A dog that had six hours of obedience training made five mistakes at a dog show, a dog that had twelve hours of obedience training made six mistakes, and a dog that had eighteen hours of obedience training made only one mistake. If we let x denote the number of hours of obedience training and y the number of mistakes, which of the two lines

$$y = 10 - \frac{1}{2}x \qquad \text{or} \qquad y = 8 - \frac{1}{3}x$$

provides a better fit to the three data points, $(6, 5)$, $(12, 6)$, and $(18, 1)$, in the sense of least squares?

16.2 With reference to Exercise 16.1, use a computer or a graphing calculator to check whether the line providing the better fit is a least-squares line.

16.3   To see whether a widely used food preservative contributes to the hyperactivity of preschool children, a dietician chose a random sample of ten four-year-olds known to be fairly hyperactive from various nursery schools and observed their behavior 45 minutes after they had eaten measured amounts of food containing the preservative. In the table that follows, x is the amount of food consumed containing the preservative (in grams) and y is a subjective rating of hyperactivity (on a scale from 1 to 20) based on a child's restlessness and interaction with other children:

x	y
36	6
82	14
45	5
49	13
21	5
24	8
58	14
73	11
85	18
52	6

(a) Draw a scattergram to judge whether a straight line might reasonably describe the overall pattern of the data.
(b) Use a ruler to draw a straight line that, judging by eye, should be fairly close to a least-squares line.
(c) Use the line drawn in part (b) to estimate the hyperactivity rating of such a child that had 65 grams of food with the preservative 45 minutes earlier.

16.4   With reference to Exercise 16.3, use appropriate software or a graphing calculator to verify that the least-squares equation for estimating y in terms of x is $\hat{y} = 1.5 + 0.16x$ rounded to two significant figures. Also, use this equation to estimate the hyperactivity rating of such a child that had 65 grams of food with the preservative 45 minutes earlier and compare the result with that of part (c) of Exercise 16.3.

*16.5   With reference to Exercise 16.3, where $\Sigma x = 525$, $\Sigma y = 100$, $\Sigma x^2 = 32,085$, and $\Sigma xy = 5,980$, set up the two normal equations and solve them by using either the method of elimination or that based on the use of determinants.

16.6   The following table shows how many weeks six persons have worked at an automobile inspection station and the number of cars each one inspected between noon and 2 P.M. on a given day:

Number of weeks employed x	Number of cars inspected y
2	13
7	21
9	23
1	14
5	15
12	21

Given that $\Sigma x = 36$, $\Sigma y = 107$, $\Sigma x^2 = 304$, and $\Sigma xy = 721$, use the computing formulas on page 466 to find a and b, and hence the equation of the least-squares line.

16.7 Use the result of Exercise 16.6 to estimate how many cars a person can be expected to inspect during the same two-hour period if he or she has been working at the inspection station for eight weeks.

16.8 Verify that the equation of the example on page 461 can be obtained by fitting a least-squares line to the following data:

Rainfall (inches)	Yield of wheat (bushels per acre)
12.9	62.5
7.2	28.7
11.3	52.2
18.6	80.6
8.8	41.6
10.3	44.5
15.9	71.3
13.1	54.4

16.9 The following data pertain to the chlorine residue in a swimming pool at various times after it has been treated with chemicals:

Number of hours	Chlorine residual (parts per million)
0	2.2
2	1.8
4	1.5
6	1.4
8	1.1
10	1.1
12	0.9

where the reading at 0 hour was taken immediately after the chemical treatment was completed.
(a) Use the computing formulas on page 466 to fit a least-squares line from which we can predict the chlorine residual in terms of the number of hours since the pool has been treated with chemicals.
(b) Use the equation of the least-squares line obtained in part (a) to estimate the chlorine residual in the pool five hours after it has been treated with chemicals.
(c) Suppose you discover that the data for this exercise were obtained on a very hot day. Explain why the results of parts (a) and (b) might be quite misleading.

16.10 Use appropriate software or a graphing calculator to rework part (a) of Exercise 16.9.

16.11 With reference to the four data points on page 464, which were (4, 6), (9, 10), (1, 2), and (6, 2), verify that the equation of the least-squares line is

$$\hat{y} = \frac{15}{17} + \frac{14}{17}x$$

Also, calculate the sum of the squares of the vertical deviations from the four points to this line, and compare the result with 44 and 26, the corresponding sums of squares obtained for the two lines shown in Figure 16.3.

*16.12 Raw material used in the production of a synthetic fiber is stored in a place that has no humidity control. Measurements of the relative humidity in the storage place and the moisture content of a sample of the raw material (both in percentages) on 12 days yielded the following results:

Humidity x	Moisture content y
46	12
53	14
37	11
42	13
34	10
29	8
60	17
44	12
41	10
48	15
33	9
40	13

(a) Draw a scattergram to verify that a straight line pretty well describes the overall relationship between the two variables.
(b) Given that $\Sigma x = 507$, $\Sigma y = 144$, $\Sigma x^2 = 22{,}265$, and $\Sigma xy = 6{,}314$, set up the two normal equations.
(c) Solve the two normal equations, using either the method of elimination or the method based on determinants.

16.13 With reference to Exercise 16.12, use the summations given in part (b) and the computing formulas on page 466 to find the equation of the least-squares line.

16.14 Use appropriate software or a graphing calculator to find the equation of the least-squares line for the relative humidity and moisture content data of Exercise 16.12.

16.15 Use the equation obtained in Exercises 16.12, 16.13, or 16.14 to estimate the moisture content when the relative humidity is 38%.

16.16 Suppose that in Exercise 16.12 we had wanted to estimate what humidity will yield a moisture content of 10%. We could substitute $\hat{y} = 10$ into the equation obtained in either of Exercises 16.12, 16.13, or 16.14, and solve for x, but this would not provide an estimate in the least-squares sense. To obtain a least-squares estimate of humidity in terms of moisture content, we would have to denote moisture content by x, humidity by y, and then fit a least-squares line to these data. Use appropriate software or a graphing calculator to obtain such a least-squares line and use it to estimate the humidity that will yield a moisture content of 10%.

16.17 When the x's are equally spaced (that is, when the differences between successive values of x are all equal), finding the equation of a least-squares line can be simplified a great deal by coding the x's by assigning them the values $\ldots, -3, -2, -1,$ $0, 1, 2, 3, \ldots$ when n is odd or $\ldots, -5, -3, -1, 1, 3, 5, \ldots$ when n is even. With this kind of coding, the sum of the coded x's, call them u's, is zero, and the computing formulas for a and b on page 466 become

$$a = \frac{\sum y}{n} \quad \text{and} \quad b = \frac{\sum uy}{\sum u^2}$$

Of course, the equation of the resulting least-squares line expresses y in terms of u, and we have to account for this when we use the equation to make estimates or predictions.

(a) During its first five years of operation, a company's gross income from sales was 1.4, 2.1, 2.6, 3.5, and 3.7 million dollars. Fit a least-squares line and, assuming that the trend continues, predict the company's gross income from sales during its sixth year of operation.

(b) At the end of eight successive years, a manufacturing company had 1.0, 1.7, 2.3, 3.1, 3.5, 3.4, 3.9, and 4.7 million dollars invested in plants and equipment. Fit a least-squares line and, assuming that the trend continues, predict the company's investment in plants and equipment at the end of the tenth year.

*16.18 Verify that if we solve the normal equations *symbolically* by using determinants, we obtain the following alternative computing formulas for a and b:

$$a = \frac{(\sum y)(\sum x^2) - (\sum x)(\sum xy)}{n(\sum x^2) - (\sum x)^2}$$

$$b = \frac{n(\sum xy) - (\sum x)(\sum y)}{n(\sum x^2) - (\sum x)^2}$$

16.19 Use the computing formulas of Exercise 16.18 to rework
(a) Exercise 16.6;
(b) Exercise 16.13.

16.3 REGRESSION ANALYSIS

In Example 16.4 we used a least-squares line to estimate, or predict, the hearing range of a person who has been exposed to the airport noise for two years as 13.5 thousand cycles per second. Even if we interpret the least-squares line correctly as a regression line (that is, treat estimates based on it as averages or expected values), there are questions that remain to be answered. For instance,

How good are the values we obtained for a and b in the least-squares equation $\hat{y} = 15.3 - 0.0175x$?

How good an estimate is $\hat{y} = 13.5$ thousand cycles per second of the average hearing range of persons who have been exposed to the airport noise for two years?

After all, $a = 15.3$ and $b = -0.0175$, as well as $\hat{y} = 13.5$, are only estimates based on sample data, and if we base our calculations on a different sample, the method of least squares would probably yield different values for a and b, and a different value for $\hat{y}$ for $x = 104$. Also, for making predictions, we might ask

> **Can we give an interval for which we can assert with some degree of confidence that it will contain the hearing range of a person who will have been exposed to the airport noise for two years?**

With regard to the first of these questions, we said that $a = 15.3$ and $b = -0.0175$ are "only estimates based on sample data," and this implies the existence of corresponding true values, usually denoted by α and β and referred to as the true **regression coefficients.** Accordingly, there is also a true regression line $\mu_{y|x} = \alpha + \beta x$, where $\mu_{y|x}$ is the true mean of y for a given value of x. To distinguish between a and α and b and β, we refer to a and b as the **estimated regression coefficients.** They are often denoted by $\hat{\alpha}$ and $\hat{\beta}$ instead of a and b.

To clarify the idea of a true regression line, let us consider Figure 16.8, where we have drawn the distributions of y for several values of x. With reference to our numerical example, these curves are the distributions of the hearing range of persons who have been exposed to the airport noise for one, two, and three weeks, and to complete the picture we can visualize similar curves for all other values of x within the range of values under consideration. Note that the means of all the distributions of Figure 16.8 lie on the true regression line $\mu_{y|x} = \alpha + \beta x$.

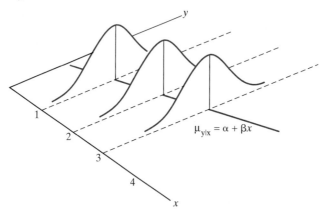

FIGURE 16.8
Distributions of y for given values of x.

In **linear regression analysis** we assume that the x's are constants, not values of random variables, and that for each value of x the variable to be predicted, y, has a certain distribution (as pictured in Figure 16.8) whose mean is $\alpha + \beta x$. In **normal regression analysis** we assume, furthermore, that these distributions are all normal distributions with the same standard deviations σ.

Based on these assumptions, it can be shown that the estimated regression coefficients a and b, obtained by the method of least squares, are values of random variables having normal distributions with the means α and β and the standard deviations

$$\sigma\sqrt{\frac{1}{n} + \frac{\bar{x}^2}{S_{xx}}} \quad \text{and} \quad \frac{\sigma}{\sqrt{S_{xx}}}$$

The estimated regression coefficients a and b are, however, not statistically independent. Note that both of these standard error formulas require that we estimate σ, the common standard deviation of the normal distributions pictured in Figure 16.8. Otherwise, since the x's are assumed to be constants, there is no problem in determining $\bar{x}$ and S_{xx}. The estimate of σ we shall use here is called the **standard error of estimate** and it is denoted by s_e. Its formula is

$$s_e = \sqrt{\frac{\sum (y - \hat{y})^2}{n - 2}}$$

where, again, the y's are the observed values of y and the $\hat{y}$'s are the corresponding values on the least-squares line. Observe that s_e^2 is the sum of the squares of the vertical deviations from the points to the line (namely, the quantity that we minimized by the method of least squares) divided by $n - 2$.

The preceding formula defines s_e, but in practice we calculate its value by means of the computing formula

*Standard error of
estimate*

$$s_e = \sqrt{\frac{S_{yy} - bS_{xy}}{n - 2}}$$

where

$$S_{yy} = \sum y^2 - \frac{1}{n} \left(\sum y \right)^2$$

analogous to the formula for S_{xx} on page 466.

EXAMPLE 16.5   Calculate s_e for the least-squares line that we fit to the data on page 462.

Solution   Since $n = 12$ and we have already shown that $S_{xy} = -668.1$, the only other quantity needed is S_{yy}. Since we gave $\sum y = 166.8$ and $\sum y^2 = 2,331.54$ in Example 16.1, it follows that

$$S_{yy} = 2,331.54 - \frac{1}{12} (166.8)^2 = 13.02$$

and, hence, that

$$s_e = \sqrt{\frac{13.02 - (-0.0175)(-668.1)}{10}}$$

$$\approx 0.3645 \qquad \blacksquare$$

Actually, all this work was not really necessary; the result is given in the computer printout of Figure 16.7, where it says $s = 0.3645$. Also, our graphing calculator could have yielded $s = 0.3644981554$, but we did not display this detail in Figure 16.6.

If we make all the assumptions of normal regression analysis, that the x's are constants and the y's are values of random variables having normal distributions with the means $\mu_{y|x} = \alpha + \beta x$ and the same standard deviation σ, inferences about the regression coefficients α and β can be based on the statistics

Statistics for inferences about regression coefficients

$$t = \frac{a - \alpha}{s_e\sqrt{\dfrac{1}{n} + \dfrac{\bar{x}^2}{S_{xx}}}}$$

$$t = \frac{b - \beta}{s_e/\sqrt{S_{xx}}}$$

whose sampling distributions are t distributions with $n - 2$ degrees of freedom. Note that the quantities in the denominators are estimates of the corresponding standard errors with s_e substituted for σ.

The example that follows illustrates how we test hypotheses about either of the regression coefficients α and β.

EXAMPLE 16.6   Suppose that it has been claimed that a person's hearing range decreases by 0.02 thousand cycles per second for each week that a person has lived near the airport directly in the flight path of departing jets, and that the data on page 462 were obtained to test this claim at the 0.05 level of significance.

Solution   For what follows, it must be assumed that all the assumptions underlying a normal regression analysis are satisfied.

1. H_0: $\beta = -0.02$
    H_A: $\beta \neq -0.02$
2. $\alpha = 0.05$
3. Reject the null hypothesis if $t \leq -2.228$ or $t \geq 2.228$, where

$$t = \frac{b - \beta}{s_e/\sqrt{S_{xx}}}$$

   and 2.228 is the value of $t_{0.025}$ for $12 - 2 = 10$ degrees of freedom; otherwise, accept the null hypothesis or reserve judgment.
4. Since we already know from Examples 16.1, 16.2, and 16.5 that $S_{xx} = 38,178.25$, $b = -0.0175$, and $s_e = 0.3645$, substitution of these values together with $\beta = -0.02$ yields

$$t = \frac{-0.0175 - (-0.02)}{0.3645/\sqrt{38,178.25}} \approx 1.340$$

5. Since $t = 1.340$ falls on the interval from -2.228 to 2.228, the null hypothesis cannot be rejected; there is no real evidence to refute the claim. ∎

Again, we could have saved ourself some work by referring to the computer printout of Figure 16.7. In the column headed **Stdev** it shows that the estimated standard error of b, the quantity that goes into the denominator of the t statistic, is 0.001865, so we can write directly

$$t = \frac{-0.0175 - (-0.02)}{0.001865} = 1.340$$

Tests concerning the regression coefficient α are performed in the same way, except that we use the first, instead of the second, of the two t statistics. In most practical applications, however, the regression coefficient α is not of much interest—it is just the y-intercept, namely, the value of y that corresponds to $x = 0$. In many cases it has no real meaning.

To construct confidence intervals for the regression coefficients α and β, we substitute for the middle term of $-t_{\alpha/2} < t < t_{\alpha/2}$ the appropriate t statistic from page 476. Then, relatively simple algebra leads to the formulas

Confidence limits for regression coefficients

$$a \pm t_{\alpha/2} \cdot s_e \sqrt{\frac{1}{n} + \frac{\bar{x}^2}{S_{xx}}}$$

and

$$b \pm t_{\alpha/2} \cdot \frac{s_e}{\sqrt{S_{xx}}}$$

where the degree of confidence is $(1 - \alpha)100\%$ and $t_{\alpha/2}$ is the entry in Table II for $n - 2$ degrees of freedom.

EXAMPLE 16.7   The following data show the average number of hours that six students spent on homework per week and their grade-point indexes for the courses they took in that semester:

Hours spent on homework x	Grade-point index y
15	2.0
28	2.7
13	1.3
20	1.9
4	0.9
10	1.7

Assuming that all the assumptions underlying a normal regression analysis are satisfied, construct a 99% confidence interval for β, the amount by which a student in the population sampled could have raised his or her grade-point in-dex by studying an extra hour per week.

```
MTB > Regress c2 1 c1;
SUBC>    Constant.



The regression equation is
C2 = 0.721 + 0.0686 C1

Predictor         Coef        Stdev      t-ratio          p
Constant        0.7209       0.2464         2.93      0.043
C1              0.06860      0.01467         4.68      0.009

s = 0.2720        R-sq = 84.5%       R-sq(adj) = 80.7%
```

FIGURE 16.9
MINITAB printout
for Example 16.7.

Solution   Using the computer printout shown in Figure 16.9, we find that $b = 0.06860$ and that the estimate of the standard error of b, by which we have to multiply $t_{\alpha/2}$, is 0.01467. Since $t_{0.005} = 4.604$ for $6 - 2 = 4$ degrees of freedom, we get $0.0686 \pm 4.604(0.01467)$ and, hence,

$$0.0011 < \beta < 0.1361$$

This confidence interval is rather wide, and this is due to two things—the very small size of the sample and the relatively large variation measured by s_e, namely, the variation among the grade-point indexes of students doing the same amount of homework. ∎

To answer the second kind of question asked on page 473, concerning the estimation, or prediction, of the average value of y for a given value of x, we use a method that is very similar to the one just discussed. With the same assumptions as before, we base our argument on another t statistic, arriving at the following $(1 - \alpha)100\%$ confidence interval for $\mu_{y|x_0}$, the mean of y when $x = x_0$:

Confidence limits for
mean of y when
x = x_0

$$(a + bx_0) \pm t_{\alpha/2} \cdot s_e \sqrt{\frac{1}{n} + \frac{(x_0 - \bar{x})^2}{S_{xx}}}$$

As before, the number of degrees of freedom is $n - 2$ and the corresponding value of $t_{\alpha/2}$ may be read from Table II.

EXAMPLE 16.8   Referring again to the data on page 462, suppose that we want to estimate the average hearing range of persons who have lived near the airport directly in the flight path of departing jets for two years. Construct a 95% confidence interval.

Solution   Assuming that all the assumptions underlying a normal regression analysis are satisfied, we substitute $n = 12$, $x_0 = 104$ weeks, $\Sigma\, x = 975$ (from Example 16.1) and hence $\bar{x} = 975/12 = 81.25$, $S_{xx} = 38,178.25$ (from Example 16.2), $a + bx_0 = 13.5$ (from Example 16.4), $s_e = 0.3645$ (from Example 16.5), and $t_{0.025} = 2.228$ for $12 - 2 = 10$ degrees of freedom into the preceding confidence-interval formula, getting

$$13.5 \pm 2.228\,(0.3645)\,\sqrt{\frac{1}{12} + \frac{(104 - 81.25)^2}{38,178.25}}$$

and hence

$$13.25 < \mu_{y|x_0} < 13.75$$

thousand cycles per second when $x = 104$ weeks. (Had we used $a = 15.32$ instead of $a = 15.3$ in Example 16.4, we would have obtained $a + bx_0 = 13.50$ instead of 13.48, which we rounded up to 13.5. So, the result would have been the same.) ■

The third question asked on pages 473 and 474 differs from the other two. It does not concern the estimation of a population parameter, but the prediction of a single future observation. The endpoints of an interval for which we can assert with a given degree of confidence that it will contain such an observation are called **limits of prediction,** and the calculation of such limits will answer the third kind of question. Basing our argument on yet another t statistic, we arrive at the following $(1 - \alpha)100\%$ limits of prediction for a value of y when $x = x_0$:

Limits of prediction

$$(a + bx_0) \pm t_{\alpha/2} \cdot s_e \sqrt{1 + \frac{1}{n} + \frac{(x_0 - \bar{x})^2}{S_{xx}}}$$

Again, the number of degrees of freedom is $n - 2$ and the corresponding value of $t_{\alpha/2}$ may be read from Table II.

Note that the only difference between these limits of prediction and the confidence limits for $\mu_{y|x_0}$ given previously is the 1 that we added to the quantity under the square root sign. Thus, it will be left to the reader to verify in Exercise 16.24 that for the hearing-range example and $x_0 = 104$ the 95% limits of prediction are 12.65 and 14.35. It should not come as a surprise that this interval is much wider than the one obtained in Example 16.8. Whereas the limits of prediction apply to a prediction for one person, the confidence limits obtained in Example 16.8 apply to the mean of all persons who live or have lived near the airport directly in the flight path of departing jets for two years.

Let us remind the reader that all these methods are based on the very stringent assumptions of normal regression analysis. Furthermore, if we base more than one inference on the same data, we will run into problems with regard to the levels of significance and/or degrees of confidence. The random variables on which the various procedures are based are clearly not independent.

■ Exercises

16.20 Assume that the data of Exercise 16.3 satisfy the assumptions required by a normal regression analysis.

(a) If the work in Exercise 16.4 was done with a computer, use the information provided by the software to test the null hypothesis $\beta = 0.15$ against the alternative hypothesis $\beta \neq 0.15$ at the 0.05 level of significance.

(b) For tests concerning the regression coefficient β, the TI-83 graphing calculator provides the value of t only for tests of the null hypothesis $\beta = 0$. Since the difference is only in the numerator, the value of t for tests of the null hypothesis $\beta = \beta_0$ may be obtained by multiplying the value of t provided by the calculator by

$$\frac{b - \beta_0}{b}$$

If the work in Exercise 16.4 was done with a graphing calculator, use this method for calculating t to test the null hypothesis $\beta = 0.15$ against the alternative hypothesis $\beta \neq 0.15$ at the 0.05 level of significance.

16.21 Assume that the data of Exercise 16.6 satisfy the assumptions required by a normal regression analysis.

(a) Use the sums given in that exercise, $\Sigma y^2 = 2{,}001$, and the result that $b = 0.898$, to calculate the value of s_e.

(b) Use the information given in part (a) as well as its result to test the null hypothesis $\beta = 1.5$ against the alternative hypothesis $\beta < 1.5$ at the 0.05 level of significance.

16.22 Use a computer or a graphing calculator to rework both parts of Exercise 16.21. If a graphing calculator is used, follow the suggestion given in part (b) of Exercise 16.20.

16.23 Assuming that the data of Exercise 16.8 satisfy the assumptions required by a normal regression analysis, use the result of that exercise and a computer or a graphing calculator to test the null hypothesis $\beta = 3.5$ against the alternative hypothesis $\beta > 3.5$ at the 0.01 level of significance. If a graphing calculator is used, follow the suggestion given in part (b) of Exercise 16.20.

16.24 With reference to the data on page 462 and the calculations in Example 16.8, show that for $x_0 = 104$ the 95% limits of prediction for the hearing range are 12.65 and 14.35 thousand cycles per second.

16.25 With reference to Exercise 16.9, use a computer or a graphing calculator to test the null hypothesis $\beta = -0.15$ against the alternative hypothesis $\beta \neq -0.15$ at the 0.01 level of significance. It must be assumed, of course, that the data of Exercise 16.9 satisfy the assumptions required by a normal regression analysis. Also, if a graphing calculator is used, follow the suggestion of part (b) of Exercise 16.20.

16.26 With reference to the preceding exercise and the same assumptions, construct a 95% confidence interval for the hourly reduction of the chlorine residual.

16.27 Assume that the data of Exercise 16.12 satisfy the assumptions required by a normal regression analysis.

(a) Use the sums given in that exercise, $\Sigma y^2 = 1{,}802$, and the result that $b = 0.272$, to calculate the value of s_e.

(b) Use the information given in part (a) as well as its result to test the null hypothesis $\beta = 0.40$ against the alternative hypothesis $\beta < 0.40$ at the 0.05 level of significance.

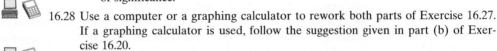

16.28 Use a computer or a graphing calculator to rework both parts of Exercise 16.27. If a graphing calculator is used, follow the suggestion given in part (b) of Exercise 16.20.

16.29 Assuming that the data of Exercise 16.12 satisfy the required assumptions for a normal regression analysis, use a computer or a graphing calculator to determine a 95% confidence interval for the mean moisture content when the humidity is 50%.

16.30 The following table shows the assessed values and the selling prices of eight houses, constituting a random sample of all the houses sold recently in a rural area:

Assessed value (thousands of dollars)	Selling price (thousands of dollars)
70.3	114.4
102.0	169.3
62.5	106.2
74.8	125.0
57.9	99.8
81.6	132.1
110.4	174.2
88.0	143.5

Assuming that these data satisfy the required assumptions for a normal regression analysis, use a computer or a graphing calculator to find
(a) a 95% confidence interval for the mean selling price of a house in this rural area that is assessed at $90,000;
(b) 95% limits of prediction for a house in this rural area that has been assessed at $90,000.

16.31 Assuming that the data of Exercise 16.3 satisfy the assumptions required by a normal regression analysis, use a computer or a graphing calculator to determine
(a) a 99% confidence interval for the mean hyperactivity rating of a four-year-old at one of the nursery schools 45 minutes after he or she has eaten 60 grams of food containing the preservative;
(b) 99% limits of prediction for the hyperactivity rating of one of these children who ate 60 grams of food with the preservative 45 minutes earlier.

16.32 Assuming that the data of Exercise 16.8 satisfy the assumptions required by a normal regression analysis, use a computer or a graphing calculator to determine
(a) a 98% confidence interval for the average yield of wheat when there are only 10 inches of rain;
(b) 98% limits of prediction for the yield of wheat when there are only 10 inches of rain.

16.33 Assuming that the data of Example 16.7 satisfy the assumptions required by a normal regression analysis, use a computer or a graphing calculator to determine
(a) a 95% confidence interval for the average grade-point index of students who average only five hours of homework per week during the semester;
(b) 95% limits of prediction for the grade-point index of a student who averages only five hours of homework per week during the semester.

*16.4 MULTIPLE REGRESSION[†]

Although there are many problems where one variable can be predicted quite accurately in terms of another, it stands to reason that predictions should improve if one considers additional relevant information. For instance, we should be able to make better predictions of the performance of newly hired teachers if we consider not only their education, but also their years of experience and their personality. Also, we should be able to make better predictions of a new textbook's success if we consider not only the quality of the work, but also the potential demand and the competition.

Many mathematical formulas can serve to express relationships among more than two variables, but most commonly used in statistics (partly for reasons of convenience) are linear equations of the form

$$y = b_0 + b_1 x_1 + b_2 x_2 + \cdots + b_k x_k$$

Here y is the variable that is to be predicted, $x_1, x_2, \ldots,$ and x_k are the k known variables on which predictions are to be based, and $b_0, b_1, b_2, \ldots,$ and b_k are numerical constants that must be determined from observed data.

To illustrate, consider the following equation, which was obtained in a study of the demand for different meats:

$$\hat{y} = 3.489 - 0.090 x_1 + 0.064 x_2 + 0.019 x_3$$

Here y denotes the total consumption of federally inspected beef and veal in millions of pounds, x_1 denotes a composite retail price of beef in cents per pound, x_2 denotes a composite retail price of pork in cents per pound, and x_3 denotes income as measured by a certain payroll index. With this equation, we can predict the total consumption of federally inspected beef and veal on the basis of specified values of x_1, x_2, and x_3.

The problem of determining a linear equation in more than two variables that best describes a given set of data is that of finding numerical values for $b_0, b_1, b_2, \ldots,$ and b_k. This is usually done by the method of least squares; that is, we minimize the sum of squares $\Sigma (y - \hat{y})^2$, where as before the y's are the observed values and the $\hat{y}$'s are the values calculated by means of the linear equation. In principle, the problem of determining the values of $b_0, b_1, b_2, \ldots,$ and b_k is the same as it is in the two-variable case, but manual solutions may be very tedious because the method of least squares requires that we solve as many normal equations as there are unknown constants $b_0, b_1, b_2, \ldots,$ and b_k. For instance, when there are two independent variables x_1 and x_2, and we want to fit the equation

$$y = b_0 + b_1 x_1 + b_2 x_2$$

[†]This section is marked optional because the calculations, while possible with a calculator for very simple problems, generally require computer software.

we must solve the three normal equations

Normal equations (two independent variables)

$$\sum y = n \cdot b_0 + b_1 \left(\sum x_1\right) + b_2 \left(\sum x_2\right)$$

$$\sum x_1 y = b_0 \left(\sum x_1\right) + b_1 \left(\sum x_1^2\right) + b_2 \left(\sum x_1 x_2\right)$$

$$\sum x_2 y = b_0 \left(\sum x_2\right) + b_1 \left(\sum x_1 x_2\right) + b_2 \left(\sum x_2^2\right)$$

Here $\sum x_1 y$ is the sum of the products obtained by multiplying each given value of x_1 by the corresponding value of y, $\sum x_1 x_2$ is the sum of the products obtained by multiplying each given value of x_1 by the corresponding value of x_2, and so on.

EXAMPLE 16.9   The following data show the number of bedrooms, the number of baths, and the prices at which eight one-family houses sold recently in a certain community:

Number of bedrooms x_1	Number of baths x_2	Price (dollars) y
3	2	143,800
2	1	109,300
4	3	158,800
2	1	109,200
3	2	154,700
2	2	114,900
5	3	188,400
4	2	142,900

Find a linear equation that will enable one to predict the average sales price of a one-family house in the given community in terms of the number of bedrooms and the number of baths.

Solution   The quantities needed for substitution into the three normal equations are $n = 8$, $\sum x_1 = 25$, $\sum x_2 = 16$, $\sum y = 1{,}122{,}000$, $\sum x_1^2 = 87$, $\sum x_1 x_2 = 55$, $\sum x_2^2 = 36$, $\sum x_1 y = 3{,}711{,}100$, and $\sum x_2 y = 2{,}372{,}700$, and we get

$$1{,}122{,}000 = 8b_0 + 25b_1 + 16b_2$$
$$3{,}711{,}100 = 25b_0 + 87b_1 + 55b_2$$
$$2{,}372{,}700 = 16b_0 + 55b_1 + 36b_2$$

We could solve these equations by the method of elimination or by using determinants, but in view of the rather tedious calculations, such work is nowadays left to computers. So, let us refer to the computer printout of Figure 16.10, where we find in the column headed "Coef" that $b_0 = 65{,}430$, $b_1 = 16{,}752$, and $b_2 = 11{,}235$. In the line immediately above the coefficients we find that the least-squares equation is

$$\hat{y} = 65{,}430 + 16{,}752 \, x_1 + 11{,}235 \, x_2$$

This tells us that (in the given community at the time the study was being made) each extra bedroom added on the average \$16,752, and each bath \$11,235, to the sales price of a house.   ■

```
MTB > Regress c3 2 c1 c2;
SUBC>    Constant.



The regression equation is
C3 = 65430 + 16752 C1 + 11235 C2

Predictor       Coef        Stdev     t-ratio        p
Constant        65430       12134       5.39      0.003
C1              16752        6636       2.52      0.053
C2              11235        9885       1.14      0.307
```

FIGURE 16.10
MINITAB printout
for Example 16.9.

EXAMPLE 16.10   Based on the result of Example 16.9, determine the average sales price of a house with three bedrooms and two baths (in the given community at the time the study was being made).

olution   Substituting $x_1 = 3$ and $x_2 = 2$ into the least-squares equation obtained in Example 16.9, we get

$$\hat{y} = 65{,}430 + 16{,}752(3) + 11{,}235(2)$$
$$= 138{,}156$$

or approximately \$138,200.   ■

We shall not go into this here, but let us point out that most computer programs like the one that produced the printout shown in Figure 16.10 provide information that makes it easy to test hypotheses about the true regression coefficients $\beta_0, \beta_1, \beta_2, \ldots$, which are estimated by $b_0, b_1, b_2, \ldots$, or to construct confidence intervals. Some further material, of no relevance at this time, was deleted from the printout.

■ E x e r c i s e s

 *16.34 The following are data on the ages and incomes of a random sample of executives working for a large multinational corporation, and the number of years they did postgraduate work at a university:

Age x_1	Years postgraduate work x_2	Income (dollars) y
38	4	181,700
46	0	173,300
39	5	189,500
43	2	179,800
32	4	169,900
52	7	212,500

(a) Use appropriate computer software to fit an equation of the form $y = b_0 + b_1x_1 + b_2x_2$ to the given data.

(b) Use the equation obtained in part (a) to estimate the average income of 39-year-old executives of the corporation who did three years of postgraduate work at a university.

 *16.35 The following data were collected to determine the relationship between two processing variables and the hardness of a certain kind of steel:

Hardness (Rockwell 30-T) y	Copper content (percent) x_1	Annealing temperature (degrees F) x_2
78.9	0.02	1,000
55.2	0.02	1,200
80.9	0.10	1,000
57.4	0.10	1,200
85.3	0.18	1,000
60.7	0.18	1,200

(a) Use appropriate computer software to fit an equation of the form $y = b_0 + b_1x_1 + b_2x_2$ to the given data.

(b) Use the equation obtained in part (a) to estimate the hardness of steel when its copper content is 0.14% and the annealing temperature is 1,100 degrees Fahrenheit.

*16.36 When the x_1's and/or the x_2's are equally spaced, the calculation of the regression coefficients can be simplified considerably by using the kind of coding described in Exercise 16.17. Rework Exercise 16.35 without a computer after coding the three x_1 values -1, 0, and 1, and the two x_2 values -1 and 1. (Note that, when coded, the 0.14% copper content becomes 0.50 and the 1,100 annealing temperature becomes 0.)

*16.37 The following are data on the percent effectiveness of a pain reliever and the amounts of three medications (in milligrams) present in each capsule:

Medication A x_1	Medication B x_2	Medication C x_3	Percent effective y
15	20	10	47
15	20	20	54
15	30	10	58
15	30	20	66
30	20	10	59
30	20	20	67
30	30	10	71
30	30	20	83
45	20	10	72
45	20	20	82
45	30	10	85
45	30	20	94

(a) Use appropriate computer software to fit an equation of the form $y = b_0 + b_1 x_1 + b_2 x_2 + b_3 x_3$ to the given data.

(b) Use the equation obtained in part (a) to estimate the average percent effectiveness of capsules that contain 12.5 milligrams of Medication A, 25 milligrams of Medication B, and 15 milligrams of Medication C.

*16.38 Rework Exercise 16.37 without a computer after coding the three x_1 values $-1, 0$, and 1, the two x_2 values -1 and 1, and the two x_3 values -1 and 1.

*16.5 NONLINEAR REGRESSION

When the pattern of a set of data points departs appreciably from a straight line, we must consider fitting some other kind of curve. In this section we shall first describe two situations where the relationship between x and y is not linear, but the method of Section 16.2 can nevertheless be employed. Then we shall give an example of **polynomial curve fitting** by fitting a parabola.

We usually plot paired data on various kinds of graph paper to see whether there are scales for which the points fall close to a straight line. Of course, when this is the case for ordinary graph paper, we proceed as in Section 16.2. If it is the case when we use **semilog paper** (with equal subdivisions for x and a logarithmic scale for y, as shown in Figure 16.11), this indicates that an **exponential curve** will provide a good fit. The equation of such a curve is

$$y = a \cdot b^x$$

or in logarithmic form

$$\log y = \log a + x \, (\log b)$$

where "log" stands for logarithm to the base 10. (Actually, we could use any base including the irrational number e, in which case the equation is often written as $y = a \cdot e^{bx}$, or in logarithmic form as $\ln y = \ln a + bx$.)

Observe that if we write A for $\log a$, B for $\log b$, and Y for $\log y$, the original equation in logarithmic form becomes $Y = A + Bx$, which is the usual equation of a straight line. Thus, to fit an exponential curve to a given set of paired data, we simply apply the method of Section 16.2 to the data points (x, Y).

EXAMPLE 16.11   The following are data on a company's net profits during the first six years that it has been in business:

Year	Net profit (thousands of dollars)
1	112
2	149
3	238
4	354
5	580
6	867

In Figure 16.11 these data are plotted on ordinary graph paper on the left and on semilog paper (with a logarithmic scale for y) on the right. As can be seen, the overall pattern is remarkably "straightened out" in the figure on the right, and this suggests that we ought to fit an exponential curve.

FIGURE 16.11
Data plotted on ordinary and semilog graph paper.

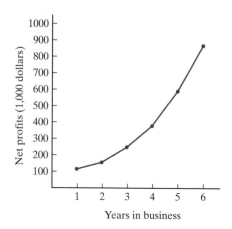 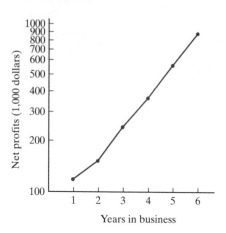

Solution   Getting the logarithms of the y's with a calculator, or perhaps from a table of logarithms, we find that

x	y	$Y = \log y$
1	112	2.0492
2	149	2.1732
3	238	2.3766
4	354	2.5490
5	580	2.7634
6	867	2.9380

Then, for these data we get $n = 6$, $\Sigma x = 21$, $\Sigma x^2 = 91$, $\Sigma Y = 14.8494$, and $\Sigma xY = 55.1664$, and hence $S_{xx} = 91 - \frac{1}{6}(21)^2 = 17.5$ and $S_{xY} = 55.1664 - \frac{1}{6}(21)(14.8494) = 3.1935$. Finally, substitution into the two formulas on page 446 yields

$$B = \frac{3.1935}{17.5} \approx 0.1825$$

$$A = \frac{14.8494 - 0.1825(21)}{6} \approx 1.8362$$

and the equation that describes the relationship is

$$\hat{Y} = 1.8362 + 0.1825x$$

Since 1.8362 and 0.1825 are the estimates corresponding to $\log a$ and $\log b$, we find by taking antilogarithms that $a = 68.58$ and $b = 1.52$. Thus, the equation of the exponential curve that best describes the relationship between the company's net profit and the number of years it has been in business is given by

$$\hat{y} = 68.58(1.52)^x$$

where $\hat{y}$ is in thousands of dollars.   ∎

FIGURE 16.12
Computer printout for
Example 16.11.

```
MTB > Regress c2 1 c1;
SUBC>    Constant.



The regression equation is
C2 = 1.84 + 0.182 C1

Predictor        Coef       Stdev     t-ratio        p
Constant      1.83620     0.02243      81.85     0.000
C1           0.182486    0.005760      31.68     0.000

s = 0.02410      R-sq = 99.6%      R-sq(adj) = 99.5%
```

Although the calculations in Example 16.11 were quite easy, we could, of course, have used a computer. Entering the values of x and Y in columns c1 and c2, we got the printout shown in Figure 16.12. As can be seen, the values we had calculated for A and B are shown in the column headed "Coef." To get the exponential equation in its final form, it would have been even easier to use a graphing calculator. After entering the original x's and y's, the command **STAT CALC ExpReg** yields the display shown in Figure 16.13. As can be seen, the constants a and b rounded to two decimals are identical with those in the exponential equation given at the end of Example 16.11.

```
ExpReg
 y=a*b^x
 a=68.57875261
 b=1.522264768
```

FIGURE 16.13
Values of a and b reproduced from display screen of TI-83 graphing calculator.

Once we fit an exponential curve to a set of paired data, we can predict a future value of y by substituting into its equation the corresponding value of x. However, it is usually much more convenient to substitute x into the logarithmic form of the equation, namely, into

$$\log \hat{y} = \log a + x(\log b)$$

EXAMPLE 16.12 With reference to Example 16.11, predict the company's net profit for the eighth year that it will have been in business.

olution Substituting $x = 8$ into the logarithmic form of the equation for the exponential curve, we get

$$\log \hat{y} = 1.8362 + 8(0.1825)$$
$$= 3.2962$$

and hence $\hat{y} = 1,980$ or $\$1,980,000$. ∎

If data points fall close to a straight line when plotted on **log-log paper** (with logarithmic scales for both x and y), this indicates that an equation of the form

$$y = a \cdot x^b$$

will provide a good fit. In the logarithmic form, the equation of such a **power function** is

$$\log y = \log a + b(\log x)$$

which is a linear equation in log x and log y. (Writing A, X, and Y for log a, log x, and log y, the equation becomes

$$Y = A + bX$$

which is the usual equation of a straight line.) For fitting a power curve, we can thus, apply the method of Section 16.2 to the problem expressed as $Y = A + bX$. The work that is required for fitting a power function is very similar to what we did in Example 16.11, and we shall not illustrate it by means of an example. However, in Exercises 16.47 and 16.49 the reader will find problems to which the method can be applied.

When the values of y first increase and then decrease, or first decrease and then increase, a **parabola** having the equation

$$y = a + bx + cx^2$$

will often provide a good fit. This equation can also be written as

$$y = b_0 + b_1 x + b_2 x^2$$

to conform with the notation of Section 16.4. Thus, it can be seen that parabolas can be looked upon as linear equations in the two unknowns $x_1 = x$ and $x_2 = x^2$, and fitting a parabola to a set of paired data is nothing new—we simply use the method of Section 16.4. If we actually wanted to use the normal equations on page 465 with $x_1 = x$ and $x_2 = x^2$, this would require that we determine Σx, Σx^2, Σx^3, Σx^4, Σy, Σxy, and $\Sigma x^2 y$, and the subsequent solution of three simultaneous linear equations. As can well be imagined, this would require a great deal of arithmetic and it is rarely done without the use of appropriate technology. In the two examples that follow, we shall first illustrate fitting a parabola with the use of a computer and then repeat the problem with the use of a graphing calculator.

EXAMPLE 16.13   The following are data on the drying time of a varnish and the amount of a certain chemical additive:

Amount of additive (grams) x	Drying time (hours) y
1	7.2
2	6.7
3	4.7
4	3.7
5	4.7
6	4.2
7	5.2
8	5.7

(a) Fit a parabola that, as would appear from Figure 16.14, is the right kind of curve to fit to the given data.

(b) Use the result of part (a) to predict the drying time of the varnish when 6.5 grams of the chemical are added.

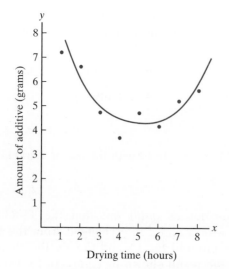

FIGURE 16.14
Scattergram of the
varnish-drying-time
data.

Solution

(a) Using the MINITAB printout shown in Figure 16.15, we find that
$b_0 = 9.2446$, $b_1 = -2.0149$, and $b_2 = 0.19940$ (in the column headed
"Coef"). Rounding to two decimals, we can, thus, write the equation
of the parabola as

$$\hat{y} = 9.24 - 2.01x + 0.20x^2$$

FIGURE 16.15
Computer printout for
fitting parabola.

```
MTB > Print c1 c2 c3.


Row   C1    C2     C3

 1    1     1     7.2
 2    2     4     6.7
 3    3     9     4.7
 4    4    16     3.7
 5    5    25     4.7
 6    6    36     4.2
 7    7    49     5.2
 8    8    64     5.7

MTB > Regress c3 2 c1 c2;
SUBC>    Constant.


The regression equation is
C3 = 9.24 - 2.01 C1 + 0.199 C2

Predictor        Coef        Stdev     t-ratio         p
Constant       9.2446       0.7645       12.09     0.000
C1            -2.0149       0.3898       -5.17     0.004
C2            0.19940      0.04228        4.72     0.005

s = 0.5480       R-sq = 85.3%     R-sq(adj) = 79.4%
```

(Note that we had entered the values of x in column c1, the values of x^2 in column c2, and the values of y in column c3.)

(b) Substituting $x = 6.5$ into the equation obtained in part (a), we get

$$\hat{y} = 9.24 - 2.01(6.5) + 0.20\,(6.5)^2$$

$$\approx 4.62 \text{ hours} \qquad \blacksquare$$

EXAMPLE 16.14   Rework part (a) of Example 16.13 with the use of a graphing calculator.

Solution   To avoid confusion, let us point out that the TI-83 uses the equation $y = ax^2 + bx + c$, with a and c interchanged from the version we gave on page 490. After we entered the x's and the y's, the command **STAT CALC QuadReg** yielded the result shown in Figure 16.16. Rounding to two decimals, we get

$$\hat{y} = 0.20x^2 - 2.01x + 9.24$$

which, except for the order of the terms, is the same as before. $\blacksquare$

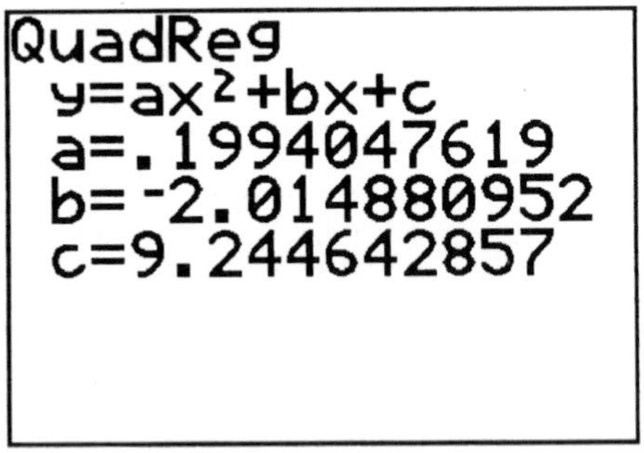

FIGURE 16.16
Fit of parabola reproduced from the display screen of a TI-83 graphing calculator.

```
QuadReg
 y=ax²+bx+c
 a=.1994047619
 b=-2.014880952
 c=9.244642857
```

On page 490 we introduced parabolas as curves that bend once—that is, their values first increase and then decrease, or first decrease and then increase. For patterns that "bend" more than once, **polynomial equations** of higher degree, such as $y = a + bx + cx^2 + dx^3$ or $y = a + bx + cx^2 + dx^3 + ex^4$ can be fitted by the technique illustrated in Example 16.13. In practice, we often use sections of such curves, especially parts of parabolas, when there is only a slight curvature in the pattern we want to describe.

Exercises

*16.39 The following data pertain to the growth of a colony of bacteria in a culture medium:

Days since inoculation x	Bacteria count (thousands) y
2	112
4	148
6	241
8	363
10	585

(a) Given that $\Sigma x = 30$, $\Sigma x^2 = 220$, $\Sigma Y = 11.9286$ (where $Y = \log y$), and $\Sigma xY = 75.2228$, set up the two normal equations for fitting an exponential curve, solve them for $\log a$ and $\log b$ by the method of elimination or by the method based on determinants, and write the equation of the exponential curve in logarithmic form.

(b) Convert the equation obtained in part (a) into the form $y = ab^x$.

(c) Use the equation obtained in part (a) to estimate the bacteria count five days after inoculation.

*16.40 Use a computer or a graphing calculator to rework Exercise 16.39.

*16.41 Use a computer or a graphing calculator to fit an exponential curve to the following data on the percentage of the radial tires made by a manufacturer that are still usable after having been driven the given numbers of miles:

Miles driven (thousands) x	Percentage usable y
1	97.2
2	91.8
5	82.5
10	64.4
20	41.0
30	29.9
40	17.6
50	11.3

*16.42 Use the equation obtained in Exercise 16.41 (if necessary converted into logarithmic form) to estimate what percentage of the tires will still be usable after having been driven for 25,000 miles.

*16.43 Use a computer or a graphing calculator to fit an exponential curve to the following data on the curing time of test samples of concrete, x, and their tensile strength, y:

x (hours)	y (psi)
1	3.54
2	8.92
3	27.5
4	78.8
5	225
6	639

*16.44 Assuming that the exponential trend continues, use the equation obtained in Exercise 16.43 (in logarithmic form) to estimate the tensile strength of a test sample of the concrete that has been cured for eight hours.

 *16.45 A small piece of a rare slow-growing cactus was grafted to another cactus with a strong root system and its height, measured annually, was as shown in the following table:

Years after grafting x	Height (millimeters) y
1	22
2	25
3	29
4	34
5	38
6	44
7	51
8	59
9	68

Use a computer or a graphing calculator to fit an exponential curve.

*16.46 Assuming that the exponential trend continues, use the equation obtained in Exercise 16.45 (in logarithmic form) to estimate the height of the graft 10 years after it had been grafted.

*16.47 On page 489 we mentioned the equation $y = a \cdot x^b$ as that of a power function (not to be confused with the kind of power function mentioned in Chapter 12), which can also be written as $\log y = \log a + b (\log x)$ or as $Y = A + bX$, where $Y = \log y$, $A = \log a$, and $X = \log x$. This means that we can fit a curve with this kind of equation using the method of Section 16.2. Use appropriate computer software to find the equation of a power function fit to the following data on the unit cost of producing certain electronic components and the number of units produced:

Lot size x	Unit cost y
50	$108
100	$53
250	$24
500	$9
1,000	$5

Also, use the result to estimate the unit cost for a lot of 300 components.

 *16.48 Finding the equation of a power function is even easier when we use a graphing calculator. We merely enter the original x's and y's and then use the command **STAT CALC PwrReg.** Thus, use a graphing calculator to verify the equation obtained in Exercise 16.47.

 *16.49 The following data pertain to the volume of a gas (in cubic inches) and its pressure (in pounds per square inch), when the gas is compressed at a constant temperature:

Volume x	Pressure y
50	16.0
30	40.1
20	78.0
10	190.5
5	532.2

Use a computer and the method described in Exercise 16.47 or a graphing calculator as described in Exercise 16.48 to find the equation of a power function fit to these data. Also, estimate the pressure of this gas when it is compressed to a volume of 15 cubic inches.

 *16.50 The following are data on the amount of fertilizer applied to the soil, x (in pounds per square yard), and the yield of a certain food crop, y (in pounds per square yard):

x	y
0.5	32.0
1.1	34.3
2.2	15.7
0.2	20.8
1.6	33.5
2.0	21.5

(a) Draw a scattergram to verify that it is reasonable to describe the overall pattern with a parabola.
(b) Use a computer or a graphing calculator to fit a parabola to the given data.
(c) Use the equation obtained in part (b) to estimate the yield when 1.5 pounds of the fertilizer are applied per square yard.

*16.51 It can be shown that $y = a + bx + cx^2$ is a maximum or a minimum when $x = -\dfrac{b}{2c}$; a maximum when c is negative and a minimum when c is positive. With reference to Exercise 16.50 and the result obtained in part (b), find the maximum yield and the amount of fertilizer that will produce this maximum yield.

 *16.52 The following data pertain to the demand for a product, y (in thousands of units), and its price, x (in dollars), in five very similar market areas:

Price	Demand
20	22
16	41
10	120
11	89
14	56

Use a computer or a graphing calculator to fit a parabola to these data, and use it to estimate the demand when the price of the product is $12.

16.6 *Checklist of Key Terms* (with page references to their definitions)

Curve fitting, 460
Data points, 462
Determinants, 466
Estimated regression coefficients, 474
Estimated regression line, 469
*Exponential curve, 486
Least-squares line, 465
Limits of prediction, 479
Linear equation, 461
Linear regression analysis, 474
*Log-log paper, 489
Method of elimination, 466
Method of least squares, 460, 463
*Multiple regression, 482
Nonlinear regression, 486

Normal equations, 465, 483, 490
Normal regression analysis, 474
*Parabola, 490
*Polynomial curve fitting, 486
*Polynomial equation, 492
*Power function, 489
Regression, 460
Regression analysis, 473, 474
Regression coefficients, 474
Regression equation, 468
Regression line, 469
*Semilog paper, 486
Standard error of estimate, 475
Scattergram, 462

16.7 *References*

Methods of deciding which kind of curve to fit to a given set of paired data may be found in books on numerical analysis and in more advanced texts in statistics. Further information about the material of this chapter may be found in

CHATTERJEE, S., and PRICE, B., *Regression Analysis by Example,* 2nd ed. New York: John Wiley & Sons, Inc., 1991.

DANIEL, C., and WOOD, F., *Fitting Equations to Data,* 2nd ed. New York: John Wiley & Sons, Inc., 1980.

DRAPER, N. R., and SMITH, H., *Applied Regression Analysis,* 2nd ed. New York: John Wiley & Sons, Inc., 1981.

EZEKIEL, M., and FOX, K. A., *Methods of Correlation and Regression Analysis,* 3rd ed. New York: John Wiley & Sons, Inc., 1959.

WEISBERG, S., *Applied Linear Regression,* 2nd ed. New York: John Wiley & Sons, Inc., 1985

WONNACOTT, T. H., and WONNACOTT, R. J., *Regression: A Second Course in Statistics.* New York: John Wiley & Sons, Inc., 1981.

17 CORRELATION

17.1   The Coefficient of Correlation 499

17.2   The Interpretation of r 506

17.3   Correlation Analysis 511

*17.4   Multiple and Partial Correlation 516

17.5   Checklist of Key Terms 519

17.6   References 519

Having learned how to fit a least-squares line to paired data, we now turn to the problem of determining how well such a line actually fits the data. Of course, we can get some idea about this by inspecting a scattergram that shows the line together with the data, but to show how we can be more objective, let us refer back to the data we used to illustrate fitting a least-squares line; namely, to the following data pertaining to the hearing range of persons exposed to the take-off noise of jets for a certain length of time:

Number of weeks x	Hearing range y
47	15.1
56	14.1
116	13.2
178	12.7
19	14.6
75	13.8

160	11.9
31	14.8
12	15.3
164	12.6
43	14.7
74	14.0

As can be seen from this table, there are substantial differences among the y's, the smallest being 11.9 and the largest being 15.3. However, we also see that the hearing range of 11.9 thousand cycles per second was that of a person who had lived at that location for 160 weeks, while the hearing range of 15.3 thousand cycles per second was that of a person who had lived there for only 12 weeks. This suggests that the differences in hearing range may well be due, at least in part, to differences in the length of time that the persons had been exposed to the airport noise. This raises the following question, which we shall answer in this chapter: Of the total variation among the y's, how much can be attributed to the relationship between the two variables x and y (that is, to the fact that the observed values of y correspond to different values of x), and how much can be attributed to chance?

In Section 17.1 we shall introduce the coefficient of correlation as a measure of the strength of the linear relationship between two variables, in Section 17.2 we shall learn how to interpret it, and in Section 17.3 we shall study related problems of inference. The problems of multiple and partial correlation will be touched upon lightly in the optional Section 17.4.

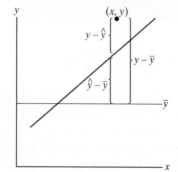

FIGURE 17.1
Illustration showing
that $y - \bar{y} =$
$(y - \hat{y}) + (\hat{y} - \bar{y})$.

17.1 THE COEFFICIENT OF CORRELATION

In reply to the question raised in the chapter opening, let us point out that we are faced here with an analysis of variance. Figure 17.1 shows what we mean. As can be seen from the diagram, the deviation of an observed value of y from the mean of all the y's, $y - \bar{y}$, can be written as the sum of two parts. One part is the deviation of $\hat{y}$ (the value on the line corresponding to an observed value of x) from the mean of all the y's, $\hat{y} - \bar{y}$; the other part is the deviation of the observed value of y from the corresponding value on the line, $y - \hat{y}$. Symbolically, we write

$$y - \bar{y} = (\hat{y} - \bar{y}) + (y - \hat{y})$$

for any observed value y, and if we square the expressions on both sides of this identity and sum over all n values of y, we find that algebraic simplifications lead to

$$\Sigma \ (y - \bar{y})^2 = \Sigma \ (\hat{y} - \bar{y})^2 + \Sigma \ (y - \hat{y})^2$$

The quantity on the left measures the total variation of the y's and we call it the **total sum of squares;** note that $\Sigma \ (y - \bar{y})^2$ is just the variance of the y's multiplied by $n - 1$. The first of the two sums on the right, $\Sigma \ (\hat{y} - \bar{y})^2$, is called the **regression sum of squares** and it measures that part of the total variation of the y's that can be ascribed to the relationship between the two variables x and y; indeed, if all the points lie on the least-squares line, then $y = \hat{y}$ and the regression sum of squares equals the total sum of squares. In practice, this is hardly, if ever, the case, and the fact that all the points do not lie on a least-squares line is an indication that there are other factors than differences among the x's that affect the values of y. It is customary to combine all these other factors under the general heading of "chance." Chance variation is thus measured by the amounts by which the points deviate from the line; specifically, it is measured by $\Sigma \ (y - \hat{y})^2$, called the **residual sum of squares,** which is the second of the two components into which we partitioned the total sum of squares.

To determine these sums of squares for the hearing-range example, we could substitute the observed values of y, $\bar{y}$, and the values of $\hat{y}$ obtained by substituting the x's into $\hat{y} = 15.3 - 0.0175x$ (see page 467), but there are simplifications. First, for $\Sigma \ (y - \bar{y})^2$ we have the computing formula

$$S_{yy} = \Sigma \ y^2 - \frac{1}{n} (\Sigma \ y)^2$$

and on page 475 we showed that it equals 13.02 for our example. Second, $\Sigma \ (y - \hat{y})^2$ is the quantity we minimized by the method of least squares, and divided by $n - 2$ it defined s_e^2 on page 475. Thus, it equals $(n - 2)s_e^2$ and $(12 - 2)(0.3645)^2 \approx 1.329$ in our example, for which we showed in Example 16.5 that $s_e = 0.3645$. Finally, by subtraction, the regression sum of squares is given by

$$\Sigma \ (\hat{y} - \bar{y})^2 = \Sigma \ (y - \bar{y})^2 - \Sigma \ (y - \hat{y})^2$$

and we get $13.02 - 1.329 = 11.69$ (rounded to two decimals) for our example.

It is of interest to note that all the sums of squares we have calculated here could have been obtained directly from the computer printout of Figure 16.7, which is reproduced in Figure 17.2. Under "Analysis of variance" in the column headed SS, we find that the total sum of squares is 13.020, the error (residual) sum of squares is 1.329, and the regression sum of squares is 11.691. The minor differences between the values shown here and calculated previously are, of course, due to rounding.

```
MTB > GPro.
MTB > Regress 'y' 1 'x';
SUBC>    Constant.




The regression equation is
y = 15.3 - 0.0175 x

Predictor          Coef        Stdev      t-ratio           p
Constant        15.3218       0.1845        83.04       0.000
x              -0.017499     0.001865        -9.38       0.000

s = 0.3645        R-sq = 89.8%      R-sq(adj) = 88.8%

Analysis of Variance

SOURCE           DF            SS           MS        F          p
Regression        1        11.691       11.691    88.00      0.000
Error            10         1.329        0.133
Total            11        13.020
```

FIGURE 17.2
Copy of Figure 16.7.

We are now ready to examine the sums of squares. Comparing the regression sum of squares with the total sum of squares, we find that

$$\frac{\sum (\hat{y} - \bar{y})^2}{\sum (y - \bar{y})^2} = \frac{11.69}{13.02} \approx 0.898$$

is the proportion of the total variation of the hearing ranges that can be attributed to the relationship with x, namely, to the differences in the length of time that the 12 persons in the sample had been exposed to the airport noise. This quantity is referred to as the **coefficient of determination** and it is denoted by r^2. Note that the coefficient of determination is also given in the printout of Figure 17.2, where it says near the middle that R-sq = 89.8%.

If we take the square root of the coefficient of determination, we get the **coefficient of correlation** that is denoted by the letter r. Its sign is chosen so that it is the same as that of the estimated regression coefficient b, and for our example, where b is negative, we get

$$r = -\sqrt{0.898} \approx -0.95$$

rounded to two decimals.

It follows that the correlation coefficient is positive when the least-squares line has an upward slope, namely, when the relationship between x and y is such that small values of y tend to go with small values of x and large values of y tend to go with large values of x. Also, the correlation coefficient is negative when the least-squares line has a downward slope, namely, when large values of y tend to go with small values of x and small values of y tend to go with large values of x. Examples of **positive** and **negative correlations** are shown in the first two diagrams of Figure 17.3.

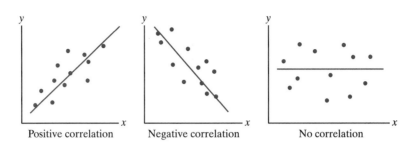

FIGURE 17.3
Types of correlation.

Positive correlation   Negative correlation   No correlation

Since part of the variation of the y's cannot exceed their total variation, $\Sigma (y - \hat{y})^2$ cannot exceed $\Sigma (y - \bar{y})^2$, and it follows from the formula defining r that correlation coefficients must lie on the interval from -1 to $+1$. If all the points actually fall on a straight line, the residual sum of squares, $\Sigma (y - \hat{y})^2$, is zero,

$$\Sigma (\hat{y} - \bar{y})^2 = \Sigma (y - \bar{y})^2$$

and the resulting value of r, -1 or $+1$, is indicative of a perfect fit. If, however, the scatter of the points is such that the least-squares line is a horizontal line coincident with $\bar{y}$ (that is, a line with slope 0 that intersects the y-axis at $a = \bar{y}$), then

$$\Sigma (y - \hat{y})^2 = \Sigma (y - \bar{y})^2 \quad \text{and} \quad r = 0$$

In that case none of the variation of the y's can be attributed to their relationship with x, and the fit is so poor that knowledge of x is of no help in predicting y. The predicted value of y is $\bar{y}$ regardless of x. An example of this is shown in the third diagram of Figure 17.3.

The formula that defines r shows clearly the nature, or essence, of the coefficient of correlation, but in actual practice it is seldom used to determine its value. Instead we use the computing formula

Computing formula
for coefficient of
correlation

$$r = \frac{S_{xy}}{\sqrt{S_{xx} \cdot S_{yy}}}$$

which has the added advantage that it automatically gives r the correct sign. The quantities needed to calculate r with this formula were previously defined, but for convenient reference, we remind the reader that

$$S_{xx} = \Sigma x^2 - \frac{1}{n}(\Sigma x)^2$$

$$S_{yy} = \Sigma y^2 - \frac{1}{n}(\Sigma y)^2$$

$$S_{xy} = \Sigma xy - \frac{1}{n}(\Sigma x)(\Sigma y)$$

EXAMPLE 17.1   Following are the scores that 12 students received in final examinations in economics and anthropology:

Economics	Anthropology
51	74
68	70
72	88
97	93
55	67
73	73
95	99
74	73
20	33
91	91
74	80
80	86

Use the computing formula to calculate r.

Solution   First calculating the necessary sums, we get $\Sigma\, x = 850$, $\Sigma\, x^2 = 65{,}230$, $\Sigma\, y = 927$, $\Sigma\, y^2 = 74{,}883$, and $\Sigma\, xy = 69{,}453$. Then, substituting these values together with $n = 12$ into the formulas for S_{xx}, S_{yy}, and S_{xy}, we find that

$$S_{xx} = 65{,}230 - \frac{1}{12}(850)^2 = 5{,}021.67$$

$$S_{yy} = 74{,}883 - \frac{1}{12}(927)^2 = 3{,}272.25$$

$$S_{xy} = 69{,}453 - \frac{1}{12}(850)(927) = 3{,}790.5$$

and

$$r = \frac{3{,}790.5}{\sqrt{(5{,}021.67)(3{,}272.25)}} = 0.935 \qquad \blacksquare$$

The quantity S_{xy} in the numerator of the formula for r is actually a computing formula for $\Sigma\,(x - \bar{x})(y - \bar{y})$, which, divided by n, is called the first **product moment.** For this reason, r is also referred to at times as the **product-moment coefficient of correlation.** Note that in $\Sigma\,(x - \bar{x})(y - \bar{y})$ we add the products obtained by multiplying the deviation of each x from $\bar{x}$ by the deviation of the corresponding y from $\bar{y}$. In this way we literally measure how the x's and y's vary together. If their relationship is such that large values of x tend to go with large values of y, and small values of x with small values of y, the deviations $x - \bar{x}$ and $y - \bar{y}$ tend to be both positive or both negative, and most of the products $(x - \bar{x})(y - \bar{y})$ will be positive. On the other hand, if the relationship is such that large values of x tend to go with small values of y, and small values of x with large values of y, the deviations $x - \bar{x}$ and $y - \bar{y}$ tend to be of opposite sign, and most of the products $(x - \bar{x})(y - \bar{y})$ will be negative. For this reason, $\Sigma (x - \bar{x})(y - \bar{y})$, divided by $n - 1$, is called the **sample covariance.**

Correlation coefficients are sometimes calculated in the analysis of $r \times c$ tables, provided that the row categories as well as the column categories are ordered. This is the kind of alternative to the chi-square analysis we suggested at the end of Section 14.4, where we pointed out that the ordering of the categories is not taken into account in the calculation of the χ^2 statistic. To use r in a problem like this, we replace the ordered categories by similarly ordered sets of numbers. As we said on page 394, such numbers are usually, though not necessarily, consecutive integers, preferably integers that will make the arithmetic as simple as possible. For three categories we might use 1, 2, and 3, or $-1, 0$, and 1; for four categories we might use 1, 2, 3, and 4, or $-1, 0, 1$, and 2, or perhaps $-3, -1, 1$, and 3. The calculation of r as a measure of the strength of the relationship between two categorical variables is illustrated by the following example.

EXAMPLE 17.2   In Example 14.7 we analyzed the following 3×3 table to see whether there is a relationship between the test scores of persons who have gone through a certain job-training program and their subsequent performance on the job:

		Performance			
		Poor	*Fair*	*Good*	
	Below average	67	64	25	156
Test score	*Average*	42	76	56	174
	Above average	10	23	37	70
		119	163	118	400

Label the test scores $x = -1, x = 0$, and $x = 1$, the performance ratings $y = -1, y = 0$, and $y = 1$, and calculate r.

Solution   Labeling the rows and the columns as indicated, we get

		y		
	-1	0	1	
-1	67	64	25	156
x 0	42	76	56	174
1	10	23	37	70
	119	163	118	400

where the row totals tell us how many times x equals $-1, 0$, and 1, and the column totals tell us how many times y equals $-1, 0$, and 1. Thus,

$$\sum x = 156(-1) + 174 \cdot 0 + 70 \cdot 1 = -86$$
$$\sum x^2 = 156(-1)^2 + 174 \cdot 0^2 + 70 \cdot 1^2 = 226$$
$$\sum y = 119(-1) + 163 \cdot 0 + 118 \cdot 1 = -1$$
$$\sum y^2 = 119(-1)^2 + 163 \cdot 0^2 + 118 \cdot 1^2 = 237$$

and for Σxy we must add the products obtained by multiplying each cell frequency by the corresponding values of x and y. Omitting all cells where either $x = 0$ or $y = 0$, we get

$$\sum xy = 67(-1)(-1) + 25(-1)1 + 10 \cdot 1(-1) + 37 \cdot 1 \cdot 1$$
$$= 69$$

Then, substitution into the formulas for S_{xx}, S_{yy}, and S_{xy} yields

$$S_{xx} = 226 - \frac{1}{400}(-86)^2 = 207.51$$

$$S_{yy} = 237 - \frac{1}{400}(-1)^2 = 237.00$$

$$S_{xy} = 69 - \frac{1}{400}(-86)(-1) = 68.78$$

all rounded to two decimals, and finally

$$r = \frac{68.78}{\sqrt{(207.51)(237.00)}} = 0.31$$ ∎

17.2 THE INTERPRETATION OF r

When r equals $+1$, -1, or 0, there is no problem about the interpretation of the coefficient of correlation. As we have already indicated, it is $+1$ or -1, when all the points actually fall on a straight line, and it is zero when the fit of the least-squares line is so poor that knowledge of x does not help in the prediction of y. In general, the definition of r tells us that $100r^2$ is the percentage of the total variation of the y's that is explained by, or is due to, their relationship with x. This itself is an important measure of the relationship between two variables; beyond this, it permits valid comparisons of the strength of several relationships.

EXAMPLE 17.3   If $r = 0.80$ in one study and $r = 0.40$ in another, would it be correct to say that the 0.80 correlation is twice as strong as the 0.40 correlation?

Solution   No! When $r = 0.80$, then $100(0.80)^2 = 64\%$ of the variation of the y's is accounted for by the relationship with x, and when $r = 0.40$, then $100(0.40)^2 = 16\%$ of the variation of the y's is accounted for by the relationship with x. Thus, in the sense of "percentage of variation accounted for" we can say that the 0.80 correlation is four times as strong as the 0.40 correlation.   ∎

In the same way, we say that a relationship for which $r = 0.60$, is nine times as strong as a relationship for which $r = 0.20$.

There are several pitfalls in the interpretation of the coefficient of correlation. First, it is often overlooked that r measures only the strength of linear relationships; second, it should be remembered that a strong correlation (a value of r close to $+1$ or -1) does not necessarily imply a cause–effect relationship.

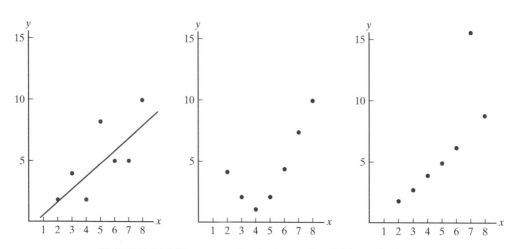

FIGURE 17.4 Three sets of paired data for which $r = 0.75$.

If *r* is calculated indiscriminately, for instance, for the three sets of data of Figure 17.4, we get $r = 0.75$ in each case, but it is a meaningful measure of the strength of the relationship only in the first case. In the second case there is a very strong curvilinear relationship between the two variables, and in the third case six of the seven points actually fall on a straight line, but the seventh point is so far off that it suggests the possibility of a gross error of measurement or an error in recording the data. Thus, before we calculate *r*, we should always plot the data to see whether there is reason to believe that the relationship is, in fact, linear.

The fallacy of interpreting a high value of *r* (that is, a value close to $+1$ or -1), as an indication of a cause–effect relationship, is best explained with a few examples. Frequently used as an illustration, is the high positive correlation between the annual sales of chewing gum and the incidence of crime in the United States. Obviously, one cannot conclude that crime might be reduced by prohibiting the sale of chewing gum; both variables depend upon the size of the population, and it is this mutual relationship with a third variable (population size) that produces the positive correlation. Another example is the strong positive correlation that was observed between the number of storks seen nesting in English villages and the number of children born in the same villages. We leave it to the reader's ingenuity to explain why there might be a strong correlation in this case in the absence of any cause–effect relationship.

■ Exercises

17.1   In Example 16.7 we gave the following data on the average number of hours that six students spent on homework per week and their grade-point indexes for the courses they took in that semester:

Hours spent on homework *x*	Grade-point index *y*
15	2.0
28	2.7
13	1.3
20	1.9
4	0.9
10	1.7

Calculate *r* and compare the result with the square root of the value of r^2 given in the printout of Figure 16.9.

17.2   In Exercise 16.3 we gave data on the amount of food containing a certain preservative that $n = 10$ four-year-olds had consumed and their hyperactivity rating 45 minutes later. Given that $\Sigma x = 525$, $\Sigma x^2 = 32{,}085$, $\Sigma y = 100$, $\Sigma y^2 = 1{,}192$, and $\Sigma xy = 5{,}980$, calculate *r*.

17.3   With reference to Exercise 17.2, what percentage of the total variation of the *y*'s (hyperactivity ratings) is accounted for by the relationship with *x* (the amount of food eaten containing the preservative)?

17.4 Following are the numbers of minutes it took $n = 12$ mechanics to assemble a piece of machinery in the morning, x, and in the late afternoon, y:

x	y
12	14
11	11
9	14
13	11
10	12
11	15
12	12
14	13
10	16
9	10
11	10
12	14

Given that $S_{xx} = 25.67$, $S_{xy} = -0.33$, and $S_{yy} = 42.67$, calculate r.

17.5 Use a computer or a graphing calculator to calculate r from the original data given in Exercise 17.4.

17.6 The following data were obtained in a study of the relationship between the resistance (ohms) and the failure time (minutes) of certain overloaded resistors:

Resistance	Failure time
48	45
28	25
33	39
40	45
36	36
39	35
46	36
40	45
30	34
42	39
44	51
48	41
39	38
34	32
47	45

Use a computer or a graphing calculator to find r. Also determine what percentage of the variation in failure times is due to differences in resistance.

17.7 Various doses of a medication were given to groups of 25 persons and they were carefully watched for showing signs of drowsiness:

Dose in mg x	Number drowsy y
5.0	3
7.5	5
10.0	5
12.5	9
15.0	13

Calculate r.

 17.8   Use a computer or a graphing calculator to rework Exercise 17.7.

17.9   Following are the elevations (feet) and average high temperatures (degrees Fahren-heit) on Labor Day of eight cities in Arizona:

Elevation	High temperature
1,418	92
6,905	70
735	98
1,092	94
5,280	79
2,372	88
2,093	90
196	96

Use a computer or a graphing calculator to find r. Also determine what percentage of the variation in the high temperatures is due to differences in elevation.

17.10 After a student calculated r for a large set of paired data, she discovered to her dismay that the variable that should have been labeled x was labeled y and the variable that should have been labeled y was labeled x. Is there any reason for being dismayed?

17.11 After a student had computed r for the heights and weights of a large number of persons, he realized that the weights were given in kilograms and the heights were given in centimeters. He had wanted to obtain r for the weights in pounds and the heights in inches. Given that there are 0.393 inch per centimeter and 2.2 pounds per kilogram, how should he correct his calculations?

17.12 If we calculate r for each of the following sets of data, should we be surprised if we get $r = 1$ for (a) and $r = -1$ for (b)? Explain your answers.

(a) x	y
6	9
14	11

(b) x	y
12	5
8	15

17.13 State in each case whether you would expect a positive correlation, a negative correlation, or no correlation:
(a) the ages of husbands and wives;
(b) the amount of rubber on tires and the number of miles they have been driven;
(c) the number of hours that golfers practice and their scores;
(d) shoe size and IQ;
(e) the weight of the load of trucks and their gasoline consumption.

17.14 State in each case whether you would expect a positive correlation, a negative correlation, or no correlation:
(a) pollen count and the sale of anti-allergy drugs;
(b) income and education;
(c) the number of sunny days in August in Detroit and the attendance at the Detroit Zoo;
(d) shirt size and sense of humor;
(e) number of persons getting flu shots and number of persons catching the flu.

17.15 If $r = 0.41$ for one set of paired data and $r = 0.29$ for another set of paired data, compare the strengths of the two relationships.

17.16 In a medical study it was found that $r = 0.70$ for the weight of babies six months old and their weight at birth, and that $r = 0.60$ for the weight of babies six months old and their average daily intake of food. Give a counterexample to show that it is not valid to conclude that weight at birth and food intake together account for

$$(0.70)^2 100\% + (0.60)^2 100\% = 85\%$$

of the variation of babies' weight when they are six months old. Can you explain why the conclusion is not valid?

17.17 Working with various socioeconomic data for recent years, a research worker got $r = 0.9225$ for the number of foreign language degrees offered by U.S. colleges and universities and the mileage of railroad track owned by U.S. railroads. Can we conclude that

$$(0.9225)^2 100\% \approx 85.1\%$$

of the variation in the foreign language degrees is accounted for by differences in the ownership of railroad track?

17.18 A student computed the correlation between height and weight for a large group of third-grade school children and obtained a value of $r = 0.32$. She was unable to decide whether she should conclude that being tall causes a child to put on more weight or that having excess weight enables a child to grow taller. Help her solve this dilemma.

17.19 In Example 17.2 we illustrated the use of the correlation coefficient in the analysis of a contingency table with ordered categories. Use the same procedure to analyze the following table reproduced from Exercise 14.38, where the relationship between the fidelity and the selectivity of radios was analyzed by means of the chi-square criterion:

	Fidelity		
	Low	*Average*	*High*
Low	7	12	31
Selectivity *Average*	35	59	18
High	15	13	0

17.20 Use the same procedure as in Example 17.2 to analyze the following table reproduced from Exercise 14.36, where the relationship between bank employees'
standard of dress and their professional advancement was analyzed by means of
the chi-square criterion:

	Speed of advancement		
	Slow	*Average*	*Fast*
Very well dressed	38	135	129
Well dressed	32	68	43
Poorly dressed	13	25	17

Note that the speed of advancement goes from low to high, whereas the standard
of dress goes from high to low.

17.3 CORRELATION ANALYSIS

When r is calculated on the basis of sample data, we may get a strong positive
or negative correlation purely by chance, even though there is actually no relationship whatever between the two variables under consideration.

Suppose, for instance, that we take a pair of dice, one red and one green,
roll them five times, and get the following results:

Red die x	Green die y
4	5
2	2
4	6
2	1
6	4

Presumably, there is no relationship between x and y, the numbers of points show-ing on the two dice. It is hard to see why large values of x should go with large values of y and small values of x with small values of y, but calculating r, we get the surprisingly high value $r = 0.66$. This raises the question of whether some-thing is wrong with the assumption that there is no relationship between x and y, and to answer it we shall have to see whether the high value of r may be attributed to chance.

When a correlation coefficient is calculated from sample data, as in the pre-ceding example, the value we get for r is only an estimate of a corresponding pa-rameter, the **population correlation coefficient,** which we denote by ρ (the Greek letter *rho*). What r measures for a sample, ρ measures for a population.

To make inferences about ρ on the basis of r, we shall have to make sev-eral assumptions about the distributions of the random variables whose values we observe. In **normal correlation analysis** we make the same assumptions as in normal regression analysis (see page 474), except that the x's are not constants, but values of a random variable having a normal distribution.

Since the sampling distribution of r is rather complicated under these assumptions, it is common practice to base inferences about ρ on the **Fisher Z transformation,** a change of scale from r to Z, which is given by

$$Z = \frac{1}{2} \cdot \ln \frac{1 + r}{1 - r}$$

Here ln denotes "natural logarithm," that is, logarithm to the base e, where $e = 2.71828. \ldots$ This transformation is named after R. A. Fisher, a prominent sta-tistician who showed that under the assumptions of normal correlation analysis and for any value of ρ, the distribution of Z is approximately normal with

$$\mu_Z = \frac{1}{2} \cdot \ln \frac{1 + \rho}{1 - \rho} \quad \text{and} \quad \sigma_Z = \frac{1}{\sqrt{n - 3}}$$

Converting Z to standard units (that is, subtracting μ_Z and then dividing by σ_Z), we arrive at the result that

Statistic for inferences about ρ

$$z = (Z - \mu_Z)\sqrt{n - 3}$$

has approximately the standard normal distribution. The application of this the-ory is greatly facilitated by the use of Table X at the end of the book, which gives the values of Z corresponding to $r = 0.00, 0.01, 0.02, 0.03, \ldots$, and 0.99. Observe that only positive values are given in this table; if r is negative, we simply look up $-r$ and take the negative of the corresponding Z. Note also that the formula for μ_Z is like that for Z with r replaced by ρ; therefore, Table X can be used to look up values of μ_Z corresponding to given values of ρ.

EXAMPLE 17.4   Using the 0.05 level of significance, test the null hypothesis of no correlation (that is, the null hypothesis $\rho = 0$) for the illustration on page 511, where we rolled a pair of dice five times and got $r = 0.66$.

Solution

1. H_0: $\rho = 0$
   H_A: $\rho \neq 0$
2. $\alpha = 0.05$
3. Since $\mu_Z = 0$ for $\rho = 0$, reject the null hypothesis if $z \leq -1.96$ or $z \geq 1.96$, where

$$z = Z \cdot \sqrt{n-3}$$

Otherwise, state that the value of r is not significant.
4. Substituting $n = 5$ and $Z = 0.793$, the value of Z corresponding to $r = 0.66$ according to Table X, we get

$$z = 0.793\sqrt{5-3}$$
$$= 1.12$$

5. Since $z = 1.12$ falls between -1.96 and 1.96, the null hypothesis cannot be rejected. In other words, the value obtained for r is not significant, as we should, of course, have expected.   ∎

An alternative way of handling this kind of problem (namely, testing the null hypothesis $\rho = 0$) is given in Exercise 17.26.

EXAMPLE 17.5   With reference to the hearing-range and airport-noise example, where we showed that $r = -0.95$ for $n = 12$, test the null hypothesis $\rho = -0.80$ against the alternative hypothesis $\rho < -0.80$ at the 0.01 level of significance.

Solution

1. H_0: $\rho = -0.80$
   H_A: $\rho < -0.80$
2. $\alpha = 0.01$
3. Reject the null hypothesis if $z \leq -2.33$, where

$$z = (Z - \mu_Z)\sqrt{n-3}$$

4. Substituting $n = 12$, $Z = -1.832$ corresponding to $r = -0.95$, and $\mu_Z = -1.099$ corresponding to $\rho = -0.80$, we get

$$z = [-1.832 - (-1.099)]\sqrt{12-3}$$
$$\approx -2.20$$

5. Since -2.20 is greater than -2.33, the null hypothesis cannot be rejected.   ∎

To construct confidence intervals for ρ, we first construct confidence intervals for μ_Z, and then convert to r and ρ by means of Table X. A confidence interval formula for μ_Z may be obtained by substituting

$$z = (Z - \mu_Z)\sqrt{n - 3}$$

for the middle term of the double inequality $-z_{\alpha/2} < z < z_{\alpha/2}$, and then manipulating the terms algebraically so that the middle term is μ_Z. This leads to the following $(1 - \alpha)100\%$ confidence interval for μ_Z:

Confidence interval for μ_Z

$$Z - \frac{z_{\alpha/2}}{\sqrt{n - 3}} < \mu_Z < Z + \frac{z_{\alpha/2}}{\sqrt{n - 3}}$$

EXAMPLE 17.6   If $r = 0.62$ for the cost estimates of two mechanics for a random sample of $n = 30$ repair jobs, construct a 95% confidence interval for the population correlation coefficient ρ.

Solution   Getting $Z = 0.725$, corresponding to $r = 0.62$, from Table X, and substituting it together with $n = 30$ and $z_{0.025} = 1.96$ into the preceding confidence interval formula for μ_Z, we find that

$$0.725 - \frac{1.96}{\sqrt{27}} < \mu_Z < 0.725 + \frac{1.96}{\sqrt{27}}$$

or

$$0.348 < \mu_Z < 1.102$$

Finally, looking up the values of r that come closest to $Z = 0.348$ and $Z = 1.102$ in Table X, we get the 95% confidence interval

$$0.33 < \rho < 0.80$$

for the true strength of the linear relationship between cost estimates made by the two mechanics.   ■

■ **E x e r c i s e s**

17.21 Assuming that the assumptions for a normal correlation analysis are met, test the null hypothesis $\rho = 0$ against the alternative hypothesis $\rho \neq 0$ at the 0.05 level of significance, given that
(a) $n = 15$ and $r = 0.59$;
(b) $n = 20$ and $r = 0.41$;
(c) $n = 40$ and $r = 0.36$.

17.22 Assuming that the assumptions for a normal correlation analysis are met, test the null hypothesis $\rho = 0$ against the alternative hypothesis $\rho \neq 0$ at the 0.01 level of significance, given that
(a) $n = 14$ and $r = 0.54$;
(b) $n = 22$ and $r = -0.61$;
(c) $n = 44$ and $r = 0.42$.

17.23 Assuming that the assumptions for a normal correlation analysis are met, test the null hypothesis $\rho = 0.40$ against the alternative hypothesis $\rho > 0.40$ at the 0.05 level of significance, given that
(a) $n = 25$ and $r = 0.57$;
(b) $n = 30$ and $r = 0.66$.

17.24 Rework Exercise 17.23, with the level of significance changed to 0.01.

17.25 Assuming that the assumptions for a normal correlation analysis are met, test the null hypothesis $\rho = -0.50$ against the alternative hypothesis $\rho > -0.50$ at the 0.01 level of significance, given that
(a) $n = 17$ and $r = -0.22$;
(b) $n = 34$ and $r = -0.43$.

17.26 Under the assumptions of normal correlation analysis, the test of the null hypothesis $\rho = 0$ may also be based on the statistic

$$t = \frac{r\sqrt{n-2}}{\sqrt{1-r^2}}$$

which has the t distribution with $n - 2$ degrees of freedom. Use this statistic to test in each case whether the value of r is significant at the 0.05 level of significance:
(a) $n = 12$ and $r = 0.77$;
(b) $n = 16$ and $r = 0.49$.

17.27 Rework Exercise 17.26 with the level of significance changed to 0.01.

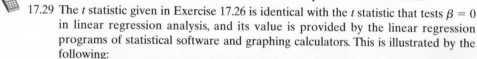 17.28 Use the t statistic of Exercise 17.26 to rework all parts of Exercise 17.21.

17.29 The t statistic given in Exercise 17.26 is identical with the t statistic that tests $\beta = 0$ in linear regression analysis, and its value is provided by the linear regression programs of statistical software and graphing calculators. This is illustrated by the following:
(a) Use the linear regression program of a statistical software package or a graphing calculator to obtain the values of t and r for the data of Exercise 16.8.
(b) Substitute the value obtained for r in part (a) and $n = 8$ into the formula for t given in Exercise 17.26.
(c) Compare the two values of t.

17.30 In a study of the relationship between the death rate from lung cancer and the per capita consumption of cigarettes 20 years earlier, data for $n = 9$ countries yielded $r = 0.73$. At the 0.05 level of significance, test the null hypothesis $\rho = 0.50$ against the alternative hypothesis $\rho > 0.50$.

17.31 In a study of the relationship between the available heat (per cord) of green wood and air-dried wood, data for $n = 13$ kinds of wood yielded $r = 0.94$. Use the 0.01 level of significance to test the null hypothesis $\rho = 0.75$ against the alternative hypothesis $\rho \neq 0.75$.

17.32 Assuming that the assumptions for a normal correlation analysis are met, use the Fisher Z transformation to construct 95% confidence intervals for ρ, given that
(a) $n = 15$ and $r = 0.80$;
(b) $n = 28$ and $r = -0.24$;
(c) $n = 63$ and $r = 0.55$.

17.33 Assuming that the assumptions for a normal correlation analysis are met, use the Fisher Z transformation to construct 99% confidence intervals for ρ, given that
(a) $n = 20$ and $r = -0.82$;
(b) $n = 25$ and $r = 0.34$;
(c) $n = 75$ and $r = 0.18$.

*17.4 MULTIPLE AND PARTIAL CORRELATION

In Section 17.1 we introduced the coefficient of correlation as a measure of the goodness of the fit of a least-squares line to a set of paired data. If predictions are to be made with an equation of the form

$$\hat{y} = b_0 + b_1 x_1 + b_2 x_2 + \cdots + b_k x_k$$

obtained by the method of least squares as in Section 16.4, we define the **multiple correlation coefficient** in the same way in which we originally defined r. We take the square root of the quantity

$$\frac{\sum (\hat{y} - \bar{y})^2}{\sum (y - \bar{y})^2}$$

which is the proportion of the total variation of the y's that can be attributed to the relationship with the x's. The only difference is that we now calculate $\hat{y}$ by means of the multiple regression equation instead of the equation $\hat{y} = a + bx$.

To give an example, let us refer to Example 16.9, where we based a multiple linear regression equation on the computer printout shown in Figure 16.10. As we pointed out at the time, part of the printout had been deleted, but since that part is needed now, the complete printout is reproduced in Figure 17.5.

```
MTB > Regress c3 2 c1 c2 ;
SUBC>    Constant.



The regression equation is
C3 = 65430 + 16752 C1 + 11235 C2

Predictor          Coef          Stdev       t-ratio          p
Constant          65430          12134          5.39      0.003
C1                16752           6636          2.52      0.053
C2                11235           9885          1.14      0.307

s = 10751          R-sq = 89.4%       R-sq(adj) = 85.2%

Analysis of Variance

SOURCE            DF            SS             MS           F          p
Regression         2      4877608448     2438804224       21.10      0.004
Error              5       577971520      115594304
Total              7      5455580160

SOURCE            DF         SEQ SS
C1                 1      4728284160
C2                 1       149324224
```

FIGURE 17.5
Complete printout for the multiple regression example.

EXAMPLE 17.7   Use the definition of the multiple correlation coefficient given above to determine its value for the data of Example 16.9, which dealt with the number of bedrooms, the number of baths, and the prices at which eight one-family houses had recently been sold.

Solution   In the analysis of variance of Figure 17.5 (in the column headed by SS), we find that the regression sum of squares is 4,877,608,448 and that the total sum of squares is 5,455,580,160. Thus, the multiple correlation coefficient equals the square root of

$$\frac{4,877,608,448}{5,455,580,160} \approx 0.894$$

Denoting it by R, we write $R \approx \sqrt{0.894} = 0.95$ rounded to two decimals, which is considered as "essentially non-negative" according to the Kendall and Buckland dictionary of statistical terms. In actual practice, R^2 is used more often than R, and it should be observed that its value, denoted by R−sq, is actually given as 89.4% in the printout of Figure 17.5.   ∎

This example also serves to illustrate that adding more independent variables in a correlation study may not be sufficiently productive to justify the extra work. As it can be shown that $r = 0.93$ for y and x_1 (number of bedrooms) alone, it is apparent that very little is gained by also considering x_2 (number of baths). The situation is quite different, though, in Exercise 17.35, where the two independent variables x_1 and x_2 together account for a much higher proportion of the total variation in y than does either x_1 or x_2 alone.

When we discussed the problem of correlation and causation, we showed that a strong correlation between two variables may be due entirely to their dependence on a third variable. We illustrated this with the examples of chewing gum sales and the crime rate, and child births and the number of storks. To give another example, let us consider the two variables x_1, the weekly amount of hot chocolate sold by a refreshment stand at a summer resort, and x_2, the weekly number of visitors to the resort. If, on the basis of suitable data, we get $r = -0.30$ for these variables, this should come as a surprise—after all, we would expect more sales of hot chocolate when there are more visitors, and vice versa, and hence a positive correlation.

However, if we think for a moment, we may surmise that the negative correlation of -0.30 may well be due to the fact that the variables x_1 and x_2 are both related to a third variable x_3, the average weekly temperature at the resort. If the temperature is high, there will be more visitors, but they will prefer cold drinks to hot chocolate; if the temperature is low, there will be fewer visitors, but they will prefer hot chocolate to cold drinks. So, let us suppose that further data yield $r = -0.70$ for x_1 and x_3, and $r = 0.80$ for x_2 and x_3. These values seem reasonable since low sales of hot chocolate should go with high temperatures and vice versa, while the number of visitors should be high when the temperature is high, and low when the temperature is low.

In the preceding example, we should really have investigated the relationship between x_1 and x_2 (hot chocolate sales and the number of visitors to the resort) when all other factors, primarily temperature, are held fixed. As it is seldom possible to control matters to such an extent, it has been found that a statistic called the **partial correlation coefficient** does a fair job of eliminating the effects of other variables. If we write the ordinary correlation coefficients for x_1 and x_2, x_1 and x_3, and x_2 and x_3, as r_{12}, r_{13}, and r_{23}, the partial correlation coefficient for x_1 and x_2 with x_3 fixed is given by

Partial correlation coefficient

$$r_{12.3} = \frac{r_{12} - r_{13} \cdot r_{23}}{\sqrt{1 - r_{13}^2}\sqrt{1 - r_{23}^2}}$$

EXAMPLE 17.8   Calculate $r_{12.3}$ for the preceding example, which dealt with the sales of hot chocolate and the number of persons who visit the resort.

Solution   Substituting $r_{12} = -0.30, r_{13} = -0.70$, and $r_{23} = 0.80$ into the formula for $r_{12.3}$, we get

$$r_{12.3} = \frac{(-0.30) - (-0.70)(0.80)}{\sqrt{1 - (-0.70)^2}\sqrt{1 - (0.80)^2}} \approx 0.607$$

This result shows that, as we expected, there is a positive relationship between the sales of hot chocolate and the number of visitors to the resort when the effect of differences in temperature is eliminated.   ∎

We have given this example primarily to illustrate what we mean by a partial correlation coefficient, but it also serves as a reminder that correlation coefficients can be very misleading unless they are interpreted with care.

■ Exercises

*17.34 In a multiple regression problem, the regression sum of squares is

$$\sum (\hat{y} - \bar{y})^2 = 45{,}225$$

and the total sum of squares is

$$\sum (y - \bar{y})^2 = 136{,}210$$

Find the value of the multiple correlation coefficient.

 *17.35 With reference to Exercise 16.34 on page 484, use the same software as before to obtain the multiple correlation coefficient. Also determine the correlation coefficients for y and x_1 (age) alone and for y and x_2 (years postgraduate work) alone, and compare them with the multiple correlation coefficient.

 *17.36 With reference to Exercise 16.35 on page 485, use the same software as before to obtain the multiple correlation coefficient.

*17.37 A team of research workers conducted an experiment to see if the height of certain rose bushes can be predicted on the basis of the amount of fertilizer and the amount of irrigation that is applied to the soil. For predicting the height on the basis of both variables, they obtained a multiple correlation coefficient of 0.58; for predicting the height on the basis of the fertilizer alone, they obtained $r = 0.66$. Comment on these results.

*17.38 With reference to Example 17.8, find
(a) the partial correlation coefficient for x_1 (sales of hot chocolate) and x_3 (temperature) when x_2 (number of visitors) remains fixed;
(b) the partial correlation coefficient for x_2 (number of visitors) and x_3 (temperature) when x_1 (sales of hot chocolate) remains fixed.

 *17.39 With reference to Exercise 16.35 on page 485, use a computer or a graphing calculator to determine the necessary correlation coefficients in order to calculate the partial correlation coefficient for hardness and annealing temperature when the copper content is held fixed.

*17.40 An experiment yielded the following results: $r_{12} = 0.80$, $r_{13} = -0.70$, and $r_{23} = 0.90$. Explain why these figures cannot all be correct.

17.5 Checklist of Key Terms *(with page references to their definitions)*

Coefficient of correlation, 501, 502
Coefficient of determination, 501
Fisher Z transformation, 512
*Multiple correlation coefficient, 516
Negative correlation, 501
Normal correlation analysis, 512
*Partial correlation coefficient, 517, 518
Population correlation coefficient, 512

Positive correlation, 501
Product moment, 503
Product-moment coefficient of correlation, 503
Regression sum of squares, 500
Residual sum of squares, 500
Sample covariance, 504
Total sum of squares 500

17.6 References

More detailed information about multiple and partial correlation may be found in

EZEKIEL, M., and FOX, K. A., *Methods of Correlation and Regression Analysis,* 3rd ed. New York: John Wiley & Sons, Inc., 1959.

HARRIS, R. J., *A Primer of Multivariate Statistics.* New York: Academic Press, Inc., 1975.

and an advanced theoretical treatment is given in Volume 2 of

KENDALL, M. G., and STUART, A., *The Advanced Theory of Statistics,* 3rd ed. New York: Hafner Press, 1973.

Volume 1 of this book provides the theoretical foundation of significance tests for r.

18

NONPARAMETRIC TESTS

18.1	The Sign Test	521
18.2	The Sign Test (Large Samples)	523
*18.3	The Signed-Rank Test	527
*18.4	The Signed-Rank Test (Large Samples)	531
18.5	The U Test	535
18.6	The U Test (Large Samples)	540
18.7	The H Test	541
18.8	Tests of Randomness: Runs	545
18.9	Tests of Randomness: Runs (Large Samples)	547
18.10	Tests of Randomness: Runs Above and Below the Median	548
18.11	Rank Correlation	551
18.12	Some Further Considerations	555
18.13	Summary	555
18.14	Checklist of Key Terms	556
18.15	References	556

Most of the tests cited in Chapters 12 through 17 require specific assumptions about the population, or populations, sampled. In many cases we must assume that the populations have roughly the shape of normal distributions, that their variances are known or are known to be equal, or that the samples are independent. Since there are many situations where it is doubtful whether all the necessary assumptions can be met, statisticians have developed alternative procedures based on less stringent assumptions, which have become known as **nonparametric tests.**

Aside from the fact that nonparametric tests can be used under more general conditions than the standard tests that they replace, they are often easier to explain and understand; moreover, in many nonparametric tests the computational burden is so light that they come under the heading of "quick and easy" or "shortcut" techniques. For all these reasons, nonparametric tests have become quite popular, and extensive literature is devoted to their theory and application. Also for these reasons, there is seldom any need to use a computer or some other specialized technology. About the only occasion when a computer or a graphing calculator comes in handy is when we have to sort samples or combinations of samples in an ascending or a descending order.

In Sections 18.1 and 18.2 we present the sign test as a nonparametric alternative to tests concerning means and to tests concerning differences between means based on paired data. Another nonparametric test that serves the same purpose but is less wasteful of information is given in Sections 18.3 and 18.4. In Sections 18.5 through 18.7 we present a nonparametric alternative to tests concerning the difference between the means of independent samples and a somewhat similar alternative to the one-way analysis of variance. In Sections 18.8 through 18.10 we shall learn how to test the randomness of a sample after the data have actually been collected; and in Section 18.11 we present a nonparametric test of the significance of a relationship between paired data. Finally, in Section 18.12 we mention some of the weaknesses of nonparametric methods, and in Section 18.13 we give a table that lists the various nonparametric tests and the "standard" test that they replace.

18.1   THE SIGN TEST

Except for the large-sample tests, all the tests concerning means that we studied in Chapter 12 were based on the assumption that the populations sampled have roughly the shape of normal distributions. When this assumption is untenable in practice, these standard tests can be replaced by any one of several nonparametric alternatives, and these are the subject matter of Sections 18.1 through 18.7. Simplest among these is the **sign test,** which we shall study in this section and in Section 18.2.

The **one-sample sign test** applies when we sample a continuous population, so that the probability of getting a sample value less than the median and the probability of getting a sample value greater than the median are both $\frac{1}{2}$. Of course, when the population is symmetrical, the mean μ and the median $\tilde{\mu}$ will coincide, and we can phrase the hypotheses in terms of either of these parameters.

To test the null hypothesis $\tilde{\mu} = \tilde{\mu}_0$ against an appropriate alternative on the basis of a random sample of size n, we replace each sample value greater than $\tilde{\mu}_0$ with a plus sign and each sample value less than $\tilde{\mu}_0$ with a minus sign. Then we test the null hypothesis that the number of plus signs are the values of a random variable having the binomial distribution with $p = \frac{1}{2}$. If any sample value actually equals $\tilde{\mu}_0$, which can easily happen when we deal with rounded data, we simply discard it.

To perform a one-sample sign test when the sample is fairly small, we refer directly to a table of binomial probabilities, such as Table V at the end of the book or the National Bureau of Standards table referred to among the references on page 232. Alternatively, we may use a computer or a calculator to obtain the required binomial probabilities. When the sample is large, however, we use the normal approximation to the binomial distribution, as is illustrated in Section 18.2.

EXAMPLE 18.1   To check a teacher's claim that the published value for the coefficient of friction for well-oiled metals, 0.050, might be too low, a science class made 18 determinations getting 0.054, 0.052, 0.044, 0.056, 0.050, 0.051, 0.055, 0.053, 0.047, 0.053,

0.052, 0.050, 0.051, 0.051, 0.054, 0.046, 0.053, and 0.043. Ordinarily, the one-sample t test would be a logical choice, but the skewness of the data suggests the use of a nonparametric alternative. Use the one-sample sign test to test the null hypothesis $\tilde{\mu} = 0.050$ against the alternative $\tilde{\mu} > 0.050$ at the 0.05 level of significance.

Solution

1. H_0: $\tilde{\mu} = 0.050$
   H_A: $\tilde{\mu} > 0.050$
2. $\alpha = 0.05$
3′. The test statistic is the number of plus signs, namely, the number of values exceeding 0.050.
4′. Replacing each value greater than 0.050 with a plus sign, each value less than 0.050 with a minus sign, and discarding the two values that equal 0.050, we get

$$+ \; + \; - \; + \; + \; + \; + \; - \; + \; + \; + \; + \; + \; - \; + \; -$$

Thus, $x = 12$, and Table V shows that for $n = 16$ and $p = 0.50$ the probability of $x \geq 12$, the p-value, is $0.028 + 0.009 + 0.002 = 0.039$.
5′. Since 0.039 is less than 0.05, the null hypothesis must be rejected. The data support the claim that the published value of the coefficient of friction is too low.   ■

Note that we used the alternative method for testing hypotheses as described on page 348. As in Section 14.1, the alternative method simplifies matters when tests are based directly on binomial tables. Although it would not seem necessary, we could have used a computer for Example 18.1. If we had done so, we would have obtained a printout like the one shown in Figure 18.1. The difference between the two p-values, 0.039 and 0.0384, is, of course, due to rounding.

FIGURE 18.1
MINITAB printout
for Example 18.1.

```
MTB > STest .050 c1;
SUBC>    Alternative 1.


Sign test of median = 0.05000 versus   G.T.   0.05000

             N   BELOW   EQUAL   ABOVE   P-VALUE    MEDIAN
C1          18     4       2      12     0.0384    0.05150
```

The sign test can also be used when we deal with paired data as in Section 12.7. In such problems, each pair of sample values is replaced with a plus sign if the first value is greater than the second, with a minus sign if the first value is smaller than the second, and pairs of equal values are discarded. For paired data, the sign test is used to test the null hypothesis that the median of the population of differences is zero. When it is used in this way, it is referred to as the **paired-sample sign test.**

EXAMPLE 18.2   In Example 12.9 we gave the following data on the average weekly losses in hours of labor due to accidents in ten industrial plants before and after the installation of an elaborate safety program:

45 and 36	73 and 60	46 and 44	124 and 119	33 and 35
57 and 51	83 and 77	34 and 29	26 and 24	17 and 11

Based on the paired-sample t test, we showed at the 0.05 level of significance that the safety program is effective. Use the paired-sample sign test to rework this example.

Solution

1. H_0: $\tilde{\mu}_D = 0$, where $\tilde{\mu}_D$ is the median of the population of differences sampled.
   H_A: $\tilde{\mu}_D > 0$
2. $\alpha = 0.05$
3'. The test statistic is the number of plus signs, namely the number of industrial plants in which the average weekly losses in hours of labor has decreased.
4'. Replacing each pair of values with a plus sign if the first value is greater than the second, and with a minus sign if the first value is smaller than the second, we get

$$+ \quad + \quad + \quad + \quad - \quad + \quad + \quad + \quad + \quad +$$

Thus, $x = 9$, and Table V shows that for $n = 10$ and $p = 0.50$ the probability of $x \geq 9$, the p-value, is $0.010 + 0.001 = 0.011$.
5'. Since 0.011 is less than 0.05, the null hypothesis must be rejected. As in Example 12.9, we conclude that the safety program is effective. ■

18.2 THE SIGN TEST (Large Samples)

When np and $n(1 - p)$ are both greater than 5, so that we can use the normal approximation to the binomial distribution, we can base the sign test on the large-sample test of Section 14.2, namely, on the statistic

$$z = \frac{x - np_0}{\sqrt{np_0(1 - p_0)}}$$

with $p_0 = 0.50$, which has approximately the standard normal distribution. When n is small, it may be wise to use the continuity correction suggested on page 379. This is true, especially if, without the continuity correction, we can *barely* reject the null hypothesis. As we have pointed out before, the continuity correction does not have to be considered when *without it* the null hypothesis cannot be rejected.

EXAMPLE 18.3   In Exercise 2.15 we gave the following scores that 50 college students obtained in a religious literacy test, in which the maximum score was 60:

35	31	54	34	41	30	38	36	43	40
50	31	36	34	44	35	49	43	39	56
26	36	30	43	36	25	40	41	39	51
25	39	48	37	29	31	33	30	43	45
46	44	38	38	53	34	51	41	36	42

Given that these students constitute a random sample of all the students attending a community college in a certain area, use the one-sample sign test to test the null hypothesis $\tilde{\mu} = 35$ (for all the students attending a community college in that area) against the alternative hypothesis $\tilde{\mu} \neq 35$ at the 0.01 level of significance. As we shall see, $n = 48$ for this example, so that $np = 48(0.50) = 24$ and $n(1 - p) = 48(0.50) = 24$ are both greater than 5 and we can use the normal approximation to the binomial distribution.

Solution

1. H_0: $\tilde{\mu} = 35$
   H_A: $\tilde{\mu} \neq 35$
2. $\alpha = 0.01$
3. Reject the null hypothesis if $z \leq -2.575$ or $z \geq 2.575$, where

$$z = \frac{x - np_0}{\sqrt{np_0(1 - p_0)}}$$

with $p_0 = 0.50$; otherwise, accept the null hypothesis or reserve judgment.

4. Counting the number of values exceeding 35 (plus signs), the number of values less than 35 (minus signs), and the number of values that are equal to 35 (and, hence, have to be discarded), we find that there are, respectively, 34, 14, and 2. Thus, $x = 34$, $n = 48$, and

$$z = \frac{34 - 48(0.50)}{\sqrt{48(0.50)(0.50)}} \approx 2.89$$

5. Since 2.89 exceeds 2.575, the null hypothesis must be rejected; in other words, the median score for all the students attending a community college in that area is not 35. (Had we used the continuity correction, we would have obtained $z = 2.74$, and the decision would have been the same.)   ∎

The next example illustrates the importance of using the continuity correction when, without it, we would barely reject the null hypothesis.

EXAMPLE 18.4   Following are two supervisors' ratings of the performance of a random sample of a large company's employees, on a scale from 0 to 100:

Supervisor 1	Supervisor 2
88	73
69	67
97	81
60	73
82	78
90	82
65	62
77	80
86	81
79	79
65	77
95	82
88	84
91	93
68	66
77	76
74	74
85	78

Use the paired-sample sign test (based on the normal approximation to the binomial distribution) to test at the 0.05 level of significance whether the differences between the two sets of ratings can be attributed to chance,

(a) without using the continuity correction;
(b) using the continuity correction.

Solution   (a) 1. H_0: $\tilde{\mu}_D = 0$, where $\tilde{\mu}_D$ is the median of the population of differences (between the supervisors' ratings) sampled.
   H_A: $\tilde{\mu}_D \neq 0$
2. $\alpha = 0.05$
3. Reject the null hypothesis if $z \leq -1.96$ or $z \geq 1.96$, where

$$z = \frac{x - np_0}{\sqrt{np_0(1 - p_0)}}$$

with $p_0 = 0.50$; otherwise, accept the null hypothesis or reserve judgment.
4. Counting the number of positive differences (plus signs), the number of negative differences (minus signs), and the number of pairs that are equal (and, hence, have to be discarded), we find that there are, respectively, 12, 4, and 2. Thus, $x = 12$ and $n = 16$, and since

$np = 16(0.50) = 8$ and $n(1 - p) = 16(0.50) = 8$ are both greater than 5, we are justified in using the normal approximation to the binomial distribution. Substituting into the formula for z, we get

$$z = \frac{12 - 16(0.50)}{\sqrt{16(0.50)(0.50)}} = 2.00$$

5. Since $z = 2.00$ *barely* exceeds 1.96, we defer making a decision until we recalculate z with the continuity correction.

(b) 4. With the continuity correction we get

$$z = \frac{11.5 - 16(0.50)}{\sqrt{16(0.50)(0.50)}} = 1.75$$

5. Since $z = 1.75$ falls between -1.96 and 1.96, we find that the null hypothesis cannot be rejected. We conclude that the differences between the supervisors' ratings can be attributed to chance. (Had we based our decision on Table V, the p-value would have exceeded 0.05, and the final decision would have been the same.)   ■

■ Exercises

18.1   On 14 occasions, a random sample, a city employee had to wait 4.5, 8.6, 7.3, 7.0, 2.5, 6.1, 8.9, 6.5, 6.3, 1.6, 5.8, 6.3, 5.9, and 9.0 minutes for the bus that he takes to work. Use the sign test based on Table V and the 0.05 level of significance to test the null hypothesis $\tilde{\mu} = 6.0$ (that his median wait is 6.0 minutes) against the alternative hypothesis $\tilde{\mu} \neq 6.0$.

18.2   Use a computer to work Exercise 18.1.

18.3   The following data, a random sample, are the weights (in grams) of 20 packages of crystalized ginger: 110.6, 113.5, 111.2, 109.8, 110.5, 111.1, 110.4, 109.7, 112.6, 110.8, 110.5, 110.0, 110.2, 111.4, 110.9, 110.5, 110.0, 109.4, 110.8, and 109.7. Use the sign test based on Table V and the 0.01 level of significance to test the null hypothesis $\tilde{\mu} = 110.0$ (that the median weight of such packages of ginger is 110.0 grams) against the alternative hypothesis $\tilde{\mu} > 110.0$.

18.4   Use a computer to work Exercise 18.3.

18.5   After playing four rounds of golf at the Padre course of the Camelback Country Club, a random sample of 15 golf professionals had total scores of 279, 281, 278, 279, 276, 280, 280, 277, 282, 278, 281, 288 (ouch), 276, 279, and 280. Use the sign test at the 0.05 level of significance to test the null hypothesis $\tilde{\mu} = 278$ (that the median score of professional golfers at that course is a two-under-par 278) against the alternative hypothesis $\tilde{\mu} > 278$. Base the test on
(a) Table V;
(b) the normal approximation to the binomial distribution.

18.6   Use a computer to rework part (a) of the preceding exercise.

18.7   Following are the miles per gallon obtained with 40 tankfuls of a certain kind of gasoline:

24.1	25.0	24.8	24.3	24.2	25.3	24.2	23.6
24.5	24.4	24.5	23.2	24.0	23.8	23.8	25.3
24.5	24.6	24.0	25.2	25.2	24.4	24.7	24.1
24.6	24.9	24.1	25.8	24.2	24.2	24.8	24.1
25.6	24.5	25.1	24.6	24.3	25.2	24.7	23.3

Use the sign test based on the normal approximation to the binomial distribution to test the null hypothesis $\tilde{\mu} = 24.2$ (that the median of the population of mileages sampled is 24.2 miles per gallon) against the alternative hypothesis $\tilde{\mu} > 24.2$. Use the 0.01 level of significance.

18.8 Following are the numbers of passengers carried on Flights 138 and 139 between Phoenix and Chicago on sixteen days: 199 and 232, 231 and 265, 236 and 250, 238 and 251, 218 and 226, 258 and 269, 253 and 247, 248 and 252, 220 and 245, 237 and 245, 239 and 235, 248 and 260, 239 and 245, 240 and 240, 233 and 239, and 247 and 236. Use the sign test based on Table V and the 0.03 level of significance to test the null hypothesis $\tilde{\mu}_D = 0$, where $\tilde{\mu}_D$ is the median of the differences for the population sampled, against the alternative hypothesis $\tilde{\mu}_D < 0$.

 18.9 Use a computer to work Exercise 18.8.

18.10 Following are the numbers of employees absent from two government agencies on 20 days: 29 and 24, 45 and 32, 38 and 38, 39 and 34, 46 and 42, 35 and 41, 42 and 36, 39 and 37, 40 and 45, 38 and 35, 31 and 37, 44 and 35, 42 and 40, 40 and 32, 42 and 45, 51 and 38, 36 and 33, 45 and 39, 33 and 28, and 32 and 38. Use the sign test at the 0.05 level of significance to test the null hypothesis $\tilde{\mu}_D = 0$, where $\tilde{\mu}_D$ is the median of the population of differences between the daily absences in the two government agencies. Use the alternative hypothesis $\tilde{\mu}_D > 0$.

18.11 Use the normal approximation to the binomial distribution to work Exercise 18.10.

18.12 Following are the numbers or artifacts dug up by two archeologists at an ancient Hohokam cliff dwelling on 30 days: 2 and 0, 4 and 1, 2 and 0, 0 and 1, 2 and 0, 3 and 1, 1 and 2, 4 and 0, 2 and 3, 3 and 2, 1 and 0, 2 and 6, 5 and 2, 3 and 2, 1 and 0, 2 and 1, 1 and 1, 4 and 2, 1 and 1, 1 and 0, 0 and 2, 3 and 1, 2 and 1, 2 and 0, 0 and 0, 1 and 3, 4 and 1, 2 and 1, 1 and 1, and 3 and 0. Use the sign test at the 0.05 level of significance to test the null hypothesis that the two archeologists are equally good at finding artifacts against the alternative hypothesis that they are not equally good.

*18.3 THE SIGNED-RANK TEST[†]

The sign test is easy to perform and has intuitive appeal, but it is wasteful of information because it utilizes only the signs of the differences between the observations and $\tilde{\mu}_0$ in the one-sample case, or the signs of the differences between the pairs of observations in the paired-sample case. It is for this reason that an alternative nonparametric test, the **signed-rank test** (also called the **Wilcoxon signed-rank test**), is often preferred.

[†]Since the signed-rank test is an alternative to the sign test, this section and Section 18.4 may be omitted without loss of continuity.

In this test, we rank the differences without regard to their signs, assigning rank 1 to the smallest numerical difference (that is, to the smallest difference in absolute value), rank 2 to the second smallest numerical difference, ..., and rank n to the largest numerical difference. Zero differences are again discarded, and if two or more differences are numerically equal, we assign each one the mean of the ranks which they jointly occupy. Then we base the test on T^+, the sum of the ranks of the positive differences, T^-, the sum of the ranks of the negative differences, or T, the smaller of the two.

The signed-rank test serves as an alternative to the one-sample sign test as well as the paired sample sign test; as such it applies when the probability of getting a value less than the median equals the probability of getting a value greater than the median. We shall illustrate it here with measurements of the octane rating of a certain brand of premium gasoline, on the basis of which we will test the null hypothesis $\tilde{\mu} = 98.5$ against the alternative hypothesis $\tilde{\mu} < 98.5$ at the 0.01 level of significance.

The actual data are shown in the lefthand column of the following table, and the middle column contains the differences obtained by subtracting 98.5 from each measurement:

Measurements	Differences	Ranks
97.5	−1.0	4
95.2	−3.3	12
97.3	−1.2	6
96.0	−2.5	10
96.8	−1.7	7
100.3	1.8	8
97.4	−1.1	5
95.3	−3.2	11
93.2	−5.3	14
99.1	0.6	2
96.1	−2.4	9
97.6	−0.9	3
98.2	−0.3	1
98.5	0.0	
94.9	−3.6	13

After we discard the zero difference, we find that the smallest numerical difference is 0.3, the next smallest numerical difference is 0.6, the next smallest after that is 0.9, ..., and the largest numerical difference is 5.3. These ranks are shown in the third column, and it follows that

$$T^+ = 8 + 2 = 10$$
$$T^- = 4 + 12 + 6 + 10 + 7 + 5 + 11 + 14 + 9 + 3 + 1 + 13$$
$$= 95$$

and, hence, $T = 10$. Since $T^+ + T^-$ equals the sum of the integers from 1 through n, namely, $\dfrac{n(n+1)}{2}$, we could have obtained T^- more easily by subtracting

$T^+ = 10$ from $\frac{14 \cdot 15}{2} = 105$. [There were no ties in rank in this example, but as we pointed out earlier, when there are ties in rank, we assign to each of the tied values (differences) the mean of the ranks that they jointly occupy.]

The close relationship among T^+, T^-, and T is also reflected by their sampling distributions, an example of which, for $n = 5$, is shown in Figure 18.2. Since there is a fifty-fifty chance for each rank to go to one of the positive differences or one of the negative differences, there are altogether 2^n possibilities, each with the probability $(\frac{1}{2})^n$. To get the probabilities associated with the various values of T^+, T^-, and T, we count the number of ways in which these values of T^+, T^-, and T can be obtained and multiply by $(\frac{1}{2})^n$. For instance, for $n = 5$ and $T^+ = 6$ there are the three possibilities 1 and 5, 2 and 4, and 1 and 2 and 3, and the probability is $3 \cdot (\frac{1}{2})^5 = \frac{3}{32}$, as shown in Figure 18.2.

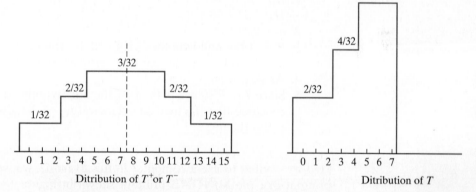

FIGURE 18.2
Distribution of T^+, T^-, and T for $n = 5$.

To simplify the construction of tables of critical values, we shall base all tests of the null hypothesis $\tilde{\mu} = \tilde{\mu}_0$ on the distribution of T and reject it for values falling into the lefthand tail. We have to be careful, though, to use the right statistic and the right critical value. When $\tilde{\mu} < \tilde{\mu}_0$ then T^+ tends to be small, so when the alternative hyothesis is $\tilde{\mu} < \tilde{\mu}_0$ we base the test on T^+; when $\tilde{\mu} > \tilde{\mu}_0$ then T^- tends to be small, so when the alternative hypothesis is $\tilde{\mu} > \tilde{\mu}_0$ we base the test on T^-; and when $\tilde{\mu} \neq \tilde{\mu}_0$ then either T^+ or T^- tends to be small, so when the alternative hypothesis is $\tilde{\mu} \neq \tilde{\mu}_0$ we base the test on T. These relationships are summarized in the following table:

Alternative hypothesis	Reject the null hypothesis if	Accept the null hypothesis or reserve judgment if
$\mu < \mu_0$	$T^+ \leq T_{2\alpha}$	$T^+ > T_{2\alpha}$
$\mu > \mu_0$	$T^- \leq T_{2\alpha}$	$T^- > T_{2\alpha}$
$\mu \neq \mu_0$	$T \leq T_{\alpha}$	$T > T_{\alpha}$

The necessary values of T_α, which are the largest values of T for which the probability of $T \leq T_\alpha$ does not exceed α, may be found in Table VI at the end of the book; the blank spaces in the table indicate that the null hypothesis cannot be rejected regardless of the value we obtain for the test statistic. Note that the same critical values serve for tests at different levels of significance depending on whether the alternative hypothesis is one sided or two sided.

EXAMPLE 18.5   With reference to the octane ratings on page 528, use the signed-rank test at the 0.01 level of significance to test the null hypothesis $\tilde{\mu} = 98.5$ against the alternative hypothesis $\tilde{\mu} < 98.5$.

Solution

1. H_0: $\tilde{\mu} = 98.5$
   H_A: $\tilde{\mu} < 98.5$
2. $\alpha = 0.01$
3. Reject the null hypothesis if $T^+ \leq 16$, where 16 is the value of $T_{0.02}$ for $n = 14$; otherwise, accept it or reserve judgment.
4. As shown on page 528, $T^+ = 10$.
5. Since $T^+ = 10$ is less than 16, the null hypothesis must be rejected. We conclude that the median octane rating of the given premium gasoline is less than 98.5.   ■

Had we wanted to use a computer in this example, we would have obtained a printout like the MINITAB printout shown in Figure 18.3. One advantage of using a computer is that we do not have to refer to a special table; Figure 18.3 gives the p-value as 0.004. This would also have led to the rejection of the null hypothesis.

```
MTB > WTest 98.5 c1;
SUBC>    Alternative -1.


TEST OF MEDIAN = 98.50 VERSUS MEDIAN L.T. 98.50

               N FOR    WILCOXON              ESTIMATED
          N    TEST    STATISTIC  P-VALUE      MEDIAN
C1       15     14       10.0      0.004        96.85
```

FIGURE 18.3
Computer printout for
Example 18.5.

The signed-rank test can also be used as an alternative to the paired-sample sign test. The procedure is exactly the same, but when we write the null hypothesis as $\tilde{\mu}_D = 0$, then $\tilde{\mu}_D$ is the median of the population of differences sampled.

EXAMPLE 18.6 Use the signed-rank test to rework Example 18.2. The actual data, the average weekly losses in hours of labor due to accidents in ten industrial plants before and after the installation of the safety program, are shown in the lefthand column of the following table. The middle column contains their differences, and, discarding signs, the ranks of the numerical differences are shown in the column on the right.

Losses in man-hours before and after	Differences	Ranks
45 and 36	9	9
73 and 60	13	10
46 and 44	2	2
124 and 119	5	4.5
33 and 35	−2	2
57 and 51	6	7
83 and 77	6	7
34 and 29	5	4.5
26 and 24	2	2
17 and 11	6	7

Thus, $T^- = 2$ and $T^+ = 53$.

Solution

1. H_0: $\tilde{\mu}_D = 0$, where $\tilde{\mu}_D$ is the median of the population of differences (between losses in hours of labor before and after the installation of the safety program) sampled.
   H_A: $\tilde{\mu}_D > 0$
2. $\alpha = 0.05$
3. Reject the null hypothesis if $T^- \leq 11$, where 11 is the value of $T_{0.10}$ for $n = 10$; otherwise, accept the null hypothesis or reserve judgment.
4. As shown previously, $T^- = 2$.
5. Since $T^- = 2$ is less than 11, the null hypothesis must be rejected. We conclude that the safety program is effective. (Had we used a computer for this example, we would have obtained a p-value of 0.005, and the conclusion would have been the same.) ∎

*18.4 THE SIGNED-RANK TEST (Large Samples)

When n is 15 or more, it is considered reasonable to assume that the distributions of T^+ and T^- can be approximated closely with normal curves. In that case we can base all tests on either T^+ or T^-, and as it does not matter which one we choose, we shall use here the statistic T^+.

Based on the assumption that each difference is as likely to be positive as negative, it can be shown that the mean and the standard deviation of the sampling distribution of T^+ are

Mean and standard deviation of T^+ statistic

$$\mu_{T^+} = \frac{n(n+1)}{4}$$

and

$$\sigma_{T^+} = \sqrt{\frac{n(n+1)(2n+1)}{24}}$$

Thus, for large samples, in this case, $n \geq 15$, we can base the signed-rank test on the statistic

Statistic for large-sample signed-rank test

$$z = \frac{T^+ - \mu_{T^+}}{\sigma_{T^+}}$$

which is a value of a random variable having approximately the standard normal distribution. When the alternative hypothesis is $\tilde{\mu} \neq \tilde{\mu}_0$ (or $\tilde{\mu}_D \neq 0$), we reject the null hypothesis if $z \leq -z_{\alpha/2}$ or $z \geq z_{\alpha/2}$; when the alternative hypothesis is $\tilde{\mu} > \tilde{\mu}_0$ (or $\tilde{\mu}_D > 0$), we reject the null hypothesis if $z \geq z_{\alpha}$; and when the alternative hypothesis is $\tilde{\mu} < \tilde{\mu}_0$ (or $\tilde{\mu}_D < 0$), we reject the null hypothesis if $z \leq -z_{\alpha}$.

EXAMPLE 18.7   The following are the weights in pounds, before and after, of 16 persons who stayed on a certain weight-reducing diet for two weeks: 169.0 and 159.9, 188.6 and 181.3, 222.1 and 209.0, 160.1 and 162.3, 187.5 and 183.5, 202.5 and 197.6, 167.8 and 171.4, 214.3 and 202.1, 143.8 and 145.1, 198.2 and 185.5, 166.9 and 158.6, 142.9 and 145.4, 160.5 and 159.5, 198.7 and 190.6, 149.7 and 149.0, and 181.6 and 183.1. Use the large-sample signed-rank test at the 0.05 level of significance to test whether the weight-reducing diet is effective.

Solution

1. H_0: $\tilde{\mu}_D = 0$, where $\tilde{\mu}_D$ is the median of the population of differences (between weights before and after) sampled.
   H_A: $\tilde{\mu}_D > 0$
2. $\alpha = 0.05$
3. Reject the null hypothesis if $z \geq 1.645$, where

$$z = \frac{T^+ - \mu_{T^+}}{\sigma_{T^+}}$$

and otherwise accept the null hypothesis or reserve judgment.

4. The original data, the differences, and the ranks of their absolute values are shown in the following table:

Weights before and after	Differences	Ranks
169.0 and 159.9	9.1	13
188.6 and 181.3	7.3	10
222.1 and 209.0	13.1	16
160.1 and 162.3	−2.2	5
187.5 and 183.5	4.0	8
202.5 and 197.6	4.9	9
167.8 and 171.4	−3.6	7
214.3 and 202.1	12.2	14
143.8 and 145.1	−1.3	3
198.2 and 185.5	12.7	15
166.9 and 158.6	8.3	12
142.9 and 145.4	−2.5	6
160.5 and 159.5	1.0	2
198.7 and 190.6	8.1	11
149.7 and 149.0	0.7	1
181.6 and 183.1	−1.5	4

It follows that

$$T^+ = 13 + 10 + 16 + 8 + 9 + 14 + 15 + 12 + 2 + 11 + 1 = 111$$

and since

$$\mu_{T^+} = \frac{16 \cdot 17}{4} = 68 \quad \text{and} \quad \sigma_{T^+} = \sqrt{\frac{16 \cdot 17 \cdot 33}{24}} \approx 19.34$$

we finally obtain

$$z = \frac{111 - 68}{19.34} \approx 2.22$$

5. Since $z = 2.22$ exceeds 1.645, the null hypothesis must be rejected; we conclude that the weight-reducing diet is effective.   ∎

■ Exercises

*18.13 On what statistic do we base our decision and for what values of the statistic do we reject the null hypothesis if we have a random sample of size $n = 10$ and are using the signed-rank test at the 0.05 level of significance to test the null hypothesis $\tilde{\mu} = \tilde{\mu}_0$ against the alternative hypothesis

(a) $\tilde{\mu} \neq \tilde{\mu}_0$;

(b) $\tilde{\mu} > \tilde{\mu}_0$;

(c) $\tilde{\mu} < \tilde{\mu}_0$?

*18.14 Rework Exercise 18.13 with the level of significance changed to 0.01.

*18.15 On what statistic do we base our decision and for what values of the statistic do we reject the null hypothesis if we have a random sample of $n = 12$ pairs of values and we are using the signed-rank test at the 0.01 level of significance to test the null hypothesis $\tilde{\mu}_D = 0$ against the alternative hypothesis
(a) $\tilde{\mu}_D \neq 0$;
(b) $\tilde{\mu}_D > 0$;
(c) $\tilde{\mu}_D < 0$?

*18.16 Rework Exercise 18.15 with the level of significance changed to 0.05.

*18.17 In a random sample of 13 issues, a newspaper listed 40, 52, 43, 27, 35, 36, 57, 39, 41, 34, 46, 32, and 37 apartments for rent. Use the signed-rank test at the 0.05 level of significance to test the null hypothesis $\tilde{\mu} = 45$ against the alternative hypothesis
(a) $\tilde{\mu} < 45$;
(b) $\tilde{\mu} \neq 45$.

*18.18 Use the signed-rank test to rework Exercise 18.1.

*18.19 Use the signed-rank test to rework Exercise 18.3.

*18.20 In a random sample taken at a public playground, it took 38, 43, 36, 29, 44, 28, 40, 50, 39, 47, and 33 minutes to play a set of tennis. Use the signed-rank test at the 0.05 level of significance to test whether or not it takes on the average 35 minutes to play a set of tennis at that public playground.

*18.21 In a random sample of 15 summer days, Casa Grande and Gila Bend reported high temperatures of 102 and 106, 103 and 110, 106 and 106, 104 and 107, 105 and 108, 102 and 109, 103 and 102, 104 and 107, 110 and 112, 109 and 110, 100 and 104, 110 and 109, 105 and 108, 111 and 114, and 105 and 106. Use the signed-rank test at the 0.05 level of significance to test the null hypothesis $\tilde{\mu}_D = 0$ against the alternative hypothesis $\tilde{\mu}_D < 0$.

*18.22 Following are the numbers of three-month and six-month certificates of deposit (CDs) that a bank sold on a random sample of 16 business days: 37 and 32, 33 and 22, 29 and 26, 18 and 33, 41 and 25, 42 and 34, 33 and 43, 51 and 31, 36 and 24, 29 and 22, 23 and 30, 28 and 29, 44 and 30, 24 and 26, 27 and 18, and 30 and 35. Test at the 0.05 level of significance whether the bank sells equally many of the two kinds of CDs against the alternative that it sells more of the three-month CDs using
(a) the signed-rank test based on Table VI;
(b) the large-sample signed-rank test.

*18.23 Use the large-sample signed-rank test to rework Exercise 18.21.

*18.24 Use the large-sample signed-rank test to rework Exercise 18.7.

*18.25 Use the large-sample signed-rank test to rework Exercise 18.10.

*18.26 Following is a random sample of the scores obtained by husbands and their wives on a spatial abilities test:

Husbands	Wives	Husbands	Wives
108	103	125	120
104	116	96	98
103	106	107	117
112	104	115	130
99	99	110	101
105	94	101	100
102	110	103	96
112	128	105	99
119	106	124	120
106	103	113	116

Use the large-sample signed-rank test at the 0.05 level of significance to test whether husbands and their wives do equally well on this test.

*18.27 Use a computer to rework Exercise 18.20.

*18.28 Use a computer to rework Exercise 18.21.

*18.29 Use a computer to rework part (a) of Exercise 18.22.

18.5 THE U TEST

We now consider a nonparametric alternative to the two-sample t test concerning the difference between two means. It is called the U **test,** the **Wilcoxon rank-sum test,** or the **Mann-Whitney test,** named after the statisticians who contributed to its development. The different names reflect the manner in which the calculations are organized; logically, they are all equivalent.

With this test we are able to check whether two independent samples come from identical populations. In particular, we can test the null hypothesis $\mu_1 = \mu_2$ without having to assume that the populations sampled have roughly the shape of normal distributions. In fact, the test requires only that the populations be continuous (to avoid ties), and even that requirement is not critical so long as the number of ties is small.

To illustrate how the U test is performed, suppose that we want to compare the grain size of sand obtained from two different locations on the moon on the basis of the following diameters (in millimeters):

Location 1: 0.37 0.70 0.75 0.30 0.45 0.16 0.62 0.73 0.33
Location 2: 0.86 0.55 0.80 0.42 0.97 0.84 0.24 0.51 0.92 0.69

The means of these two samples are 0.49 and 0.68, and their difference seems large, but it remains to be seen whether that is significant.

To perform the U test we first arrange the data jointly, as if they comprise one sample, in an increasing order of magnitude. For our data we get

0.16	0.24	0.30	0.33	0.37	0.42	0.45	0.51	0.55	0.62
1	2	1	1	1	2	1	2	2	1

0.69	0.70	0.73	0.75	0.80	0.84	0.86	0.92	0.97
2	1	1	1	2	2	2	2	2

where we indicated for each value whether it came from location 1 or location 2. Assigning the data, in this order, the ranks 1, 2, 3, . . . , and 19, we find that the values of the first sample (location 1) occupy ranks 1, 3, 4, 5, 7, 10, 12, 13, and 14, while those of the second sample (location 2) occupy ranks, 2, 6, 8, 9, 11, 15, 16, 17, 18, and 19. There are no ties here between values in different samples, but if there are, we assign each of the tied observations the mean of the ranks that they jointly occupy. For instance, if the third and fourth values are the same, we assign each the rank $\frac{3 + 4}{2} = 3.5$, and if the ninth, tenth, and eleventh values are the same, we assign each the rank $\frac{9 + 10 + 11}{3} = 10$. When there are ties among values belonging to the same sample, it does not matter how they are ranked. For instance, if the third and fourth values are the same but belong to the same sample, it does not matter which one is ranked 3 and which one is ranked 4.

Now, if there is an appreciable difference between the means of the two populations, most of the lower ranks are likely to go to the values of one sample while most of the higher ranks are likely to go to the values of the other sample. The test of the null hypothesis that the two samples come from identical populations may thus be based on W_1, the sum of the ranks of the values of the first sample, or on W_2, the sum of the ranks of the values of the second sample. In practice, it does not matter which sample we refer to as sample 1 and which sample we refer to as sample 2, and whether we base the test on W_1 or W_2. (When the sample sizes are unequal, we usually let the smaller of the two samples be sample 1; however, this is not required for the work in this book.)

If the sample sizes are n_1 and n_2, the sum of W_1 and W_2 is simply the sum of the first $n_1 + n_2$ positive integers, which is known to be

$$\frac{(n_1 + n_2)(n_1 + n_2 + 1)}{2}$$

This formula enables us to find W_2 if we know W_1, and vice versa. For our illustration we get

$$W_1 = 1 + 3 + 4 + 5 + 7 + 10 + 12 + 13 + 14 = 69$$

and since the sum of the first 19 positive integers is $\frac{19 \cdot 20}{2} = 190$, it follows that

$$W_2 = 190 - 69 = 121$$

(This value is the sum of the ranks 2, 6, 8, 9, 11, 15, 16, 17, 18, and 19.)

When the use of **rank sums** was first proposed as a nonparametric alternative to the two-sample t test, the decision was based on W_1 or W_2. Nowadays, it is more common to base the decision on either of the related statistics

U_1 and U_2 statistics | or

$$U_1 = W_1 - \frac{n_1(n_1 + 1)}{2}$$

$$U_2 = W_2 - \frac{n_2(n_2 + 1)}{2}$$

or on the statistic U, which always equals the smaller of the two. The resulting tests are equivalent to those based on W_1 or W_2, but they have the advantage that they lend themselves more readily to the construction of tables of critical values. Not only do U_1 and U_2 take on values on the same interval from 0 to $n_1 n_2$—indeed, their sum is always equal to $n_1 n_2$—but they have identical distributions that are symmetrical about $\frac{n_1 n_2}{2}$. The relationship between the sampling distributions of U_1, U_2, and U is pictured in Figure 18.4 for the special case where $n_1 = 3$ and $n_2 = 3$.

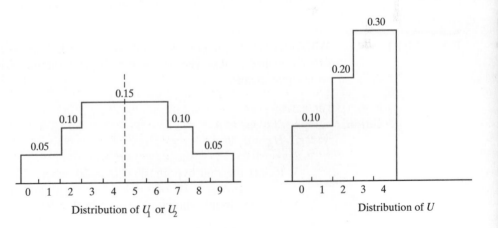

FIGURE 18.4
Distributions of U_1, U_2, and U for $n_1 = 3$ and $n_2 = 3$.

Distribution of U_1 or U_2

Distribution of U

As we have said earlier, it is assumed that we are dealing with independent random samples from identical populations, but we care mainly whether $\mu_1 = \mu_2$. All the tests will be based on the sampling distribution of one and the same statistic, as in Section 18.3. However, here it is the statistic U, and the null hypothesis will be rejected for values falling into its lefthand tail. Again, we have to be careful, though, to use the right statistic and the right critical value. When $\mu_1 < \mu_2$ then U_1 will tend to be small, so when the alternative hypothesis is $\mu_1 < \mu_2$ we base the test on U_1; when $\mu_1 > \mu_2$ then U_2 will tend to be small, so when the alternative hypothesis is $\mu_1 > \mu_2$ we base the test on U_2; and when $\mu_1 \neq \mu_2$ then either U_1 or U_2 will tend to be small, so when the alternative hypothesis is $\mu_1 \neq \mu_2$ we base the test on U. All this is summarized in the following table:

Alternative hypothesis	Reject the null hypothesis if	Accept the null hypothesis or reserve judgment if
$\mu_1 < \mu_2$	$U_1 \leq U_{2\alpha}$	$U_1 > U_{2\alpha}$
$\mu_1 > \mu_2$	$U_2 \leq U_{2\alpha}$	$U_2 > U_{2\alpha}$
$\mu_1 \neq \mu_2$	$U \leq U_\alpha$	$U > U_\alpha$

The necessary values of U_α, which are the largest values of U for which the probability of $U \leq U_\alpha$ does not exceed α, may be found in Table VII at the end of the book; the blank spaces in the table indicate that the null hypothesis cannot be rejected regardless of the value we obtain for the test statistic. Note that the same critical values serve for tests at different levels of significance depending on whether the alternative hypothesis is one sided or two sided.

EXAMPLE 18.8   With reference to the grain-size data on page 535, use the U test at the 0.05 level of significance to test whether or not the two samples come from populations with equal means.

Solution
1. H_0: $\mu_1 = \mu_2$
   H_A: $\mu_1 \neq \mu_2$
2. $\alpha = 0.05$
3. Reject the null hypothesis if $U \leq 20$, where 20 is the value of $U_{0.05}$ for $n_1 = 9$ and $n_2 = 10$; otherwise, reserve judgment.
4. Having already shown on page 536 that $W_1 = 69$ and $W_2 = 121$, we get

$$U_1 = 69 - \frac{9 \cdot 10}{2} = 24$$

$$U_2 = 121 - \frac{10 \cdot 11}{2} = 66$$

and, hence, $U = 24$. Note that $U_1 + U_2 = 24 + 66 = 90$, which equals $n_1 n_2 = 9 \cdot 10$.

5. Since $U = 24$ is greater than 20, the null hypothesis cannot be rejected; in other words, we cannot conclude that there is a difference in the mean grain size of sand from the two locations on the moon.   ■

Had we wanted to use a computer in this example, MINITAB would have yielded the printout shown in Figure 18.5. It is based on the statistic W, which we called W_1, and it gives the *p*-value as 0.0942. Since the *p*-value exceeds 0.05, we conclude (as before) that the null hypothesis cannot be rejected.

FIGURE 18.5
Computer printout for
Example 18.8.

```
MTB > Mann-Whitney 95.0 c1 c2;
SUBC>    Alternative 0.


C1              N =   9      Median =       0.4500
C2              N =  10      Median =       0.7450

W = 69.0
Test of ETA1 = ETA2   vs.   ETA1 ~= ETA2 is significant at 0.0942

Cannot reject at alpha = 0.05
```

EXAMPLE 18.9   The following are the burning times (rounded to the nearest tenth of a minute) of random samples of two kinds of emergency flares:

| Brand 1: | 17.2 | 18.1 | 19.3 | 21.1 | 14.4 | 13.7 | 18.8 | 15.2 | 20.3 | 17.5 |
| Brand 2: | 13.6 | 19.1 | 11.8 | 14.6 | 14.3 | 22.5 | 12.3 | 13.5 | 10.9 | 14.8 |

Use the *U* test at the 0.05 level of significance to test whether it is reasonable to say that on the average brand 1 flares are better (last longer) than brand 2 flares.

Solution

1. H_0: $\mu_1 = \mu_2$
   H_A: $\mu_1 > \mu_2$
2. $\alpha = 0.05$
3. Reject the null hypothesis if $U_2 \leq 27$, where 27 is the value of $U_{0.10}$ for $n_1 = 10$ and $n_2 = 10$; otherwise, accept it or reserve judgment.
4. Ranking the data jointly according to size, we find that the values of the second sample occupy ranks 5, 16, 2, 9, 7, 20, 3, 4, 1, and 10, so that

$$W_2 = 5 + 16 + 2 + 9 + 7 + 20 + 3 + 4 + 1 + 10 = 77$$

and

$$U_2 = 77 - \frac{10 \cdot 11}{2} = 22$$

5. Since $U_2 = 22$ is less than 27, the null hypothesis must be rejected; we conclude that brand 1 flares are, indeed, better than brand 2 flares.   ■

18.6   THE *U* TEST (Large Samples)

The large-sample *U* test may be based on either U_1 or U_2 as defined on page 537, but since the resulting tests are equivalent and it does not matter which sample we denote sample 1 and which sample we denote sample 2, we shall use here the statistic U_1.

Based on the assumption that the two samples come from identical continuous populations, it can be shown that the mean and the standard deviation of the sampling distribution of U_1 are[†]

Mean and standard deviation of U_1 statistic

$$\mu_{U_1} = \frac{n_1 n_2}{2}$$

and

$$\sigma_{U_1} = \sqrt{\frac{n_1 n_2 (n_1 + n_2 + 1)}{12}}$$

Observe that these formulas remain the same when we interchange the subscripts 1 and 2, but this should not come as a surprise—as we pointed out on page 537, the distributions of U_1 and U_2 are the same.

Furthermore, if n_1 and n_2 are both greater than 8, the sampling distribution of U_1 can be approximated closely by a normal distribution. Thus, we base the test of the null hypothesis $\mu_1 = \mu_2$ on the statistic

Statistic for large-sample U test

$$z = \frac{U_1 - \mu_{U_1}}{\sigma_{U_1}}$$

which has approximately the standard normal distribution. When the alternative hypothesis is $\mu_1 \neq \mu_2$, we reject the null hypothesis if $z \leq -z_{\alpha/2}$ or $z \geq z_{\alpha/2}$; when the alternative hypothesis is $\mu_1 > \mu_2$, we reject the null hypothesis if $z \geq z_\alpha$; and when the alternative hypothesis is $\mu_1 < \mu_2$, we reject the null hypothesis if $z \leq -z_\alpha$.

EXAMPLE 18.10   The following are the weight gains (in pounds) of two random samples of young turkeys fed two different diets but otherwise kept under identical conditions:

Diet 1:   16.3   10.1   10.7   13.5   14.9   11.8   14.3   10.2
             12.0   14.7   23.6   15.1   14.5   18.4   13.2   14.0

Diet 2:   21.3   23.8   15.4   19.6   12.0   13.9   18.8   19.2
             15.3   20.1   14.8   18.9   20.7   21.1   15.8   16.2

[†]When there are ties in rank the formula for the standard deviation provides only an approximation, but unless the number of ties is very large, there is rarely any need to make an adjustment.

Use the large-sample U test at the 0.01 level of significance to test the null hypothesis that the two populations sampled are identical against the alternative hypothesis that on the average the second diet produces a greater gain in weight.

Solution

1. H_0: $\mu_1 = \mu_2$ (populations are identical)
   H_A: $\mu_1 < \mu_2$
2. $\alpha = 0.01$
3. Reject the null hypothesis if $z \leqslant -2.33$, where

$$z = \frac{U_1 - \mu_{U_1}}{\sigma_{U_1}}$$

and otherwise accept the null hypothesis or reserve judgment.

4. Ranking the data jointly according to size, we find that the values of the first sample occupy ranks 21, 1, 3, 8, 15, 4, 11, 2, 5.5, 13, 31, 16, 12, 22, 7, and 10. (The fifth and sixth values are both equal to 12.0, so we assign each the rank 5.5.) Thus

$$W_1 = 1 + 2 + 3 + 4 + 5.5 + 7 + 8 + 10 + 11 + 12 + 13$$
$$+ 15 + 16 + 21 + 22 + 31$$
$$= 181.5$$

and

$$U_1 = 181.5 - \frac{16 \cdot 17}{2} = 45.5$$

Since $\mu_{U_1} = \frac{16 \cdot 16}{2} = 128$ and $\sigma_{U_1} = \sqrt{\frac{16 \cdot 16 \cdot 33}{12}} \approx 26.53$, it follows that

$$z = \frac{45.5 - 128}{26.53} \approx -3.11$$

5. Since $z = -3.11$ is less than -2.33, the null hypothesis must be rejected; we conclude that on the average the second diet produces a greater gain in weight. ∎

18.7 THE H TEST

The **H test**, or the **Kruskal-Wallis test**, is a rank-sum test that serves to test the assumption that k independent random samples come from identical populations, and in particular the null hypothesis $\mu_1 = \mu_2 = \cdots = \mu_k$, against the alternative hypothesis that these means are not all equal. Unlike the standard test that it replaces, the one-way analysis of variance of Section 15.3, it does not require the assumption that the populations sampled have, at least approximately, normal distributions.

As in the U test, the data are ranked jointly from low to high as though they constitute a single sample. Then, if R_i is the sum of the ranks assigned to the n_i values of the ith sample and $n = n_1 + n_2 + \cdots + n_k$, the H test is based on the statistic

Statistic for H test

$$H = \frac{12}{n(n+1)} \sum_{i=1}^{k} \frac{R_i^2}{n_i} - 3(n+1)$$

If the null hypothesis is true and each sample has at least five observations, it is generally considered reasonable to approximate the sampling distribution of H with a chi-square distribution having $k - 1$ degrees of freedom. Consequently, we reject the null hypothesis $\mu_1 = \mu_2 = \cdots = \mu_k$ and accept the alternative hypothesis that these means are not all equal, when the value we get for H exceeds or equals χ_α^2 for $k - 1$ degrees of freedom.

EXAMPLE 18.11   Students are randomly assigned to groups that are taught Spanish by three different methods: (1) classroom instruction and language laboratory, (2) only classroom instruction, and (3) only self-study in language laboratory. Following are the final examination scores of samples of students from the three groups:

Method 1:	94	88	91	74	86	97	
Method 2:	85	82	79	84	61	72	80
Method 3:	89	67	72	76	69		

Use the H test at the 0.05 level of significance to test the null hypothesis that the populations sampled are identical against the alternative hypothesis that their means are not all equal.

Solution

1. H_0: $\mu_1 = \mu_2 = \mu_3$ (The populations are identical)
   H_A: $\mu_1, \mu_2,$ and μ_3 are not all equal.
2. $\alpha = 0.05$
3. Reject the null hypothesis if $H \geqslant 5.991$, which is the value of $\chi_{0.05}^2$ for $3 - 1 = 2$ degrees of freedom; otherwise, accept it or reserve judgment.
4. Arranging the data jointly according to size, we get 61, 67, 69, 72, 72, 74, 76, 79, 80, 82, 84, 85, 86, 88, 89, 91, 94, and 97. Assigning the data, in this order, the ranks 1, 2, 3, . . . , and 18, we find that

$$R_1 = 6 + 13 + 14 + 16 + 17 + 18 = 84$$
$$R_2 = 1 + 4.5 + 8 + 9 + 10 + 11 + 12 = 55.5$$
$$R_3 = 2 + 3 + 4.5 + 7 + 15 = 31.5$$

and it follows that

$$H = \frac{12}{18 \cdot 19} \left(\frac{84^2}{6} + \frac{55.5^2}{7} + \frac{31.5^2}{5} \right) - 3 \cdot 19$$
$$\approx 6.67$$

5. Since $H = 6.67$ exceeds 5.991, the null hypothesis must be rejected; we conclude that the three methods of instruction are not all equally effective.   ∎

Had we used a computer for this example, we would have found that the p-value corresponding to $H = 6.67$ is 0.036, and that the p-value adjusted for the tie is also 0.036. Since 0.036 is less than 0.05, we would have concluded, as before, that the null hypothesis must be rejected.

■ Exercises

18.30 On what statistic do we base the decision and for what values of the statistic do we reject the null hypothesis $\mu_1 = \mu_2$ if we have random samples of size $n_1 = 9$ and $n_2 = 9$ and are using the U test based on Table VII and the 0.05 level of significance to test the null hypothesis against the alternative hypothesis
(a) $\mu_1 > \mu_2$;
(b) $\mu_1 \neq \mu_2$;
(c) $\mu_1 < \mu_2$?

18.31 Rework Exercise 18.30 with the level of significance changed to 0.01.

18.32 On what statistic do we base the decision and for what values of the statistic do we reject the null hypothesis $\mu_1 = \mu_2$ if we have random samples of size $n_1 = 10$ and $n_2 = 14$ and are using the U test based on Table VII and the 0.01 level of significance to test the null hypothesis against the alternative hypothesis
(a) $\mu_1 > \mu_2$;
(b) $\mu_1 \neq \mu_2$;
(c) $\mu_1 < \mu_2$?

18.33 Rework Exercise 18.32 with the level of significance changed to 0.05.

18.34 On what statistic do we base the decision and for what values of the statistic do we reject the null hypothesis $\mu_1 = \mu_2$ against the alternative hypothesis $\mu_1 \neq \mu_2$ if we are using the U test based on Table VII and the 0.05 level of significance, and
(a) $n_1 = 4$ and $n_2 = 6$;
(b) $n_1 = 9$ and $n_2 = 8$;
(c) $n_1 = 5$ and $n_2 = 12$;
(d) $n_1 = 7$ and $n_2 = 3$?

18.35 Rework Exercise 18.34 with the alternative hypothesis changed to $\mu_1 > \mu_2$.

18.36 Explain why there is no value in Table VII for $U_{0.05}$ corresponding to $n_1 = 3$ and $n_2 = 3$. (*Hint*: Refer to Figure 18.4.)

18.37 Following are the scores that random samples of students from two minority groups obtained on a current events test:

Minority group 1:	73	82	39	68	91	75
	89	67	50	86	57	65

Minority group 2:	51	42	36	53	88	59
	49	66	25	64	18	76

Use the U test based on Table VII to test at the 0.05 level of significance whether or not students from the two minority groups can be expected to score equally well on this test.

18.38 Use a computer to rework Exercise 18.37.

18.39 The following are the number of minutes taken by random samples of 15 men and 12 women to complete a written test given for the renewal of their driver's licenses:

Men:	9.9	7.4	8.9	9.1	7.7	9.7	11.8	7.5
	9.2	10.0	10.2	9.5	10.8	8.0	11.0	
Women:	8.6	10.9	9.8	10.7	9.4	10.3		
	7.3	11.5	7.6	9.3	8.8	9.6		

Use the U test based on Table VII to test at the 0.05 level of significance whether or not $\mu_1 = \mu_2$, where μ_1 and μ_2 are the average amounts of time it takes men and women to complete the test.

18.40 Use the large-sample U test to rework Exercise 18.39.

18.41 The following are the Rockwell hardness numbers obtained for six aluminum die castings randomly selected from production lot A and eight from production lot B:

Production lot A:	75	56	63	70	58	74		
Production lot B:	63	85	77	80	86	76	72	82

Use the U test based on Table VII to test at the 0.05 level of significance whether the castings of production lot B are on the average harder than those of production lot A.

 18.42 Use a computer to rework Exercise 18.41.

18.43 Use the large-sample U test to rework Example 18.8.

18.44 Use the large-sample U test to rework Example 18.9.

 18.45 Use a computer to rework Example 18.10.

18.46 Following are data for the breaking strength (in pounds) of random samples of two kinds of 2-inch cotton ribbons:

Type I ribbon:	133	144	165	169	171	176	180	181
	182	183	186	187	194	197	198	200
Type II ribbon:	134	154	159	161	164	164	164	169
	170	172	175	176	185	189	194	198

(For convenience, the values have been arranged in ascending order.) Use the large-sample U test at the 0.05 level of significance to test the claim that type I ribbon is, on the average, stronger than type II ribbon.

18.47 Use a computer to rework Exercise 18.46.

18.48 Following are the miles per gallon that a test driver got in random samples of six tankfuls of each of three kinds of gasoline:

Gasoline 1:	15	24	27	29	30	32
Gasoline 2:	17	20	22	28	32	33
Gasoline 3:	18	19	22	23	25	32

(For convenience, the values have been arranged in ascending order.) Use the H test at the 0.05 level of significance to test the claim that there is no difference in the true average mileage yield of the three kinds of gasoline.

18.49 Use the *H* test at the 0.01 level of significance to rework Exercise 15.15.

18.50 Use the *H* test at the 0.01 level of significance to rework Exercise 15.17.

18.51 To compare four bowling balls, a professional bowler bowls five games with each ball and gets the following scores:

Bowling ball D:	221	232	207	198	212
Bowling ball E:	202	225	252	218	226
Bowling ball F:	210	205	189	196	216
Bowling ball G:	229	192	247	220	208

Use the *H* test at the 0.05 level of significance to test the null hypothesis that on the average the bowler performs equally well with the four bowling balls.

 18.52 Use a computer to rework Exercise 18.51.

18.53 Three groups of guinea pigs, injected with 0.5, 1.0, and 1.5 mg, respectively, of a tranquilizer fell asleep in the following number of seconds:

0.5 mg dose:	7.8	8.2	10.0	10.2	
	10.9	12.7	13.7	14.0	
1.0 mg dose:	7.5	7.9	8.8	9.7	10.5
	11.0	12.5	12.9	13.1	13.3
1.5 mg dose:	7.2	8.0	8.5	9.0	
	9.4	11.3	11.5	12.0	

(For convenience, the values have been arranged in increasing order.) Use the *H* test at the 0.01 level of significance to test the null hypothesis that the differences in dosage have no effect on the length of time it takes guinea pigs to fall asleep.

 18.54 Use a computer to rework Exercise 18.53.

18.8 TESTS OF RANDOMNESS: RUNS

All the methods of inference studied in this book are based on the assumption that our samples are random; yet there are many applications where it is difficult to determine whether this assumption is justifiable. This is true, particularly, when we have little or no control over the selection of the data, as is the case, for example, when we rely on whatever records are available to make long-range predictions of the weather, when we use whatever data are available to estimate the mortality rate of a disease, or when we use sales records for past months to make predictions of a department store's future sales. None of this information constitutes a random sample in the strict sense.

There are several methods of judging the randomness of a sample on the basis of the order in which the observations are obtained; they enable us to decide, after the data have been collected, whether patterns that look suspiciously nonrandom may be attributed to chance. The technique we describe here and in the next two sections, the *u* **test,** is based on the **theory of runs.**

A **run** is a succession of identical letters (or other kinds of symbols) that is followed and preceded by different letters or no letters at all. To illustrate, consider the following arrangement of healthy, H, and diseased, D, elm trees that were planted many years ago along a country road:

$$\underline{H\,H\,H\,H}\,\underline{D\,D\,D}\,\underline{H\,H\,H\,H\,H\,H\,H}\,\underline{D\,D}\,\underline{H\,H}\,\underline{D\,D\,D\,D}$$

Using underlines to combine the letters that constitute the runs, we find that there is first a run of four H's, then a run of three D's, then a run of seven H's, then a run of two D's, then a run of two H's, and finally a run of four D's.

The **total number of runs** appearing in an arrangement of this kind is often a good indication of a possible lack of randomness. If there are too few runs, we might suspect a definite grouping or clustering, or perhaps a trend; if there are too many runs, we might suspect some sort of repeated alternating, or cyclical pattern. In the preceding example there seems to be a definite clustering—the diseased trees seem to come in groups—but it remains to be seen whether this is significant or whether it can be attributed to chance.

If there are n_1 letters of one kind, n_2 letters of another kind, and u runs, we base this kind of decision on the following criterion:

Reject the null hypothesis of randomness if

$$u \leq u'_{\alpha/2} \qquad \text{or} \qquad u \geq u_{\alpha/2}$$

where $u'_{\alpha/2}$ and $u_{\alpha/2}$ are given in Table VIII for values of n_1 and n_2 through 15, and $\alpha = 0.05$ and $\alpha = 0.01$.

In the construction of Table VIII, $u'_{\alpha/2}$ is the largest value of u for which the probability of $u \leq u'_{\alpha/2}$ does not exceed $\alpha/2$ and $u_{\alpha/2}$ is the smallest value of u for which the probability of $u \geq u_{\alpha/2}$ does not exceed $\alpha/2$; the blank spaces in the table indicate that the null hypothesis cannot be rejected for values in that tail of the sampling distribution regardless of the value we obtain for u.

EXAMPLE 18.12   With reference to the arrangement of healthy and diseased elm trees cited previously, use the u test at the 0.05 level of significance to test the null hypothesis of randomness against the alternative hypothesis that the arrangement is not random.

Solution

1. H_0: Arrangement is random.
   H_A: Arrangement is not random.
2. $\alpha = 0.05$.
3. Reject the null hypothesis if $u \leq 6$ or $u \geq 17$, where 6 and 17 are the values of $u'_{0.025}$ and $u_{0.025}$ for $n_1 = 13$ and $n_2 = 9$; otherwise, accept it or reserve judgment.
4. $u = 6$ by inspection of the data.

5. Since $u = 6$ equals the value of $u'_{0.025}$, the null hypothesis must be rejected; we conclude that the arrangement of healthy and diseased elm trees is not random. There are fewer runs than might have been expected, and it appears that the diseased trees come in clusters. ∎

18.9 TESTS OF RANDOMNESS: RUNS (Large Samples)

Under the null hypothesis that n_1 letters of one kind and n_2 letters of another kind are arranged at random, it can be shown that the mean and the standard deviation of u, the total number of runs, are

Mean and standard deviation of u

$$\mu_u = \frac{2n_1 n_2}{n_1 + n_2} + 1$$

and

$$\sigma_u = \sqrt{\frac{2n_1 n_2 (2n_1 n_2 - n_1 - n_2)}{(n_1 + n_2)^2 (n_1 + n_2 - 1)}}$$

Furthermore, if neither n_1 nor n_2 is less than 10, the sampling distribution of u can be approximated closely by a normal distribution. Thus, we base the test of the null hypothesis of randomness on the statistic

Statistic for large-sample u test

$$z = \frac{u - \mu_u}{\sigma_u}$$

which has approximately the standard normal distribution. If the alternative hypothesis is that the arrangement is not random, we reject the null hypothesis for $z \leqslant -z_{\alpha/2}$ or $z \geqslant z_{\alpha/2}$; if the alternative hypothesis is that there is a clustering or a trend, we reject the null hypothesis for $z \leqslant -z_\alpha$; and if the alternative hypothesis is that there is an alternating, or cyclical, pattern, we reject the null hypothesis for $z \geqslant z_\alpha$.

EXAMPLE 18.13   The following is an arrangement of men, M, and women, W, lined up to purchase tickets for a rock concert:

$M\ W\ M\ W\ M\ M\ M\ W\ M\ W\ M\ M\ M\ W\ W\ M\ M\ M\ M\ W\ W\ M\ W\ M$

$M\ M\ W\ M\ M\ M\ W\ W\ W\ M\ W\ M\ M\ M\ W\ M\ W\ M\ M\ M\ M\ W\ W\ M$

Test for randomness at the 0.05 level of significance.

Solution

1. H_0: Arrangement is random.
   H_A: Arrangement is not random.
2. $\alpha = 0.05$
3. Reject the null hypothesis if $z \leqslant -1.96$ or $z \geqslant 1.96$, where

$$z = \frac{u - \mu_u}{\sigma_u}$$

and otherwise accept the null hypothesis or reserve judgment.

4. Since $n_1 = 30$, $n_2 = 18$, and $u = 27$, we get

$$\mu_u = \frac{2 \cdot 30 \cdot 18}{30 + 18} + 1 = 23.5$$

$$\sigma_u = \sqrt{\frac{2 \cdot 30 \cdot 18(2 \cdot 30 \cdot 18 - 30 - 18)}{(30 + 18)^2(30 + 18 - 1)}} = 3.21$$

and, hence,

$$z = \frac{27 - 23.5}{3.21} \approx 1.09$$

5. Since $z = 1.09$ falls between -1.96 and 1.96, the null hypothesis cannot be rejected; in other words, there is no real evidence to suggest that the arrangement is not random.   ∎

18.10 TESTS OF RANDOMNESS: RUNS ABOVE AND BELOW THE MEDIAN

The u test is not limited to testing the randomness of sequences of attributes, such as the H's and D's, or M's and W's, of our examples. Any sample consisting of numerical measurements or observations can be treated similarly by using the letters a and b to denote values falling above and below the median of the sample. Numbers equal to the median are omitted. The resulting sequence of a's and b's (representing the data in their original order) can then be tested for randomness on the basis of the total number of runs of a's and b's, namely, the total number of **runs above and below the median.** Depending on the size of n_1 and n_2, we use Table VIII or the large-sample test of Section 18.9.

EXAMPLE 18.14   On 24 successive runs between two cities, a bus carried

| 24 | 19 | 32 | 28 | 21 | 23 | 26 | 17 | 20 | 28 | 30 | 24 |
| 13 | 35 | 26 | 21 | 19 | 29 | 27 | 18 | 26 | 14 | 21 | 23 |

passengers. Use the total number of runs above and below the median to test at the 0.01 level of significance whether it is reasonable to treat these data as if they constitute a random sample.

Solution   Since the median of the data is 23.5, we get the following arrangement of values above and below the median:

$$a\ b\ a\ a\ b\ b\ a\ b\ b\ a\ a\ a\ b\ a\ a\ b\ b\ a\ a\ b\ a\ b\ b\ b$$

1. H_0:   Arrangement is random.
   H_A: Arrangement is not random.
2. $\alpha = 0.01$
3. Reject the null hypothesis if $u \leq 6$ or $u \geq 20$, where 6 and 20 are the values of $u'_{0.005}$ and $u_{0.005}$ for $n_1 = 12$ and $n_2 = 12$; otherwise, accept the null hypothesis or reserve judgment.
4. $u = 14$ by inspection of the preceding arrangement of a's and b's.
5. Since $u = 14$ falls between 6 and 20, the null hypothesis cannot be rejected; in other words, there is no real evidence to indicate that the data do not constitute a random sample.   ∎

■ Exercises

18.55 Following is the order in which a broker received 25 orders to buy, B, and sell, S, shares of a certain stock:

$$S\ S\ S\ B\ B\ B\ B\ B\ B\ B\ B\ S\ B\ S\ S\ S\ S\ S\ B\ B\ B\ B\ S\ S$$

Use Table VIII to test for randomness at the 0.05 level of significance.

18.56 Use the large-sample test to rework Exercise 18.55.

18.57 A driver buys gasoline either at a Chevron station, C, or an Arco station, A, and the following arrangement shows the order of the stations from which he made 29 purchases of gasoline over a recent period of time:

$$A\ CA\ C\ CA\ CA\ CA\ C\ CA\ A\ A\ CA\ C\ C\ CA\ C\ A\ CA\ A\ A\ C\ CA\ C$$

Use Table VIII to test for randomness at the 0.01 level of significance.

18.58 Use the large-sample test to rework Exercise 18.57.

18.59 Test at the 0.05 level of significance whether the following arrangement of defective, D, and nondefective, N, engines coming off an assembly line may be regarded as random:

$$N\ N\ N\ N\ N\ N\ N\ D\ N\ N\ N\ N\ N\ D\ D\ D\ N\ N\ N\ D\ D\ D\ N\ N$$

18.60 The following arrangement indicates whether 60 consecutive persons interviewed by a pollster are for, F, or against, A, an increase of half a cent in the state sales tax to build a new football stadium:

$$A\ F\ F\ F\ A\ A\ A\ A\ F\ A\ A\ F\ F\ A\ A\ A\ A\ F\ F\ F\ A\ F\ A\ A\ A\ F\ F\ F\ A\ A$$
$$A\ A\ F\ F\ F\ F\ A\ A\ F\ A\ A\ F\ F\ F\ F\ F\ A\ F\ A\ A\ A\ A\ F\ F\ A\ F\ F\ F\ F\ A$$

Test for randomness at the 0.01 level of significance.

18.61 To test whether a radio signal contains a message or constitutes random noise, an interval of time is subdivided into a number of very short intervals and for each of these it is determined whether the signal strength exceeds, E, or does not exceed,

N, a certain level of background noise. Test at the 0.05 level of significance whether the following arrangement, thus obtained, may be regarded as random, and hence that the signal contains no message and may be regarded as random noise:

$$E\,N\,N\,N\,N\,E\,N\,E\,N\,N\,N\,E\,E\,N\,N\,N\,E\,E\,N\,E\,N\,N\,N\,E\,E\,N\,N\,N$$
$$N\,N\,E\,E\,N\,E\,N\,N\,E\,N\,N\,N\,E\,E\,N\,N\,N\,E\,N\,E\,N\,N\,N\,N\,N\,E\,N$$

18.62 Flip a coin 50 times and test at the 0.05 level of significance whether the resulting sequence of *H*'s and *T*'s (heads and tails) may be regarded as random.

18.63 Record whether 60 consecutive cars arriving at an intersection from the north have local license plates, *L*, or out-of-state plates, *O*. Test for randomness at the 0.05 level of significance.

18.64 In the beginning of Section 11.1 we gave the following data on the number of minutes it took 36 persons to assemble an "easy to assemble" toy: 17, 13, 18, 19, 17, 21, 29, 22, 16, 28, 21, 15, 26, 23, 24, 20, 8, 17, 17, 21, 32, 18, 25, 22, 16, 10, 20, 22, 19, 14, 30, 22, 12, 24, 28, and 11. Test for randomness at the 0.05 level of significance.

18.65 In Exercise 12.47 we gave the following weights of 40 mature dogs of a certain breed: 66.2, 59.2, 70.8, 58.0, 64.3, 50.7, 62.5, 58.4, 48.7, 52.4, 51.0, 35.7, 62.6, 52.3, 41.2, 61.1, 52.9, 58.8, 64.1, 48.9, 74.3, 50.3, 55.7, 55.5, 51.8, 55.8, 48.9, 51.8, 63.1, 44.6, 47.0, 49.0, 62.5, 45.0, 78.6, 54.2, 72.2, 52.4, 60.5, and 46.8 ounces. Given that the median of these weights is 54.85 ounces, test for randomness at the 0.05 level of significance.

18.66 Following are the examination grades of 42 students in the order in which they finished an examination:

75	95	77	93	89	83	69	77	92	88	62	64	91	72
76	83	50	65	84	67	63	54	58	76	70	62	65	41
63	55	32	58	61	68	54	28	35	49	82	60	66	57

Test for randomness at the 0.05 level of significance.

18.67 The total number of retail stores opening for business and also quitting business within the calendar years 1970–1999 in a large city were

107	125	142	147	122	116	153	144	106	138
126	125	129	134	137	143	150	148	152	145
112	162	139	132	122	143	148	155	146	158

Making use of the fact that the median is 140.5, test at the 0.05 level of significance whether there is a significant trend.

18.68 The following are six year's quarterly sales (in millions of dollars) of a manufacturer of heavy machinery:

83.8	102.5	121.0	90.5	106.6	104.8	114.7	93.6
98.9	96.9	122.6	85.6	103.2	96.9	118.0	92.1
100.5	92.9	125.6	79.2	110.8	95.1	125.6	86.7

Making use of the fact that the median is 99.7, test at the 0.05 level of significance whether there is a real cyclical pattern.

18.11 RANK CORRELATION

Since the significance test for r of Section 17.3 is based on very stringent assumptions, we sometimes use a nonparametric alternative that can be applied under much more general conditions. This test of the null hypothesis of no correlation is based on the **rank-correlation coefficient,** often called **Spearman's rank-correlation coefficient** and denoted by r_S.

   To calculate the rank-correlation coefficient for a given set of paired data, we first rank the x's among themselves from low to high or high to low; then we rank the y's in the same way, find the sum of the squares of the differences, d, between the ranks of the x's and the y's, and substitute into the formula

Rank-correlation coefficient

$$r_S = 1 - \frac{6\left(\sum d^2\right)}{n(n^2 - 1)}$$

where n is the number of pairs of x's and y's. When there are ties in rank, we proceed as before and assign to each of the tied observations the mean of the ranks that they jointly occupy.

EXAMPLE 18.15   The following are the numbers of hours that ten students studied for an examination and the scores that they obtained:

Number of hours studied x	Scores in examination y
9	56
5	44
11	79
13	72
10	70
5	54
18	94
15	85
2	33
8	65

Calculate r_S.

Solution   Ranking the x's among themselves from low to high and also the y's, we get the ranks shown in the first two columns of the following table:

Rank of x	Rank of y	d	d^2
5	4	1.0	1.00
2.5	2	0.5	0.25
7	8	−1.0	1.00
8	7	1.0	1.00
6	6	0.0	0.00
2.5	3	−0.5	0.25
10	10	0.0	0.00
9	9	0.0	0.00
1	1	0.0	0.00
4	5	−1.0	1.00
			4.50

Note that the second and third smallest values among the x's are both equal to 5, so we assign each of them the rank $\dfrac{2 + 3}{2} = 2.5$. Then, determining the d's (differences between the ranks) and their squares, and substituting $n = 10$ and $\Sigma\, d^2 = 4.50$ into the formula for r_S, we get

$$r_S = 1 - \frac{6\,(4.50)}{10\,(10^2 - 1)} = 0.97 \qquad \blacksquare$$

As can be seen from this example, r_S is easy to compute manually, and this is why it is sometimes used instead of r when no calculator is available. When there are no ties, r_S actually equals the correlation coefficient r calculated for the two sets of ranks; when ties exist there may be a small (but usually negligible) difference. Of course, by using ranks instead of the original data we lose some information, but this is usually offset by the rank-correlation coefficient's computational ease. It is of interest to note that if we had calculated r for the original x's and y's in the preceding example, we would have obtained 0.96 instead of 0.97; at least in this case, the difference between r and r_S is very small.

The main advantage in using r_S is that we can test the null hypothesis of no correlation without having to make any assumptions about the populations sampled. Under the null hypothesis of no correlation—indeed, the null hypothesis that the x's and the y's are randomly matched—the sampling distribution of r_S has the mean 0 and the standard deviation

$$\sigma_{r_S} = \frac{1}{\sqrt{n - 1}}$$

Since this sampling distribution can be approximated with a normal distribution even for relatively small values of n, we base the test of the null hypothesis on the statistic

Statistic for testing
significance of r_S

$$z = \frac{r_S - 0}{1/\sqrt{n-1}} = r_S\sqrt{n-1}$$

which has approximately the standard normal distribution.

EXAMPLE 18.16   With reference to Example 18.15, where we had $n = 10$ and $r_S = 0.97$, test the null hypothesis of no correlation at the 0.01 level of significance.

olution

1. H_0: $\rho = 0$ (no correlation)
   H_A: $\rho \neq 0$
2. $\alpha = 0.01$
3. Reject the null hypothesis if $z \leqslant -2.575$ or $z \geqslant 2.575$, where

$$z = r_S\sqrt{n-1}$$

and otherwise, accept it or reserve judgment.
4. For $n = 10$ and $r_S = 0.97$ we get

$$z = 0.97\sqrt{10 - 1} = 2.91$$

5. Since $z = 2.91$ exceeds 2.575, the null hypothesis must be rejected; we conclude that there is a relationship between study time and scores in the population sampled.   ■

■ **Exercises**

18.69 Calculate r_S for the following sample data representing the number of minutes it took 12 mechanics to assemble a piece of machinery in the morning, x, and in the late afternoon, y:

x	y
10.8	15.1
16.6	16.8
11.1	10.9
10.3	14.2
12.0	13.8
15.1	21.5
13.7	13.2
18.5	21.1
17.3	16.4
14.2	19.3
14.8	17.4
15.3	19.0

18.70 Test at the 0.05 level of significance whether the value obtained for r_S in Exercise 18.69 is significant.

18.71 With reference to Exercise 16.3, calculate r_S and test whether it is significant at the 0.05 level of significance.

18.72 With reference to Exercise 16.6, calculate r_S and test whether it is significant at the 0.05 level of significance.

18.73 With reference to Exercise 16.12, calculate r_S and test whether it is significant at the 0.01 level of significance.

18.74 If a sample of $n = 37$ pairs of data yielded $r_S = 0.39$, is this rank-correlation coefficient significant at the 0.01 level of significance?

18.75 If a sample of $n = 50$ pairs of data yielded $r_S = 0.31$, is this rank-correlation coefficient significant at the 0.05 level of significance?

18.76 The following table shows how a panel of nutrition experts and a panel of heads of household ranked 15 breakfast foods on their palatability:

Breakfast food	Nutrition experts	Heads of household
I	7	5
II	3	4
III	11	8
IV	9	14
V	1	2
VI	4	6
VII	10	12
VIII	8	7
IX	5	1
X	13	9
XI	12	15
XII	2	3
XIII	15	10
XIV	6	11
XV	14	13

Calculate r_S as a measure of the consistency of the two ratings.

18.77 The following are the rankings which three judges gave to the work of ten artists:

Judge A:	5	8	4	2	3	1	10	7	9	6
Judge B:	3	10	1	4	2	5	6	7	8	9
Judge C:	8	5	6	4	10	2	3	1	7	9

Calculate r_S for each pair of rankings and decide
(a) which two judges are most alike in their opinions about these artists;
(b) which two judges differ the most in their opinions about these artists.

18.12 SOME FURTHER CONSIDERATIONS

Although nonparametric tests have a great deal of intuitive appeal and are widely applicable, it should not be overlooked that they are usually **less efficient** than the standard tests that they replace. To illustrate what we mean here by "less efficient," let us refer to Example 10.12, where we showed that the mean of a random sample of size $n = 128$ is as reliable an estimate of the mean of a symmetrical population as a median of a random sample of size $n = 200$. Thus, the median requires a larger sample than the mean, and this is what we mean when we say that it is "less efficient."

Put another way, nonparametric tests tend to be wasteful of information. The one-sample sign test and the paired-sample sign test are especially wasteful, whereas the other procedures introduced in this chapter are wasteful to a lesser degree. Above all, nonparametric tests should not be used indiscriminately when the assumptions underlying the corresponding standard tests are satisfied.

In actual practice, nonparametric procedures are often used to confirm conclusions based on standard tests when there is some uncertainty about the validity of the assumptions that underly the standard tests. Nonparametric tests are indispensable when sample sizes are too small to form an opinion one way or the other about the validity of assumptions.

18.13 SUMMARY

The table that follows summarizes the various nonparametric tests we have discussed (except for the tests of randomness based on runs) and the corresponding standard tests that they replace. In each case we list the section or sections of the book where they are discussed.

Null hypothesis	Standard tests	Nonparametric tests
$\mu = \mu_0$	One-sample t test (Section 12.4) or one-sample z test (Section 12.3)	One-sample sign test (Sections 18.1 and 18.2) or signed-rank test (Sections 18.3 and 18.4)
$\mu_1 = \mu_2$ (independent samples)	Two-sample t test (Section 12.6) or two-sample z test (Section 12.5)	U test (Sections 18.5 and 18.6)
$\mu_1 = \mu_2$ (paired data)	Paired-sample t test or paired-sample z test (Section 12.7)	Paired-sample sign test (Sections 18.1 and 18.2) or signed-rank test (Sections 18.3 and 18.4)

(continued on page 556)

Null hypothesis	Standard tests	Nonparametric tests
$\mu_1 = \mu_2 = \cdots = \mu_k$	One-way analysis of variance (Section 15.3)	H test (Section 18.7)
$\rho = 0$	Test based on Fisher Z transformation (Section 17.3)	Test based on rank-correlation coefficient (Section 18.11)

Tests of randomness are discussed in Sections 18.8, 18.9, and 18.10, but there are no corresponding standard tests.

18.14 Checklist of Key Terms *(with page references to their definitions)*

Efficiency, 555
H test, 541, 542
Kruskal-Wallis test, 541
Mann-Whitney test, 535
Nonparametric tests, 520
One-sample sign test, 521
Paired-sample sign test, 523
Rank-sum tests, 535, 541
Rank-correlation coefficient, 551
Runs, 546
Runs above and below the median, 548

Sign test, 521, 523
Signed-rank test, 527, 532
Spearman's rank-correlation coefficient, 551
Theory of runs, 545
Total number of runs, 546
u test, 545, 546, 547
U test, 535, 538, 540
Wilcoxom signed-rank test, 527
Wilcoxon rank-sum test, 535

18.15 References

Further information about the nonparametric tests discussed in this chapter and many others may be found in

CONOVER, W. J., *Practical Nonparametric Statistics*. New York: John Wiley & Sons, Inc., 1971.

DANIEL, W. W., *Applied Nonparametric Statistics*. Boston: Houghton Mifflin Company, 1978.

GIBBONS, J. D., *Nonparametric Statistical Inference*. New York: Marcel Dekker, 1985.

LEHMANN, E. L., *Nonparametrics: Statistical Methods Based on Ranks*. San Francisco: Holden-Day, Inc., 1975.

MOSTELLER, F., and ROURKE, R. E. K., *Sturdy Statistics, Nonparametrics and Order Statistics*. Reading, Mass.: Addison-Wesley Publishing Company, Inc., 1973.

NOETHER, G. E., *Introduction to Statistics: The Nonparametric Way*. New York: Springer-Verlag, 1990.

RANDLES, R., and WOLFE, D., *Introduction to the Theory of Nonparametric Statistics*. New York: John Wiley & Sons, Inc., 1979.

SIEGEL, S., *Nonparametric Statistics for the Behavioral Sciences*. New York: McGraw-Hill Book Company, 1956.

R.183 To find the best arrangement of instruments on a control panel of an airplane, three different arrangements were tested by simulating emergency conditions and observing the reaction time required to correct the condition. The reaction times (in tenths of a second) of 12 pilots (randomly assigned to the different arrangements) were as follows:

Arrangement 1:   8   15   10   11
Arrangement 2:   16   11   14   19
Arrangement 3:   12   7   13   8

(a) Calculate $n \cdot s_{\bar{x}}^2$ for these data, and also the mean of the variances of the three samples and the value of F.

(b) Assuming that the necessary assumptions can be met, test at the 0.01 level of significance whether the differences among the three sample means can be attributed to chance.

R.184 Recent government statistics show that for couples with 0, 1, 2, 3, or 4 children, the relationship between the number of children, x, and family income in dollars, y, is fairly well described by the least-squares line $\hat{y} = 38{,}600 + 3{,}500x$. If a childless couple has twins, will this increase their income by $2(3{,}500) = \$7{,}000$?

R.185 Following are the numbers of hours that ten persons (interviewed as part of a sample survey) spent watching television, x, and reading books or magazines, y, per week:

x	y
18	7
25	5
19	1
12	5
12	10
27	2
15	3
9	9
12	8
18	4

For these data, $\Sigma x = 167$, $\Sigma x^2 = 3{,}101$, $\Sigma y = 54$, $\Sigma y^2 = 374$, and $\Sigma xy = 798$.

(a) Fit a least-squares line that will enable us to predict y in terms of x.

(b) If a person spends 22 hours watching television per week, predict how many hours he or she will spend reading books or magazines.

R.186 Calculate r for the data of Exercise R.185.

R.187 Following are the prices (in dollars) charged for a certain camera in a random sample of 15 discount stores: 57.25, 58.14, 54.19, 56.17, 57.21, 55.38, 54.75, 57.29, 57.80, 54.50, 55.00, 56.35, 54.26, 60.23, and 53.99. Use the sign test based on Table V to test at the 0.05 level of significance whether or not the median price charged for such cameras in the population sampled is $55.00.

R.188 Following are the numbers of minutes that patients had to wait for their appointments with four doctors:

Doctor A:	18	26	29	22	16
Doctor B:	9	11	28	26	15
Doctor C:	20	13	22	25	10
Doctor D:	21	26	39	32	24

Use the H test at the 0.05 level of significance to test the null hypothesis that the four samples come from identical populations against the alternative hypothesis that the means of the four populations are not all equal.

R.189 The following sequence shows whether a certain senator was present, P, or absent, A, at 30 consecutive meetings of an appropriations committee:

$$P\ P\ P\ P\ P\ P\ P\ A\ A\ P\ P\ P\ P\ P\ P\ A\ A\ P\ P\ P\ P\ P\ A\ P\ P\ P\ P\ A\ A\ P$$

At the 0.01 level of significance, is there any real indication of a lack of randomness?

R.190 A school has seven department heads who are assigned to seven different committees, as shown in the following table:

Committee	Department heads			
Textbooks	Dodge,	Fleming,	Griffith,	Anderson
Athletics	Bowman,	Evans,	Griffith,	Anderson
Band	Bowman,	Carlson,	Fleming,	Anderson
Dramatics	Bowman,	Carlson,	Dodge,	Griffith
Tenure	Carlson,	Evans,	Fleming,	Griffith
Salaries	Bowman,	Dodge,	Evans,	Fleming
Discipline	Carlson,	Dodge,	Evans,	Anderson

(a) Verify that this arrangement is a balanced incomplete block design.

(b) If Dodge, Bowman, and Carlson are (in that order) appointed chairpersons of the first three committees, how will the chairpersons of the other four committees have to be chosen so that each of the department heads is chairperson of one of the committees?

R.191 If $k = 6$ and $n = 9$ in a two-way analysis of variance without interaction, what are the degrees of freedom for treatments, blocks, and error?

R.192 The following data pertain to a study of the effects of environmental pollution on wildlife; in particular, the relationship between DDT and the thickness of the eggshells of certain birds:

DDT residue in yolk lipids (parts per million)	Thickness of eggshells (millimeters)
117	0.49
65	0.52
303	0.37
98	0.53
122	0.49
150	0.42

If x denotes the DDT residue and y denotes the eggshell thickness, then $S_{xx} = 34{,}873.50$, $S_{xy} = -23.89$, and $S_{yy} = 0.0194$. Calculate the coefficient of correlation.

R.193   With reference to Exercise R.192, use the 0.05 level of significance to test whether the value obtained for r is significant.

 R.194   Following are the numbers of computer modem cards produced by four assembly lines on 12 workdays:

Line 1	Line 2	Line 3	Line 4
904	835	873	839
852	857	803	849
861	822	855	913
770	796	851	840
877	808	856	843
929	832	857	892
955	777	873	841
836	830	830	807
870	808	921	875
843	862	886	898
847	843	834	976
864	802	939	822

(a) Perform an analysis of variance and, assuming that the required assumptions can be met, test at the 0.05 level of significance whether the differences obtained for the means of the four samples, 867.33, 822.67, 864.83, and 866.25, can be attributed to chance.

(b) Use the studentized-range method at the 0.05 level of significance to analyze the performance of the four assembly lines.

 *R.195   Following are data on the average weekly profit (in $1,000) of five restaurants, their seating capacities, and the average daily traffic (in thousands of cars) that pass their locations:

Seating capacity x_1	Traffic count x_2	Weekly net profit y
120	19	23.8
200	8	24.2
150	12	22.0
180	15	26.2
240	16	33.5

(a) Use the method of least squares to fit an equation of the form $y = b_0 + b_1 x_1 + b_2 x_2$ to these data.

(b) Use the equation obtained in part (a) to predict the average weekly profit of a restaurant with a seating capacity of 210 at a location where the daily traffic count averages 14,000 cars.

R.196 The sample data in the following table are the grades in a statistics test obtained by nine college students from three majors who were taught by three different instructors:

	Instructor A	Instructor B	Instructor C
Marketing	77	88	71
Finance	88	97	81
Insurance	85	95	72

Assuming that the necessary assumptions can be met, use the 0.05 level of significance to analyze this two-factor experiment.

R.197 Following are the numbers of persons who attended a "singles only" dance on 12 Saturdays: 172, 208, 169, 232, 123, 165, 197, 178, 221, 195, 209, and 182. Use the sign test based on Table V to test at the 0.05 level of significance whether or not the median of the population sampled is $\tilde{\mu} = 169$.

R.198 Use the signed-rank test based on Table VI to rework Exercise R.197.

R.199 Use the large-sample sign test to rework Exercise R.197.

R.200 On what statistic do we base our decision and for what values of the statistic do we reject the null hypothesis $\mu_1 = \mu_2$ if we have random samples of size $n_1 = 8$ and $n_2 = 11$ and are using the U test based on Table VII at the 0.05 level of significance to test the null hypothesis against the alternative hypothesis

(a) $\mu_1 \neq \mu_2$;

(b) $\mu_1 < \mu_2$;

(c) $\mu_1 > \mu_2$?

R.201 Assuming that the conditions underlying normal correlation analysis are met use the Fisher Z transformation to construct approximate 99% confidence intervals for ρ when

(a) $r = 0.45$ and $n = 18$;

(b) $r = -0.32$ and $n = 38$.

*R.202 An experiment yielded $r_{12} = 0.40$, $r_{13} = -0.90$, and $r_{23} = 0.90$. Explain why these figures cannot all be correct.

*R.203 The figures in the following 5×5 Latin square are the numbers of minutes engines E_1, E_2, E_3, E_4, and E_5, tuned up by mechanics M_1, M_2, M_3, M_4, and M_5, ran with a gallon of fuel A, B, C, D, or E:

	E_1	E_2	E_3	E_4	E_5
M_1	A 31	B 24	C 20	D 20	E 18
M_2	B 21	C 27	D 23	E 25	A 31
M_3	C 21	D 27	E 25	A 29	B 21
M_4	D 21	E 25	A 33	B 25	C 22
M_5	E 21	A 37	B 24	C 24	D 20

Analyze this Latin square, using the 0.01 level of significance for each of the tests of significance.

R.204 Following are data on the percentage kill of two kinds of insecticides used against mosquitos:

Insecticide X: 41.9   46.9   44.6   43.9   42.0   44.0
                  41.0   43.1   39.0   45.2   44.6   42.0

Insecticide Y: 45.7   39.8   42.8   41.2   45.0   40.2
                  40.2   41.7   37.4   38.8   41.7   38.7

Use the U test based on Table VII to test at the 0.05 level of significance whether or not the two insecticides are on the average equally effective.

R.205 Use the large-sample U test to rework Exercise R.204.

R.206 Working a homework assignment, a student obtained $S_{xx} = 145.22$, $S_{xy} = -210.58$, and $S_{yy} = 287.45$ for a given set of paired data. Explain why there must be an error in these calculations.

R.207 Following are the numbers of burglaries committed in two cities on 22 days: 87 and 81, 83 and 80, 98 and 87, 114 and 86, 112 and 120, 77 and 102, 103 and 94, 116 and 81, 136 and 95, 156 and 158, 83 and 127, 105 and 104, 117 and 102, 86 and 100, 150 and 108, 119 and 124, 111 and 91, 137 and 103, 160 and 153, 121 and 140, 143 and 105, and 129 and 129. Use the large-sample sign test at the 0.05 level of significance to test whether or not $\tilde{\mu}_D = 0$, where $\tilde{\mu}_D$ is the median of the population of differences sampled.

*R.208 Use the large-sample signed-rank test to rework the preceding exercise.

R.209 To study the earnings of faculty members in statistics and economics from speeches, writing, and consulting, a research worker interviewed four male assistant professors of economics, four male professors of economics, four female professors of statistics, four male associate professors of economics, four female assistant professors of statistics, and four female associate professors of statistics. If he combines the first and fifth groups, the second and third groups, and the fourth and sixth groups, performs an analysis of variance with $k = 3$ and $n = 8$, and gets a significant value of F, to what source of variation (sex, rank, or subject) can this be attributed?

R.210 With reference to Exercise R.209, explain why there is no way in which the research worker can use the data to test whether there is a significant difference that can be attributed to sex.

R.211 The manager of a restaurant wants to determine whether the sales of chicken dinners depend on how this entree is described on the menu. He has three kinds of menus printed, listing chicken dinners among the other entrees or featuring them as "Chef's Special," or as "Gourmet's Delight," and he intends to use each kind of menu on six different Sundays. Actually, the manager collects only the following data, showing the number of chicken dinners sold on 12 Sundays:

Listed among
other entrees: 76   94   85   77

Featured as
Chef's Special: 109   117   102   92   115

Featured as
Gourmet's Delight: 100   83   102

Perform a one-way analysis of variance at the 0.05 level of significance.

R.212 Following are the batting averages, x, and home runs hit, y, by a random sample of 15 major league baseball players during the first half of the season:

x	y
0.252	12
0.305	6
0.299	4
0.303	15
0.285	2
0.191	2
0.283	16
0.272	6
0.310	8
0.266	10
0.215	0
0.211	3
0.272	14
0.244	6
0.320	7

Calculate the rank-correlation coefficient and test at the 0.01 level of significance whether it is statistically significant.

*R.213 In the years 1986–1992 the profits of stockholders in petroleum and coal products corporations were 6.1, 7.7, 14.9, 14.6, 12.8, 7.7, and 2.4 percent of equity. Code the years $-3, -2, -1, 0, 1, 2,$ and 3, and fit a parabola by the method of least squares.

R.214 The following are the closing prices of a commodity (in dollars) on 20 consecutive trading days: 378, 379, 379, 378, 377, 376, 374, 374, 373, 373, 374, 375, 376, 376, 376, 375, 374, 374, 373, and 374. Test for randomness at the 0.01 level of significance.

R.215 The following are the scores of 16 golfers on the first two days of a tournament: 68 and 71, 73 and 76, 70 and 73, 74 and 71, 69 and 72, 72 and 74, 67 and 70, 72 and 68, 71 and 72, 73 and 74, 68 and 69, 70 and 72, 73 and 70, 71 and 75, 67 and 69, and 73 and 71. Use the sign test at the 0.05 level of significance to test whether on the average the hundreds of golfers participating in the tournament scored equally well on the first two days or whether they tended to score lower on the first day.

R.216 If $r = 0.28$ for the ages of a group of college students and their knowledge of foreign affairs, what percentage of the variation of their knowledge of foreign affairs can be attributed to differences in age?

R.217 State in each case whether you would expect a positive correlation, a negative correlation, or no correlation:

(a) Family expenditures on restaurant meals and family expenditures on property taxes.

(b) Daily low temperatures and closing prices of shares of an electronics firm.

(c) The number of hours that basketball players practice and their free-throw percentages.

(d) The number of passengers on a cruise ship and the number of empty cabins.

*R.218 In a multiple regression problem, the residual sum of squares is 926 and the total sum of squares is 1,702. Find the value of the multiple correlation coefficient.

R.219 Following are the number of inquiries a car dealer received in eight weeks about cars for sale, x, and cars for lease, y:

x	y
325	29
212	20
278	22
167	14
201	17
265	23
305	26
259	19

Calculate r.

R.220 With reference to Exercise R.219, calculate 95% confidence limits for ρ.

R.221 Following are the high school averages, x, and the first-year college grade-point averages, y, of seven students:

x	y
2.7	2.5
3.6	3.8
3.0	2.8
2.4	2.1
2.4	2.5
3.1	3.2
3.5	2.9

Fit a least-squares line that will enable us to predict y in terms of x, and use it to predict y for a student with $x = 2.8$.

R.222 With reference to Exercise R.221, construct a 95% confidence interval for the regression coefficient β.

*R.223 The following data pertain to the cosmic-ray doses measured at various altitudes:

Altitude (hundreds of feet) x	Dose rate (mrem/year) y
0.5	28
4.5	30
7.8	32
12.0	36
48.0	58
53.0	69

Use a computer or a graphing calculator to fit an exponential curve and use it to estimate the cosmic radiation at 6,000 feet.

*R.224 Market research shows that weekly sales of a new candy bar will be related to its price as follows:

Price (cents)	Weekly sales (number of bars)
50	232,000
55	194,000
60	169,000
65	157,000

Finding that the parabola $\hat{y} = 1{,}130{,}000 - 28{,}000x + 200x^2$ provides an excellent fit, the person conducting the study substitutes $x = 85$, gets $\hat{y} = 195{,}000$, and predicts that weekly sales will total 195,000 bars if the candy is priced at 85 cents. Comment on this argument.

R.225 If $r = 0.41$ for one set of data and $r = -0.92$ for another, compare the strengths of the two relationships.

R.226 The following sequence shows whether a television news program had at least 25% of a city's viewing audience, A, or less than 25%, L, on 36 consecutive weekday evenings:

$$L\,L\,L\,L\,A\,A\,L\,L\,L\,A\,L\,L\,L\,A\,A\,A\,A\,L$$

$$A\,L\,L\,L\,A\,A\,L\,L\,L\,L\,L\,A\,L\,L\,L\,L\,L\,A$$

Test for randomness at the 0.05 level of significance.

STATISTICAL TABLES

I   Normal-Curve Areas 569

II   Critical Values of t 571

III   Critical Values of χ^2 573

IV   Critical Values of F 575

V   Binomial Probabilities 577

VI   Critical Values of T 582

VII   Critical Values of U 583

VIII   Critical Values of u 585

IX   Critical Values of q_α 587

X   Values of $Z = \dfrac{1}{2} \cdot \ln \dfrac{1+r}{1-r}$ 589

XI   Binomial Coefficients 590

XII   Values of e^{-x} 591

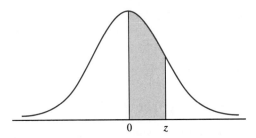

The entries in Table I are the probabilities that a random variable having the standard normal distribution will take on a value between 0 and z; they are given by the area of the tinted region in the figure shown above.

TABLE I Normal-Curve Areas

z	.00	.01	.02	.03	.04	.05	.06	.07	.08	.09
0.0	.0000	.0040	.0080	.0120	.0160	.0199	.0239	.0279	.0319	.0359
0.1	.0398	.0438	.0478	.0517	.0557	.0596	.0636	.0675	.0714	.0753
0.2	.0793	.0832	.0871	.0910	.0948	.0987	.1026	.1064	.1103	.1141
0.3	.1179	.1217	.1255	.1293	.1331	.1368	.1406	.1443	.1480	.1517
0.4	.1554	.1591	.1628	.1664	.1700	.1736	.1772	.1808	.1844	.1879
0.5	.1915	.1950	.1985	.2019	.2054	.2088	.2123	.2157	.2190	.2224
0.6	.2257	.2291	.2324	.2357	.2389	.2422	.2454	.2486	.2517	.2549
0.7	.2580	.2611	.2642	.2673	.2704	.2734	.2764	.2794	.2823	.2852
0.8	.2881	.2910	.2939	.2967	.2995	.3023	.3051	.3078	.3106	.3133
0.9	.3159	.3186	.3212	.3238	.3264	.3289	.3315	.3340	.3365	.3389
1.0	.3413	.3438	.3461	.3485	.3508	.3531	.3554	.3577	.3599	.3621
1.1	.3643	.3665	.3686	.3708	.3729	.3749	.3770	.3790	.3810	.3830
1.2	.3849	.3869	.3888	.3907	.3925	.3944	.3962	.3980	.3997	.4015
1.3	.4032	.4049	.4066	.4082	.4099	.4115	.4131	.4147	.4162	.4177
1.4	.4192	.4207	.4222	.4236	.4251	.4265	.4279	.4292	.4306	.4319
1.5	.4332	.4345	.4357	.4370	.4382	.4394	.4406	.4418	.4429	.4441
1.6	.4452	.4463	.4474	.4484	.4495	.4505	.4515	.4525	.4535	.4545
1.7	.4554	.4564	.4573	.4582	.4591	.4599	.4608	.4616	.4625	.4633
1.8	.4641	.4649	.4656	.4664	.4671	.4678	.4686	.4693	.4699	.4706
1.9	.4713	.4719	.4726	.4732	.4738	.4744	.4750	.4756	.4761	.4767
2.0	.4772	.4778	.4783	.4788	.4793	.4798	.4803	.4808	.4812	.4817
2.1	.4821	.4826	.4830	.4834	.4838	.4842	.4846	.4850	.4854	.4857
2.2	.4861	.4864	.4868	.4871	.4875	.4878	.4881	.4884	.4887	.4890
2.3	.4893	.4896	.4898	.4901	.4904	.4906	.4909	.4911	.4913	.4916
2.4	.4918	.4920	.4922	.4925	.4927	.4929	.4931	.4932	.4934	.4936
2.5	.4938	.4940	.4941	.4943	.4945	.4946	.4948	.4949	.4951	.4952
2.6	.4953	.4955	.4956	.4957	.4959	.4960	.4961	.4962	.4963	.4964
2.7	.4965	.4966	.4967	.4968	.4969	.4970	.4971	.4972	.4973	.4974
2.8	.4974	.4975	.4976	.4977	.4977	.4978	.4979	.4979	.4980	.4981
2.9	.4981	.4982	.4982	.4983	.4984	.4984	.4985	.4985	.4986	.4986
3.0	.4987	.4987	.4987	.4988	.4988	.4989	.4989	.4989	.4990	.4990

Also, for $z = 4.0, 5.0$, and 6.0, the areas are $0.49997, 0.4999997$, and 0.499999999.

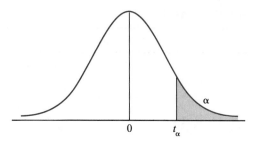

The entries in Table II are values for which the area to their right under the t distribution with given degrees of freedom (the area of the tinted region under the curve shown above) is equal to α.

TABLE II   Critical Values of $t^{\dagger}$

d.f.	$t_{.100}$	$t_{.050}$	$t_{.025}$	$t_{.010}$	$t_{.005}$	d.f.
1	3.078	6.314	12.706	31.821	63.657	1
2	1.886	2.920	4.303	6.965	9.925	2
3	1.638	2.353	3.182	4.541	5.841	3
4	1.533	2.132	2.776	3.747	4.604	4
5	1.476	2.015	2.571	3.365	4.032	5
6	1.440	1.943	2.447	3.143	3.707	6
7	1.415	1.895	2.365	2.998	3.499	7
8	1.397	1.860	2.306	2.896	3.355	8
9	1.383	1.833	2.262	2.821	3.250	9
10	1.372	1.812	2.228	2.764	3.169	10
11	1.363	1.796	2.201	2.718	3.106	11
12	1.356	1.782	2.179	2.681	3.055	12
13	1.350	1.771	2.160	2.650	3.012	13
14	1.345	1.761	2.145	2.624	2.977	14
15	1.341	1.753	2.131	2.602	2.947	15
16	1.337	1.746	2.120	2.583	2.921	16
17	1.333	1.740	2.110	2.567	2.898	17
18	1.330	1.734	2.101	2.552	2.878	18
19	1.328	1.729	2.093	2.539	2.861	19
20	1.325	1.725	2.086	2.528	2.845	20
21	1.323	1.721	2.080	2.518	2.831	21
22	1.321	1.717	2.074	2.508	2.819	22
23	1.319	1.714	2.069	2.500	2.807	23
24	1.318	1.711	2.064	2.492	2.797	24
25	1.316	1.708	2.060	2.485	2.787	25
26	1.315	1.706	2.056	2.479	2.779	26
27	1.314	1.703	2.052	2.473	2.771	27
28	1.313	1.701	2.048	2.467	2.763	28
29	1.311	1.699	2.045	2.462	2.756	29
inf.	1.282	1.645	1.960	2.326	2.576	inf.

†From Richard A. Johnson and Dean W. Wichern, *Applied Multivariate Statistical Analysis*, © 1982, p. 582. Adapted by permission of Prentice-Hall, Inc., Upper Saddle River, N.J.

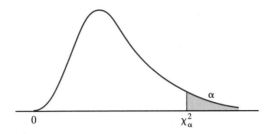

The entries in Table III are values for which the area to their right under the chi-square distribution with given degrees of freedom (the area of the tinted region under the curve shown above) is equal to α.

TABLE III Critical Values of $\chi^{2\dagger}$

d.f.	$\chi^2_{.995}$	$\chi^2_{.99}$	$\chi^2_{.975}$	$\chi^2_{.95}$	$\chi^2_{.05}$	$\chi^2_{.025}$	$\chi^2_{.01}$	$\chi^2_{.005}$	d.f.
1	.0000393	.000157	.000982	.00393	3.841	5.024	6.635	7.879	1
2	.0100	.0201	.0506	.103	5.991	7.378	9.210	10.597	2
3	.0717	.115	.216	.352	7.815	9.348	11.345	12.838	3
4	.207	.297	.484	.711	9.488	11.143	13.277	14.860	4
5	.412	.554	.831	1.145	11.070	12.832	15.086	16.750	5
6	.676	.872	1.237	1.635	12.592	14.449	16.812	18.548	6
7	.989	1.239	1.690	2.167	14.067	16.013	18.475	20.278	7
8	1.344	1.646	2.180	2.733	15.507	17.535	20.090	21.955	8
9	1.735	2.088	2.700	3.325	16.919	19.023	21.666	23.589	9
10	2.156	2.558	3.247	3.940	18.307	20.483	23.209	25.188	10
11	2.603	3.053	3.816	4.575	19.675	21.920	24.725	26.757	11
12	3.074	3.571	4.404	5.226	21.026	23.337	26.217	28.300	12
13	3.565	4.107	5.009	5.892	22.362	24.736	27.688	29.819	13
14	4.075	4.660	5.629	6.571	23.685	26.119	29.141	31.319	14
15	4.601	5.229	6.262	7.261	24.996	27.488	30.578	32.801	15
16	5.142	5.812	6.908	7.962	26.296	28.845	32.000	34.267	16
17	5.697	6.408	7.564	8.672	27.587	30.191	33.409	35.718	17
18	6.265	7.015	8.231	9.390	28.869	31.526	34.805	37.156	18
19	6.844	7.633	8.907	10.117	30.144	32.852	36.191	38.582	19
20	7.434	8.260	9.591	10.851	31.410	34.170	37.566	39.997	20
21	8.034	8.897	10.283	11.591	32.671	35.479	38.932	41.401	21
22	8.643	9.542	10.982	12.338	33.924	36.781	40.289	42.796	22
23	9.260	10.196	11.689	13.091	35.172	38.076	41.638	44.181	23
24	9.886	10.856	12.401	13.848	36.415	39.364	42.980	45.558	24
25	10.520	11.524	13.120	14.611	37.652	40.646	44.314	46.928	25
26	11.160	12.198	13.844	15.379	38.885	41.923	45.642	48.290	26
27	11.808	12.879	14.573	16.151	40.113	43.194	46.963	49.645	27
28	12.461	13.565	15.308	16.928	41.337	44.461	48.278	50.993	28
29	13.121	14.256	16.047	17.708	42.557	45.722	49.588	52.336	29
30	13.787	14.953	16.791	18.493	43.773	46.979	50.892	53.672	30

$\dagger$Based on Table 8 of *Biometrika Tables for Statisticians, Volume I* (Cambridge University Press, 1954), by permission of the *Biometrika* trustees.

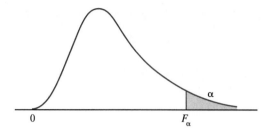

The entries in Table IV are values for which the area to their right under the F distribution with given degrees of freedom (the area of the tinted region under the curve shown above) is equal to α.

TABLE IV Critical Values of F†

Values of $F_{0.05}$

	Degrees of freedom for numerator																		
	1	2	3	4	5	6	7	8	9	10	12	15	20	24	30	40	60	120	∞
1	161	200	216	225	230	234	237	239	241	242	244	246	248	249	250	251	252	253	254
2	18.5	19.0	19.2	19.2	19.3	19.3	19.4	19.4	19.4	19.4	19.4	19.4	19.4	19.5	19.5	19.5	19.5	19.5	19.5
3	10.1	9.55	9.28	9.12	9.01	8.94	8.89	8.85	8.81	8.79	8.74	8.70	8.66	8.64	8.62	8.59	8.57	8.55	8.53
4	7.71	6.94	6.59	6.39	6.26	6.16	6.09	6.04	6.00	5.96	5.91	5.86	5.80	5.77	5.75	5.72	5.69	5.66	5.63
5	6.61	5.79	5.41	5.19	5.05	4.95	4.88	4.82	4.77	4.74	4.68	4.62	4.56	4.53	4.50	4.46	4.43	4.40	4.37
6	5.99	5.14	4.76	4.53	4.39	4.28	4.21	4.15	4.10	4.06	4.00	3.94	3.87	3.84	3.81	3.77	3.74	3.70	3.67
7	5.59	4.74	4.35	4.12	3.97	3.87	3.79	3.73	3.68	3.64	3.57	3.51	3.44	3.41	3.38	3.34	3.30	3.27	3.23
8	5.32	4.46	4.07	3.84	3.69	3.58	3.50	3.44	3.39	3.35	3.28	3.22	3.15	3.12	3.08	3.04	3.01	2.97	2.93
9	5.12	4.26	3.86	3.63	3.48	3.37	3.29	3.23	3.18	3.14	3.07	3.01	2.94	2.90	2.86	2.83	2.79	2.75	2.71
10	4.96	4.10	3.71	3.48	3.33	3.22	3.14	3.07	3.02	2.98	2.91	2.85	2.77	2.74	2.70	2.66	2.62	2.58	2.54
11	4.84	3.98	3.59	3.36	3.20	3.09	3.01	2.95	2.90	2.85	2.79	2.72	2.65	2.61	2.57	2.53	2.49	2.45	2.40
12	4.75	3.89	3.49	3.26	3.11	3.00	2.91	2.85	2.80	2.75	2.69	2.62	2.54	2.51	2.47	2.43	2.38	2.34	2.30
13	4.67	3.81	3.41	3.18	3.03	2.92	2.83	2.77	2.71	2.67	2.60	2.53	2.46	2.42	2.38	2.34	2.30	2.25	2.21
14	4.60	3.74	3.34	3.11	2.96	2.85	2.76	2.70	2.65	2.60	2.53	2.46	2.39	2.35	2.31	2.27	2.22	2.18	2.13
15	4.54	3.68	3.29	3.06	2.90	2.79	2.71	2.64	2.59	2.54	2.48	2.40	2.33	2.29	2.25	2.20	2.16	2.11	2.07
16	4.49	3.63	3.24	3.01	2.85	2.74	2.66	2.59	2.54	2.49	2.42	2.35	2.28	2.24	2.19	2.15	2.11	2.06	2.01
17	4.45	3.59	3.20	2.96	2.81	2.70	2.61	2.55	2.49	2.45	2.38	2.31	2.23	2.19	2.15	2.10	2.06	2.01	1.96
18	4.41	3.55	3.16	2.93	2.77	2.66	2.58	2.51	2.46	2.41	2.34	2.27	2.19	2.15	2.11	2.06	2.02	1.97	1.92
19	4.38	3.52	3.13	2.90	2.74	2.63	2.54	2.48	2.42	2.38	2.31	2.23	2.16	2.11	2.07	2.03	1.98	1.93	1.88
20	4.35	3.49	3.10	2.87	2.71	2.60	2.51	2.45	2.39	2.35	2.28	2.20	2.12	2.08	2.04	1.99	1.95	1.90	1.84
21	4.32	3.47	3.07	2.84	2.68	2.57	2.49	2.42	2.37	2.32	2.25	2.18	2.10	2.05	2.01	1.96	1.92	1.87	1.81
22	4.30	3.44	3.05	2.82	2.66	2.55	2.46	2.40	2.34	2.30	2.23	2.15	2.07	2.03	1.98	1.94	1.89	1.84	1.78
23	4.28	3.42	3.03	2.80	2.64	2.53	2.44	2.37	2.32	2.27	2.20	2.13	2.05	2.01	1.96	1.91	1.86	1.81	1.76
24	4.26	3.40	3.01	2.78	2.62	2.51	2.42	2.36	2.30	2.25	2.18	2.11	2.03	1.98	1.94	1.89	1.84	1.79	1.73
25	4.24	3.39	2.99	2.76	2.60	2.49	2.40	2.34	2.28	2.24	2.16	2.09	2.01	1.96	1.92	1.87	1.82	1.77	1.71
30	4.17	3.32	2.92	2.69	2.53	2.42	2.33	2.27	2.21	2.16	2.09	2.01	1.93	1.89	1.84	1.79	1.74	1.68	1.62
40	4.08	3.23	2.84	2.61	2.45	2.34	2.25	2.18	2.12	2.08	2.00	1.92	1.84	1.79	1.74	1.69	1.64	1.58	1.51
60	4.00	3.15	2.76	2.53	2.37	2.25	2.17	2.10	2.04	1.99	1.92	1.84	1.75	1.70	1.65	1.59	1.53	1.47	1.39
120	3.92	3.07	2.68	2.45	2.29	2.18	2.09	2.02	1.96	1.91	1.83	1.75	1.66	1.61	1.55	1.50	1.43	1.35	1.25
∞	3.84	3.00	2.60	2.37	2.21	2.10	2.01	1.94	1.88	1.83	1.75	1.67	1.57	1.52	1.46	1.39	1.32	1.22	1.00

†Reproduced from M. Merrington and C. M. Thompson, "Tables of percentage points of the inverted beta (F) distribution," *Biometrika*, vol. 33 (1943), by permission of the *Biometrika* trustees.

Degrees of freedom for denominator

TABLE IV Critical Values of F† (Continued)

Values of $F_{0.01}$

Degrees of freedom for numerator

Denominator	1	2	3	4	5	6	7	8	9	10	12	15	20	24	30	40	60	120	∞
1	4,052	5,000	5,403	5,625	5,764	5,859	5,928	5,982	6,023	6,056	6,106	6,157	6,209	6,235	6,261	6,287	6,313	6,339	6,366
2	98.5	99.0	99.2	99.2	99.3	99.3	99.4	99.4	99.4	99.4	99.4	99.4	99.4	99.5	99.5	99.5	99.5	99.5	99.5
3	34.1	30.8	29.5	28.7	28.2	27.9	27.7	27.5	27.3	27.2	27.1	26.9	26.7	26.6	26.5	26.4	26.3	26.2	26.1
4	21.2	18.0	16.7	16.0	15.5	15.2	15.0	14.8	14.7	14.5	14.4	14.2	14.0	13.9	13.8	13.7	13.7	13.6	13.5
5	16.3	13.3	12.1	11.4	11.0	10.7	10.5	10.3	10.2	10.1	9.89	9.72	9.55	9.47	9.38	9.29	9.20	9.11	9.02
6	13.7	10.9	9.78	9.15	8.75	8.47	8.26	8.10	7.98	7.87	7.72	7.56	7.40	7.31	7.23	7.14	7.05	6.97	6.88
7	12.2	9.55	8.45	7.85	7.46	7.19	6.99	6.84	6.72	6.62	6.47	6.31	6.16	6.07	5.99	5.91	5.82	5.74	5.65
8	11.3	8.65	7.59	7.01	6.63	6.37	6.18	6.03	5.91	5.81	5.67	5.52	5.36	5.28	5.20	5.12	5.03	4.95	4.86
9	10.6	8.02	6.99	6.42	6.06	5.80	5.61	5.47	5.35	5.26	5.11	4.96	4.81	4.73	4.65	4.57	4.48	4.40	4.31
10	10.0	7.56	6.55	5.99	5.64	5.39	5.20	5.06	4.94	4.85	4.71	4.56	4.41	4.33	4.25	4.17	4.08	4.00	3.91
11	9.65	7.21	6.22	5.67	5.32	5.07	4.89	4.74	4.63	4.54	4.40	4.25	4.10	4.02	3.94	3.86	3.78	3.69	3.60
12	9.33	6.93	5.95	5.41	5.06	4.82	4.64	4.50	4.39	4.30	4.16	4.01	3.86	3.78	3.70	3.62	3.54	3.45	3.36
13	9.07	6.70	5.74	5.21	4.86	4.62	4.44	4.30	4.19	4.10	3.96	3.82	3.66	3.59	3.51	3.43	3.34	3.25	3.17
14	8.86	6.51	5.56	5.04	4.70	4.46	4.28	4.14	4.03	3.94	3.80	3.66	3.51	3.43	3.35	3.27	3.18	3.09	3.00
15	8.68	6.36	5.42	4.89	4.56	4.32	4.14	4.00	3.89	3.80	3.67	3.52	3.37	3.29	3.21	3.13	3.05	2.96	2.87
16	8.53	6.23	5.29	4.77	4.44	4.20	4.03	3.89	3.78	3.69	3.55	3.41	3.26	3.18	3.10	3.02	2.93	2.84	2.75
17	8.40	6.11	5.19	4.67	4.34	4.10	3.93	3.79	3.68	3.59	3.46	3.31	3.16	3.08	3.00	2.92	2.83	2.75	2.65
18	8.29	6.01	5.09	4.58	4.25	4.01	3.84	3.71	3.60	3.51	3.37	3.23	3.08	3.00	2.92	2.84	2.75	2.66	2.57
19	8.19	5.93	5.01	4.50	4.17	3.94	3.77	3.63	3.52	3.43	3.30	3.15	3.00	2.92	2.84	2.76	2.67	2.58	2.49
20	8.10	5.85	4.94	4.43	4.10	3.87	3.70	3.56	3.46	3.37	3.23	3.09	2.94	2.86	2.78	2.69	2.61	2.52	2.42
21	8.02	5.78	4.87	4.37	4.04	3.81	3.64	3.51	3.40	3.31	3.17	3.03	2.88	2.80	2.72	2.64	2.55	2.46	2.36
22	7.95	5.72	4.82	4.31	3.99	3.76	3.59	3.45	3.35	3.26	3.12	2.98	2.83	2.75	2.67	2.58	2.50	2.40	2.31
23	7.88	5.66	4.76	4.26	3.94	3.71	3.54	3.41	3.30	3.21	3.07	2.93	2.78	2.70	2.62	2.54	2.45	2.35	2.26
24	7.82	5.61	4.72	4.22	3.90	3.67	3.50	3.36	3.26	3.17	3.03	2.89	2.74	2.66	2.58	2.49	2.40	2.31	2.21
25	7.77	5.57	4.68	4.18	3.86	3.63	3.46	3.32	3.22	3.13	2.99	2.85	2.70	2.62	2.53	2.45	2.36	2.27	2.17
30	7.56	5.39	4.51	4.02	3.70	3.47	3.30	3.17	3.07	2.98	2.84	2.70	2.55	2.47	2.39	2.30	2.21	2.11	2.01
40	7.31	5.18	4.31	3.83	3.51	3.29	3.12	2.99	2.89	2.80	2.66	2.52	2.37	2.29	2.20	2.11	2.02	1.92	1.80
60	7.08	4.98	4.13	3.65	3.34	3.12	2.95	2.82	2.72	2.63	2.50	2.35	2.20	2.12	2.03	1.94	1.84	1.73	1.60
120	6.85	4.79	3.95	3.48	3.17	2.96	2.79	2.66	2.56	2.47	2.34	2.19	2.03	1.95	1.86	1.76	1.66	1.53	1.38
∞	6.63	4.61	3.78	3.32	3.02	2.80	2.64	2.51	2.41	2.32	2.18	2.04	1.88	1.79	1.70	1.59	1.47	1.32	1.00

Degrees of freedom for denominator

†Reproduced from M. Merrington and C. M. Thompson, "Tables of percentage points of the inverted beta (F) distribution," *Biometrika*, vol. 33 (1943), by permission of the *Biometrika* trustees.

TABLE V   Binomial Probabilities

n	x	0.05	0.1	0.2	0.3	0.4	0.5	0.6	0.7	0.8	0.9	0.95
2	0	0.902	0.810	0.640	0.490	0.360	0.250	0.160	0.090	0.040	0.010	0.002
	1	0.095	0.180	0.320	0.420	0.480	0.500	0.480	0.420	0.320	0.180	0.095
	2	0.002	0.010	0.040	0.090	0.160	0.250	0.360	0.490	0.640	0.810	0.902
3	0	0.857	0.729	0.512	0.343	0.216	0.125	0.064	0.027	0.008	0.001	
	1	0.135	0.243	0.384	0.441	0.432	0.375	0.288	0.189	0.096	0.027	0.007
	2	0.007	0.027	0.096	0.189	0.288	0.375	0.432	0.441	0.384	0.243	0.135
	3		0.001	0.008	0.027	0.064	0.125	0.216	0.343	0.512	0.729	0.857
4	0	0.815	0.656	0.410	0.240	0.130	0.062	0.026	0.008	0.002		
	1	0.171	0.292	0.410	0.412	0.346	0.250	0.154	0.076	0.026	0.004	
	2	0.014	0.049	0.154	0.265	0.346	0.375	0.346	0.265	0.154	0.049	0.014
	3		0.004	0.026	0.076	0.154	0.250	0.346	0.412	0.410	0.292	0.171
	4			0.002	0.008	0.026	0.062	0.130	0.240	0.410	0.656	0.815
5	0	0.774	0.590	0.328	0.168	0.078	0.031	0.010	0.002			
	1	0.204	0.328	0.410	0.360	0.259	0.156	0.077	0.028	0.006		
	2	0.021	0.073	0.205	0.309	0.346	0.312	0.230	0.132	0.051	0.008	0.001
	3	0.001	0.008	0.051	0.132	0.230	0.312	0.346	0.309	0.205	0.073	0.021
	4			0.006	0.028	0.077	0.156	0.259	0.360	0.410	0.328	0.204
	5				0.002	0.010	0.031	0.078	0.168	0.328	0.590	0.774
6	0	0.735	0.531	0.262	0.118	0.047	0.016	0.004	0.001			
	1	0.232	0.354	0.393	0.303	0.187	0.094	0.037	0.010	0.002		
	2	0.031	0.098	0.246	0.324	0.311	0.234	0.138	0.060	0.015	0.001	
	3	0.002	0.015	0.082	0.185	0.276	0.312	0.276	0.185	0.082	0.015	0.002
	4		0.001	0.015	0.060	0.138	0.234	0.311	0.324	0.246	0.098	0.031
	5			0.002	0.010	0.037	0.094	0.187	0.303	0.393	0.354	0.232
	6				0.001	0.004	0.016	0.047	0.118	0.262	0.531	0.735
7	0	0.698	0.478	0.210	0.082	0.028	0.008	0.002				
	1	0.257	0.372	0.367	0.247	0.131	0.055	0.017	0.004			
	2	0.041	0.124	0.275	0.318	0.261	0.164	0.077	0.025	0.004		
	3	0.004	0.023	0.115	0.227	0.290	0.273	0.194	0.097	0.029	0.003	
	4		0.003	0.029	0.097	0.194	0.273	0.290	0.227	0.115	0.023	0.004
	5			0.004	0.025	0.077	0.164	0.261	0.318	0.275	0.124	0.041
	6				0.004	0.017	0.055	0.131	0.247	0.367	0.372	0.257
	7					0.002	0.008	0.028	0.082	0.210	0.478	0.698
8	0	0.663	0.430	0.168	0.058	0.017	0.004	0.001				
	1	0.279	0.383	0.336	0.198	0.090	0.031	0.008	0.001			
	2	0.051	0.149	0.294	0.296	0.209	0.109	0.041	0.010	0.001		
	3	0.005	0.033	0.147	0.254	0.279	0.219	0.124	0.047	0.009		
	4		0.005	0.046	0.136	0.232	0.273	0.232	0.136	0.046	0.005	
	5			0.009	0.047	0.124	0.219	0.279	0.254	0.147	0.033	0.005
	6			0.001	0.010	0.041	0.109	0.209	0.296	0.294	0.149	0.051
	7				0.001	0.008	0.031	0.090	0.198	0.336	0.383	0.279
	8					0.001	0.004	0.017	0.058	0.168	0.430	0.663

All values omitted in this table are 0.0005 or less.

TABLE V Binomial Probabilities (*Continued*)

n	x	0.05	0.1	0.2	0.3	0.4	0.5	0.6	0.7	0.8	0.9	0.95
9	0	0.630	0.387	0.134	0.040	0.010	0.002					
	1	0.299	0.387	0.302	0.156	0.060	0.018	0.004				
	2	0.063	0.172	0.302	0.267	0.161	0.070	0.021	0.004			
	3	0.008	0.045	0.176	0.267	0.251	0.164	0.074	0.021	0.003		
	4	0.001	0.007	0.066	0.172	0.251	0.246	0.167	0.074	0.017	0.001	
	5		0.001	0.017	0.074	0.167	0.246	0.251	0.172	0.066	0.007	0.001
	6			0.003	0.021	0.074	0.164	0.251	0.267	0.176	0.045	0.008
	7				0.004	0.021	0.070	0.161	0.267	0.302	0.172	0.063
	8					0.004	0.018	0.060	0.156	0.302	0.387	0.299
	9						0.002	0.010	0.040	0.134	0.387	0.630
10	0	0.599	0.349	0.107	0.028	0.006	0.001					
	1	0.315	0.387	0.268	0.121	0.040	0.010	0.002				
	2	0.075	0.194	0.302	0.233	0.121	0.044	0.011	0.001			
	3	0.010	0.057	0.201	0.267	0.215	0.117	0.042	0.009	0.001		
	4	0.001	0.011	0.088	0.200	0.251	0.205	0.111	0.037	0.006		
	5		0.001	0.026	0.103	0.201	0.246	0.201	0.103	0.026	0.001	
	6			0.006	0.037	0.111	0.205	0.251	0.200	0.088	0.011	0.001
	7			0.001	0.009	0.042	0.117	0.215	0.267	0.201	0.057	0.010
	8				0.001	0.011	0.044	0.121	0.233	0.302	0.194	0.075
	9					0.002	0.010	0.040	0.121	0.268	0.387	0.315
	10						0.001	0.006	0.028	0.107	0.349	0.599
11	0	0.569	0.314	0.086	0.020	0.004						
	1	0.329	0.384	0.236	0.093	0.027	0.005	0.001				
	2	0.087	0.213	0.295	0.200	0.089	0.027	0.005	0.001			
	3	0.014	0.071	0.221	0.257	0.177	0.081	0.023	0.004			
	4	0.001	0.016	0.111	0.220	0.236	0.161	0.070	0.017	0.002		
	5		0.002	0.039	0.132	0.221	0.226	0.147	0.057	0.010		
	6			0.010	0.057	0.147	0.226	0.221	0.132	0.039	0.002	
	7			0.002	0.017	0.070	0.161	0.236	0.220	0.111	0.016	0.001
	8				0.004	0.023	0.081	0.177	0.257	0.221	0.071	0.014
	9				0.001	0.005	0.027	0.089	0.200	0.295	0.213	0.087
	10					0.001	0.005	0.027	0.093	0.236	0.384	0.329
	11							0.004	0.020	0.086	0.314	0.569
12	0	0.540	0.282	0.069	0.014	0.002						
	1	0.341	0.377	0.206	0.071	0.017	0.003					
	2	0.099	0.230	0.283	0.168	0.064	0.016	0.002				
	3	0.017	0.085	0.236	0.240	0.142	0.054	0.012	0.001			
	4	0.002	0.021	0.133	0.231	0.213	0.121	0.042	0.008	0.001		
	5		0.004	0.053	0.158	0.227	0.193	0.101	0.029	0.003		
	6			0.016	0.079	0.177	0.226	0.177	0.079	0.016		
	7			0.003	0.029	0.101	0.193	0.227	0.158	0.053	0.004	
	8			0.001	0.008	0.042	0.121	0.213	0.231	0.133	0.021	0.002
	9				0.001	0.012	0.054	0.142	0.240	0.236	0.085	0.017
	10					0.002	0.016	0.064	0.168	0.283	0.230	0.099
	11						0.003	0.017	0.071	0.206	0.377	0.341
	12							0.002	0.014	0.069	0.282	0.540

TABLE V Binomial Probabilities (*Continued*)

n	x	0.05	0.1	0.2	0.3	0.4	0.5	0.6	0.7	0.8	0.9	0.95
13	0	0.513	0.254	0.055	0.010	0.001						
	1	0.351	0.367	0.179	0.054	0.011	0.002					
	2	0.111	0.245	0.268	0.139	0.045	0.010	0.001				
	3	0.021	0.100	0.246	0.218	0.111	0.035	0.006	0.001			
	4	0.003	0.028	0.154	0.234	0.184	0.087	0.024	0.003			
	5		0.006	0.069	0.180	0.221	0.157	0.066	0.014	0.001		
	6		0.001	0.023	0.103	0.197	0.209	0.131	0.044	0.006		
	7			0.006	0.044	0.131	0.209	0.197	0.103	0.023	0.001	
	8			0.001	0.014	0.066	0.157	0.221	0.180	0.069	0.006	
	9				0.003	0.024	0.087	0.184	0.234	0.154	0.028	0.003
	10				0.001	0.006	0.035	0.111	0.218	0.246	0.100	0.021
	11					0.001	0.010	0.045	0.139	0.268	0.245	0.111
	12						0.002	0.011	0.054	0.179	0.367	0.351
	13							0.001	0.010	0.055	0.254	0.513
14	0	0.488	0.229	0.044	0.007	0.001						
	1	0.359	0.356	0.154	0.041	0.007	0.001					
	2	0.123	0.257	0.250	0.113	0.032	0.006	0.001				
	3	0.026	0.114	0.250	0.194	0.085	0.022	0.003				
	4	0.004	0.035	0.172	0.229	0.155	0.061	0.014	0.001			
	5		0.008	0.086	0.196	0.207	0.122	0.041	0.007			
	6		0.001	0.032	0.126	0.207	0.183	0.092	0.023	0.002		
	7			0.009	0.062	0.157	0.209	0.157	0.062	0.009		
	8			0.002	0.023	0.092	0.183	0.207	0.126	0.032	0.001	
	9				0.007	0.041	0.122	0.207	0.196	0.086	0.008	
	10				0.001	0.014	0.061	0.155	0.229	0.172	0.035	0.004
	11					0.003	0.022	0.085	0.194	0.250	0.114	0.026
	12					0.001	0.006	0.032	0.113	0.250	0.257	0.123
	13						0.001	0.007	0.041	0.154	0.356	0.359
	14							0.001	0.007	0.044	0.229	0.488
15	0	0.463	0.206	0.035	0.005							
	1	0.366	0.343	0.132	0.031	0.005						
	2	0.135	0.267	0.231	0.092	0.022	0.003					
	3	0.031	0.129	0.250	0.170	0.063	0.014	0.002				
	4	0.005	0.043	0.188	0.219	0.127	0.042	0.007	0.001			
	5	0.001	0.010	0.103	0.206	0.186	0.092	0.024	0.003			
	6		0.002	0.043	0.147	0.207	0.153	0.061	0.012	0.001		
	7			0.014	0.081	0.177	0.196	0.118	0.035	0.003		
	8			0.003	0.035	0.118	0.196	0.177	0.081	0.014		
	9			0.001	0.012	0.061	0.153	0.207	0.147	0.043	0.002	
	10				0.003	0.024	0.092	0.186	0.206	0.103	0.010	0.001
	11				0.001	0.007	0.042	0.127	0.219	0.188	0.043	0.005
	12					0.002	0.014	0.063	0.170	0.250	0.129	0.031
	13						0.003	0.022	0.092	0.231	0.267	0.135
	14							0.005	0.031	0.132	0.343	0.366
	15								0.005	0.035	0.206	0.463

TABLE V   Binomial Probabilities (*Continued*)

n	x	0.05	0.1	0.2	0.3	0.4	0.5	0.6	0.7	0.8	0.9	0.95
16	0	0.440	0.185	0.028	0.003							
	1	0.371	0.329	0.113	0.023	0.003						
	2	0.146	0.275	0.211	0.073	0.015	0.002					
	3	0.036	0.142	0.246	0.146	0.047	0.009	0.001				
	4	0.006	0.051	0.200	0.204	0.101	0.028	0.004				
	5	0.001	0.014	0.120	0.210	0.162	0.067	0.014	0.001			
	6		0.003	0.055	0.165	0.198	0.122	0.039	0.006			
	7			0.020	0.101	0.189	0.175	0.084	0.019	0.001		
	8			0.006	0.049	0.142	0.196	0.142	0.049	0.006		
	9			0.001	0.019	0.084	0.175	0.189	0.101	0.020		
	10				0.006	0.039	0.122	0.198	0.165	0.055	0.003	
	11				0.001	0.014	0.067	0.162	0.210	0.120	0.014	0.001
	12					0.004	0.028	0.101	0.204	0.200	0.051	0.006
	13					0.001	0.009	0.047	0.146	0.246	0.142	0.036
	14						0.002	0.015	0.073	0.211	0.275	0.146
	15							0.003	0.023	0.113	0.329	0.371
	16								0.003	0.028	0.185	0.440
17	0	0.418	0.167	0.023	0.002							
	1	0.374	0.315	0.096	0.017	0.002						
	2	0.158	0.280	0.191	0.058	0.010	0.001					
	3	0.041	0.156	0.239	0.125	0.034	0.005					
	4	0.008	0.060	0.209	0.187	0.080	0.018	0.002				
	5	0.001	0.017	0.136	0.208	0.138	0.047	0.008	0.001			
	6		0.004	0.068	0.178	0.184	0.094	0.024	0.003			
	7		0.001	0.027	0.120	0.193	0.148	0.057	0.009			
	8			0.008	0.064	0.161	0.185	0.107	0.028	0.002		
	9			0.002	0.028	0.107	0.185	0.161	0.064	0.008		
	10				0.009	0.057	0.148	0.193	0.120	0.027	0.001	
	11				0.003	0.024	0.094	0.184	0.178	0.068	0.004	
	12				0.001	0.008	0.047	0.138	0.208	0.136	0.017	0.001
	13					0.002	0.018	0.080	0.187	0.209	0.060	0.008
	14						0.005	0.034	0.125	0.239	0.156	0.041
	15						0.001	0.010	0.058	0.191	0.280	0.158
	16							0.002	0.017	0.096	0.315	0.374
	17								0.002	0.023	0.167	0.418
18	0	0.397	0.150	0.018	0.002							
	1	0.376	0.300	0.081	0.013	0.001						
	2	0.168	0.284	0.172	0.046	0.007	0.001					
	3	0.047	0.168	0.230	0.105	0.025	0.003					
	4	0.009	0.070	0.215	0.168	0.061	0.012	0.001				
	5	0.001	0.022	0.151	0.202	0.115	0.033	0.004				
	6		0.005	0.082	0.187	0.166	0.071	0.015	0.001			
	7		0.001	0.035	0.138	0.189	0.121	0.037	0.005			
	8			0.012	0.081	0.173	0.167	0.077	0.015	0.001		
	9			0.003	0.039	0.128	0.185	0.128	0.039	0.003		
	10			0.001	0.015	0.077	0.167	0.173	0.081	0.012		
	11				0.005	0.037	0.121	0.189	0.138	0.035	0.001	
	12				0.001	0.015	0.071	0.166	0.187	0.082	0.005	

TABLE V Binomial Probabilities (*Continued*)

n	x	0.05	0.1	0.2	0.3	0.4	0.5	0.6	0.7	0.8	0.9	0.95
	13					0.004	0.033	0.115	0.202	0.151	0.022	0.001
	14					0.001	0.012	0.061	0.168	0.215	0.070	0.009
	15						0.003	0.025	0.105	0.230	0.168	0.047
	16						0.001	0.007	0.046	0.172	0.284	0.168
	17							0.001	0.013	0.081	0.300	0.376
	18								0.002	0.018	0.150	0.397
19	0	0.377	0.135	0.014	0.001							
	1	0.377	0.285	0.068	0.009	0.001						
	2	0.179	0.285	0.154	0.036	0.005						
	3	0.053	0.180	0.218	0.087	0.017	0.002					
	4	0.011	0.080	0.218	0.149	0.047	0.007	0.001				
	5	0.002	0.027	0.164	0.192	0.093	0.022	0.002				
	6		0.007	0.095	0.192	0.145	0.052	0.008	0.001			
	7		0.001	0.044	0.153	0.180	0.096	0.024	0.002			
	8			0.017	0.098	0.180	0.144	0.053	0.008			
	9			0.005	0.051	0.146	0.176	0.098	0.022	0.001		
	10			0.001	0.022	0.098	0.176	0.146	0.051	0.005		
	11				0.008	0.053	0.144	0.180	0.098	0.017		
	12				0.002	0.024	0.096	0.180	0.153	0.044	0.001	
	13				0.001	0.008	0.052	0.145	0.192	0.095	0.007	
	14					0.002	0.022	0.093	0.192	0.164	0.027	0.002
	15					0.001	0.007	0.047	0.149	0.218	0.080	0.011
	16						0.002	0.017	0.087	0.218	0.180	0.053
	17							0.005	0.036	0.154	0.285	0.179
	18							0.001	0.009	0.068	0.285	0.377
	19								0.001	0.014	0.135	0.377
20	0	0.358	0.122	0.012	0.001							
	1	0.377	0.270	0.058	0.007							
	2	0.189	0.285	0.137	0.028	0.003						
	3	0.060	0.190	0.205	0.072	0.012	0.001					
	4	0.013	0.090	0.218	0.130	0.035	0.005					
	5	0.002	0.032	0.175	0.179	0.075	0.015	0.001				
	6		0.009	0.109	0.192	0.124	0.037	0.005				
	7		0.002	0.055	0.164	0.166	0.074	0.015	0.001			
	8			0.022	0.114	0.180	0.120	0.035	0.004			
	9			0.007	0.065	0.160	0.160	0.071	0.012			
	10			0.002	0.031	0.117	0.176	0.117	0.031	0.002		
	11				0.012	0.071	0.160	0.160	0.065	0.007		
	12				0.004	0.035	0.120	0.180	0.114	0.022		
	13				0.001	0.015	0.074	0.166	0.164	0.055	0.002	
	14					0.005	0.037	0.124	0.192	0.109	0.009	
	15					0.001	0.015	0.075	0.179	0.175	0.032	0.002
	16						0.005	0.035	0.130	0.218	0.090	0.013
	17						0.001	0.012	0.072	0.205	0.190	0.060
	18							0.003	0.028	0.137	0.285	0.189
	19								0.007	0.058	0.270	0.377
	20								0.001	0.012	0.122	0.358

TABLE VI   Critical Values of $T^{\dagger}$

n	$T_{0.10}$	$T_{0.05}$	$T_{0.02}$	$T_{0.01}$
4				
5	1			
6	2	1		
7	4	2	0	
8	6	4	2	0
9	8	6	3	2
10	11	8	5	3
11	14	11	7	5
12	17	14	10	7
13	21	17	13	10
14	26	21	16	13
15	30	25	20	16
16	36	30	24	19
17	41	35	28	23
18	47	40	33	28
19	54	46	38	32
20	60	52	43	37
21	68	59	49	43
22	75	66	56	49
23	83	73	62	55
24	92	81	69	61
25	101	90	77	68

†From F. Wilcoxon and R. A. Wilcox, *Some Rapid Approximate Statistical Procedures,* American Cyanamid Company, Pearl River, N.Y., 1964. Reproduced with permission of American Cyanamid Company.

TABLE VII   Critical Values of $U^{\dagger}$

Values of $U_{0.10}$

n_1 \ n_2	2	3	4	5	6	7	8	9	10	11	12	13	14	15
2				0	0	0	1	1	1	1	2	2	3	3
3		0	0	1	2	2	3	4	4	5	5	6	7	7
4		0	1	2	3	4	5	6	7	8	9	10	11	12
5	0	1	2	4	5	6	8	9	11	12	13	15	16	18
6	0	2	3	5	7	8	10	12	14	16	17	19	21	23
7	0	2	4	6	8	11	13	15	17	19	21	24	26	28
8	1	3	5	8	10	13	15	18	20	23	26	28	31	33
9	1	4	6	9	12	15	18	21	24	27	30	33	36	39
10	1	4	7	11	14	17	20	24	27	31	34	37	41	44
11	1	5	8	12	16	19	23	27	31	34	38	42	46	50
12	2	5	9	13	17	21	26	30	34	38	42	47	51	55
13	2	6	10	15	19	24	28	33	37	42	47	51	56	61
14	3	7	11	16	21	26	31	36	41	46	51	56	61	66
15	3	7	12	18	23	28	33	39	44	50	55	61	66	72

Values of $U_{0.05}$

n_1 \ n_2	2	3	4	5	6	7	8	9	10	11	12	13	14	15
2							0	0	0	0	1	1	1	1
3				0	1	1	2	2	3	3	4	4	5	5
4			0	1	2	3	4	4	5	6	7	8	9	10
5		0	1	2	3	5	6	7	8	9	11	12	13	14
6		1	2	3	5	6	8	10	11	13	14	16	17	19
7		1	3	5	6	8	10	12	14	16	18	20	22	24
8	0	2	4	6	8	10	13	15	17	19	22	24	26	29
9	0	2	4	7	10	12	15	17	20	23	26	28	31	34
10	0	3	5	8	11	14	17	20	23	26	29	33	36	39
11	0	3	6	9	13	16	19	23	26	30	33	37	40	44
12	1	4	7	11	14	18	22	26	29	33	37	41	45	49
13	1	4	8	12	16	20	24	28	33	37	41	45	50	54
14	1	5	9	13	17	22	26	31	36	40	45	50	55	59
15	1	5	10	14	19	24	29	34	39	44	49	54	59	64

†This table is based on Table 11.4 of D. B. Owen, *Handbook of Statistical Tables,* © 1962, U.S. Department of Energy. Published by Addison-Wesley Publishing Company, Inc., Reading Mass. Reprinted with permission of the publisher.

TABLE VII   Critical Values of U (*Continued*)

n_1 \ n_2	2	3	4	5	6	7	8	9	10	11	12	13	14	15
Values of $U_{0.02}$														
2												0	0	0
3						0	0	1	1	1	2	2	2	3
4				0	1	1	2	3	3	4	5	5	6	7
5			0	1	2	3	4	5	6	7	8	9	10	11
6			1	2	3	4	6	7	8	9	11	12	13	15
7		0	1	3	4	6	7	9	11	12	14	16	17	19
8		0	2	4	6	7	9	11	13	15	17	20	22	24
9		1	3	5	7	9	11	14	16	18	21	23	26	28
10		1	3	6	8	11	13	16	19	22	24	27	30	33
11		1	4	7	9	12	15	18	22	25	28	31	34	37
12		2	5	8	11	14	17	21	24	28	31	35	38	42
13	0	2	5	9	12	16	20	23	27	31	35	39	43	47
14	0	2	6	10	13	17	22	26	30	34	38	43	47	51
15	0	3	7	11	15	19	24	28	33	37	42	47	51	56

n_1 \ n_2	3	4	5	6	7	8	9	10	11	12	13	14	15
Values of $U_{0.01}$													
3							0	0	0	1	1	1	2
4			0	0	1	1	2	2	3	3	4	5	
5		0	1	1	2	3	4	5	6	7	7	8	
6	0	1	2	3	4	5	6	7	9	10	11	12	
7	0	1	3	4	6	7	9	10	12	13	15	16	
8	1	2	4	6	7	9	11	13	15	17	18	20	
9	0	1	3	5	7	9	11	13	16	18	20	22	24
10	0	2	4	6	9	11	13	16	18	21	24	26	29
11	0	2	5	7	10	13	16	18	21	24	27	30	33
12	1	3	6	9	12	15	18	21	24	27	31	34	37
13	1	3	7	10	13	17	20	24	27	31	34	38	42
14	1	4	7	11	15	18	22	26	30	34	38	42	46
15	2	5	8	12	16	20	24	29	33	37	42	46	51

TABLE VIII Critical Values of $u^\dagger$

	Values of $u_{0.025}$											
n_1 \ n_2	4	5	6	7	8	9	10	11	12	13	14	15
4		9	9									
5	9	10	10	11	11							
6	9	10	11	12	12	13	13	13	13			
7		11	12	13	13	14	14	14	14	15	15	15
8		11	12	13	14	14	15	15	16	16	16	16
9			13	14	14	15	16	16	16	17	17	18
10			13	14	15	16	16	17	17	18	18	18
11			13	14	15	16	17	17	18	19	19	19
12			13	14	16	16	17	18	19	19	20	20
13				15	16	17	18	19	19	20	20	21
14				15	16	17	18	19	20	20	21	22
15				15	16	18	18	19	20	21	22	22

	Values of $u'_{0.025}$												
n_1 \ n_2	2	3	4	5	6	7	8	9	10	11	12	13	14	15
2											2	2	2	2
3					2	2	2	2	2	2	2	2	2	3
4				2	2	2	3	3	3	3	3	3	3	3
5			2	2	3	3	3	3	3	4	4	4	4	4
6		2	2	3	3	3	3	4	4	4	4	5	5	5
7		2	2	3	3	3	4	4	5	5	5	5	5	6
8		2	3	3	3	4	4	5	5	5	6	6	6	6
9		2	3	3	4	4	5	5	5	6	6	6	7	7
10		2	3	3	4	5	5	5	6	6	7	7	7	7
11		2	3	4	4	5	5	6	6	7	7	7	8	8
12	2	2	3	4	4	5	6	6	7	7	7	8	8	8
13	2	2	3	4	5	5	6	6	7	7	8	8	9	9
14	2	2	3	4	5	5	6	7	7	8	8	9	9	9
15	2	3	3	4	5	6	6	7	7	8	8	9	9	10

$\dagger$This table is adapted, by permission, from F. S. Swed and C. Eisenhart, "Tables for testing randomness of grouping in a sequence of alternatives," *Annals of Mathematical Statistics,* Vol. 14.

TABLE VIII   Critical Values of u (*Continued*)

n_1 \ n_2	5	6	7	8	9	10	11	12	13	14	15
5		11									
6	11	12	13	13							
7		13	13	14	15	15	15				
8		13	14	15	15	16	16	17	17	17	
9			15	15	16	17	17	18	18	18	19
10			15	16	17	17	18	19	19	19	20
11			15	16	17	18	19	19	20	20	21
12				17	18	19	19	20	21	21	22
13				17	18	19	20	21	21	22	22
14				17	18	19	20	21	22	23	23
15					19	20	21	22	22	23	24

Values of $u_{0.005}$

n_1 \ n_2	3	4	5	6	7	8	9	10	11	12	13	14	15
3										2	2	2	2
4						2	2	2	2	2	2	2	3
5				2	2	2	2	3	3	3	3	3	3
6			2	2	2	3	3	3	3	3	3	4	4
7			2	2	3	3	3	3	4	4	4	4	4
8		2	2	3	3	3	3	4	4	4	5	5	5
9		2	2	3	3	3	4	4	5	5	5	5	6
10		2	3	3	3	4	4	5	5	5	5	6	6
11		2	3	3	4	4	5	5	5	6	6	6	7
12	2	2	3	3	4	4	5	5	6	6	6	7	7
13	2	2	3	3	4	5	5	5	6	6	7	7	7
14	2	2	3	4	4	5	5	6	6	7	7	7	8
15	2	3	3	4	4	5	6	6	7	7	7	8	8

Values of $u'_{0.005}$

TABLE IX Values of $q_\alpha{}^\dagger$

$\alpha = 0.05$

Degrees of freedom	\multicolumn{20}{c}{Number of treatments}
	3	4	5	6	7	8	9	10	11	12	13	14	15	16	20	24	28	32	36	40
1	27.0	32.8	37.1	40.4	43.1	45.4	47.4	49.1	50.6	52.0	53.2	54.3	55.4	56.3	59.6	62.1	64.2	66.0	67.6	68.9
2	8.33	9.80	10.9	11.7	12.4	13.0	13.5	14.0	14.4	14.8	15.1	15.4	15.7	15.9	16.8	17.5	18.0	18.5	18.9	19.3
3	5.91	6.83	7.50	8.04	8.48	8.85	9.18	9.46	9.72	9.95	10.2	10.4	10.5	10.7	11.2	11.7	12.1	12.4	12.6	12.9
4	5.04	5.76	6.29	6.71	7.05	7.35	7.60	7.83	8.03	8.21	8.37	8.53	8.66	8.79	9.23	9.58	9.88	10.1	10.3	10.5
5	4.60	5.22	5.67	6.03	6.33	6.58	6.80	7.00	7.17	7.32	7.47	7.60	7.72	7.83	8.21	8.51	8.76	8.98	9.17	9.33
6	4.34	4.90	5.31	5.63	5.90	6.12	6.32	6.49	6.65	6.79	6.92	7.03	7.14	7.24	7.59	7.86	8.09	8.28	8.45	8.60
7	4.17	4.68	5.06	5.36	5.61	5.82	6.00	6.16	6.30	6.43	6.55	6.66	6.76	6.85	7.17	7.42	7.63	7.81	7.97	8.11
8	4.04	4.53	4.89	5.17	5.40	5.60	5.77	5.92	6.05	6.18	6.29	6.39	6.48	6.57	6.87	7.11	7.31	7.48	7.63	7.76
9	3.95	4.42	4.76	5.02	5.24	5.43	5.60	5.74	5.87	5.98	6.09	6.19	6.28	6.36	6.64	6.87	7.06	7.22	7.36	7.49
10	3.88	4.33	4.65	4.91	5.12	5.31	5.46	5.60	5.72	5.83	5.94	6.03	6.11	6.19	6.47	6.69	6.87	7.02	7.16	7.28
11	3.82	4.26	4.57	4.82	5.03	5.20	5.35	5.49	5.61	5.71	5.81	5.90	5.98	6.06	6.33	6.54	6.71	6.86	6.99	7.11
12	3.77	4.20	4.51	4.75	4.95	5.12	5.27	5.40	5.51	5.62	5.71	5.80	5.88	5.95	6.21	6.41	6.59	6.73	6.86	6.97
13	3.74	4.15	4.45	4.69	4.89	5.05	5.19	5.32	5.43	5.53	5.63	5.71	5.79	5.86	6.11	6.31	6.48	6.62	6.74	6.85
14	3.70	4.11	4.41	4.64	4.83	4.99	5.13	5.25	5.36	5.46	5.55	5.64	5.71	5.79	6.03	6.22	6.39	6.53	6.65	6.75
15	3.67	4.08	4.37	4.60	4.78	4.94	5.08	5.20	5.31	5.40	5.49	5.57	5.65	5.72	5.96	6.15	6.31	6.45	6.56	6.67
16	3.65	4.05	4.33	4.56	4.74	4.90	5.03	5.15	5.26	5.35	5.44	5.52	5.59	5.66	5.90	6.08	6.24	6.37	6.49	6.59
17	3.63	4.02	4.30	4.52	4.71	4.86	4.99	5.11	5.21	5.31	5.39	5.47	5.54	5.61	5.84	6.03	6.18	6.31	6.43	6.53
18	3.61	4.00	4.28	4.50	4.67	4.82	4.96	5.07	5.17	5.27	5.35	5.43	5.50	5.57	5.79	5.98	6.13	6.26	6.37	6.47
19	3.59	3.98	4.25	4.47	4.65	4.79	4.92	5.04	5.14	5.23	5.32	5.39	5.46	5.53	5.75	5.93	6.08	6.21	6.32	6.42
20	3.58	3.96	4.23	4.45	4.62	4.77	4.90	5.01	5.11	5.20	5.28	5.36	5.43	5.49	5.71	5.89	6.04	6.17	6.28	6.37
24	3.53	3.90	4.17	4.37	4.54	4.68	4.81	4.92	5.01	5.10	5.18	5.25	5.32	5.38	5.59	5.76	5.91	6.03	6.13	6.23
30	3.49	3.85	4.10	4.30	4.46	4.60	4.72	4.82	4.92	5.00	5.08	5.15	5.21	5.27	5.48	5.64	5.77	5.89	5.99	6.08
40	3.44	3.79	4.04	4.23	4.39	4.52	4.64	4.74	4.82	4.90	4.98	5.04	5.11	5.16	5.36	5.51	5.64	5.75	5.85	5.93
60	3.40	3.74	3.98	4.16	4.31	4.44	4.55	4.65	4.73	4.81	4.88	4.94	5.00	5.06	5.24	5.39	5.51	5.62	5.71	5.79
120	3.36	3.69	3.92	4.10	4.24	4.36	4.47	4.56	4.64	4.71	4.78	4.84	4.90	4.95	5.13	5.27	5.38	5.48	5.57	5.64
∞	3.31	3.63	3.86	4.03	4.17	4.29	4.39	4.47	4.55	4.62	4.69	4.74	4.80	4.85	5.01	5.14	5.25	5.35	5.43	5.50

†SOURCE: Adapted from H. L. Harter (1969), *Order Statistics and Their Use in Testing and Estimation. Vol 1: Tests Based on Range and Studentized Range of Samples From a Normal Population*, Aerospace Research Laboratories, U.S. Air Force.

TABLE IX Values of q_α (*Continued*)

$\alpha = 0.01$

Degrees of freedom	Number of treatments																			
	3	4	5	6	7	8	9	10	11	12	13	14	15	16	20	24	28	32	36	40
1	135	164	186	202	216	227	237	246	253	260	266	272	277	282	298	311	321	330	338	345
2	19.0	22.3	24.7	26.6	28.2	29.5	30.7	31.7	32.6	33.4	34.1	34.9	35.4	36.0	38.0	39.5	40.8	41.8	42.8	43.6
3	10.6	12.2	13.3	14.2	15.0	15.6	16.2	16.7	17.1	17.5	17.9	18.3	18.5	18.8	19.8	20.6	21.2	21.7	22.2	22.6
4	8.12	9.17	9.96	10.6	11.1	11.6	11.9	12.3	12.6	12.8	13.1	13.4	13.5	13.7	14.4	15.0	15.4	15.8	16.1	16.4
5	6.98	7.80	8.42	8.91	9.32	9.67	9.97	10.2	10.5	10.7	10.9	11.1	11.2	11.4	11.9	12.4	12.7	13.0	13.3	13.5
6	6.33	7.03	7.56	7.97	8.32	8.61	8.87	9.10	9.30	9.49	9.65	9.81	9.95	10.1	10.5	11.0	11.2	11.5	11.7	11.9
7	5.92	6.54	7.01	7.37	7.68	7.94	8.17	8.37	8.55	8.71	8.86	9.00	9.12	9.24	9.65	9.97	10.2	10.5	10.7	10.9
8	5.64	6.20	6.63	6.96	7.24	7.47	7.68	7.86	8.03	8.18	8.31	8.44	8.55	8.66	9.03	9.33	9.57	9.78	9.96	10.1
9	5.43	5.96	6.35	6.66	6.92	7.13	7.33	7.50	7.65	7.78	7.91	8.03	8.13	8.23	8.57	8.85	9.08	9.27	9.44	9.59
10	5.27	5.77	6.14	6.43	6.67	6.88	7.06	7.21	7.36	7.49	7.60	7.72	7.81	7.91	8.23	8.49	8.70	8.88	9.04	9.19
11	5.15	5.62	5.97	6.25	6.48	6.67	6.84	6.99	7.13	7.25	7.36	7.47	7.56	7.65	7.95	8.20	8.40	8.58	8.73	8.86
12	5.05	5.50	5.84	6.10	6.32	6.51	6.67	6.81	6.94	7.06	7.17	7.27	7.36	7.44	7.73	7.97	8.16	8.33	8.47	8.60
13	4.96	5.40	5.73	5.98	6.19	6.37	6.53	6.67	6.79	6.90	7.01	7.11	7.19	7.27	7.55	7.78	7.96	8.12	8.26	8.39
14	4.90	5.32	5.63	5.88	6.09	6.26	6.41	6.54	6.66	6.77	6.87	6.97	7.05	7.13	7.40	7.62	7.79	7.95	8.08	8.20
15	4.84	5.25	5.56	5.80	5.99	6.16	6.31	6.44	6.56	6.66	6.76	6.85	6.93	7.00	7.26	7.48	7.65	7.80	7.93	8.05
16	4.79	5.19	5.49	5.72	5.92	6.08	6.22	6.35	6.46	6.56	6.66	6.75	6.82	6.90	7.15	7.36	7.53	7.67	7.80	7.92
17	4.74	5.14	5.43	5.66	5.85	6.01	6.15	6.27	6.38	6.48	6.57	6.66	6.73	6.81	7.05	7.26	7.42	7.56	7.69	7.80
18	4.70	5.09	5.38	5.60	5.79	5.94	6.08	6.20	6.31	6.41	6.50	6.58	6.66	6.73	6.97	7.17	7.33	7.47	7.59	7.70
19	4.67	5.05	5.33	5.55	5.74	5.89	6.02	6.14	6.25	6.34	6.43	6.51	6.59	6.65	6.89	7.09	7.24	7.38	7.50	7.61
20	4.64	5.02	5.29	5.51	5.69	5.84	5.97	6.09	6.19	6.29	6.37	6.45	6.52	6.59	6.82	7.02	7.17	7.30	7.42	7.52
24	4.55	4.91	5.17	5.37	5.54	5.69	5.81	5.92	6.02	6.11	6.19	6.27	6.33	6.39	6.61	6.79	6.94	7.06	7.17	7.27
30	4.46	4.80	5.05	5.24	5.40	5.54	5.65	5.76	5.85	5.93	6.01	6.08	6.14	6.20	6.41	6.58	6.71	6.83	6.93	7.02
40	4.37	4.70	4.93	5.11	5.27	5.39	5.50	5.60	5.69	5.76	5.84	5.90	5.96	6.02	6.21	6.37	6.49	6.60	6.70	6.78
60	4.28	4.60	4.82	4.99	5.13	5.25	5.36	5.45	5.53	5.60	5.67	5.73	5.79	5.84	6.02	6.16	6.28	6.38	6.47	6.55
120	4.20	4.50	4.71	4.87	5.01	5.12	5.21	5.30	5.38	5.44	5.51	5.57	5.61	5.66	5.83	5.96	6.07	6.16	6.24	6.32
∞	4.12	4.40	4.60	4.76	4.88	4.99	5.08	5.16	5.23	5.29	5.35	5.40	5.45	5.49	5.65	5.77	5.87	5.95	6.03	6.09

TABLE X   Values of $Z = \dfrac{1}{2} \cdot \ln \dfrac{1 + r}{1 - r}$

r	.00	.01	.02	.03	.04	.05	.06	.07	.08	.09
0.0	0.000	0.010	0.020	0.030	0.040	0.050	0.060	0.070	0.080	0.090
0.1	0.100	0.110	0.121	0.131	0.141	0.151	0.161	0.172	0.182	0.192
0.2	0.203	0.213	0.224	0.234	0.245	0.255	0.266	0.277	0.288	0.299
0.3	0.310	0.321	0.332	0.343	0.354	0.365	0.377	0.388	0.400	0.412
0.4	0.424	0.436	0.448	0.460	0.472	0.485	0.497	0.510	0.523	0.536
0.5	0.549	0.563	0.576	0.590	0.604	0.618	0.633	0.648	0.662	0.678
0.6	0.693	0.709	0.725	0.741	0.758	0.775	0.793	0.811	0.829	0.848
0.7	0.867	0.887	0.908	0.929	0.950	0.973	0.996	1.020	1.045	1.071
0.8	1.099	1.127	1.157	1.188	1.221	1.256	1.293	1.333	1.376	1.422
0.9	1.472	1.528	1.589	1.658	1.738	1.832	1.946	2.092	2.298	2.647

For negative values of r put a minus sign in front of the corresponding Z's, and vice versa.

TABLE XI   Binomial Coefficients

n	$\binom{n}{0}$	$\binom{n}{1}$	$\binom{n}{2}$	$\binom{n}{3}$	$\binom{n}{4}$	$\binom{n}{5}$	$\binom{n}{6}$	$\binom{n}{7}$	$\binom{n}{8}$	$\binom{n}{9}$	$\binom{n}{10}$
0	1										
1	1	1									
2	1	2	1								
3	1	3	3	1							
4	1	4	6	4	1						
5	1	5	10	10	5	1					
6	1	6	15	20	15	6	1				
7	1	7	21	35	35	21	7	1			
8	1	8	28	56	70	56	28	8	1		
9	1	9	36	84	126	126	84	36	9	1	
10	1	10	45	120	210	252	210	120	45	10	1
11	1	11	55	165	330	462	462	330	165	55	11
12	1	12	66	220	495	792	924	792	495	220	66
13	1	13	78	286	715	1287	1716	1716	1287	715	286
14	1	14	91	364	1001	2002	3003	3432	3003	2002	1001
15	1	15	105	455	1365	3003	5005	6435	6435	5005	3003
16	1	16	120	560	1820	4368	8008	11440	12870	11440	8008
17	1	17	136	680	2380	6188	12376	19448	24310	24310	19448
18	1	18	153	816	3060	8568	18564	31824	43758	48620	43758
19	1	19	171	969	3876	11628	27132	50388	75582	92378	92378
20	1	20	190	1140	4845	15504	38760	77520	125970	167960	184756

For $r > 10$ it may be necessary to make use of the identity $\binom{n}{r} = \binom{n}{n-r}$.

TABLE XII Values of e^{-x}

x	e^{-x}	x	e^{-x}	x	e^{-x}	x	e^{-x}
0.0	1.0000	2.5	0.082085	5.0	0.006738	7.5	0.00055308
0.1	0.9048	2.6	0.074274	5.1	0.006097	7.6	0.00050045
0.2	0.8187	2.7	0.067206	5.2	0.005517	7.7	0.00045283
0.3	0.7408	2.8	0.060810	5.3	0.004992	7.8	0.00040973
0.4	0.6703	2.9	0.055023	5.4	0.004517	7.9	0.00037074
0.5	0.6065	3.0	0.049787	5.5	0.004087	8.0	0.00033546
0.6	0.5488	3.1	0.045049	5.6	0.003698	8.1	0.00030354
0.7	0.4966	3.2	0.040762	5.7	0.003346	8.2	0.00027465
0.8	0.4493	3.3	0.036883	5.8	0.003028	8.3	0.00024852
0.9	0.4066	3.4	0.033373	5.9	0.002739	8.4	0.00022487
1.0	0.3679	3.5	0.030197	6.0	0.002479	8.5	0.00020347
1.1	0.3329	3.6	0.027324	6.1	0.002243	8.6	0.00018411
1.2	0.3012	3.7	0.024724	6.2	0.002029	8.7	0.00016659
1.3	0.2725	3.8	0.022371	6.3	0.001836	8.8	0.00015073
1.4	0.2466	3.9	0.020242	6.4	0.001662	8.9	0.00013639
1.5	0.2231	4.0	0.018316	6.5	0.001503	9.0	0.00012341
1.6	0.2019	4.1	0.016573	6.6	0.001360	9.1	0.00011167
1.7	0.1827	4.2	0.014996	6.7	0.001231	9.2	0.00010104
1.8	0.1653	4.3	0.013569	6.8	0.001114	9.3	0.00009142
1.9	0.1496	4.4	0.012277	6.9	0.001008	9.4	0.00008272
2.0	0.1353	4.5	0.011109	7.0	0.000912	9.5	0.00007485
2.1	0.1225	4.6	0.010052	7.1	0.000825	9.6	0.00006773
2.2	0.1108	4.7	0.009095	7.2	0.000747	9.7	0.00006128
2.3	0.1003	4.8	0.008230	7.3	0.000676	9.8	0.00005545
2.4	0.0907	4.9	0.007447	7.4	0.000611	9.9	0.00005017

ANSWERS TO ODD NUMBERED EXERCISES

In exercises involving extensive calculations, the reader may well get answers differing somewhat from those given here due to rounding at various intermediate stages.

CHAPTER 1

1.1 The following are possibilities:
(a) It has been claimed that more than 70% of all persons over 35 have some form of life insurance. If 15 of 18 such persons selected at random have some form of life insurance, test the claim at the 0.05 level of significance.
(b) It has been claimed that more than 70% of all persons planning a trip to Europe will include a stopover in London. If 15 of 18 such persons selected at random will include a stopover in London, test the claim at the 0.05 level of significance.

1.3 (a) The results may be misleading because "Xerox copier" is often used as a generic term.
(b) Since Concorde flights are very expensive, persons disembarking from such flights can hardly be described as average vacationers.

1.5 (a) Many persons are reluctant to give honest answers about their health habits.
(b) Successful graduates are more likely to return the questionnaire than graduates that have not done so well.

1.7 (a) The statement is purely descriptive.
(b) The statement requires a generalization.
(c) The statement requires a generalization.
(d) The statement requires a generalization.

1.9 (a) The statement requires a generalization.
(b) The statement is a generalization.
(c) The statement is purely descriptive.
(d) The statement is purely descriptive.

1.11 (a) This statement is nonsense, of course.
(b) This is a generalization.

1.13 The data are nominal data.

1.15 (a) Interval data; (b) nominal data.

CHAPTER 2

2.1 The numbers of dots corresponding to 0, 1, 2, 3, and 6 are, respectively, 1, 6, 1, 1, and 1.

2.3 On 2, 3, 7, 11, 9, 5, and 3 days the pharmacy filled 4, 5, 6, 7, 8, 9, and 10 prescriptions for the medication.

2.5 The persons named 5 afghans, 2 bassets, 8 beagles, 1 bloodhound, 8 dachshunds, and 6 greyhounds.

2.7 There are 8 dots for A, 5 dots for B, 2 dots for C, 1 dot for D, and 1 for E.

2.9 There were 16 faulty connections, 9 missing parts, 5 broken parts, 3 paint defects, and 1 other defective.

2.11 (a) 36, 31, 37, 35, and 32;
(b) 415, 438, 450, and 477;
(c) 254, 254, 250, 253, and 259.

2.13
```
5 | 8  6
6 | 5  6  4  0  7
7 | 9  7  8  1  2  1  3  5
8 | 6  4  3  8  1  1  5  9  0
9 | 5
```

2.15
```
2 | 5 5 6 9
3 | 0 0 0 1 1 1 3 4 4 4 4 5 5 6 6 6 6 6 7 8 8 8 9 9 9
4 | 0 0 1 1 1 2 3 3 3 3 4 4 5 6 8 9
5 | 0 1 1 3 6
```

2.17
```
6 | 55  75  32
7 | 84  83  60  60  18
8 | 34  65  39  88  31  86  42  54  26  66  65
9 | 19  12  39  61  54  01
```

2.19
```
1.3 | 7
1.4 | 2  4  6  9
1.5 | 0  2  3  3  4  4  8  8  9
1.6 | 0  2  3  6  8
1.7 | 2
```

2.21
```
 8 | 4  8
 9 | 2  3  6  7  7  9
10 | 1  3  3  3  4  5  5  6  8  9
11 | 0  3  5
12 | 2  4  7
```

593

2.23 A convenient choice would be 260–279, 280–299, 300–319, 320–339, 340–359, 360–379, and 380–399.

2.25 Convenient choices would be 25.00–49.99, 50.00–74.99, 75.00–99.99, 100.00–124.99, 125.00–149.99, and 150.00–174.99, or 20.00–39.99, 40.00–59.99, 60.00–79.99, 80.00–99.99, 100.00–119.99, 120.00–139.99, 140.00–159.99, and 160.00–179.99.

2.27 (a) 5.0, 20.0, 35.0, 50.0, 65.0, and 80.0;
(b) 19.9, 34.9, 49.9, 64.9, 79.9, and 94.9;
(c) 4.95, 19.95, 34.95, 49.95, 64.95, 79.95, and 94.95;
(d) 15, 15, 15, 15, 15, and 15.

2.29 The values from 70.00 to 79.99 appear in two classes, and there is no provision for values from 50.00 to 59.99

2.31 There is no provision, for example, for cookies or Jello. Also, there is ambiguity about classifying, say, fruit cake, pie and ice cream, fruit with ice cream, etc.

2.33 (a) 20–24, 25–29, 30–34, 35–39, and 40–44;
(b) 22, 27, 32, 37, and 42;
(c) 5, 5, 5, 5, and 5.

2.35 (a) 60.0–74.9, 75.0–89.9, 90.0–104.9, 105.0–119.9, and 120.0–134.9;
(b) 67.45, 82.45, 97.45, 112.45, and 127.45.

2.37 The respective percentages are 2.5, 5.0, 37.5, 40.0, 10.0, and 5.0.

2.39 The respective class frequencies are 13, 14, 16, 12, 4, and 1.

2.41 The cumulative percentages corresponding to 19 or less, 24 or less, 29 or less, 34 or less, 39 or less, 44 or less, and 49 or less are, respectively, 0, 21.67, 45.00, 71.67, 91.67, 98.33, and 100.00.

2.43 The cumulative frequencies more than 0.49, more than 0.59, more than 0.69, more than 0.79, more than 0.89, more than 0.99, more than 1.09, more than 1.19, more than 1.29, more than 1.39, and more than 1.49 are, respectively, 120, 118, 112, 100, 62, 36, 23, 16, 8, 3, and 0.

2.45 The frequencies corresponding to 30–39, 40–49, 50–59, 60–69, 70–79, 80–89, and 90–99 are, respectively, 3, 7, 31, 56, 59, 16 and 8.

2.47 The cumulative percentages corresponding to 30 or more, 40 or more, 50 or more, 60 or more, 70 or more, 80 or more, 90 or more, and 100 or more are, respectively, 100.00, 98.33, 94.44, 77.22, 46.11, 13.33, 4.44, and 0.00.

2.49 The cumulative "less than" frequencies corresponding to 0, 2, 4, 6, 8, 10, 12, and 14 are, respectively, 0, 23, 41, 53, 62, 67, 69, and 70.

2.57 It might easily give a misleading impression because we tend to compare the areas of rectangles rather than their heights. Since the 80–99 class is twice as wide as the others, we could make the areas of the four rectangles proportional to the class frequencies by dividing the height of the 80–99 rectangle by 2.

2.59 The central angles are 106.1, 134.9, 21.5, 32.2, 8.6, and 56.7 degrees.

2.61 The central angles corresponding to the six classes are 31.2, 57.6, 120, 74.4, 40.8, and 36 degrees.

2.63 Since each dimension is doubled, the area is quadrupled, and we tend to compare areas. We can correct for this by making each dimension of the larger figure $\sqrt{2}$ (or approximately 1.41) times that of the smaller figure.

2.65 The class frequencies are 1, 5, 8, 33, 40, 30, 20, 11, and 2.

2.67 There is an upward linear trend, but the points are fairly widely scattered.

2.69 There is no distinct pattern.

2.71 The class frequencies are 2, 3, 1, 1, and 0 for the first row; 1, 4, 3, 1 and 1 for the second row; 0, 2, 4, 3, and 1 for the third row; 0, 1, 0, 3 and 1 for the fourth row; and 0, 0, 0, 3, and 1 for the fifth row.

2.73 The frequencies are 2, 3, 2, and 0 for the first row; 1, 4, 6, and 1 for the second row; and 0, 2, 5, and 4 for the third row.

CHAPTER 3

3.1 (a) The figures would constitute a population if we are interested only in the number of home runs hit in each game of the 1999 College World Series.
(b) The figures would constitute a sample if we are interested in the number of home runs hit in each game of the College World Series during the last ten years.

3.3 The population might consist of the lengths of the delays of all TWA flights departing from LAX during that period of time. Alternatively, the population might consist of the lengths of the delays of all flights departing from LAX during that period of time.

3.5 $\bar{x} = 97.5$.

3.7   On the average, the calibration is off by 0.04.

3.9 $\bar{x} = 0.9608$ rounded to four decimals.

3.11   (a) $\bar{x} = 75.3$;
(b) $\bar{x} = 75.3$. If we subtract a constant from each value, calculate the mean of the differences, and then add the same constant to the mean of the differences, we obtain the mean of the original values.

3.13   (a) At most 0.67;   (b) at most 0.86.

3.15   (a) 18;   (b) 6;   (c) 2; 96 and 192.

3.17 $\bar{x}_w = \$10,514.32$.

3.19   The weighted mean is 4.69% rounded to two decimals. This is the investor's actual return on the three investments.

3.21   The combined average score was 20.14 rounded to two decimals.

3.23   (a) 13;   (b) 9.5.

3.25   The median is 55.

3.27   142 minutes.

3.29   With 238 replaced by 832, the median becomes 336, so that the error is only $336 - 298.5 = 37.5$

3.31   The median is 118.5 grams.

3.35   Since the three midranges are 29.8, 30.0, and 30.3, the manufacturers of car C can use the midrange to substantiate the claim that their car performed best.

3.37   (a) The median, Q_1, and Q_3 positions are 16, 8, and 24;
(b) the median, Q_1, and Q_3 positions are 14.5, 7.5, and 21.5.

3.39   There are eight values to the left of the Q_1 position, eight values between the Q_1 and the median positions, eight values between the median and the Q_3 positions, and eight values to the right of the Q_3 position.

3.41   The smallest value is 41 and the largest value is 66.

3.43   The smallest value is 405 and the largest value is 440.

3.45   The smallest value is 33 and the largest value is 118.

3.47   The smallest value is 82 and the largest value is 148.

3.49   The smallest value is 0.53 and the largest value is 1.47.

3.53   The two hinges are 76.5 and 88.

3.55   (a) The mode is 7;
(b) the mode does not exist;
(c) the mode is 13.

3.57   The mode is 0.

3.59   The data become bimodal, with modes at 0 and 5.

3.61   There are two modes: beagle and dachshund.

3.63   The mean is 7.8 and the median is 6.83 rounded to two decimals.

3.65   The mean is 4.883 rounded to three decimals and the median is 4.89.

3.67   The mean is 47.638 rounded to three decimals and the median is 46.2.

3.69   Both rounded to three decimals, the mean is 50.167 and the median is 50.571.

3.73   (a) $x_1 + x_2 + x_3 + x_4 + x_5 + x_6$;
(b) $y_1 + y_2 + y_3 + y_4 + y_5$;
(c) $x_1 y_1 + x_2 y_2 + x_3 y_3$;
(d) $x_1 f_1 + x_2 f_2 + x_3 f_3 + x_4 f_4 + x_5 f_5 + x_6 f_6 + x_7 f_7 + x_8 f_8$;
(e) $x_3^2 + x_4^2 + x_5^2 + x_6^2 + x_7^2$;
(f) $(x_1 + y_1) + (x_2 + y_2) + (x_3 + y_3) + (x_4 + y_4)$.

3.75   (a) 16;   (b) 72.

3.77   (a) 4;   (b) 6;   (c) 30;   (d) 46;
(e) 37.

3.79   (a) 30;   (b) 30.

3.81   The equality is not true in general.

CHAPTER 4

4.1   (a) 1.7;   (b) 0.83 rounded to two decimals.

4.3   The range, 0.05, is fairly close to $2(0.0216) = 0.0432$.

4.5   The range, 8, is not very close to $3(2.183) = 6.549$.

4.7   28.

4.9   46.

4.11   (a) 1.75 rounded to two decimals;
(b) 1.75 rounded to two decimals.

4.13   (a) 0.245 rounded to three decimals;
(b) 0.245 rounded to three decimals.

4.15   (a) 0.207 rounded to three decimals;
(b) 20.7 rounded to one decimal. If each value is multiplied by the same constant, the value obtained for s must be divided by that constant.

4.19 11.625 rounded to three decimals.

4.21 (a) $\frac{35}{36}$;   (b) $\frac{143}{144}$.

4.23 (a) At least 75%;   (b) at least 96%.

4.25 At most 4% of the time.

4.27 (a) The applicant is in a relatively better position with respect to the first branch.
(b) The applicant is in a relatively better position with respect to the second branch.

4.29 The first golfer is relatively more consistent.

4.31 The second person's blood glucose level was relatively more variable.

4.33 (a) Approximately 9.90;
(b) approximately 9.90.

4.35 0.51 rounded to two decimals.

4.37 −0.076 rounded to three decimals.

4.39 $SK = -0.370$ rounded to three decimals. The distribution has a slight negative skewness.

4.41 The accident data are negatively skewed.

4.43 The data are positively skewed.

4.45 The distribution is a reverse J-shaped distribution; it is highly positively skewed.

4.47 The distribution is U-shaped.

REVIEW EXERCISES

R.1 The numbers 23 and 24 can go into the third and fourth classes; the distribution does not accommodate the numbers 36, 37, 38, and 39.

R.3 125, 123, 137, 130, 134, 138, 141, 144, 146, 146, 149, 143, 152, 152, 158, 150, 155, 167, and 161.

R.7 (a) 7.3125;   (b) 6;
(c) 5.70 rounded to two decimals;
(d) 0.69 rounded to two decimals.

R.9 (a) 0.04 | 5 6 7 8 9 9
      0.05 | 0 2 2 4 4 4 5 5 6 7 7 8 8 8
      0.06 | 0 1 2 2 3 3 5 6 7 8
      0.07 | 2 2
(b) 0.057, 0.052, and 0.0625;
(c) data are positively skewed.

R.11 (a) The data would constitute a population if the meteorologist wants to determine the average number of days in June that the maximum temperature in Palm Springs, California, exceeded 110 degrees in these ten years.
(b) The data would constitute a sample if the meteorologist wants to predict for future years in how many days in June the maximum temperature will exceed 110 degrees.

R.13 The following are possibilities:
(a) It has been claimed that more than 70% of all suits against health insurance companies are settled before they come to trial. If 15 of 18 such suits are settled before they came to trial, test the claim at the 0.05 level of significance.
(b) It has been claimed that more than 70% of all successful mystery novels are made into movies. If 15 of 18 such mystery novels selected at random were made into movies, test the claim at the 0.05 level of significance.

R.15 At least 88.9% rounded to one decimal.

R.17 (a) 11;   (b) 44;
(c) cannot be determined;
(d) cannot be determined.

R.19 42.55% rounded to two decimals.

R.21 The histograms are very similar, but the one of Figure 2.3 looks slightly more skewed.

R.23 (a) 9.5, 29.5, 49.5, 69.5, 89.5, and 109.5;
(b) 19.5, 39.5, 59.5, 79.5, and 99.5;
(c) 20.

R.25 (a) 17.1;   (b) 87.45;   (c) 292.41.

R.27 6.24.

R.29 There are other kinds of fibers and also shirts made of combinations of fibers.

R.31 (a) Cannot be determined;
(b) yes, the number in the fourth class;
(c) yes, the sum of the numbers in the second and third classes;
(d) cannot be determined.

R.33 (a) Referring to the practice as "unfair" is begging the question.
(b) The difference in opinion between persons having telephones and persons not having telephones may affect the result.

R.35 (a) 14.5, 29.5, 44.5, 59.5, 74.5, 89.5, 104.5, and 119.5;
(b) 22, 37, 52, 67, 82, 97, and 112;
(c) 15.

R.37 The cumulative "less than" frequencies are 0, 3, 17, 35, 61, 81, 93, and 100.

R.39

12	4
13	0 0 5
14	2 6 9
15	1 3 4 5 6 8 9
16	2 2 2 5
17	2 3
18	2
19	
20	4

R.41 At least 93.75%.

R.43 The frequencies are 11, 9, 5, 3, 1, and 1.

R.45 Approximately 25.57%.

R.47 It is assumed that the difference between A and B is as important as the differences between B and C, C and D, and D and F.

CHAPTER 5

5.1 AL can win 4 to 1, 4 to 2, or 4 to 3, and NL can win 4 to 3.

5.3 (a) In three cases; (b) in two cases.

5.5 There are four ways in which the store can sell two or three of the cheesecakes on the two days.

5.7 (a) In 6 ways; (b) in 9 ways.

5.9 24.

5.11 128.

5.13 (a) 4;   (b) 16;   (c) 12.

5.15 480.

5.17 4,096.

5.19 (a) True;   (b) false;   (c) true;
(d) false;   (e) true;   (f) false.

5.21 720.

5.23 5,040.

5.25 30,240.

5.27 (a) 24;   (b) 1,152.

5.29 (a) 60;   (b) 20;   (c) 120;
(d) If the n objects were all distinct, the number of arrangements would be $n!$. The duplicate objects cause overcounting by $r!$, and the overall number must be $n!/r!$.

5.31 66.

5.33 117,600.

5.35 (a) 55;   (b) 165.

5.37 (a) 6;   (b) 30;
(c) 20;   (d) $6 + 30 + 20 = 56$.

5.39 (a) 11,440;   (b) 1,287;
(c) 11,628;   (d) 1,365.

5.43 (a) 1/52;   (b) 3/13;
(c) 1/2;   (d) 3/13;   (e) 1/4.

5.45 (a) 11/850;   (b) 6/5,525;   (c) 16/5,525.

5.47 (a) 1/12;   (b) 1/9;   (c) 2/9.

5.49 (a) 5/24;   (b) 25/36;
(c) 7/72;   (d) 35/72.

5.51 (a) 37/75;   (b) 1/5;   (c) 16/75.

5.53 (a) 24/91;   (b) 45/91;   (c) 2/91.

5.55 1/2.

5.57 (a) 1/4;   (b) 1/28.

5.59 4/9.

5.61 0.75.

5.63 0.6875.

5.69 In either case there is a fifty-fifty chance of being right.

5.71 What happens in a single event cannot prove or refute a probability.

5.73 A minor surgery at hospital B is a better risk than a major surgery at hospital A.

5.75 If one person claims that the probability for rain is 0.60 and another person claims that the probability is 0.75, then rain would confirm both probabilities. Since 0.60 and 0.75 cannot both be right, this would make the whole concept of probability meaningless.

CHAPTER 6

6.1 (a) $U' = \{a, c, d, f, g\}$, the scholarship is awarded to Ms. Adams, Miss Clark, Mrs. Daily, Ms. Fuentes, or Ms. Gardner;
(b) $U \cap V = \{e, h\}$, the scholarship is awarded to Mr. Earl or Mr. Hall;
(c) $U \cup V' = \{a, b, c, d, e, h\}$, the scholarship is not awarded to Ms. Fuentes or Ms. Gardner.

6.3 (a) $\{(0, 2), (1, 1), (2, 0)\}$;   (b) $\{(0, 0), (1, 1)\}$;
(c) $\{(1, 1), (2, 1), (1, 2)\}$.

6.5 (a) There is one fewer professor than there are assistants.
(b) Altogether there are four professors and assistants.

(c) There are two assistants. K and L are mutually exclusive; K and M are not mutually exclusive; and L and M are not mutually exclusive.

6.7 (a) $T \cap U = \{(4, 1), (3, 2)\}$;
(b) $U \cap V = \{(4, 3)\}$;
(c) $V \cap T' = \{(3, 3), (4, 3)\}$.

6.9 (a) $K' = \{(0, 0), (1, 0), (2, 0), (3, 0), (0, 1),$ $(1, 1), (2, 1)\}$, at most one boat is rented out for the day;
(b) $L \cap M = \{(2, 1), (3, 0)\}$, at least two boats are in dry dock and any boat not in dry dock is rented out for the day.

6.11 (a) $\{A, D\}$;   (b) $\{C, E\}$;   (c) $\{B\}$.

6.15 (a) Regions 1 and 2 together represent the event that a graduate student speaks very good French.
(b) Regions 2 and 4 together represent the event that a graduate student is not studying at the Sourbonne.

6.17 (a) Regions 1 and 3 together represent the event that a flight arrives in Holland on time.
(b) Regions 3 and 4 together represent the event that a flight does not leave Denmark on time.
(c) Regions 2, 3, and 4 together represent the event that either a flight does not leave Denmark on time or does not arrive in Holland on time.

6.19 33.

6.21 (a) The car needs an engine overhaul, transmission repairs, and new tires.
(b) The car needs transmission repairs and new tires, but no engine overhaul.
(c) The car needs an engine overhaul but neither transmission repairs nor new tires.
(d) The car needs an engine overhaul and new tires.
(e) The car needs transmission repairs but no new tires.
(f) The car does not need an engine overhaul.

6.23 (a) They will run into bad weather and have problems with local authorities but no difficulties with their photographic equipment.
(b) They will have difficulties with their photographic equipment but will not run into bad weather nor have problems with local authorities.
(c) They will run into bad weather and have problems with local authorities.

(d) They will not run into bad weather, have no problems with local authorities, and have no difficulties with their photographic equipment.
(e) They will run into bad weather but have no problems with local authorities.
(f) They will have no problems with local authorities.

6.25 (a) 24;   (b) 16.

6.27 (a) The probability that there will not be enough capital for expansion.
(b) The probability that there will not be adequate transportation.
(c) The probability that there will be enough capital for expansion and/or adequate transportation.
(d) The probability that there will be enough capital for expansion and adequate transportation.

6.29 (a) $P(T')$;   (b) $P(T \cup U)$;
(c) $P(T' \cap U')$.

6.31 (a) Postulate 1;   (b) Postulate 2;
(c) Postulate 2;   (d) Postulate 3.

6.35 (a) $P(A \cap B') = 0.30$, $P(A' \cap B) = 0.30$, and $P(A' \cap B') = 0.40$;
(b) $P(A \cap B') = 0.50$ and $P(A' \cap B) = 0.50$.

6.37 (a) 0.36;   (b) 0.85;   (c) 0.15.

6.39 (a) The odds are 11 to 5;
(b) the odds are 7 to 2 against it.

6.41 (a) The probability is $\frac{34}{55}$;
(b) the probability is $\frac{1}{6}$ that we will get a meaningful word.

6.43 No.

6.45 (a) He feels that the probability is 1/4;
(b) he feels that the probability is less than 1/4.

6.47 The probabilities are consistent.

6.51 (a) 2/9;   (b) 1/9;   (c) 1/2.

6.53 (a) 0.38;   (b) 0.62;   (c) 0.47;   (d) 0.93.

6.55 (a) 7/9;   (b) 2/9;   (c) 2/15.

6.57 1/32, 5/32, 10/32, 10/32, 5/32, and 1/32.

6.59 0.41.

6.61 The odds are 7 to 3.

6.63 (a) The probability that either or both will be fired would be 1.07, which is an impossible value.
(b) The third of the probabilities cannot exceed the second.

6.65 (a) The probability that an astronaut who is a well-trained scientist is a member of the armed services.
(b) The probability that an astronaut who was not a test pilot is not a member of the armed services.
(c) The probability that an astronaut who was once a test pilot is a well-trained scientist but not a member of the armed services.
(d) The probability that an astronaut who is a well-trained scientist and was once a test pilot is a member of the armed services.

6.67 (a) The probability that a student who rates high on the social adjustment scale will score high in intelligence.
(b) The probability that a student who does not display neurotic tendencies will not rate high on the social adjustment scale.
(c) The probability that a student who does not rate high on the social adjustment scale will score high in intelligence but have neurotic tendencies.
(d) The probability that a student who scores high in intelligence and rates high on the social adjustment scale will not have neurotic tendencies.

6.69 (a) $P(M) = 0.6$ is the probability that the chosen applicant will be married.
(b) $P(E') = 0.7$ is the probability that the chosen applicant will have had no experience.
(c) $P(M \cap E) = 0.2$ is the probability that the chosen applicant will be married and have had some experience.
(d) $P(M' \cap E') = 0.3$ is the probability that the chosen applicant will not be married and not have had any experience.
(e) $P(E \mid M) = 1/3$ is the probability that the chosen applicant, who is married, will have had some experience.
(f) $P(M'|E') = 3/7$ is the probability that the chosen applicant, who has not had any experience, will not be married.

6.71 (a) 0.6;   (b) 0.5;   (c) 0.4;
(d) 0.1;   (e) 2/3;   (f) 0.2.

6.73 (a) 0.90;   (b) 0.96.

6.75 0.60.

6.77 (a) 1/16;   (b) 1/17.

6.79 Events A and C are independent.

6.81 (a) 0.42, 0.18, and 0.12.

6.83 Approximately 0.508.

6.85 0.03528.

6.87 Approximately 0.71.

6.89 0.447 rounded to three decimals.

6.91 0.679 rounded to three decimals.

6.93 3/4.

6.95 (a) 0.3999;
(b) 0.810 rounded to three decimals.

6.97 (a) $P(M) = 5P(Y) - 2$;   (b) 0.12.

CHAPTER 7

7.1 $0.25.

7.3 (a) $0.27 rounded to the nearest cent;
(b) no.

7.5 (a) $210,000 and $210,000;
(b) $228,000 and $192,000.

7.7 107 guests rounded up to the nearest integer.

7.9 $1,260.

7.11 1.87.

7.13 The probability is less than 0.25.

7.15 The expected number of mowings is greater than 10.

7.17 $89.

7.19 Since the average price per grab-bag is $3.50, it is not worthwhile to pay $4.

7.21 Her expectation is $120, which is not worth her time.

7.23 (a) He should go first to the barn.
(b) It does not matter where he goes first.

7.27 (a) To minimize the maximum loss, the tests should be discontinued.
(b) It does not matter where he goes first.

7.29 (a) The operation should be continued.
(b) The operation should not be continued.

7.31 (a) Continue the tests.
(b) Vote against it.

7.33 The mode, 17.

7.35 The midrange, $2,695.

REVIEW EXERCISES

R.49 (a) 0.60;   (b) 0.20;
(c) 0.92;   (d) 0.12.

R.51 (a) 0.38;   (b) 0.42;   (c) 0.50.

R.55 $\frac{3}{4} \le p < \frac{4}{5}$.

R.57 1,365.

R.59 0.63.

R.61 (a) $\frac{6}{7}$; 0.234 rounded to three decimals.

R.63 65,536.

R.65 (a) 165;   (b) 168.

R.67 0.014 rounded to three decimals.

R.69 (a) The probability is 7/8 that the driver will lose the race.
(b) The probability is 11/16 that at most two of the cards will be black.

R.71 (a) Events A and B are mutually exclusive;
(b) events A and B are not independent.

R.73 $39.

R.75 (a) 30, the mode;   (b) 31, the mean.

R.77 (a) 120;   (b) 360;   (c) 180.

R.79 The probabilities are not consistent.

R.81 (a) 0.187 rounded to three decimals;
(b) 0.998 rounded to three decimals.

R.83 Many persons would prefer a guaranteed 4.5% to a potentially risky 6.2%.

R.85 $P(A) = 3/4$.

R.87 0.153 rounded to three decimals.

R.89 (a) 24;   (b) 120.

CHAPTER 8

8.1 (a) No;   (b) no;   (c) yes.

8.3 (a) yes;   (b) no;   (c) yes.

8.5 .421875.

8.7 0.6561, compared to 0.656 in Table V.

8.9 (a) 0.250822656;   (b) 0.251.

8.11 (a) 0.649;   (b) 0.047.

8.13 (a) 0.984;   (b) 0.358;   (c) 0.639.

8.15 (a) 0.718;   (b) 0.069;   (c) 0.014.

8.17 0.000, 0.000, 0.000, 0.000, 0.000, 0.000, 0.001, 0.003, 0.014, 0.043, 0.103, 0.188, 0.250, 0.231, 0132, and 0.035.

8.19 (a) 0.1205;   (b) 0.1205.

8.21 (a) 0.7021;   (b) 0.0342.

8.23 (a) 0.0864;
(b) 0.0791 rounded to four decimals;
(c) 0.063.

8.25 0.509 rounded to three decimals.

8.27 0.317 rounded to three decimals.

8.29 (a) 33/112;   (b) 55/112;
(c) 22/112;   (d) 2/112.

8.31 (a) 5/6;   (b) 5/11;   (c) 17/22.

8.33 0.0922 rounded to four decimals.

8.35 (a) 0.280 rounded to three decimals;
(b) 0.276 rounded to three decimals.

8.37 (a) Conditions are not satisfied;
(b) conditions are satisfied;
(c) conditions are not satisfied.

8.39 0.024 rounded to three decimals.

8.41 0.02025 rounded to five decimals.

8.43 0.0228 rounded to four decimals.

8.45 Yes.

8.47 0.1820 rounded to four decimals.

8.49 0.00166 rounded to five decimals.

8.51 0.0790 rounded to four decimals.

8.53 0.0403 rounded to four decimals.

8.55 (a) 0.1798 rounded to four decimals;
(b) same.

8.57 (a) 1.1;   (b) 0.69 and 0.83;
(c) 0.69 and 0.83.

8.59 1.37 rounded to two decimals.

8.61 $\mu = 2$ and $\sigma = 1$.

8.63 (a) $\mu = 7.990$ and $\sigma = 1.292$, both rounded to three decimals;
(b) $\mu = 8$ and $\sigma = 1.265$ rounded to three decimals.

8.65 $\mu = 4.86$ and $\sigma = 1.884$ rounded to three decimals.

8.67 $\mu = 1.5$.

8.69 $\mu = \frac{15}{16}$.

8.71 $\mu = 2.5$ and $\mu = 2.5$.

8.73 4/7.

8.75 (a) $\frac{11}{12}$;   (b) $\frac{15}{44}$;   (c) $\frac{143}{256}$.

8.77 $\mu = 2.501$, which is very close to $\lambda = 2.5$.

8.79 $\sigma = 1.578$ rounded to three decimals, which is very close to $\sqrt{2.5} = 1.581$ rounded to three decimals.

8.81 (a) The probability is at least $\frac{143}{144}$ that there will be between 30 and 246 days with above 100 degree temperature.
(b) The probability is at least 0.91.

CHAPTER 9

9.1 (a) $P(S)$ cannot equal 0.75;
(b) $f(x)$ is negative for $x < \frac{7}{4}$.

9.3 (a) 0.75;   (b) 0;   (c) 0.8.

9.5 (a) $\frac{2}{9}$;   (b) $\frac{3}{4}$;   (c) $\frac{17}{25}$.

9.7 (a) First area is bigger;
(b) second area is bigger;
(c) first area is bigger;
(d) first area is bigger;
(e) the areas are equal;
(f) second area is bigger;
(g) the areas are equal;
(h) first area is bigger.

9.9 (a) 0.2171;   (b) 0.8427;   (c) 0.5959.

9.11 (a) First area is bigger;
(b) first area is bigger;
(c) the areas are equal.

9.13 (a) 1.92;   (b) 2.22;
(c) −0.50;   (d) 1.44 or −1.44.

9.15 (a) 0.6826;   (b) 0.9544;
(c) 0.9974;   (d) 0.99994;   (e) 0.9999994.

9.17 (a) 0.9332;   (b) 0.7734;
(c) 0.2957;   (d) 0.9198.

9.19 $\sigma = 20$.

9.21 (a) 0.3297;   (b) 0.1999;   (c) 0.2019.

9.23 (a) 0.1353;   (b) 0.049787;   (c) 0.3935.

9.33 The mean is about 78 and the standard deviation is about $\frac{92 - 63}{2} = 14.5$.

9.35 71.4 minutes, rounded to one decimal.

9.37 (a) 0.0401;   (b) 0.7734;   (c) 0.5859.

9.39 Supplier A.

9.41 Above 2.08 inches, rounded to two decimals.

9.43 Approximately 1.7 million men would be discharged.

9.45 (a) 0.0985;   (b) 0.1335.

9.47 Approximately 0.42%.

9.49 (a) Conditions are satisfied;
(b) conditions are not satisfied;
(c) conditions are satisfied.

9.51 (a) 0.1492;   (b) 0.3669.

9.53 (a) 0.0502;   (b) 0.1002;   (c) 0.4840.

9.57 Approximately 0.25%.

CHAPTER 10

10.1 (a) 3;   (b) 120;
(c) 300;   (d) 4,950.

10.3 (a) $\frac{1}{1,820}$;   (b) $\frac{1}{91,390}$.

10.5 *uvw, uvx, uvy, uvz, uwx, uwy, uwz, uxy, uxz, uyz, vwx, vwy, vwz, vxy, vxz, vyz, wxy, wxz, wyz,* and *xyz.*

10.7 0.2.

10.9 (a) 0.1;   (b) 0.6;   (c) 0.3.

10.13 3406, 3591, 3383, 3554, 3513, 3439, 3707, 3416, 3795, and 3329.

10.15 264, 429, 437, 419, 418, 252, 326, 443, 410, 472, 446, and 318.

10.17 6094, 2749, 0160, 0081, 0662, 5676, 6441, 6356, 2269, 4341, 0922, 6762, 5454, 7323, 1522, 1615, 4363, 3019, 3743, 5173, 5186, 4030, 0276, 7845, 5025, 0792, 0903, 5667, 4814, 3676, 1435, 5552, 7885, 1186, 6769, 5006, 0165, 1380, 0831, 3327, 0279, 7607, 3231, 5015, 4909, 6100, 0633, 6299, 3350, and 3597.

10.23 16.8, 24.0, 20.1, 21.9, 15.8, 22.1, 20.9, 21.3, 18.8, and 18.5; 27.6, 15.9, 24.2, 15.2, 20.4, 21.1, 15.7, 25.0, 16.9, and 25.0; 25.4, 16.9, 19.4, 17.2, 19.0, 25.8, 16.8, 12.9, 21.1, and 13.2; 16.6, 19.4, 16.6, 16.1, 16.9, 18.0, 20.5, 15.0, 27.9, and 16.7; 22.6, 17.0, 17.0, 18.6, 18.4, 15.5, 17.0, 15.8, 14.7, and 24.2.

10.25 The sixth sample contains all the December figures that are, of course, above average.

10.27 (a) 115 and 125, 115 and 135, 115 and 185, 115 and 195, 115 and 205, 125 and 135, 125 and 185, 125 and 195, 125 and 205, 135 and 185, 135 and 195, 135 and 205, 185 and 195, 185 and 205, and 195 and 205; the means are 120, 125, 150, 155, 160, 130, 155, 160, 165, 160, 165, 170, 190, 195, and 200; the probability is 8/15.
(b) 115 and 185, 115 and 195, 115 and 205, 125 and 185, 125 and 195, 125 and 205, 135 and 185, 135 and 195, and 135 and 205; the means are 150, 155, 160, 155, 160, 165, 160, 165, and 170; the probability is 2/9.
(c) 125 and 115, 125 and 195, 125 and 205, 135 and 115, 135 and 195, 135 and 205, 185 and 115, 185 and 195, and 185 and 205; the means are 120, 160, 165, 125, 165, 170, 150, 190, and 195; the probability is 2/3.
(d) The relevant stratification in (b) greatly improves on the simple random sampling of (a). The stratification in (c) actually has a negative (destructive) effect.

10.29 (a) 151,470;   (b) 323,136.

10.31 10, 24, 4, and 2.

10.35 $n_1 = 50$, $n_2 = 24$, and $n_3 = 10$.

10.37 (a) $81 and $510; $216.47.

10.39 (a) 0.52;   (b) 0.76.

10.41 (a) 12 and 12, 12 and 12, 12 and 12, 12 and 12, 12 and 12, 12 and 12, 12 and 12, 12 and 12, 12 and 12, 12 and 12, 12 and 14, 12 and 14, 12 and 14, 12 and 14, 12 and 14, 12 and 20, 12 and 20, 12 and 20, 12 and 20, 12 and 20, 12 and 42, 12 and 42, 12 and 42, 12 and 42, 12 and 42, 14 and 20, 14 and 42, 20 and 42;

(b) 12, 12, 12, 12, 12, 12, 12, 12, 12, 12, 13, 13, 13, 13, 13, 16, 16, 16, 16, 16, 27, 27, 27, 27, 27, 17, 28, and 31;

(c)

mean	frequency
12	10
13	5
16	5
17	1
27	5
28	1
31	1

(d) 17 and 6.414 rounded to three decimals.

10.43 (a) 4, 5, and 6; 4, 5, and 7; 4, 5, and 8; 4, 5, and 9; 4, 6, and 7; 4, 6, and 8; 4, 6, and 9; 4, 7, and 8; 4, 7, and 9; 4, 8, and 9; 5, 6, and 7; 5, 6, and 8; 5, 6, and 9; 5, 7, and 8; 5, 7, and 9; 5, 8, and 9; 6, 7, and 8; 6, 7, and 9; 6, 8, and 9; 7, 8, and 9;

(b) 5, $5\frac{1}{3}$, $5\frac{2}{3}$, 6, $5\frac{2}{3}$, 6, $6\frac{1}{3}$, $6\frac{1}{3}$, $6\frac{2}{3}$, 7, 6, $6\frac{1}{3}$, $6\frac{2}{3}$, $6\frac{2}{3}$, 7, $7\frac{1}{3}$, 7, $7\frac{1}{3}$, $7\frac{2}{3}$, 8;

(c) The frequencies corresponding to 5, $5\frac{1}{3}$, . . . , $7\frac{2}{3}$, and 8 are 1, 1, 2, 3, 3, 3, 3, 2, 1, and 1;

(d) 0.80.

10.45 The medians are 5, 5, 5, 5, 6, 6, 6, 7, 7, 8, 6, 6, 6, 7, 7, 8, 7, 7, 8, and 8; the frequencies corresponding to 5, 6, 7, and 8 are 4, 6, 6, and 4. The probability is 0.60 compared to 0.80.

10.47 (a) Divided by 3;
(b) multiplied by 2;
(c) divided by 1.5.

10.49 (a) Standard error is reduced by 35.4%; standard error is reduced by 83.2%.

10.51 (a) 0.990 rounded to three decimals;
(b) 0.985 rounded to three decimals;
(c) 0.988 rounded to three decimals.

10.53 (a) 2.091 rounded to three decimals;
(b) 12.316 rounded to three decimals.

10.57 $\sigma_{\bar{x}} = 35$, which is greater than the other two.

10.59 (a) The probability is at least 0.84;
(b) the probability is 0.9876.

10.61 0.8904.

10.67 The medians are 11, 15, 19, 17, 14, 13, 14, 17, 18, 14, 14, 19, 16, 18, 17, 18, 19, 17, 14, 12, 11, 16, 16, 19, 18, 17, 17, 15, 13, 14, 15, 16, 14, 15, 12, 16, 16, 17, 14, and 17; their standard deviation is 2.20 and the corresponding theoretical value is 2.24.

10.69 The percentage error is 3.76%.

10.71 The standard deviation is 0.620 compared to 0.625.

REVIEW EXERCISES

R.91 (a) $\frac{132}{323}$;   (b) $\frac{99}{323}$.

R.93 $\alpha = 0.12$.

R.95 (a) Condition is satisfied;
(b) condition is not satisfied;
(c) condition is not satisfied.

R.97 $\frac{1}{341,055}$.

R.99 (a) 3.203;   (b) 3.20.

R.101 (a) Conditions are not satisfied;
(b) conditions are satisfied;
(c) conditions are satisfied.

R.103 0.7888.

R.105 0.066 rounded to three decimals.

R.107 (a) 5/42;   (b) 5/14.

R.109 The probability is a maximum for $N = 11$ and $N = 12$.

R.111 (c) 0.04.

R.113 (a) 364;   (b) 42,504.

R.115 (a) 0.1108;   (b) 0.6227.

R.117 (a) Yes;   (b) no;   (c) yes.

R.119 (a) 0.8088;   (b) 0.1867;   (c) 0.0186.

R.121 1.4974 rounded to four decimals.

R.123 (a) 1 and 3, 1 and 5, 1 and 7, 3 and 5, 3 and 7, and 5 and 7; their means are 2, 3, 4, 4, 5, and 6.
(b) The probabilities corresponding to 2, 3, 4, 5, and 6 are 1/6, 1/6, 2/6, 1/6, and 1/6;
(c) 4 and $\sqrt{5/3}$.

R.125 (a) 1.5;   (b) 1.43.

R.127 (a) 1.464 rounded to three decimals;
(b) 1.470 rounded to three decimals.

R.129 (a) 0.1502;   (b) 0.4144;   (c) 0.0304.

R.131 (a) Approximately 2.03%;
(b) approximately 1.54%;
(c) approximately 1.69%.

R.133 $\frac{1}{43,680}$.

R.135 $\frac{21}{110}$.

CHAPTER 11

11.1 $E = 41.84$ rounded to two decimals.

11.3 $E = 1.80$ mm rounded to two decimals.

11.5 $E = \$106$ rounded to the nearest dollar; $\$1,363 < \mu < \$1,513$.

11.7 The degree of confidence is approximately 91%.

11.9 $n = 52$ rounded up to the nearest integer.

11.11 $n = 37$ rounded up to the nearest integer.

11.13 (a) 27;   (b) 26.

11.15 (a) $81.7 < \mu < 85.1$;   (b) same.

11.17 (a) $1.97 < \mu < 2.71$ micrograms;
(b) $E = 0.54$ micrograms rounded to two decimals.

11.19 $E = 1,174$ rounded to the nearest pound.

11.21 $0.403 < \mu < 0.409$ cm.

11.23 (a) It is reasonable to treat the data as a sample from a normal population;
(b) $30.07 < \mu < 33.15$.

11.27 $9.55 < \mu < 12.45$ mistakes.

11.29 $0.122 < \mu < 0.178$ inches rounded to three decimals.

11.31 $E = 1.67$ fillings rounded to two decimals.

11.33 $0.0088 < \sigma < 0.0189$ rounded to four decimals.

11.35 $1.76 < \sigma < 5.65$ rounded to two decimals.

11.37 $0.052 < \sigma^2 < 1.200$ rounded to three decimals.

11.39 (a) $0.018 < \sigma < 0.236$ rounded to three decimals;
(b) $1.19 < \sigma < 3.82$ rounded to two decimals.

11.41 (a) $4.4 < \sigma < 7.1$ rounded to one decimal;
(b) $0.035 < \sigma < 0.063$ rounded to three decimals.

11.43 The estimate is 13.24 rounded to two decimals, which is of the same general magnitude as 14.35.

11.45 (a) $n > 125$;   (b) $n > 62.5$.

11.47 $E = 0.088$ rounded to three decimals.

11.49 $E = 0.039$ rounded to three decimals.

11.51 $0.724 < p < 0.836$ rounded to three decimals.

11.53 $E = 0.0625$ rounded to four decimals.

11.55 $E = 6.2\%$ rounded to one decimal.

11.57 $E = 0.130$ rounded to three decimals.

11.59 (a) $n = 151$ rounded up to the nearest integer;
(b) $n = 2,401$.

11.61 (a) $n = 260$ rounded up to the nearest integer;
(b) $n = 369$ rounded up to the nearest integer;
(c) $n = 637$ rounded up to the nearest integer.

CHAPTER 12

12.1 (a) $\mu < \mu_0$ and buy the new ambulances only if the null hypothesis can be rejected;
(b) $\mu > \mu_0$ and buy the new ambulances unless the null hypothesis can be rejected.

12.3 (a) Since the null hypothesis is true and accepted, the psychologist will not be making an error;
(b) since the null hypothesis is false and accepted, the psychologist will be making a Type II error.

12.5 If it erroneously rejects the null hypothesis, the testing service will be committing a Type I error; if it erroneously accepts the null hypothesis, the testing service will be committing a Type II error.

12.7 Use the null hypothesis that the antipollution device is not effective.

12.9 (a) 0.1056;   (b) 0.8413;   (c) 0.2743.

12.11 (a) It is reduced from 0.11 to 0.05;
(b) it is increased from 0.21 to 0.34.

12.13 The term "statistically significant" applies to statistics calculated from sample data and not to population parameters.

12.15 No! It means that the probability is 0.05 that the null hypothesis will be rejected even though it is true.

12.17 (a) Three rejections are well within expectation even though the production process is under control. There is no reason for concern.
(b) Seven rejections are quite a bit more than what can reasonably be expected if the production process is under control. There is reason for concern.

12.19 With the alternative hypothesis $\mu > 2.6$, the efficiency expert would use a one-tailed test.

12.21 With the alternative hypothesis $\mu > 20$, the medical team would use a one-tailed test.

12.23 A result may be statistically significant without being of any practical significance. In this case,

24.02 is so close to 24 that expensive adjustments may not be required.

12.25 (a) 0.878;   (b) 0.608.

12.27 Since the p-value is $0.2112 > 0.05$, the null hypothesis cannot be rejected.

12.29 Since $z = 2.42$ falls between -2.575 and 2.575, the null hypothesis cannot be rejected.

12.31 Since $z = 2.73 > 2.33$, the null hypothesis must be rejected.

12.33 Since $z = 1.89 < 2.33$, the null hypothesis $\mu = 0.87$ cannot be rejected.

12.35 $n = 306$; the null hypothesis will be rejected for $\bar{x} \leq 531.7$ rounded to one decimal.

12.37 $n = 71$; the null hypothesis will be rejected for $\bar{x} \leq 644$ or $\bar{x} \geq 656$.

12.39 Since $t = 4.00 \geq 2.896$, the null hypothesis must be rejected.

12.41 Since $t = -1.56 \leq -1.341$, the null hypothesis $\mu = 4.50$ must be rejected.

12.43 (a) The normal probability plot is not linear; (b) the test cannot be performed.

12.45 Since $t = 2.89 > 2.539$, the null hypothesis must be rejected.

12.47 Since $t = -1.39 > -1.685$, the null hypothesis cannot be rejected.

12.49 Since $z = 2.36 \geq 1.96$, the null hypothesis must be rejected. The difference is significant.

12.51 Since $z = -2.13$ falls between -2.33 and 2.33, the null hypothesis cannot be rejected. The difference is not significant.

12.53 The p-value is 0.034.

12.55 Since $z = 2.78 \geq 2.33$, the null hypothesis must be rejected.

12.57 $-14.1 < \delta < -0.6$.

12.59 Since $t = -1.88$ falls between -2.074 and 2.074, the null hypothesis $\delta = 0$ cannot be rejected.

12.61 Since $t = 1.29$ falls between -3.143 and 3.143, the null hypothesis cannot be rejected.

12.63 Since $t = 2.01$ falls between -2.228 and 2.228, the null hypothesis $\delta = 2.0$ cannot be rejected.

12.65 The p-value is $0.0570 > 0.05$, so that the null hypothesis cannot be rejected.

12.67 The p-value is $0.0720 \leq 0.10$, so that the null hypothesis must be rejected.

12.69 The p-value is $0.0011 \leq 0.025$, so that the null hypothesis must be rejected.

12.71 Since $t = 0.77$ falls between -2.120 and 2.120, the null hypothesis cannot be rejected.

12.73 The p-value is $0.006 \leq 0.05$, so that the null hypothesis must be rejected.

12.75 The p-value is $0.0135 > 0.01$, so that the null hypothesis could not have been rejected at the 0.01 level of significance.

12.77 $6.63 < \delta < 10.33$.

12.79 The p-value is $0.0550 > 0.05$, so that the null hypothesis could not have been rejected at the 0.05 level of significance.

12.81 Since $t = 1.66$ falls between -2.069 and 2.069, the null hypothesis cannot be rejected and, hence, will have to be accepted.

CHAPTER 13

13.1 Since $\chi^2 = 6.66 > 3.325$, the null hypothesis cannot be rejected.

13.3 The p-value is $0.0184 \leq 0.03$, so that the null hypothesis must be rejected at the 0.03 level of significance.

13.5 The p-value is $0.0389 > 0.015$, so that the null hypothesis could not have been rejected at the 0.015 level of significance.

13.7 Since $\chi^2 = 9 < 9.488$, the null hypothesis cannot be rejected.

13.9 Since $z = -1.88$ falls between -1.96 and 1.96, the null hypothesis cannot be rejected.

13.11 The p-value is $0.0228 > 0.01$, so that the null hypothesis cannot be rejected at the 0.01 level of significance.

13.13 Since $F = 2.86 \geq 2.72$, the null hypothesis must be rejected. The first lighting technique is less variable than the second.

13.15 Since $F = 1.16 < 5.35$, the null hypothesis cannot be rejected. There is no reason to doubt the assumption.

13.17 Since $F = 1.93 < 6.99$, the null hypothesis cannot be rejected. There is no reason to doubt the assumption.

CHAPTER 14

14.1 Since the p-value is $0.028 \leq 0.05$, the null hypothesis must be rejected.

14.3 Since the p-value is $0.116 > 0.10$, the null hypothesis cannot be rejected.

14.5 Since the p-value is $0.1788 > 0.01$, the null hypothesis cannot be rejected. This supports the state highway official's claim.

14.7 The p-values corresponding to $x = 16, 17,$ and 18 are 0.061, 0.015, and 0.002. To reject the null hypothesis, at least 17 of the students would have to be opposed to the plan.

14.9 (a) Since $z = -2.00 \leq -1.96$, the null hypothesis must be rejected.
(b) Since $z = -2.00$ falls between -2.575 and 2.575, the null hypothesis cannot be rejected.

14.11 (a) Since $z = -1.68 \leq -1.645$, the null hypothesis must be rejected.
(b) Since $z = -1.62 > -1.645$, the null hypothesis cannot be rejected.

14.13 Since $z = -0.053$ falls between -1.96 and 1.96, the null hypothesis cannot be rejected. There is no evidence of any lack of randomness.

14.15 Since $z = 0.81 < 1.645$, the null hypothesis cannot be rejected.

14.17 Since $z = -0.94$ falls between -1.96 and 1.96, the null hypothesis cannot be rejected.

14.19 Since $z = -1.11$ falls between -1.96 and 1.96, the null hypothesis cannot be rejected. The difference between the two sample proportions is not significant.

14.21 Since $z = -0.47 > -2.33$, the null hypothesis cannot be rejected. The trend is not statistically significant.

14.23 Since $z = 1.81 \geq 1.645$, the null hypothesis must be rejected. The test item is good.

14.25 $\chi^2 = 40.90$ rounded to two decimals.

14.27 H_0: The probabilities for the three response categories (part-time employment, full-time employment, no employment) are the same regardless of the number of children.
H_A: The probabilities for at least one of the response categories are not all the same.

14.29 Since $\chi^2 = 1.41 < 5.991$, the null hypothesis cannot be rejected.

14.31 The p-value is $0.00000108 \leq 0.01$, so that the null hypothesis must be rejected.

14.33 Since $\chi^2 = 20.72 \geq 16.812$, the null hypothesis must be rejected. There are differences of opinion among the four groups concerning the curriculum change.

14.35 Since $\chi^2 = 26.77 \geq 9.488$, the null hypothesis must be rejected. There is a relationship be-

tween students' interest and ability in studying a foreign language.

14.37 For Exercise 14.35 the p-value is $0.000022 \leq 0.01$, so that the null hypothesis must be rejected at the 0.01 level of significance. For Exercise 14.36 the p-value is $0.0139 > 0.01$, so that the null hypothesis cannot be rejected at the 0.01 level of significance.

14.39 The p-value is $0.0000000001 \leq 0.01$, so that the null hypothesis must be rejected.

14.41 Since $\chi^2 = 1.39 < 5.991$, the null hypothesis cannot be rejected. There is no evidence that the handicaps affect the performance.

14.43 $C = 0.33$ rounded to two decimals.

14.45 Since $\chi^2 = 4.86 < 9.210$, the null hypothesis cannot be rejected.

14.47 Since $\chi^2 = 0.88 < 3.841$, the null hypothesis cannot be rejected. Also, $z^2 = (-0.94)^2 = 0.8836$ approximately equals $\chi^2 = 0.88$.

14.49 Since $\chi^2 = 8.36 \geq 6.635$, the null hypothesis must be rejected. Also, $z^2 = (2.89)^2 = 8.3521$ approximately equals $\chi^2 = 8.36$.

14.53 Since $\chi^2 = 5.07 < 7.815$, the null hypothesis cannot be rejected. This substantiates the claim.

14.55 Since $\chi^2 = 0.41 < 6.635$, the null hypothesis cannot be rejected. The binomial distribution with $n = 3$ and $p = 0.2$ provides a very close fit.

14.57 Since $\chi^2 = 6.81 \geq 5.991$, the null hypothesis must be rejected. Data may be looked upon as a random sample from a binomial population.

14.59 Since $\chi^2 = 10.465 < 15.507$, the null hypothesis cannot be rejected. Daily number of power failures is a random variable having the Poisson distribution with $\lambda = 3.2$.

14.61 (c) Since $\chi^2 = 1.77 < 3.841$, the null hypothesis cannot be rejected. Data can be looked upon as a random sample from a normal population.

REVIEW EXERCISES

R.137 Normal probability plot justifies the assumption that the data constitute a sample from a normal population.

R.139 $n = 784$.

R.141 $32.9 < \mu < 36.7$ micrograms per liter (rounded to one decimal).

R.143 Since $z = -3.85 \leq -2.575$, the null hypothesis must be rejected. The data refute the claim.

R.145 (a) 0.031;   (b) 0.851;
(c) 0.604;   (d) 0.308.

R.147 Since $t = 6.96 > 3.747$, the null hypothesis must be rejected.

R.149 The pattern on the normal probability plot is definitely not linear.

R.151 $n = 423$.

R.153 Since $z = 2.88 \geq 2.575$, the null hypothesis must be rejected. The difference between the two sample proportions is significant.

R.155 Since $F = 2.47 < 5.05$, the null hypothesis cannot be rejected.

R.157 Since $t = -2.11 > -2.201$, the null hypothesis cannot be rejected.

R.159 Since $\chi^2 = 0.91 < 5.991$, the null hypothesis cannot be rejected.

R.161 $0.618 < p < 0.782$ rounded to three decimals.

R.163 $0.0125 < p < 0.9875$.

R.165 Since $z = -10.94 \leq -1.96$, the null hypothesis must be rejected. The difference between the means is significant.

R.169 Erroneously accepting the hypothesis that the athlete is physically fit to play is a Type II error. Erroneously rejecting the hypothesis that the athlete is physically fit to play is a Type I error.

R.171 $E = 77.04$ pieces rounded to two decimals.

R.173 $2.41 < \sigma < 6.39$ rounded to two decimals.

R.175 (c) Since $\chi^2 = 1.27 < 5.991$, the null hypothesis cannot be rejected. Data can be looked upon as a random sample from a normal population.

R.177 $1.37 < \sigma < 2.45$ grams rounded to two decimals.

R.179 (a) $n = 267$;   (b) $n = 257$.

R.181 Since $z = -3.52 \leq -2.575$, the null hypothesis must be rejected. The difference between the two sample means is significant.

CHAPTER 15

15.1 (a) 1.86, 3.215, and 0.58.
(b) Since $F = 0.58 < 3.68$, the null hypothesis cannot be rejected. The differences among the sample means can be attributed to chance.

15.3 (a) 5 and 54;   (b) 4 and 35.

15.5 The scores for School 2 are much more variable than those for the other two schools.

15.7 The three kinds of tulips should have been assigned at random to the 12 locations in the flower bed. If the soil is sloped, for example, different rows of the flower bed might get different amounts of irrigation.

15.9 This is controversial, and statisticians argue about the appropriateness of discarding a proper randomization because it happens to possess some undesirable property. Most practitioners will examine the results of their randomizations to check for odd patterns. For the situation described here, it is likely that the analysis of variance will not be done.

15.15 Since $F = 10.78 \geq 4.94$, the null hypothesis must be rejected. The differences among the sample means cannot be attributed to chance.

15.17 Since $F = 11.70 \geq 5.01$, the null hypothesis must be rejected.

15.19 Since $F = 3.19 < 3.89$, the null hypothesis cannot be rejected. The differences among the three sample means are not significant.

15.21 Since $F = 9.28 \geq 2.25$, the null hypothesis must be rejected. The eight spraying schedules are not equally effective.

15.23 $\underline{\text{32 in. 28 in. 24 in. 20 in.}}$ For 20 in. and 24 in. the average yield is significantly higher than for 32 in. Otherwise, the average yields for 20 in., 24 in., and 28 in., are not significantly different.

15.25 The average sales for Smith and Black do not differ significantly; the average sales for Smith differ significantly from those of Brown and Jones; the average sales of Black differ significantly from those of Brown.

15.29 She might consider only programs of the same length, or she might use the program lengths as blocks and perform a two-way analysis of variance.

15.31 Since $F = 6.63 < 10.92$, the null hypothesis for treatments (diet foods) cannot be rejected. Since $F = 4.90 < 9.78$, the null hypothesis for blocks (laboratories) cannot be rejected.

15.33 Since $F = 8.31 \geq 3.26$, the null hypothesis for treatments (threads) must be rejected. Since $F = 0.58 < 3.49$, the null hypothesis for blocks (measuring instruments) cannot be rejected.

15.35 Since $F = 6.58 < 6.93$, the null hypothesis for treatments (operators) cannot be rejected. Since $F = 4.65 < 5.95$, the null hypothesis for blocks (bonders) cannot be rejected. Since $F = 3.01 < 4.82$, the null hypothesis for interaction cannot be rejected.

15.37 Since $F = 0.08 < 3.89$, the null hypothesis for recipes cannot be rejected. Since $F = 1.62 < 3.49$, the null hypothesis for mixes cannot be rejected. Since $F = 2.65 < 3.00$, the null hypothesis for interaction cannot be rejected.

15.39 180.

15.41 $A_L B_L C_L D_L$, $A_L B_L C_L D_H$, $A_L B_L C_H D_L$, $A_L B_L C_H D_H$, $A_L B_H C_L D_L$, $A_L B_H C_L D_H$, $A_L B_H C_H D_L$, $A_L B_H C_H D_H$, $A_H B_L C_L D_L$, $A_H B_L C_L D_H$, $A_H B_L C_H D_L$, $A_H B_L C_H D_H$, $A_H B_H C_L D_L$, $A_H B_H C_L D_H$, $A_H B_H C_H D_L$, and $A_H B_H C_H D_H$.

15.43 Since $F = 15.63 \geq 4.76$, the null hypothesis for rows (golf pros) must be rejected. Since $F = 0.82 < 4.76$, the null hypothesis for columns (drivers) cannot be rejected. Since $F = 45.80 \geq 4.76$, the null hypothesis for treatments (brands of golf balls) must be rejected.

15.45 The doctor who is a Westerner is a Republican.

15.47 4, 3, and 7 on Tuesday; 2, 5, and 7 on Thursday; and 5, 3, and 6 on Saturday.

CHAPTER 16

16.1 The second line provides a better fit.

16.3 (a) The overall pattern is linear, although the points are fairly widely scattered.

16.5 $100 = 10a + 525b$
$5,980 = 525a + 32,085b$
$a = 1.526$ and $b = 0.161$ rounded to three decimals.

16.7 19.6 cars rounded to one decimal.

16.9 (a) $\hat{y} = 2.039 - 0.102x$;
(b) $\hat{y} = 2.039 - (0.102)5 = 1.529$.
(c) On a very hot day the chlorine would dissipate much faster than usual.

16.11 The sum of the squares of the vertical deviations from the points to the line is 20.94 rounded to two decimals.

16.13 $\hat{y} = 0.49 + 0.27x$.

16.15 $\hat{y} = 10.75$ rounded to two decimals.

16.17 (a) $\hat{y} = 2.66 + 0.6u$; $\hat{y} = \$4.46$ million.
(b) $\hat{y} = 2.95 + 0.242u$; $\hat{y} = \$5.612$ million.

16.19 (a) $a = 12.45$ and $b = 0.898$;
(b) $a = 0.49$ and $b = 0.27$.

16.21 (a) $s_e = 2.34$;   (b) since $t = -2.41 \leq -2.132$, the null hypothesis must be rejected.

16.23 Since $t = 2.68 < 3.143$, the null hypothesis cannot be rejected.

16.25 Since $t = 4.17 \geq 4.032$, the null hypothesis must be rejected.

16.27 (a) $s_e = 1.070$;
(b) since $t = -3.48 \leq -1.812$, the null hypothesis must be rejected.

16.29 $13.16 < \mu_{y|50} < 15.02$ rounded to two decimals.

16.31 (a) $7.78 < \mu_{y|60} < 14.64$ rounded to two decimals;
(b) $0.43 - 21.99$ rounded to two decimals.

16.33 (a) $0.554 < \mu_{y|5} < 1.574$ rounded to three decimals;
(b) $0.736 - 1.392$ rounded to three decimals.

16.35 (a) $\hat{y} = 198 + 37.2x_1 - 0.120x_2$;
(b) $71.2°$ rounded to one decimal.

10.37 (a) $\hat{y} = -2.33 + 0.90x_1 + 1.27x_2 + 0.90x_3$;
(b) $\hat{y} = 54.17$ rounded to two decimals.

16.39 (a) $\log \hat{y} = 1.8379 + 0.0913x$;
(b) $\hat{y} = 68.9 (1.234)^x$;
(c) $\hat{y} = 197.0$.

16.41 $\hat{y} = (101.17)(0.9575)^x$.

16.43 $\hat{y} = (1.178)(2.855)^x$.

16.45 $\hat{y} = (18.99)(1.152)^x$.

16.47 $\hat{y} = 6,481x^{-1.0399}$; $\hat{y} = \$17.20$.

16.49 $\hat{y} = 6,114x^{-1.4938}$; $\hat{y} = 107.0$ psi.

16.51 Maximum yield of 38.8 pounds per square yard is produced by 1.11 pounds of fertilizer per square yard.

CHAPTER 17

17.1 $r = 0.92$; $\sqrt{.845} = 0.919$ rounded to three decimals.

17.3 60.8%.

17.5 $r = -0.01$.

17.7 $r = 0.95$.

17.9 $r = -0.99$; $(-0.99)^2 100 = 98.0\%$.

17.11 No correction is needed.

17.13 (a) Positive;   (b) negative;
(c) negative;   (d) none;   (e) positive.

17.15 $(0.41/0.29)^2 = 2.00$; first relationship is twice as strong as the second.

17.17 For one thing, correlation does not necessarily imply causation. Actually, both variables (foreign language degrees and mileage of railroad track) depend on other variables, such as population growth and economic conditions on general.

17.19 $r = -0.45$.

17.21 (a) Since $z = 2.35 \geq 1.96$, the null hypothesis must be rejected;
(b) since $z = 1.80$ falls between -1.96 and 1.96, the null hypothesis cannot be rejected;
(c) Since $z = 2.29 \geq 1.96$, the null hypothesis must be rejected.

17.23 (a) Since $z = 1.05 < 1.645$, the null hypothesis cannot be rejected;
(b) since $z = 1.92 \geq 1.645$, the null hypothesis must be rejected.

17.25 (a) Since $z = 1.22 < 2.33$, the null hypothesis cannot be rejected;
(b) since $z = 0.50 < 2.33$, the null hypothesis cannot be rejected.

17.27 (a) Since $t = 3.82 \geq 3.169$, the null hypothesis must be rejected;
(b) since $t = 2.10$ falls between -2.977 and 2.977, the null hypothesis cannot be rejected.

17.29 (a) $t = 12.8823$ and $r = 0.9824$;
(b) $t = 12.88$;
(c) except for rounding, the two values of t are the same.

17.31 Since $z = 2.42$ falls between -2.575 and 2.575, the null hypothesis cannot be rejected.

17.33 (a) $-0.945 < \rho < -0.49$;
(b) $-0.19 < \rho < 0.72$;
(c) $-0.12 < \rho < 0.45$.

17.35 $R = 0.992$; $r = 0.699$ for y and x_1 and $r = 0.770$ for y and x_2.

17.37 Impossible. The multiple correlation coefficient based on both variables cannot be numerically less than the ordinary correlation coefficient based on only one of the two variables.

17.39 $r_{23 \cdot 1} = -1.00$.

CHAPTER 18

18.1 Since the p-value is $0.424 > 0.05$, the null hypothesis cannot be rejected.

18.3 Since the p-value is $0.016 > 0.01$, the null hypothesis cannot be rejected.

18.5 (a) Since the p-value is $0.047 \leq 0.05$, the null hypothesis must be rejected.
(b) Without the continuity correction $z = 1.94 \geq 1.645$, so that the null hypothesis must be rejected. With the continuity correction $z = 1.66 \geq 1.645$, so that the null hypothesis must be rejected.

18.7 Without the continuity correction $z = 2.333 \geq 2.33$, so that the null hypothesis should be rejected. With the continuity correction $z = 2.17 < 2.33$, so that the null hypothesis should not have been rejected.

18.9 Since the p-value is $0.01758 \leq 0.03$, the null hypothesis must be rejected.

18.11 Since $z = 2.065 \geq 1.645$, the null hypothesis must be rejected; with the continuity correction $z = 1.84$, and the conclusion is the same.

18.13 (a) Reject the null hypothesis if $T \leq 8$;
(b) reject the null hypothesis if $T^- \leq 11$;
(c) reject the null hypothesis if $T^+ \leq 11$.

18.15 (a) Reject the null hypothesis if $T \leq 7$;
(b) reject the null hypothesis if $T^- \leq 10$;
(c) reject the null hypothesis if $T^+ \leq 10$.

18.17 (a) Since $T^+ = 18 \leq 21$, the null hypothesis must be rejected;
(b) since $T = 18 > 17$, the null hypothesis cannot be rejected.

18.19 Since $T^- = 18 \leq 33$, the null hypothesis must be rejected.

18.21 Since $T^+ = 5 \leq 26$, the null hypothesis must be rejected.

18.23 Since $z = -2.98 \leq -1.645$, the null hypothesis must be rejected. With the continuity correction $z = -2.95$ and the conclusion is the same.

18.25 Since $z = -1.75 \leq -1.645$, the null hypothesis must be rejected. With the continuity correction $z = -1.73$ and the conclusion is the same.

18.31 (a) Reject the null hypothesis if $U_2 \leq 14$;
(b) reject the null hypothesis if $U \leq 11$;
(c) reject the null hypothesis if $U_1 \leq 14$.

18.33 (a) Reject the null hypothesis if $U_2 \leq 41$;
(b) reject the null hypothesis if $U \leq 36$;
(c) reject the null hypothesis if $U_1 \leq 41$.

18.35 (a) Reject the null hypothesis if $U_2 \leq 3$;
(b) reject the null hypothesis if $U_2 \leq 18$;
(c) reject the null hypothesis if $U_2 \leq 13$;
(d) reject the null hypothesis if $U_2 \leq 2$.

18.37 Since $U = 34 \leq 37$, the null hypothesis must be rejected. Students from the two minority groups cannot be expected to score equally well on the test.

18.39 Since $U = 88 > 49$, the null hypothesis cannot be rejected.

18.41 Since $U_1 = 5.5 \leq 10$, the null hypothesis must be rejected.

18.43 Since $z = -1.71$ falls between -1.96 and 1.96, the null hypothesis cannot be rejected. (Continuity correction not needed.)

18.49 Since $H = 11.96 \geq 11.345$, the null hypothesis must be rejected.

18.51 Since $H = 4.51 < 7.815$, the null hypothesis cannot be rejected.

18.53 Since $H = 1.53 < 9.210$, the null hypothesis cannot be rejected. The differences in dosage have no effect on the length of time it takes the guinea pigs to fall asleep.

18.55 Since $u = 7 \leq 8$, the null hypothesis of randomness must be rejected.

18.57 Since $u = 20$ falls between 8 and 23, the null hypothesis of randomness cannot be rejected.

18.59 Without the continuity correction $z = -2.00 \leq -1.96$ and the null hypothesis should be rejected. With the continuity correction $z = -1.74$ (between -1.96 and 1.96) and the null hypothesis cannot be rejected.

18.61 Without the continuity correction $z = 0.38$ (between -1.96 and 1.96) and the null hypothesis of randomness cannot be rejected. There is no need to check the continuity correction.

18.65 Without the continuity correction $z = 1.60$ (between -1.96 and 1.96), so that the null hypothesis of randomness cannot be rejected. There is no need to try the continuity correction.

18.67 Without the continuity correction $z = -2.23 \leq -1.645$, so that the null hypothesis must be rejected. (The continuity correction yields $z = -2.04$ and the conclusion is the same.) There is a significant trend.

18.69 $r_S = 0.65$.

18.71 $r_S = 0.80$ and since $z = 2.40 \geq 1.96$, it is significant.

18.73 $r_S = 0.89$ and since $z = 2.95 \geq 2.575$, it is significant.

18.75 Since $z = 2.17 \geq 1.96$, $r_S = 0.31$ for $n = 50$ is significant.

18.77 (a) Judges A and B are most alike;
(b) judges B and C are least alike.

REVIEW EXERCISES

R.183 (a) 28, $\frac{86}{9}$, and 2.93 rounded to two decimals.
(b) Since $F = 2.93 < 8.02$, the null hypothesis cannot be rejected. The differences among the sample means can be attributed to chance.

R.185 (a) $\hat{y} = 10.95 - 0.333x$;
(b) $\hat{y} = 3.62$ rounded to two decimals.

R.187 Since the p-value is $0.424 > 0.05$, the null hypothesis cannot be rejected.

R.189 Without the continuity correction $z = -1.44$ (between -2.575 and 2.575), so that the null hypothesis of randomness cannot be rejected. There is no need to check the continuity correction.

R.191 5, 8, and 40.

R.193 Since $z = -2.75 \leq -1.96$, the null hypothesis must be rejected. The value obtained for r is significant.

R.195 (a) $\hat{y} = -0.63 + 0.0972x_1 + 0.662x_2$;
(b) $\hat{y} = \$29.05$ thousand rounded to two decimals.

R.197 Since the p-value is $0.064 > 0.05$, the null hypothesis cannot be rejected.

R.199 With the continuity correction $z = 1.81$ (between -1.96 and 1.96) and the null hypothesis cannot be rejected. (Not using the continuity correction would have led to the rejection of the null hypothesis.)

R.201 (a) $-0.18 < \rho < 0.82$;
(b) $-0.645 < \rho < 0.103$.

R.203 Since $F = 2.31 < 5.41$, the null hypothesis for rows (mechanics) cannot be rejected; since $F = 8.24 \geq 5.41$, the null hypothesis for columns (engines) must be rejected; since $F = 31.28 \geq 5.41$, the null hypothesis for treatments (fuels) must be rejected.

R.205 Since $z = -2.14 \leq -1.96$, the null hypothesis must be rejected. (With the continuity correction, conclusion would have been the same.)

R.207 Without continuity correction $z = 1.53$ falls between -1.96 and 1.96, so that the null hypothesis cannot be rejected. No need to try the continuity correction.

R.209 A significant value of F could be attributed to differences in rank.

R.211 Since $F = 6.83 \geq 4.26$, the null hypothesis must be rejected. The descriptions do affect the sales.

R.213 $\hat{y} = 14.2 - 0.471x - 1.18x^2$, where the x's are the coded years.

R.215 Since the p-value is $0.039 \leq 0.05$, the null hypothesis must be rejected.

R.217 (a) Positive correlation;
(b) no correlation;
(c) positive correlation;
(d) negative correlation.

R.219 $r = 0.94$.

R.221 $\hat{y} = -0.069 + 0.980x$; $\hat{y} = 2.675$.

R.223 $\hat{y} = 28.286 \, (1.016)^x$; $\hat{y} = 74.2$ mrem/year rounded to one decimal.

R.225 The second relationship is five times as strong as the first.

INDEX

A

Addition rules of probability, 156, 157
α *(alpha)*, probability of Type I error, 335
α *(alpha)*, regression coefficient, 474
  confidence interval for, 477
  test for, 476
Alternative hypothesis, 332
  one-sided, 339
  two-sided, 339
Analysis of an $r \times c$ table, 387
Analysis of variance, 417, 424
  F test, 420, 426
  Latin square, 450
  one-way, 424, 429
  table, 426, 439, 452
  two-way, 437, 438
ANOVA *(see* Analysis of variance)
Area sampling, 273
Arithmetic mean, 49
Arithmetic probability paper, 247
Average *(see* Mean)

B

Balanced incomplete block design, 454
Bar chart, 15
Bayesian analysis, 185
Bayes' theorem, 172
Bell-shaped distribution, 87, 95
β *(beta)*, probability of Type II error, 336
β *(beta)*, regression coefficient, 474
  confidence interval for, 477
  test for, 476
Betting odds, 149
Biased estimator, 83
Binomial coefficient, 117, 122
  table, 590
Binomial distribution, 203
  and hypergeometric distribution, 214
  mean, 224
  normal approximation, 253

Binomial distribution *(cont.)*
  Poisson approximation, 216
  standard deviation, 227
  table, 577
Binomial population, 208
Block, 437
  effect, 438
  sum of squares, 439
Blocking, 437
Blocks, complete, 437
Boundary, class, 24
Box-and-whisker plot, 64
Boxplot, 64

C

Categorical data, 7
Categorical distribution, 21
Cell frequency:
  expected 389, 390
  observed, 389, 391
Central limit theorem, 283
Central location, measures of, 48
Chebyshev's theorem, 85, 227
Chi-square distribution, 317, 390, 542
  degrees of freedom, 317, 390, 542
  table, 573
χ^2 *(chi-square)* statistic, 317, 367, 390, 542
Class:
  boundary, 24
  frequency, 24
  interval, 25
  limit, 24
  mark, 25
  open, 22
  real limit, 24
Classical probability concept, 124
Cluster sampling, 273
Coding, 76, 473, 485
Coefficient of correlation, 501, 502
  confidence interval for, 514

Coefficient of correlation, *(cont.)*
   test for, 512
   Z transformation, 512
Coefficient of determination, 501
Coefficient of quartile variation, 92
Coefficient of skewness, Pearsonian, 95
Coefficient of variation, 88
Coefficients, binomial, 117, 122
   table, 590
Column sum of squares, 451
Combinations, 116
Complement, 137
Complete block design, 437
Complete blocks, 437
Completely randomized design, 423
Composite hypothesis, 337
Computer simulation, 127, 278, 279, 291
Conditional probability, 160, 162
Confidence, 305
   and probability, 305
   degree of, 307
Confidence interval, 307
   correlation coefficient, 514
   mean, 307, 309
   mean of y for given x, 478
   proportion, 324
   regression coefficients, 477
   standard deviation, 318, 321
   variance, 318
Confidence limits, 307
Consistency criterion, 150
Contingency coefficient, 399
Contingency table, 388
Continuity correction, 252
Continuous distribution, 234, 235
Continuous random variable, 233
Controlled experiment, 422
Correction factor, finite population, 281
Correlation:
   multiple, 516
   negative, 501
   partial, 517
   positive, 501
   rank, 551
Correlation analysis, normal, 512
Correlation coefficient:
   computing formula, 502

Correlation coefficient: *(cont.)*
   definition, 501
   confidence interval for, 514
   interpretation, 506
   multiple, 516
   partial, 517, 518
   population, 512
   product-moment, 503
   rank, 551
   test for, 512
Count data, 323
Countably infinite sample space, 216
Covariance, sample, 504
Critical value, 345
Cross stratification, 272
Cumulative distribution, 26
Cumulative percentage distribution, 26
Curve fitting, 460
   exponential, 486
   linear, 465
   parabolic, 490
   polynomial, 486
   power function, 489

D

Data:
   count, 323
   interval, 7
   nominal, 7
   numerical, 7
   ordinal, 7
   paired, 37, 359, 459, 498
   ratio, 7
   raw, 20
Data point, 38, 462
Decile, 62, 73
Decision theory, 188
Degree of confidence, 307
Degrees of freedom:
   chi-square distribution, 317, 390, 542
   F distribution, 371, 420, 426, 429, 439, 451
   t distribution, 308, 349, 357, 476
Denominator degrees of freedom, 371, 420, 426, 429, 439
Density, probability, 235
Dependent events, 163

Descriptive statistics, 4
Design, sample, 269
Design of experiments, 418
    balanced incomplete block design, 454
    complete block design, 437
    completely randomized design, 421
    incomplete block design, 450
    randomized block design, 438
Determinants, 466
Deviation from mean, 82
Difference between means:
    confidence interval for, 361
    paired data, 359
    standard error, 354
    test for, 356, 357
Difference between proportions:
    standard error, 383
    test for, 383, 386, 394
Discrete random variable, 200
Distribution (*see also* Frequency distribution,
        Probability density, Probability distribution,
        and by name the various special
        distributions)
    bell-shaped, 87, 95
    categorical, 21
    continuous, 234, 235
    cumulative, 26
    numerical, 21
    percentage, 25
    qualitative, 21
    quantitative, 21
    reverse J-shaped, 98
    skewed, 95
    symmetrical, 98
    U-shaped, 98
Dot diagram, 14
Double-stem display, 17
Double summation, 77

Equality of two standard deviations, 376
Equitable game, 180
Error:
    experimental, 424
    grouping, 71
    percentage, 254
    probable, 285
    standard, 280, 285, 324, 354
    Type I, 335
    Type II, 336
Error mean square, 426
Error sum of squares, 424
    Latin square, 451
    one-way analysis of variance, 424
    two-way analysis of variance, 437, 438
Estimate:
    biased, 83
    interval, 307
    point, 303
    unbiased, 83
Estimated regression coefficients, 474
Estimated regression line, 469
Estimation, 302
Event, 136
Events:
    dependent and independent, 163
    mutually exclusive, 137
Expectation, mathematical, 178, 180
Expected cell frequencies, 389, 390
Expected value of a random variable, 223
Experiment, 135
    controlled, 422
    two-factor, 438, 443
Experimental design, 418
Experimental error, 424
Exploratory data analysis, 6
Exponential curve fitting, 486
Exponential distribution, 246

E

Effect:
    block, 438
    treatment, 418
Elimination, rule of, 172
Empirical rule, 87
Empty set, 136

F

Factor, 438, 443
Factorial notation, 114
Fair game, 180
Fair odds, 149
F distribution, 371, 420
    degrees of freedom, 371

F statistic, 370, 420
Fisher *Z* transformation, 512
F test, 372, 426
Finite population, 262
  correction factor, 281
Finite sample space, 136
Fractile, 62
Frequency distribution, 20
  bell-shaped, 87, 95
  categorical, 21
  class boundary, 24
  class frequency, 24
  class interval, 25
  class limit, 24
  class mark, 25
  cumulative, 26
  deciles, 73
  histogram, 30
  mean, 70
  median, 71
  ogive, 32
  percentage, 25
  percentiles, 62, 73
  qualitative, 21
  quantitative, 21
  quartiles, 62, 73
  reverse J-shaped, 98
  skewed, 95
  standard deviation, 82
  symmetry, 98
  two-way, 40
  U-shaped, 98
Frequency interpretation of probability, 126
Frequency polygon, 32

G

Game:
  equitable, 180
  fair, 180
General addition rule, 156
General multiplication rule, 163, 164
Generalized addition rule, 157
Geometric distribution, 211
Geometric mean, 57
Goodness of fit, 401
Grand mean, 55, 418, 424
Grand total, 388, 427

Graphical presentation, 20
  bar chart, 15
  frequency polygon, 32
  histogram, 30
  ogive, 32
  pictogram, 33
  pie chart, 33
Grouping, 13
Grouping error, 71

H

Harmonic mean, 57
Hinge, 69
Histogram, 30
  three-dimensional, 40
H test, 541, 542
Hypergeometric distribution, 212
  and binomial distribution, 214
  mean, 225
  multivariate, 220
  variance, 231
Hypothesis:
  alternative, 332
    one-sided, 339
    two-sided, 339
  composite, 337
  null, 332
  simple, 337
  statistical, 332
Hypothesis testing (*see* Tests of hypotheses)

I

Incomplete block design, 454
  balanced, 454
Independent events, 163
Independent samples, 354
Inference, statistical, 5
Infinite population, 262
  random sample from, 262, 266
Infinite sample space, 136
Interaction, 438, 443
Interpretation of *r*, 506
Interquartile range, 81, 323
Intersection, 137
Interval, class, 25
Interval data, 7
Interval estimate, 307

J

J-shaped distribution, reverse, 98
Judgment sample, 272

K

Kruskal-Wallis test, 541

L

Large numbers, law of, 127
Large-sample confidence interval:
　for mean, 307
　for proportion, 324
　for standard deviation, 321
　for variance, 318
Large-sample test:
　for mean, 345
　for difference between means, 356
　for difference between proportions, 383
　for standard deviation, 369
Large-sample nonparametric test:
　for sign test, 523
　for signed-rank test, 532
　for u test, 547
　for U test, 540
Latin square, 450
Law of large numbers, 127
Leaf, 17
Lease squares, method of, 460, 463
Least-squares line, 465
Level of significance, 341
Limit, class, 24
Limits of prediction, 479
Linear equation, 461
Linear regression analysis, 474
Listing, 13
Location, measures of, 47
Log-log paper, 489
Lower class boundary, 24
Lower class limit, 24
Lower hinge, 69

M

Mann-Whitney test, 535
Mark, class, 25
Markov's theorem, 57

Mathematical expectation, 178, 180
Maximax criterion, 187
Maximin criterion, 187
Maximum error:
　estimation of mean, 304
　estimation of proportion, 325, 326
Mean:
　arithmetic, 49
　binomial distribution, 224
　combined data, 55
　confidence interval for, 307, 309
　deviation from, 82
　geometric, 57
　grand, 55, 418, 424
　grouped data, 70
　harmonic 57
　hypergeometric distribution, 225
　maximum error of estimate, 304
　Poisson distribution, 225
　population, 50
　probability density, 237
　probability distribution, 223
　probable error, 288
　sample, 50
　standard error, 280
　test for, 345, 349
　weighted, 54
Mean deviation, 82
Mean square:
　blocks, 441
　error, 426, 440, 451
　treatments, 425, 440
Measures of central location, 48
Measures of location, 47
Measures of relative variation, 88
Measures of skewness, 95
Measures of variation, 47, 81
Median, 58
　grouped data, 71
　population, 60
　position, 60
　sample, 60
　standard error, 285
Method of elimination, 466
Method of least squares, 460, 463
Midquartile, 64
Midrange, 68

Minimax criterion, 187
Minimin criterion, 187
Mode, 66
Model, statistical, 3
μ *(mu)*, mean:
   population, 50
   probability density, 237
   probability distribution, 223
Multiple comparison, 431
Multinomial distribution, 220
Multivariate hypergeometric distribution, 220
Multiple correlation coefficient, 516
Multiple regression, 482
Multiplication of choices, 110, 112
Multiplication rules of probability, 163, 164
Mutually exclusive events, 137

N

Negative correlation, 501
Negatively skewed distribution, 95
Nominal data, 7
Nonlinear regression, 486
Nonparametric tests, 520
   efficiency, 555
   H test, 541, 542
   Kruskal-Wallis test, 542
   Mann-Whitney test, 535
   randomness, 545, 546
   rank correlation, 551
   runs test, 546
   signed-rank test, 527, 532
   sign test, 521, 523
   u test, 545, 546, 547
   U test, 535, 538, 540
   Wilcoxon rank-sum test, 535
   Wilcoxon signed-rank test, 527
Normal correlation analysis, 512
Normal-curve areas, 238
   table, 569
Normal distribution, 234, 237
   binomial approximation, 253
   check for normality, 247
   standard, 238
Normal equations, 465, 483, 490
Normal population, 247
Normal probability paper, 247
Normal probability plot, 247

Normal regression analysis, 474
Null hypothesis, 332
Number of degrees of freedom, 308, 317
Numerator degrees of freedom, 371, 420, 429, 439
Numerical data, 7
Numerical distribution, 21

O

Observed cell frequencies, 389, 390
OC-curve, 336
Odds, 124, 148
   betting, 149
   fair, 149
   and probability, 148, 150
Ogive, 32
One-sample sign test, 521
One-sample *t* test, 349
One-sample *z* test, 345
One-sided alternative, 339
One-sided criterion, 342
One-sided test, 342
One-tailed test, 342
One-way analysis of variance, 424
   unequal sample sizes, 429
Open class, 22
Operating characteristic curve, 336
Optimum allocation, 275
Ordinal data, 7
Outcome of an experiment, 135

P

Paired data, 37, 359, 459, 498
Paired-sample sign test, 523
Paired-sample *t* test, 360
Parabola, 490
Parameter, 50
Pareto diagram, 19
Partial correlation coefficient, 517, 518
Pascal's triangle, 123
Pearsonian coefficient of skewness, 95
Percentage distribution, 25
   cumulative, 26
Percentage error, 254
Percentile, 62, 73
Permutations, 113
   indistinguishable objects, 121

Personal probability, 129
Pictogram, 33
Pie chart, 33
Point estimate, 303
Poisson distribution, 21, 219
    and binomial distribution, 216
    mean, 225
    standard deviation, 231
Polygon, frequency, 32
Polynomial curve fitting, 486
Polynomial equation, 492
Pooled standard deviation, 358
Pooling, 383
Population, 48
    binomial, 203
    correlation coefficient, 512
    finite, 262
    infinite, 262
    mean, 50
    median, 60
    normal, 247
    size, 50
    standard deviation, 83
    uniform, 243
    variance, 83
Positive correlation, 501
Positively-skewed distribution, 95
Postulates of probability, 145, 146
    generalized, 216
Power function, 336, 489
Prediction, limits of, 479
Probability, 108, 124
    addition rules, 153, 157
    and area under curve, 235
    and confidence, 305
    and odds, 148, 150
    Bayes' theorem, 172
    classical concept, 124
    conditional, 160, 162
    and confidence, 305
    consistency criterion, 150
    empty set, 146
    equally likely events, 124
    frequency interpretation, 126
    general addition rule, 157
    generalized addition rule, 153
    general multiplication rule, 163, 164

Probability (cont.)
    independence, 163, 164
    multiplication rules, 163, 164
    and odds, 148, 150
    paper, arithmetic, 247
    paper, normal, 247
    personal, 129, 149
    postulates, 145, 146
    special addition rule, 157
    special multiplication rule, 164
    subjective, 129, 149
    total, 172
Probability density, 235
    chi-square, 317
    exponential, 246
    F, 371, 420
    mean, 236, 237
    normal, 234, 237
    standard deviation, 236, 237
    t, 308, 349, 357, 476
Probability distribution, 199, 210
    binomial, 203
    geometric, 211
    hypergeometric, 212
    mean, 223
    multinomial, 220
    Poisson, 216, 219
    standard deviation, 226
Probability paper:
    arithmetic, 247
    normal, 247
Probability theory, 5, 108
Probable error of mean, 288
Product moment, 503
Product-moment coefficient of correlation,
    503
Proportion:
    confidence interval for, 324
    maximum error of estimate, 325, 326
    sample, 323
    standard error, 324
    test for, 378, 379
Proportional allocation, 271
Proportions:
    difference between, 383
    differences among, 394
p-value, 347, 348

Q

Qualitative data, 7
Qualitative distribution, 21
Quantitative data, 7
Quantitative distribution, 21
Quartile deviation, 81
Quartile variation, coefficient of, 92
Quartiles, 62, 73
Quota sampling, 272

R

r (*see* Correlation coefficient)
Randomization, 423
Randomized block design, 438
Randomized design, completely, 423
Randomized response technique, 174
Randomness, tests of, 545, 546, 547, 548
Random numbers, 264, 265
Random sample, 261
  finite population, 263
  infinite population, 266
  simple, 263
Random selection, 125
Random variable, 199
  continuous, 233
  discrete, 200
Range, 81
  interquartile, 81
  semi-interquartile, 81
Rank correlation coefficient, 551
  standard error, 552
  test for, 553
Rank sums, 535
Ratio data, 7
Raw data, 20
$r \times c$ table, 387
  expected cell frequencies, 389, 390
Real class limits, 24
Regression, 460
  analysis, 473, 474
    linear, 474
    normal, 474
  coefficients, 474
    confidence interval for, 476
    estimated, 474
    tests for, 474

Regression (*cont.*)
  equation, 468
  line, 469
    estimated, 469
  multiple, 482
  nonlinear, 486
Regression sum of squares, 500
Relative variation, measures of, 97
Replication, 443
Residual sum of squares, 500
Reverse J-shaped distribution, 98
ρ (*rho*), population correlation coefficient, 512
Robust, 373
Root-mean-square deviation, 82
Row sum of squares, 451
Rule of elimination, 172
Rule of total probability, 172
Replication, 443
Runs, 546
  above and below the median, 548
  theorey of, 545

S

Sample, 48
  covariance, 504
  design, 269
  mean, 50
  proportion, 323
  random, 261, 263, 264
  range, 81
  size, 50
  standard deviation, 82
  variance, 83
Samples, independent, 354
Sample space, 135
  countably infinite, 216
  finite, 136
  infinite, 136
Sampling:
  area, 273
  cluster, 273
  judgment, 272
  quota, 272
  random, 261, 263, 264
  with replacement, 168
  without replacement, 168
  stratified, 270

Sampling: *(cont.)*
  optimum allocation, 275
  proportional allocation, 271
  systematic, 269
Sampling distribution, 262
  difference between means, 354, 357
  difference between proportions, 383
  mean, 278, 283
  median, 285
Sampling frame, 268
scaling, 11
Scattergram, 38, 462
Scores, standard, 238
Self weighting, 276
Semi-interquartile range, 81
Semilog paper, 486
σ *(sigma)*, standard deviation:
  population, 83
  probability density, 236, 237
  probability distribution, 226
Σ *(sigma)*, summation sign, 50, 76
Signed-rank test, 527, 532
Significance, level of, 341
Significance, statistical, 338
Significance test, 338
Sign test, 521, 523
  large sample, 523
  one sample, 521
  paired-sample, 523
Simple hypothesis, 337
Simple random sample, 263
Simulation, 290
Size:
  population, 50
  sample, 50
Skewed distribution, 95
Skewness, 95
  measures of, 95
  Pearsonian coefficient, 95
  positive and negative, 95
Slope, 461
Spearman's rank correlation coefficient, 551
Special addition rule, 157
Special multiplication rule, 164
Standard deviation, 82
  binomial distribution, 227
  confidence interval for, 318, 321

Standard deviation *(cont.)*
  grouped data, 93
  hypergeometric distribution, 231
  Poisson distribution, 231
  pooled, 358
  population, 83
  probability density, 236, 237
  probability distribution, 226
    computing formula, 226
  sample, 82
    computing formula, 84
  test for, 367
Standard deviations, equality of, 370
Standard error:
  difference between means, 354
  difference between proportions, 383
  mean, 280
  median, 285
  proportion, 324
  rank correlation coefficient, 552
  standard deviation, 320
Standard error of estimate, 475
Standard normal distribution, 238
Standard scores, 238
Standard units, 86, 238
States of Nature, 185
Statistic, 50
Statistical hypothesis, 332
Statistical inference, 5
Statistical model, 3, 205
Statistically significant, 338
Statistics, descriptive, 4
Stem, 17
Stem-and-leaf display, 16
  double stem, 17
Stem label, 17
Strata, 270
Stratified sampling, 270
  cross stratification, 272
  optimum allocation, 275
  proportional allocation, 271
Student's *t* distribution, 308
Studentized range, 432
Studentizing, 432
Subjective probability, 129, 149
  consistency criterion, 150
Subset, 136

Summation, 50, 76
  double, 77
Sum of squares:
  block, 439
  column, 451
  error, 424, 437, 438, 451
  interaction, 445
  regression, 500
  residual, 500
  row, 451
  total, 424, 500
  treatment, 424
Sums of squares, computing formulas, 427, 429,
    439, 451
Symmetrical distribution, 98
Systematic sampling, 269

T

Table, analysis of variance:
  Latin square, 452
  one-way, 426
  two-way, 439
Tail probability (*see* p-value)
t distribution, 308
  degrees of freedom, 308
  table, 571
t interval, 309
t statistic, 308
Test statistic, 342
Tests of hypotheses:
  alternative hypothesis, 332
  analysis of variance, 429, 439, 452
  correlation coefficient, 512
    rank, 553
  differences among proportions, 394
  differences between means, 354, 356
  differences between proportions, 383
  equality of standard deviations, 370
  goodness of fit, 401
  H test, 541, 542
  Kruskal-Wallis test, 541, 542
  Level of significance, 341
  Mann-Whitney test, 535
  mean, 343, 349
  nonparametric, 520
  null hypothesis, 332

Tests of hypotheses: *(cont.)*
  one-sample t test, 349
  one-sample z test, 345
  one-sided, 342
  one-tailed, 342
  operating characteristic curve, 336
  paired-sample sign test, 523
  paired-sample t test, 360
  proportion, 378, 379
  randomness, 545, 546, 547, 548
  rank correlation coefficient, 553
  regression coefficients, 474
  runs, 546, 547, 548
  significance test, 338
  sign test, 521, 523
  signed-rank test, 527, 529, 532
  standard deviation, 367
  two-sample t test, 357
  two-sample z test, 356
  two-sided, 342
  two-tailed, 342
Treatments, 425
Two-way analysis of variance, 437, 438
  with interaction, 443
  without interaction, 438
Two-way frequency distribution, 40
Type I error, 335
Type II error, 335

U

Unbiased estimator, 83
Uniform distribution, 243
Union, 137
Units, standard, 96, 238
Upper class boundary, 24
Upper class limit, 24
U-shaped distribution, 98
U statistics, 537
  mean, 540
  standard deviation, 540
u statistic, 546
  mean, 547
  standard deviation, 547
U test, 535, 538, 540
u test, 545, 546, 547

V

Variable, random, 199, 200, 233
Variance:
  binomial distribution, 227
  confidence interval for, 318, 321
  hypergeometric distribution, 231
  Poisson distribution, 231
  population, 83
  probability distribution, 225
    computing formula, 226
  random variable, 225
  sample, 83
    computing formula, 84
Variance, analysis of (*see* Analysis of variance)
Variance ratio, 371, 420
Variation:
  coefficient, 88
  coefficient of quartile, 92
  measures of, 47
Venn diagram, 138

W

Weights, 54
Weighted mean, 54
Wilcoxon rank-sum test, 535
Wilcoxon signed-rank test, 527

Z

z scores, 88, 238
z test:
  one-sample, 345, 346
  two-sample, 354, 356
z interval, 307
Z transformation, 512

PROBLEMS OF ESTIMATION

Confidence Interval for

Mean (σ known)

$$\bar{x} - z_{\alpha/2} \cdot \frac{\sigma}{\sqrt{n}} < \mu < \bar{x} + z_{\alpha/2} \cdot \frac{\sigma}{\sqrt{n}}$$

Mean (σ unknown)

$$\bar{x} - t_{\alpha/2} \cdot \frac{s}{\sqrt{n}} < \mu < \bar{x} + t_{\alpha/2} \cdot \frac{s}{\sqrt{n}}$$

Proportion (large sample)

$$\hat{p} - z_{\alpha/2} \cdot \sqrt{\frac{\hat{p}(1 - \hat{p})}{n}} < p < \hat{p} + z_{\alpha/2} \cdot \sqrt{\frac{\hat{p}(1 - \hat{p})}{n}}$$

where $\hat{p} = \dfrac{x}{n}$

Maximum Error

Estimation of mean

$$E = z_{\alpha/2} \cdot \frac{\sigma}{\sqrt{n}}$$

Estimation of proportion

$$E = z_{\alpha/2} \cdot \sqrt{\frac{\hat{p}(1 - \hat{p})}{n}}$$

Sample Size

Estimation of mean

$$n = \left[\frac{z_{\alpha/2} \cdot \sigma}{E} \right]^2$$

Estimation of proportion

$$n = p(1 - p)\left[\frac{z_{\alpha/2}}{E} \right]^2 \qquad \text{or} \qquad n = \frac{1}{4}\left[\frac{z_{\alpha/2}}{E} \right]^2$$

TESTS OF HYPOTHESES

Statistics for Tests Concerning

Difference between means (σ's known)

$$z = \frac{\bar{x}_1 - \bar{x}_2 - \delta}{\sqrt{\dfrac{\sigma_1^2}{n_1} + \dfrac{\sigma_2^2}{n_2}}}$$

Difference between means (σ's unknown)

$$t = \frac{\bar{x}_1 - \bar{x}_2}{s_p\sqrt{\dfrac{1}{n_1} + \dfrac{1}{n_2}}} \qquad \text{where} \qquad s_p = \sqrt{\frac{(n_1 - 1)s_1^2 + (n_2 - 1)s_2^2}{n_1 + n_2 - 2}}$$

Difference between proportions (large samples)

$$z = \frac{\dfrac{x_1}{n_1} - \dfrac{x_2}{n_2}}{\sqrt{\hat{p}(1 - \hat{p})\left(\dfrac{1}{n_1} + \dfrac{1}{n_2}\right)}} \qquad \text{with} \qquad \hat{p} = \frac{x_1 + x_2}{n_1 + n_2}$$

Differences among proportions, contingency tables, or goodness of fit

$$\chi^2 = \sum \frac{(o - e)^2}{e}$$

Mean (σ known)

$$z = \frac{\bar{x} - \mu_0}{\sigma/\sqrt{n}}$$

Mean (σ unknown)

$$t = \frac{\bar{x} - \mu_0}{s/\sqrt{n}}$$

Proportion (large sample)

$$z = \frac{x - np_0}{\sqrt{np_0(1 - p_0)}}$$